Red Biotechnology

Red Biotechnology

Edited by **Giorgio Salati**

SYRAWOOD
PUBLISHING HOUSE
New York

Published by Syrawood Publishing House,
750 Third Avenue, 9th Floor,
New York, NY 10017, USA
www.syrawoodpublishinghouse.com

Red Biotechnology
Edited by Giorgio Salati

International Standard Book Number: 978-1-68286-210-0 (Hardback)

Printed in China.

Contents

Preface

This book has been an outcome of determined endeavour from a group of educationists in the field. The primary objective was to involve a broad spectrum of professionals from diverse cultural background involved in the field for developing new researches. The book not only targets students but also scholars pursuing higher research for further enhancement of the theoretical and practical applications of the subject.

Red biotechnology is a distinct branch of biotechnology that deals with the therapeutic and pharmaceutical application of biotechnology. This book brings forth some of the most innovative concepts and elucidates the unexplored aspects in the field of red biotechnology. Antibiotics and metabolites, enzymes, lipids, toxins, genetic engineering and therapeutic proteins are some of the concepts related to this field included in this book. It includes contributions of experts and scientists which will provide innovative insights and will prove to be immensely beneficial to students and researchers in this field.

It was an honour to edit such a profound book and also a challenging task to compile and examine all the relevant data for accuracy and originality. I wish to acknowledge the efforts of the contributors for submitting such brilliant and diverse chapters in the field and for endlessly working for the completion of the book. Last, but not the least; I thank my family for being a constant source of support in all my research endeavours.

Editor

1

Integrative gene transfer in the truffle *Tuber borchii* by *Agrobacterium tumefaciens*-mediated transformation

Andrea Brenna[1], Barbara Montanini[2], Eleonora Muggiano[1], Marco Proietto[1], Patrizia Filetici[3], Simone Ottonello[2] and Paola Ballario[1,3*]

Abstract

Agrobacterium tumefaciens-mediated transformation is a powerful tool for reverse genetics and functional genomic analysis in a wide variety of plants and fungi. *Tuber* spp. are ecologically important and gastronomically prized fungi ("truffles") with a cryptic life cycle, a subterranean habitat and a symbiotic, but also facultative saprophytic lifestyle. The genome of a representative member of this group of fungi has recently been sequenced. However, because of their poor genetic tractability, including transformation, truffles have so far eluded in-depth functional genomic investigations. Here we report that *A. tumefaciens* can infect *Tuber borchii* mycelia, thereby conveying its transfer DNA with the production of stably integrated transformants. We constructed two new binary plasmids (pABr1 and pABr3) and tested them as improved transformation vectors using the green fluorescent protein as reporter gene and hygromycin phosphotransferase as selection marker. Transformants were stable for at least 12 months of in vitro culture propagation and, as revealed by TAIL- PCR analysis, integration sites appear to be heterogeneous, with a preference for repeat element-containing genome sites.

Keywords: *Tuber* spp; Truffles; *Agrobacterium tumefaciens*-mediated transformation; T-DNA; binary plasmid; Green fluorescent protein; Hygromycin phosphotransferase; TAIL-PCR

Introduction

The ascomycete *Tuber borchii* is a hypogeous fungus (genus, *Tuber*; family Tuberaceae; order Pezizales) that establishes a beneficial mutualistic interaction ('ectomycorrhiza') with the roots of many tree species. *T. borchii* has also a pronounced saprobiotic capacity and can be grown in vitro (albeit quite slowly) as vegetative free-living mycelium in the absence of a plant host. This mixed symbiotic/saprophytic lifestyle (Hebe et al. 1999), together with the lack of asexual spores amenable to in vitro culture has hampered the genetic manipulation of this fungus and obscured our understanding of its complex life cycle. *Tuber* fruitbodies ('truffles') lack an active system for the dispersal of spores, which are disseminated by the action of mycophagous animals (Pegler et al. 1993). Another peculiarity of certain *Tuber* spp. is the high commercial value of their fruitbodies which are prized as gourmet food. *Tuber melanosporum* and *T. borchii* are the two most studied truffle species. Sequencing and annotation of the *T. melanosporum* genome, the first symbiotic Ascomycete to be sequenced, revealed many aspects of truffles' biology and genetic organization, including their massive content of repeated transposable elements and their heterotallic mode of conjugation and ability to outcross through the action of two distinct mating type loci (Martin et al. 2010). *T. borchii* is the truffle that can be more easily handled under laboratory conditions.

With the acquisition of new genome sequence data, functional genetic studies are all the more necessary to decipher gene function. Reverse genetics in *Tuber* has been hindered, however, by the absence of a well-established stable transformation system. An effective approach in this direction is represented by *Agrobacterium tumefaciens*-mediated (ATM) transformation of intact hyphae. This technique was first successfully applied to fungi

* Correspondence: paola.ballario@uniroma1.it
[1]Pasteur Cenci Bolognetti Foundation, c/o Department of Biology and Biotechnology "Charles Darwin", Sapienza University, Piazzale A. Moro 5, 00185 Rome, Italy
[3]Institute of Biology and Molecular Pathology, CNR, 00185 Rome, Italy
Full list of author information is available at the end of the article

19 years ago (Bundock et al. 1995) by exploiting the natural ability of *A. tumefaciens* to transfer a portion of its DNA to a foreign infected organism, most notably dicotyledonous plants. To date, ATM transformation is applied to the study of a variety of fungal species, including Ascomycetes, Basidiomycetes and Zygomycetes. The transfer DNA (T-DNA), located on a >200 kb tumor-inducing (Ti) plasmid, is flanked by two 25 bp direct imperfect repeats, known as left border (LB) and right border (RB) sequences, which also include the *vir* genes encoding the genetic factors required for transfer (Zupan et al. 2000). Recent improvements of ATM transformation include the optimization of temperature and co-cultivation conditions, and the development of new selection markers (Michielse et al. 2005). We previously described an ATM transformation procedure for *T. borchii* (Grimaldi et al. 2005), which due to the insertion of transgenic DNA driven by the T-DNA, and consequent lack of stability, could not be effectively exploited for functional genomics studies. Building upon these earlier attempts, in the present work we improved transformation conditions through the development of two new vectors and also gained insight into the fate of the transferred T-DNA, which was found to be randomly integrated, with a preference for repeat element-containing genome sites. These results represent an important technical advancement for the molecular biological investigation of truffles, which will be instrumental to the construction of mutant strain collections.

Materials and methods

Strains and media

T. borchii Vittad. mycelia (isolate ATCC 95640) were grown and propagated in the dark at 24°C on potato-dextrose medium with agar 39 g/l (PDA: 0.2% peptone, 0.2% yeast extract, 1.8% glucose, 0.5% potato starch, 1.5% agar; Liofilchem), as described (Ambra et al. 2004). For easy medium replacement, mycelial cultures were usually grown on sterile dialysis membranes. For liquid culture replacement, mycelial cultures were inoculated in potato-dextrose liquid medium (PDB, without agar). *A. tumefaciens* was grown and propagated in Luria-Bertani (LB) (1% tryptone, 0.5% yeast extract, 1.5% agar) solid or liquid (w/o agar) medium at 28°C in the presence of appropriate marker selection supplements. After co-cultivation, mycelia were transferred onto Evans minimal medium (1x Vogel's salts, 1% glucose, 2 mg/l vitamin B1, 1.5% agar). The hypervirulent *Agrobacterium* strain AGL-1 [(C58 pTiBo542) recA::bla, T-region deleted Mop (+) Cb (R)] was kindly provided by Peter Romaine (Department of Plant Pathology, Pennsylvania State University, University Park, PA, USA) (Lazo et al. 1991). Carbenicillin (Sigma-Aldrich, St Louis, MO) 100 μg/ml was used for AGL-1 propagation. The GV3101 strain (background C58; Ti-plasmid cured) was kindly provided by Paolo Costantino (Department of Biology and Biotechnology "La Sapienza" University, Rome, Italy). Rifampicin (25 μg/ml) and gentamicin (25 μg/ml), both provided by Duchefa Biochemie, were used for GV3101 propagation.

Plasmids

The binary plasmid vector pBGgHg (Chen et al. 2000) consists of a pCAMBIA 1300 (CAMBIA, Canberra, Australia), backbone containing the *Escherichia coli* hygromycin B phosphotransferase (*hph*) gene and the enhanced green fluorescent proteins (*egfp*) gene, both under control of the strong and constitutive glyceraldehyde-3- phosphate dehydrogenase (*gpd*) promoter from *Agaricus bisporus* and the *CaMV35S* terminator. The plasmid was kindly provided by Peter Romaine (Department of Plant Pathology, Pennsylvania State University). The binary vector pABr1 (Figure 1a) was derived from pCAMBIA 1302 (CAMBIA, Canberra, Australia) through the incorporation of a region extracted from pCT74 (Ciuffetti et al. 1997; Lorang et al. 2001). This region contains a synthetic green fluorescent protein (*sgfp)* gene under control of the strong and constitutive *ToxA* promoter from *Pyrenophora tritici-repentis* and a hygromycin B phosphotransferase (*hph*) gene under control of the constitutive promoter *pdgp* from *Agaricus bisporus.* To generate pABr1, both starting plasmids were subjected to *Eco*RI-*Xho*I digestion, which removed the T-DNA region of pCAMBIA 1302 containing the plant *hph* gene under control of the *CaMV35S* promoter, while the *GFP* gene controlled by the *CaMV35S* promoter was retained. The binary plasmid vector pABr3 is a shortened version of pABr1 from which the region containing the *GFP* gene under control of the *CaMV35S* promoter used for plant transformation was deleted.

ATM transformation

T. borchii mycelia were pre-grown in the dark at 24°C for 7 days on cellophane-overlaid PDA plates and then transferred to PDA medium containing 200 μM acetosyringone (AS) to stimulate *Agrobacterium* virulence (outlined in Additional file 1: Figure S1). Three days before the start of co-cultivation, pre-transformed *Agrobacterium* cells were grown at 28°C for 2 days in the presence of the appropriate selection marker antibiotics (100 μg/ml carbenicillin for AGL-1; 25 μg/ml each of rifampicin and gentamycin for GV3101; 30 μg/ml kanamycin for both vectors). On the day of co-cultivation, a fresh overnight bacterial culture with an optical density at 600 nm (OD_{600nm}) of 1.2-1.5 was diluted to an OD_{600nm} of 0.075 with antibiotic-supplemented medium also containing 200 μM AS and grown for 4 h to an OD_{600nm} of 0.3. After reaching this cell density, 50 μl of bacteria in LB (1.5×10^3 cells/μl) were overlaid onto *T. borchii* mycelia and co-cultivation was continued for 3 days at 22-24°C. Following co-cultivation, mycelia

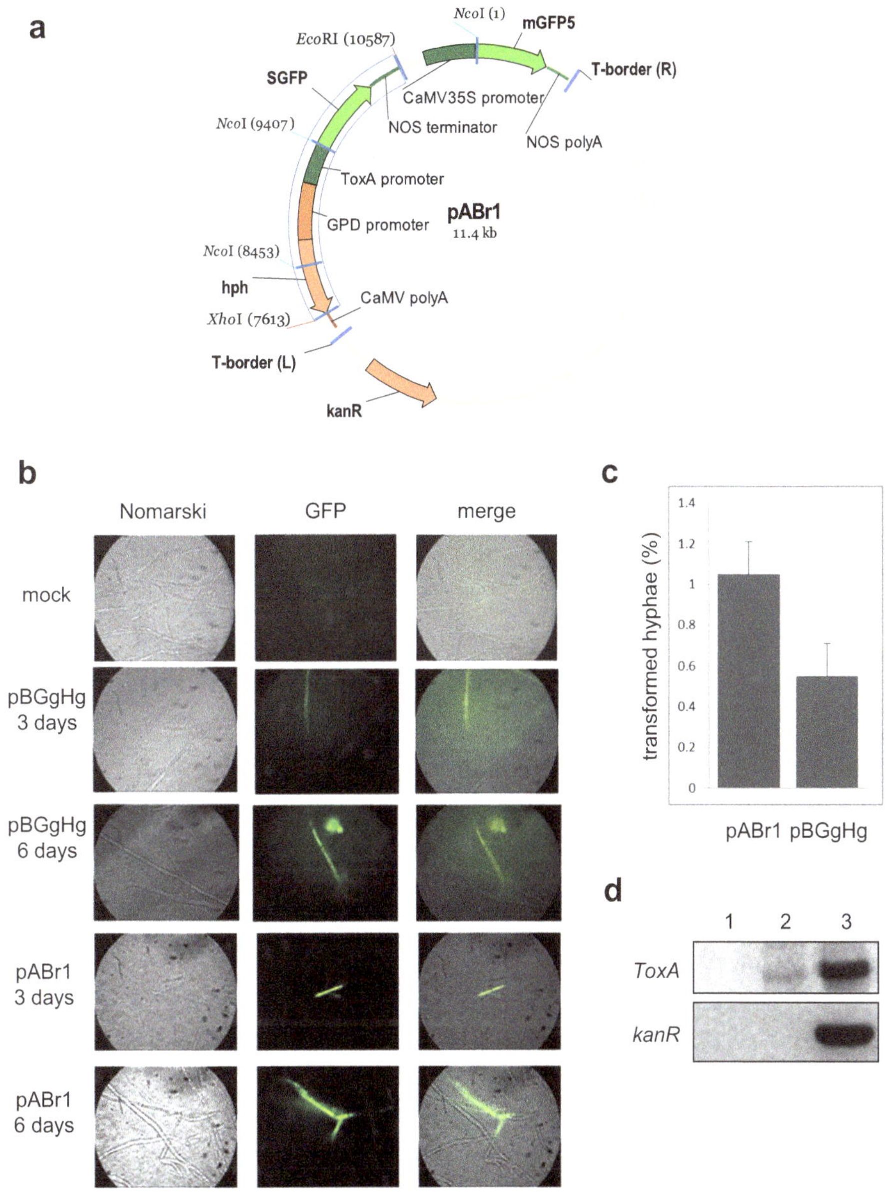

Figure 1 ATM-transformation with the pABr1 vector. a) pABr1 (11.4 kb) vector map: *Hygromycin (R)*, hygromycin phosphotransferase gene (*hph*); Kanamicin resistance gene (*kanR*); *ToxA promoter*, promoter sequence from *P. tritici-repentis*; *SGFP*, the S65T variant of the green fluorescent protein (GFP); *CAMV35S* cauliflower mosaic virus promoter; *mGFP5*, folding-enhanced green fluorescent protein variant; *NOS polyA*, Nopaline synthase terminator sequence; *T-border (R)* and *T-border (L)*, right (RB) and left (LB) borders of *A. tumefaciens* T-DNA. The blue highlighted box is the plasmid region derived from pCT74. **b)** Images (100x magnification) of *T. borchii* hyphae transformed with either the pABr1 or the pBGgHg vector, plus an untransformed control (*mock*), obtained by phase-contrast (*Nomarski*) and GFP fluorescence microscopy (*GFP*). The type of treatment, vector and co-cultivation times are indicated on the left; a merge of the Nomarski and GFP images is shown in the rightmost panels. **c)** Quantification of transformed hyphae obtained with pBGgHg or pABr-1, expressed as percentage with respect to the total number of hyphae (~4500) present in each analyzed section. Data are the mean ± s.e.m. of at least five independent experiments. **d)** PCR amplification with *ToxA*-specific primers of total DNA extracted from mock-infected mycelia (*lane 1*), pABr1-transformed mycelia (*lane 2*), and pABr1/AGL-1 bacterial cells (*lane 3*) is shown in the *upper panel*. The results of PCR amplification of the same samples with *kanR*-specific primers are shown in the *lower panel*.

were washed with sterile ddH_2O and 400 μM cefotaxime, and mycelial samples were sliced and examined by confocal microscopy. The remaining mycelia were transferred to new plates of Evans minimal medium containing 400 μM cefotaxime (to inhibit bacterial growth) and 15 μg/ml hygromycin (to aid in transformant hyphae isolation). Additional mycelial samples were analyzed after 6 days of co-cultivation, and the remaining mycelia were monitored over time for hygromycin B resistance and GFP expression.

Fluorescence and confocal microscopy

Sections (3 × 3 mm) of either control, untransformed *T. borchii* mycelia or mycelia co-cultivated with *A. tumefaciens* for 3 days were washed 5 times with distilled water and cefotaxime (400 μM) prior to microscopic analysis. An Axioskop 2 microscope (Carl Zeiss International, Oberkochen, Germany) equipped with a 100X oil immersion objective (Plan Neofluar), a Zeiss filter set (green excitation filter 450–490 HB) and a Zeiss AttoArc2 HB-100 W mercury lamp were used for fluorescence microscopy analysis. Images were captured with a Microcolor camera (RGB-MS-C; CRI, Boston, MA) and processed with Diffraction Micro CCD software and Adobe Photoshop (Adobe System, San Jose, CA). Typical exposure times were 30–60 ms for the SGFP and Nomarski images. Confocal laser scanning microscopy was performed with a Leica microscope equipped with a TCS SP2 Laser (Ar/Kr, Gre/Ne, He/Ne; standard phase contrast plus Normaski contrast). SGFP was excited with a 488-nm laser line and detected at 515-530 nm.

DNA extraction and PCR analysis

Mycelia to be analyzed by confocal microscopy were washed 5 times with ddH_2O and 400 μM cefotaxime and then disrupted by freezing in liquid nitrogen. DNA was extracted from frozen mycelia using the DNeasy Plant Minikit (QIAGEN N. V., Hilden, Germany) as per manufacturer's instructions. The amplification protocols described in (Grimaldi et al. 2005) were employed for PCR analysis of the aminoglycoside 3'-phosphotransferase (*kanR*) and *sgfp* genes, using the oligonucleotide primers listed below.

kanR primers (to amplify a region comprised within the binary plasmid vector):

kanR_Fw: 5′-GGTCATGCATTCTAGGTACT-3′; kanR_Rw: 5′-AATGGCTAAAATGAGAATAT-3′.

Sgfp primers (to amplify a region comprised within the T-DNA):

sgfp_Fw: 5′- CACATGAAGCAGCACGACTT-3′; sgfp_Rw: 5′-TGCTCAGGTAGTGGTTGTCG-3′;

ToxA promoter region (also comprised within the T-DNA):

ToxA_Fw: 5′-GGGGAATTCTAGTGGAACTGATTGGAATGCA-3′;

ToxA_Rw: 5′-GGGGATAGAACCCATGGCCTATAT-3′

DNA blot analysis

Genomic DNA from transformed and untransformed (negative control) *T. borchii* mycelia (10 μg each) plus the same amount of pABr1 plasmid DNA (positive control) were digested with *Nco*I. A separate sample of pABr1 was digested in parallel in order to generate a pair of DNA fragments to be used as probes. Four fragments were obtained from *Nco*I-*Xho*I digestion, including an 840 bp fragment that was used as probe. This fragment comprises a region (located around the T-DNA junction site) that encompasses the site of recombination with genomic DNA. The second probe was obtained by *Nco*I digestion, which produced three different fragments. The fragment that was utilized as probe was 954 bp in length (from the middle of the hygromycin phosphotransferase cassette to the *ToxA* promoter). This probe matches the region of the T-DNA that is transferred to *T. borchii* hyphae. Radioactive labeling with approximately 20 ng of purified templates, blotting, hybridization and washing were performed as described (Green and Sambrook 2012; Montanini et al. 2003).

TAIL-PCR amplification, cloning and sequencing of T-DNA integration sites

Genomic DNA sequences flanking the T-DNA were amplified by Thermal Asymmetric Interlaced PCR (TAIL-PCR) as described in (Wang and Li 2008). The following oligonucleotides were used as primers to amplify the Left Border (LB) and the Right Border (RB) regions, in combination with the degenerate, genome-sequence targeting primer AD1.

LB1 5′-GGGTTCCTATAGGGTTTCGCTCATG-3′ LB2 5′-CATGTGTTGAGCATATAAGAAACCCT-3′ LB3 5′-GAATTAATTCGGCGTTAATTCAGT-3′

RB1 5′-GGCATGGCCGTCGTTTTACAAC-3′ RB2 5′-AACGTCGTGACTGGGAAAACCCT-3′ RB3 5′-CCCTTCCCAACAGTTGCGCA-3′

AD1 5′-WGTGNAGWANCANAGA-3′

DNA amplified by TAIL-PCR was cloned into the pDrive A/T vector (QIAGEN) according to the manufacturer's instructions, followed by DNA sequence analysis of plasmids isolated from transformant colonies.

Other procedures

Mycelial sections (3 × 3 mm) containing ~4500 hyphal apexes, as estimated by counting with a Burker chamber, were used for transformation efficiency determinations. The fraction (%) of transformed hyphae was obtained by dividing the number of fluorescent hyphae (average of multiple determinations) by 4500; data were calculated as the mean ± standard error of the mean (s.e.m.) of at least five independent replicates.

The Vector NTI® Software (Life Technologies) was used for drawing plasmid maps. BLAST searches were conducted against the *T. melanosporum* genome sequence at the MycorWeb site (http://mycor.nancy.inra.fr).

Results

ATM transformation of *T. borchii* mycelia using a *Pyrenophora ToxA* promoter-bearing binary vector

We previously documented the production of transformed, GFP-expressing *T. borchii* mycelia with the use of ATM and the pBGgHg plasmid (Grimaldi et al. 2005). However, transformation efficiency was rather low (0.2-0.5%) with a fairly weak GFP signal that was lost after a few weeks, possibly due to the lack of stable integration. In order to improve transformation efficiency and to achieve higher GFP expression levels as well as stable transformants, a new series of cloning vectors for ATM transformation was constructed. We first created a binary vector, named pABr1 (Figure 1a), with the *sgfp* gene under control of the *ToxA* promoter from the ascomycete *P. tritici-repentis* (Ciuffetti et al. 1997), inserted within the T-DNA region, which also contains the hygromycin phosphotransferase *(hph)* gene utilized for positive selection of transformants. The *sgfp* gene, which codes for the S65T variant of GFP, is a commonly utilized transformation marker in fungi (Niwa 2003) and use of the *ToxA* promoter has previously been shown to enhance the expression of reporter genes in various fungi including *Neurospora crassa* (Lorang et al. 2001). Our goal was to couple the enhanced strength of the *ToxA* promoter with the increased stability of SGFP in order to obtain a stronger fluorescence signal and thus an improved visualization of transformed hyphae. To this end, we performed ATM transformation by treating mycelia with the AGL-1 strain, carrying either pABr1 or pBGgHg, and comparing the transformation efficiency of these two vectors. After 3 days of co-cultivation (outlined in Additional file 1: Figure S1), a subset of mycelia was transferred to new antibiotic-containing plates (ampicillin, cefotaxime and hygromycin B) and samples were taken for hyphal section visualization by confocal microscopy. This allowed the identification of fluorescent, GFP-producing hyphae, by comparison with the background signal associated with control mycelia transformed with the empty AGL-1 strain (Figure 1b). The diffuse fluorescence background present in mock-transformed samples, contrasted with the localized signal observed in the apical hyphae of *T. borchii* transformed with AGL-1 carrying the binary vector. Transformation efficiency obtained with the pABr1 vector after 3 days of co-cultivation was consistently found to be at least two-fold higher (≥1%) than that obtained with pBGgHg using the same AGL-1 *A. tumefaciens* strain (Figure 1c). This was calculated as the ratio between fluorescent hyphae and the total number of hyphae per section (~4500). When the analysis was repeated at 6 days, i.e. after 3 additional days since the end of co-cultivation, the hyphae appeared more fluorescent, likely due to SGFP accumulation (Figure 1b), but without any appreciable change in the relative number of fluorescent hyphae (data not shown). Still, the SGFP signal in pABr1-transformed hyphae appeared stronger than that of hyphae transformed with the pBGgHg vector, thus facilitating the identification of transformants grown over time. Transformation was also verified by PCR amplification of the *ToxA* promoter using genomic-DNA extracted from transformed mycelia as template (Figure 1d). To confirm that *ToxA* amplification did indeed result from the mobilized T-DNA rather than from residual bacteria, we performed a parallel PCR analysis using a pair of primers annealing to the kanamycin resistance (*kanR*) gene region that is present in the plasmid outside of the T-DNA. The absence of a *kanR* amplification signal ruled out a false positive result caused by residual contaminating bacteria. We then used a different *Agrobacterium* strain (GV3101) to verify that the improved transformation efficiency obtained with pABr1 was a genuine effect of the plasmid vector, rather than an indirect effect of the specific *A. tumefaciens* strain utilized for transformation. The results obtained with GV3101 confirmed that also in this strain the transformation efficiency yielded by pABr1 was at least two-fold higher than that obtained with pBGgHg (Additional file 1: Figure S2a,b). However, global efficiency with GV3101-mediated transformation was generally lower than that obtained with AGL-1 (Additional file 1: Figure S2c), suggesting that not all *A. tumefaciens* strains are equally effective for *Tuber* transformation. Also in this case, the data were confirmed by PCR analysis, which revealed an *sgfp* amplicon in the case of DNA extracted from transformed mycelia, but not in the case of mock-infected mycelia (Additional file 1: Figure S2d). Furthermore, a stronger GFP signal was systematically observed in pABr1 transformants compared to transformants obtained with the pBGgHg vector utilized in previous studies (Grimaldi et al. 2005). Altogether the data confirm the superior performance of the newly designed pABr1 vector.

T-DNA transformation occurs by random integration in the *Tuber* genome

As described previously (Bundock and Hooykaas 1996), the T-DNA can either be integrated into the host genome or transferred and incorporated as stable, extra-chromosomal T-DNA elements that can form circular structures in plants (Singer et al. 2012). To investigate this point, we used a phenotype-based approach. We hypothesized that if the T-DNA was integrated in the *Tuber* genome, the same transformant section analyzed at different times after transformation should show a progressive increase in the fluorescence of neighboring hyphae, resulting from multiple nuclear divisions of hyphae containing the integrated transgene. We thus isolated regions of AGL-1/pABr1-transformed mycelia containing one or very few transformed hyphae (visualized by confocal microscopy after 6 days of co-cultivation; see Figure 2a "6d") and, after sterile washing, put them back to liquid medium for an additional 15 days. Subsequent analysis by confocal microscopy revealed a four-fold increase in the

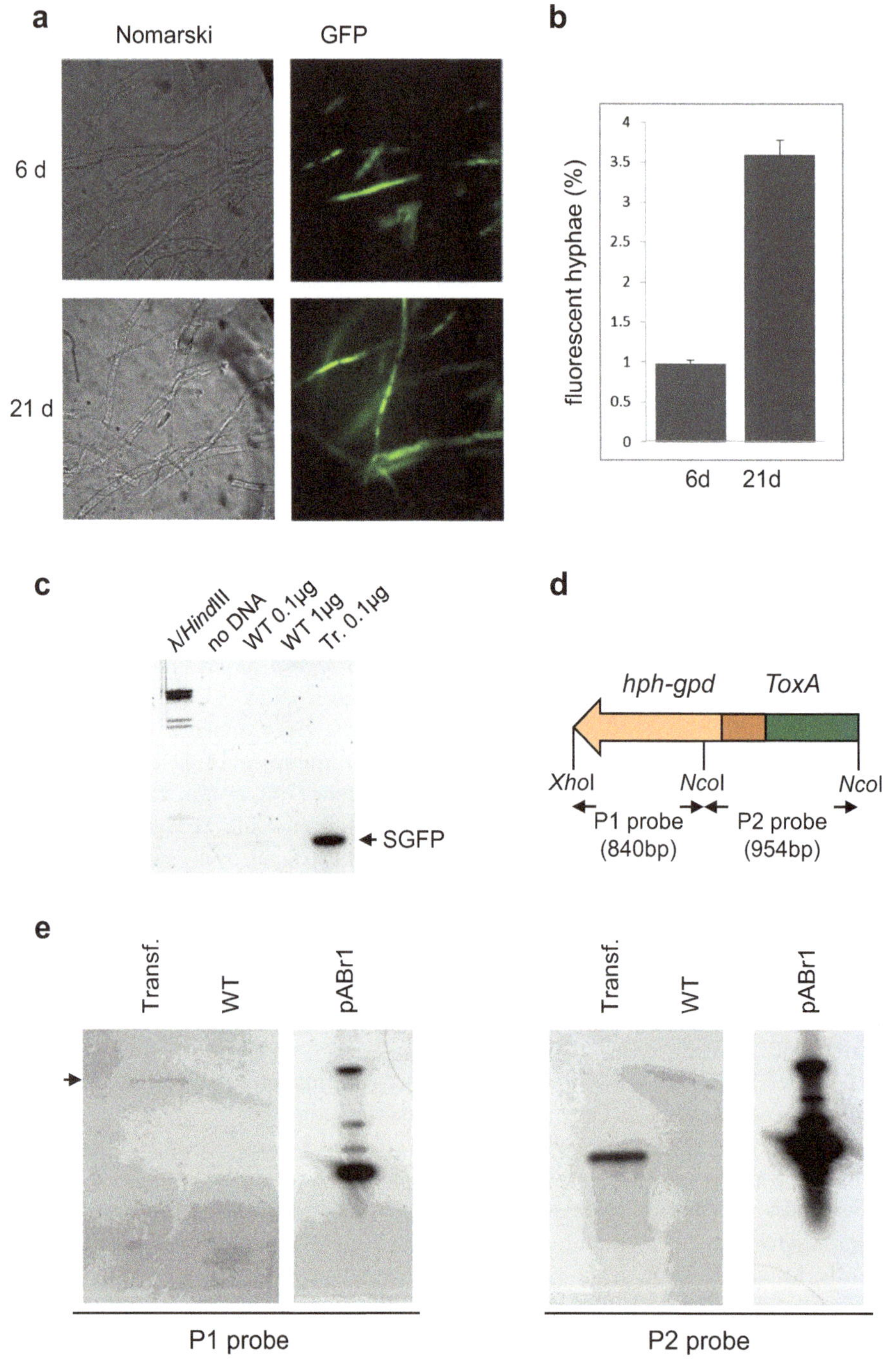

Figure 2 Following the fate of the T-DNA in pABr1-transformed hyphae. a) Sections (3 x 3 mm) of transformed *T. borchii* mycelia analyzed by confocal microscopy (GFP wavelength) at the indicated times after *Agrobacterium* infection. **b)** Histogram representation of cumulative data produced by the experiments in panel **a)** expressed as percentage of fluorescent hyphae with respect to the total number of hyphae present in each section (~4500). Data are the mean ± s.e.m. of at least five independent experiments. **c)** PCR amplification with *sgfp* specific-primers of genomic DNA obtained from mock-transformed mycelia (*WT*) using 0.1 µg (*lane 3*) and 1 µg (*lane 4*) of template DNA, and from AGL-1/pABr1-transformed mycelia (0.1 µg of template DNA; *lane5*). A PCR negative control (*no DNA*) and size markers (*λ/HIII*) are shown in the two leftmost lanes. **d)** Schematic representation of the probes (P1 and P2) utilized for DNA blot analysis. **e)** DNA blot analysis of DNA extracted from AGL-1/pABr1-transformed mycelia (*"transformant"*, *lane 1*), mock-transformed mycelia (*WT*, lane 2), and AGL-1/ABr-1 bacterial cells (*pABr1*, lane 3), hybridized with the P1 probe, which detects T-DNA integration regions. After stripping, the same membrane was hybridized with the P2 control probe.

number of contiguous fluorescent hyphae (see Figure 2a "21d" and b), in keeping with the hypothesis that transformation mainly occurs via stable integration rather than through the incorporation of extra-chromosomal elements. Transformation was confirmed at the molecular level by the detection of an *sgfp* PCR amplification product in transformed mycelia and the lack of a similar product when using increasing amounts of genomic DNA from mock infected wild-type mycelia as template (Figure 2c). To demonstrate the presence of integrated transgene copies, we further investigated the fate of the transgene by DNA blot analysis. Since very little amounts of genomic DNA can usually be obtained from *Tuber* mycelia, and the number of transformed nuclei is far less than that of untransformed nuclei, unique regions of integration were difficult to visualize. To overcome this problem, also considering that random, non-sequence specific T-DNA integration has been shown to be the main outcome of ATM-transformation in other fungi (Bundock et al. 2002), we mapped T- DNA insertions at the genomic level by hybridization analysis conducted with two specific probes (named P1 and P2).

The P1 probe was obtained by using an 840 bp sequence derived from *Nco*I-*Xho*I digestion of pABr1 as template (Figure 2d). P1 recognizes the T-DNA region between the left border and the *hph* sequence. This region presumably recombines with the *Tuber* genome as described in other fungi (Thomas and Jones 2007). The other probe (P2) was generated by using as template, a 954 bp region of the T-DNA obtained by *Nco*I digestion (Figure 2d). This probe recognizes an internal sequence comprising part of the *hph* gene, the *pdgp* promoter and the entire *ToxA* promoter sequence, that is present regardless of the integrated or extra-chromosomal state of the T-DNA. The P2 probe was thus used as a positive control for hybridization experiments, to verify that DNA blotting had been performed correctly. Total genomic DNA from transformed and untransformed mycelia was then subjected to *Nco*I digestion, which would generate a smear of fragments recognized by the P1 probe only in the case of T-DNA integration. As shown in Figure 2e (left panel), a weak, P1 probe-positive band was only observed in transformed mycelia. The fact that the size of this band (~10,000 bp) was much higher than expected (840 bp) strongly suggests the occurrence of an integration event. In the same sample, a stronger signal resulting from the sum of integrated and extra chromosomal T-DNA fragments of the same size, was observed with the P2-probe, which hybridized with a DNA fragment, located within the T-DNA region, having an expected size of 954 bp (Figure 2e, right panel). Also in this case, the P2-associated signal was only observed in the case of DNA derived from transformed, but not mock-infected, mycelia. Altogether these data suggest the integration of at least some copies of the pABr1-derived T-DNA in the *Tuber* genome.

Improved transformation efficiency of the shortened pABr3 vector

pABr1 is a binary vector containing the T-DNA sequence interposed between the left border and the right border. It also contains an unmodified *gfp* gene, under control of the *CaMV35S* promoter and a *NOS* terminator (necessary for plant transformation), between the *Hin*dIII site and the right border of the T-DNA (Figure 1a). While contributing to, and considerably expanding the size of the T-DNA, this region is presumably not necessary for *Tuber* transformation and it was thus eliminated. The T-DNA of the resulting vector, designated as pABr3, only contained sequences derived from the pCT74 plasmid (*hph* gene, *gpd* promoter, *ToxA* promoter and the *sgfp* reporter gene). We hypothesized that the smaller size of the T-DNA sequence harbored by pABr3, corresponding to approximately 1/3 of the sequence contained in pABr1, may result in a higher transformation efficiency, thus increasing the number of transferred T-DNA copies. Indeed, as shown in Figure 3, a comparison of pABr1- and pABr3-transformed hyphae (obtained after 3 days of co-cultivation with the AGL-1/pABr1 and AGL-1/pABr3 *Agrobacterium* strains) revealed an approximately 6-fold higher transformation efficiency with the latter binary vector, likely resulting from the combined effect of a smaller sized T-DNA and an enhanced expression of the SGFP reporter. Moreover, as revealed by confocal microscopy, the SGFP signal was found to be strikingly stable and diffuse even 1 year after the initial transformation, further indicating the increased performance of this optimized-transformation procedure.

Since it was previously impossible to obtain portions of *Tuber* mycelia capable of growing on media containing more than 15 μg/ml hygromycin B, we also analyzed the ability of AGL-1/pABr3-transformed mycelia to grow in the presence of increasing amounts of hygromycin B. To this end, transformed mycelia were transferred to plates containing 15 μg/ml hygromycin B and the outline of the mycelium was marked so to allow an easy visualization of mycelial clumps overgrowth over time. After 60 days (during which mycelia were transferred to fresh antibiotic-containing medium every 10 days), full-bodied mycelial clumps were found to propagate on medium containing 15 μg/ml of hygromycin B (Figure 3c, top panel). Following isolation and transfer of hygromycin-resistant mycelial slices to media containing increasing concentrations of hygromycin B, we found that transformants were able to propagate in the presence of hygromycin B concentrations as high as 300 μg/ml (Figure 3c, bottom panel). Mycelia subjected to this high-stringency

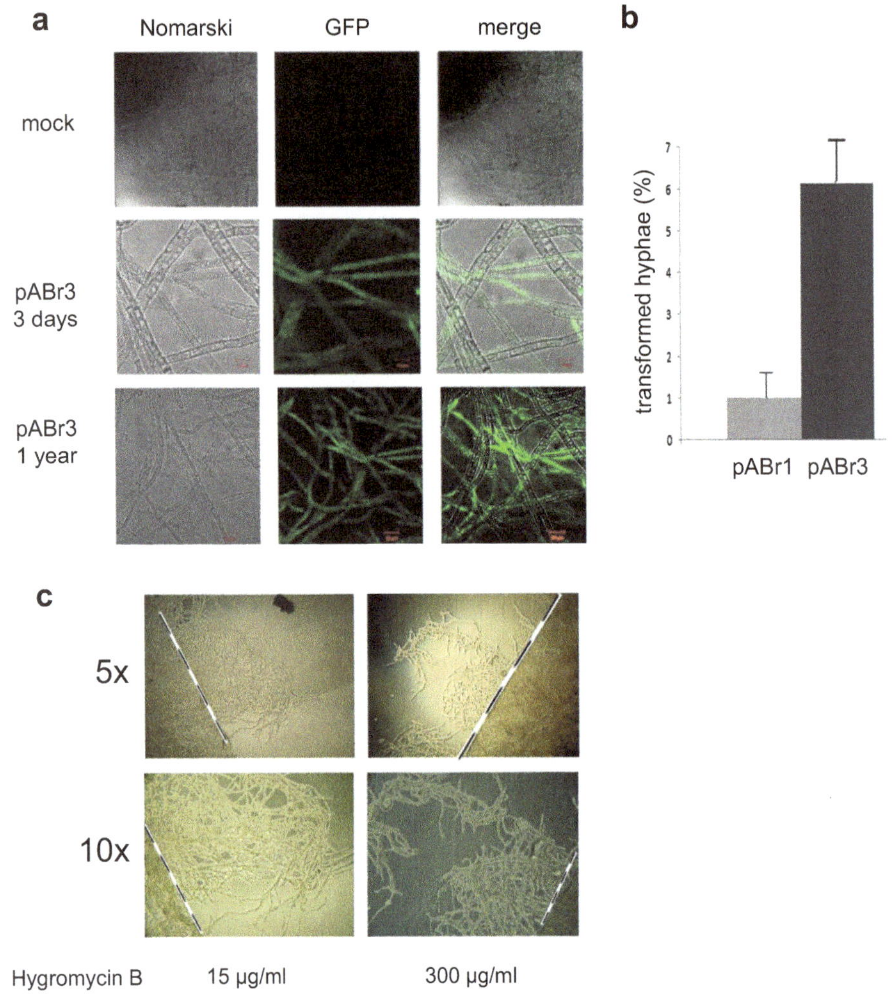

Figure 3 pABr3 transformation, comparison of transformation efficiency with pABr1, and hygromycin transformant amplification. **a)** Images (100x magnification) of pABr3-transformed hyphae, at 3 days and 1 year after transformation as indicated, obtained by phase-contrast (*Nomarski*) and SGFP fluorescence microscopy. A negative, untransformed control (*mock*) is shown at the top; a merge of the two types of images is shown in the rightmost panels. **b)** Comparison of the transformation efficiencies obtained with pABr1 and pABr3. Data are the mean ± s.e.m. of at least five independent experiments. **c)** Transformed mycelia were grown on hygromycin (15 µg/ml)-supplemented medium for 60 days and then transferred to solid medium containing the indicated concentrations of hygromycin B. After 7 days, mycelia were visualized under a stereo-microscope at the indicated magnifications to detect the presence of tufts grown beyond the dotted line.

selection procedure were used as starting material for further molecular analyses (see below).

Integration site identification by TAIL-PCR

The above results suggest that *A. tumefaciens* transforms *T. borchii* via genomic integration of the T-DNA. We further investigated this point by isolating T-DNA integration sites comprised of unknown *Tuber* genome sequences flanking known sequences of the T-DNA borders. To overcome potential difficulties deriving from the syncytium organization of *Tuber* hyphae, which upon transformation become a mosaic with only one transformed nucleus surrounded by a multitude of untransformed nuclei, we used TAIL-PCR as a sensitive tool to recover flanking-DNA sequences (Liu and Whittier 1995). This consisted of three subsequent PCR cycles amplifying a region interposed between the T-DNA border and the flanking genomic DNA using different, LB or RB, site-specific primers and a degenerate primer. The high specificity of this reaction makes TAIL-PCR a powerful tool to verify T-DNA integration in fungal cells after ATM transformation (Rolland et al. 2003). DNA obtained from mycelia transformed with AGL-1/pABr3 and subjected to high-stringency hygromycin selection, together with DNA from control untransformed mycelia was used for TAIL-PCR analysis. While the first cycle of TAIL-PCR, conducted on the left border, generated non-specific amplification products, the last cycle yielded specific products that were only detectable in genomic DNA from transformed mycelia (Figure 4a). The same approach applied to

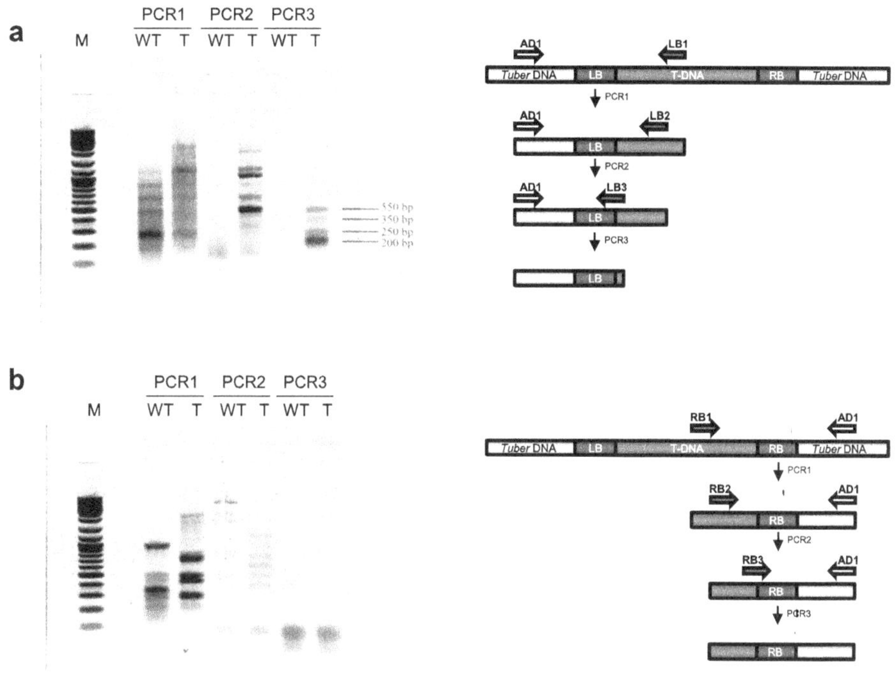

c

>Clone 2 (~250bp)
GTGATTAGTGCAGAACCAGAGAGAACCCCGCCAAGCAGCTCATATCGCAA
ACGAGCTATCCATTCTCCTTACATCTACCAAAACCCACGATCCTATTATA
TAAAAACATCCACCTGATCAATTAGTCAGTACAGTAGGGCACAGTCCACG
ATTAACTTAATAACACATTGCGGACGTTTT**TAATGTACTGAATTAACGCC**
GAATTAATTC

>Clone 8 (~250bp)
GTGATTAGTGCAGAACCACAGAGAACCCCGCCAAGCAGCTCATATCGCAA
ACGAGCTATCCATTCTCTTTACATCTACCAAAACCCACGATCCTATTATA
TAAAAACATTCACCTGATCAATTAGTCAGTACAGTAGGGCACAGTCCACG
ATTAACTTAATAACACATTGCGGACGTTTT**TAATGTACTGAATTAACGCC**
GAATTAATTC

>Clone 10 (~200bp)
AGATTTGTGGAGTAGCAGAGATAATAGTAGATTCCGTTACAGCCTCCTTG
CTCAACCAAAGCTTCCTCTTTACCCGGAAAACCCCAAACTTTAACTCCCT
TAGTGCCCTCTGACACATGCAGCTCCCGGAATCAC

>Clone 11 (~550bp)
GTGATTAGTGGAGAAGCAGAGAGGCGGGAGGCGACAATCCTCTCGAGGAT
CTTGGAGACCGTTTCCAGAAGGACGATAATTCGAAAAGAGGCCGGCGAGT
CATAGGAGGGCTTCCCAGGCTTGTCGAGAACTACGCCGTGCGCTCGTTTG
AGAGAGGCGGGGTGGTGGCCGTATACCAGGAGCGGTGAAAGAAGGGTGGT
CAGAATGTGGGGGCAGGCTTTATGCAGGGTCTTTCAAACAGAGTAGGGAA
TAGTATCCGGACCAGGGGCGGACGACGGAGAGCATTTTGAGAGGGCAGCG
GCTATCTCGCTCTGTTTCCTGCTCATTCTCATACTTCATACTTATCACCT
GTACCTACCACTCTCCAGCCTGTACCTATATATTGTGGTGTAAACAAATT
GACGCTTAGACAACTTAATAACACATTGCGGACGTTTT**TAATGTACTGAA**
TTAACGCCGAATTAATTC

>Clone 12 (~350bp)
AGATAGTGGAGAAGCAGAGAGAAGAGGAAAACCCCACCGGCCCTACCACA
CTCAAGAGGAAGAACAGGACACCCTTACAGTCACAAATACTCCCAACCAT
GTAGGCTCATCTGGGACTGCCAACAGAGCCCTTGAAGCCTACATTACAGT
CAATGGCCACAAAGCGAAGGCCTTATTCGATACAGGAACTATGAGAAATA
ACCTAATCTCTGGAAAATTCATCTCCACCTTCCT**TAATGTACTGAATTAA**
CGCCGAATTAATTC

Figure 4 TAIL-PCR and integration site identification. a) TAIL-PCR amplification of DNA fragments covering pABr3 integration sites. Representative results obtained from three amplification runs (PCR1, PCR2, PCR3) carried out with three nested primers annealing with the left border (LB) of the T-DNA (LB1, LB2 and LB3) and a degenerate primer annealing with the neighboring genome sequence (AD1) (see explanatory scheme on the right and "Materials and Methods" for details). Template genomic DNA was extracted from a pABr3-transformed mycelium (*T*) and from a mock-transformed mycelium (*WT*); size markers (*M*) are in the leftmost lane. **b)** Amplification of the right border (RB). **c)** Integration site sequences retrieved from TAIL-PCR analysis of transformants #2, #8, #10, #11 and #12; The underlined and highlighted in bold sequences represent the portion of the T-DNA adjacent to the unknown sequence obtained by PCR.

the right border did not yield any amplicon (Figure 4b). This is a common problem previously described in various plant studies (Mazars et al. 1991) and it is likely caused by illegitimate recombination events that also occur during fungal transformation (Choi et al. 2007). Sequence analysis of transformant-specific LB amplification products revealed T-DNA/genomic DNA junction sites (Figure 4c). The non-T-DNA portions of the sequences thus retrieved were used as queries for a BLAST search performed against the heterologous, but likely closely related, *T. melanosporum* genome (Martin et al. 2010). Two sequences (#10 and # 11) yielded positive matches, corresponding, respectively, to a region between two closely located transposable elements and to a retrotransposable element gene. The apparent lack of positive matches for the other non-T-DNA sequences retrieved from TAIL-PCR may be explained by the relatively short length of such sequences compared to the extremely high number and variability of transposons and other repeated elements in the *T. melanosporum* genome utilized as reference for BLAST search. TAIL-PCR analysis thus provided final proof of the integrative transfer of the T-DNA region into the *Tuber* genome.

Discussion

Tuber spp. are ecologically and economically important filamentous Ascomycetes. These fungi, whose fruitbodies are commonly known as truffles (Pegler et al. 1993), have a subterranean habitat and establish a symbiotic relationship with several trees and shrubs via ectomycorrhiza formation (Hebe et al. 1999). Crucial for the success of the ectomycorrhizal interaction is the mutualistic exchange of nutrients between the symbiont and the host plant. Indeed, inoculation of plants with *Tuber* mycelia leads to an increased plant growth, indicating that the fungus is a mutualistic symbiont (Giovannetti and Fontana 1982). Recently, the genome sequence of a representative *Tuber* species, the black truffle *T. melanosporum*, has been determined (Martin et al. 2010). Although a large fraction of genes has been functionally defined by sequence homology, a consistent number of them (>40%) remains functionally unknown. Reverse genetic analysis will thus be required in order to assign a role to these orphan genes. In *Tuber*, conventional genetic transformation procedures failed due to inefficient protoplast regeneration (Poma et al. 2006). We sought to circumvent this problem with the use of *A. tumefaciens*-mediated transformation, a foreign DNA transfer strategy largely used in plants (Sheng and Citovsky 1996) and fungi (Michielse et al. 2005).

Our first attempt toward ATM-transformation in *T. borchii* was based on the pBGgHg binary vector (Grimaldi et al. 2005). Although the T-DNA entered host cells, transformation was transient and the organization of T-DNA (extrachromosomal vs. integrated) was not clear. Moreover, transformation efficiency was rather low (0.2-0.5%) and the EGFP fluorescence signal as well as hygromycin resistance were lost quite early, typically after 4–5 weeks of in vitro culture. To overcome these limitations, while trying to achieve integrative transformation, we built two new binary vectors (pABr1 and pABr3). In both pABr1 and pABr3, GFP expression is driven by the strong *ToxA* promoter from the pathogenic fungus *Phirenophora trittici repentis*, which has been employed successfully for transgene overexpression in other filamentous fungi (Freitag et al. 2001; Lorang et al. 2001). Transformation with AGL-1/pABr1 (Figure 1a) produced stable transformants with a strong and long-lasting SGFP expression (Figure 1b). Also, transformation efficiency increased by about two-fold compared to AGL-1/pBGgHg (Figure 1c). This difference in transformation efficiency was confirmed with GV3101, an *Agrobacterium* strain less virulent than AGL-1 (Additional file 1: Figure S2).

A further increase in transformation efficiency (~6-fold compared to pBGgHg) was obtained with pABr3 (Figure 3a, b), a shortened derivative of pABr1, lacking the unnecessary plant promoter/GFP/NOS terminator region (Figure 1a). This suggests that a smaller T-DNA size facilitates DNA transfer into host cells. Another important goal in order to strengthen selection efficiency was to maximize the concentration of hygromycin B that could be tolerated by transformed mycelia. Unlike transformation with the other AGL-1/binary vector combinations, AGL-1/pABr3 allowed the amplification of transformed mycelial clumps capable of growing in the presence of hygromycin B concentrations as high as 300 μg/ml (Figure 3c). This enabled the selective isolation and propagation of transformed mycelial regions. This is important, especially considering the lack of conidia-like elements in *Tuber* spp., the syncytial nature of *Tuber* hyphae, and thus the fact that transformants are mosaics, likely bearing more than one transformed locus. We noted, however, that transformed hyphae are not randomly distributed within mycelia, but appear to be concentrated at the periphery of mycelial clumps, the only region from which the fungus could be propagated. This suggests that T-DNA may require a metabolically active tissue for activation and efficient functioning.

With pAbr3, the SGFP signal not only increased in intensity and distribution, but most of the analyzed mycelial sections appeared to be enriched in transformed hyphae even after 1 year from the initial transformation. In fact, in transformed mycelia, the GFP signal progressively became more diffuse and increasingly spread to contiguous hyphae (Figure 3a). This can be explained if the T-DNA is integrated in the *Tuber* genome, so that following duplication of a transformed nucleus the amount of produced/accumulated GFP gradually increases over time.

Integrative transfer was indeed demonstrated by PCR amplification of the *ToxA* (Figure 1d) and the *sgfp* (Additional file 1: Figure S2d) sequences, as opposed to the negative amplification results obtained with the *kanR* gene region, which is located outside of the T-DNA (Figure 1d and Additional file 1: Figure S2d). This was further corroborated by the results of hybridization analyses conducted with region-specific probes. Most notably, the fact that the size of the fragment recognized by a probe hybridizing with the left T-DNA border (i.e., a region that is usually involved in the integration process (Thomas and Jones 2007)) increased from an expected size of 854 bp to ~10,000 bp.

Lastly, we used transformed, hygromycin B-selected mycelia for the identification of the genomic sites of integration by TAIL-PCR (Liu and Whittier 1995), a widely used approach for the analysis of transformants in systems where homokaryotic colonies are hard to obtain (Poma et al. 2006). We sequenced T-DNA/*Tuber* genomic DNA flanking site amplicons obtained from transformed mycelia and found that two of them matched transposon-related elements (Figure 4). In *Tuber*, repeated elements, including retrotransposons, account for about 60% of the genome (Martin et al. 2010), suggesting that integrative transformation mediated by heterologous recombination is likely driven by repetitive DNA sequences.

The highly improved genetic transformation procedure described in this work represents a first, but crucial step toward a functional genomic analysis of *Tuber*, and perhaps other transformation-recalcitrant filamentous fungi as well, with the ultimate goal of producing a comprehensive site-specific mutant collection as in *Aspergillus nidulans*, *N. crassa* and *Magnaporthe oryzae* (http://www.fgsc.net/). Although whole genome sequence data are only available for *T. melanosporum* so far, we used *T. borchii* as an experimental organism to set up this improved transformation procedure because of the easier and more efficient propagation of its mycelium under axenic culture conditions. However, ongoing work, including the setting up of more efficient mycelial growth conditions, indicates that the optimized AGL-1/pABr3-protocol here applied to *T. borchii* also works well in *T. melanosporum*, with the production of well detectable levels of GFP and resistance to hygromycin concentrations as high as 50 μg/ml (A.B and P.B., unpublished).

The next step will be the disruption of non-homologous end joining (NHEJ) mediated by the Ku70 and Ku80 proteins in order to favor integration at homologous sites. The NHEJ pathway, discovered for the first time in mammals (Jeggo et al. 1991) and so named because it does not require homologous sequences to join DNA ends, repairs double-strand breaks in the absence of a homologous donor (Moore and Haber 1996). As shown in *N. crassa* (Ninomiya et al. 2004), disruption of the ku70/ku80 genes strongly enhances homologous recombination and thus the production of site-specific mutants. The presence of ku70/ku80 homologs in the *T. melanosporum* genome (GSTUMT00005220001 and GSTUMT000019 28001 gene models; http://mycor.nancy.inra.fr) suggests that such an approach may also be feasible in *Tuber*.

Additional file

Additional file 1: Figure S1. Outline of the ATM-transformation protocol; **Figure S2.** ATM-transformation with the pABr1 vector using GV3101 as recipient strain.

Abbreviations

ATM: *Agrobacterium tumefaciens*-mediated; egfp: Enhanced green fluorescent protein; gpd: Glyceraldehyde-3-phosphate dehydrogenase; hph: Hygromycin B phosphotransferase; kanR: Kanamycin resistance gene, aminoglycoside 3'-phosphotransferase; LB: Left border; PDA: Potato-dextrose agar; RB: Right border; sgfp: Synthetic green fluorescent protein; TAIL-PCR: Thermal Asymmetric Interlaced PCR; T-DNA: Transfer DNA; Ti: Tumor-inducing.

Competing interests

The authors declare that they have no competing interests.

Authors' contributions

AB was in charge of plasmid vector construction and confocal microscopy analysis, set up the transformation protocol and prepared the first draft of the manuscript; BM contributed to plasmid vector construction, confocal microscopy analysis, final editing of the manuscript and figure assembly; EM performed Tail-PCR experiments; MP was in charge of integration site analyses and T. melanosporum experiments; PF contributed to data analysis; SO proposed the project, designed part of the experiments and wrote the last draft of the manuscript; PB coordinated the project, designed part of the experiments and contributed to manuscript drafting.

Acknowledgments

We are grateful to P. Romaine and to L. Ciuffetti for the kind gift of the BGgHg and CT74 plasmids, respectively. We thank R. Gerace for help in maintaining the *Tuber* strains, G. Buglia for skillful assistance with microscopy analysis and P. Costantino's group for the GV3101 *A. tumefaciens* strain. P.B. thanks La Sapienza University, Rome (Fondi di Ateneo) for partial financial support. A.B. was supported by a Teresa Ariaudo fellowship.

Author details

[1]Pasteur Cenci Bolognetti Foundation, c/o Department of Biology and Biotechnology "Charles Darwin", Sapienza University, Piazzale A. Moro 5, 00185 Rome, Italy. [2]Department of Life Sciences, Biochemistry and Molecular Biology Unit, Laboratory of Functional Genomics and Protein Engineering, University of Parma, 43124 Parma, Italy. [3]Institute of Biology and Molecular Pathology, CNR, 00185 Rome, Italy.

References

Ambra R, Grimaldi B, Zamboni S, Filetici P, Macino G, Ballario P (2004) Photomorphogenesis in the hypogeous fungus *Tuber borchii*: isolation and characterization of Tbwc-1, the homologue of the blue-light photoreceptor of *Neurospora crassa*. Fungal Genet Biol 41(7):688–697

Bundock P, Hooykaas PJ (1996) Integration of *Agrobacterium tumefaciens* T-DNA in the *Saccharomyces cerevisiae* genome by illegitimate recombination. Proc Natl Acad Sci USA 93(26):15272–15275

Bundock P, den Dulk-Ras A, Beijersbergen A, Hooykaas PJ (1995) Trans-kingdom T-DNA transfer from *Agrobacterium tumefaciens* to *Saccharomyces cerevisiae*. The EMBO journal 14(13):3206–3214

Bundock P, van Attikum H, den Dulk-Ras A, Hooykaas PJ (2002) Insertional mutagenesis in yeasts using T-DNA from *Agrobacterium tumefaciens*. Yeast 19 (6):529–536

Chen X, Stone M, Schlagnhaufer C, Romaine CP (2000) A fruiting body tissue method for efficient *Agrobacterium*-mediated transformation of *Agaricus bisporus*. Appl Environ Microbiol 66(10):4510–4513

Choi J, Park J, Jeon J, Chi MH, Goh J, Yoo SY, Park J, Jung K, Kim H, Park SY, Rho HS, Kim S, Kim BR, Han SS, Kang S, Lee YH (2007) Genome-wide analysis of T-DNA integration into the chromosomes of *Magnaporthe oryzae*. Mol Microbiol 66(2):371–382

Ciuffetti LM, Tuori RP, Gaventa JM (1997) A single gene encodes a selective toxin causal to the development of tan spot of wheat. Plant Cell 9(2):135–144

Freitag M, Ciuffetti LM, Selker EU (2001) Expression and visualization of Green Fluorescent Protein (GFP) in *Neurospora crassa*. Fungal Genet Newsletter 48:15–19

Giovannetti G, Fontana A (1982) Mycorrhizal synthesis between *Cistaceae* and *Tuberaceae*. New Phytol 92(4):533–537

Green MR, Sambrook J (2012) Molecular cloning: a laboratory manual, 4th edn. Laboratory Press, Cold Spring Harbor, NY

Grimaldi B, de Raaf MA, Filetici P, Ottonello S, Ballario P (2005) *Agrobacterium*-mediated gene transfer and enhanced green fluorescent protein visualization in the mycorrhizal ascomycete *Tuber borchii*: a first step towards truffle genetics. Curr Genet 48(1):69–74

Hebe G, Hager A, Salzer P (1999) Initial signalling processes induced by elicitors of ectomycorrhiza-forming fungi in spruce cells can also be triggered by G-protein-activating mastoparan and protein phosphatase-inhibiting cantharidin. Planta 207(3):418–425

Jeggo PA, Tesmer J, Chen DJ (1991) Genetic analysis of ionising radiation sensitive mutants of cultured mammalian cell lines. Mutat Res 254(2):125–133

Lazo GR, Stein PA, Ludwig RA (1991) A DNA transformation-competent Arabidopsis genomic library in *Agrobacterium*. Biotechnology 9(10):963–967

Liu YG, Whittier RF (1995) Thermal asymmetric interlaced PCR: automatable amplification and sequencing of insert end fragments from P1 and YAC clones for chromosome walking. Genomics 25(3):674–681

Lorang JM, Tuori RP, Martinez JP, Sawyer TL, Redman RS, Rollins JA, Wolpert TJ, Johnson KB, Rodriguez RJ, Dickman MB, Ciuffetti LM (2001) Green fluorescent protein is lighting up fungal biology. Appl Environ Microbiol 67(5):1987–1994

Martin F, Kohler A, Murat C, Balestrini R, Coutinho PM, Jaillon O, Montanini B, Morin E, Noel B, Percudani R, Porcel B, Rubini A, Amicucci A, Amselem J, Anthouard V, Arcioni S, Artiguenave F, Aury JM, Ballario P, Bolchi A, Brenna A, Brun A, Buee M, Cantarel B, Chevalier G, Couloux A, Da Silva C, Denoeud F, Duplessis S, Ghignone S et al (2010) Perigord black truffle genome uncovers evolutionary origins and mechanisms of symbiosis. Nature 464(7291):1033–1038

Mazars GR, Moyret C, Jeanteur P, Theillet CG (1991) Direct sequencing by thermal asymmetric PCR. Nucleic Acids Res 19(17):4783

Michielse CB, Hooykaas PJJ, van den Hondel CAMJJ, Ram AFJ (2005) *Agrobacterium*-mediated transformation as a tool for functional genomics in fungi. Curr Genet 48(1):1–17

Montanini B, Betti M, Marquez AJ, Balestrini R, Bonfante P, Ottonello S (2003) Distinctive properties and expression profiles of glutamine synthetase from a plant symbiotic fungus. Biochem J 373(Pt 2):357–368

Moore JK, Haber JE (1996) Cell cycle and genetic requirements of two pathways of nonhomologous end-joining repair of double-strand breaks in *Saccharomyces cerevisiae*. Mol Cell Biol 16(5):2164–2173

Ninomiya Y, Suzuki K, Ishii C, Inoue H (2004) Highly efficient gene replacements in *Neurospora* strains deficient for nonhomologous end-joining. Proc Natl Acad Sci U S A 101(33):12248–12253

Niwa Y (2003) A synthetic green fluorescent protein gene for plant biotechnology. Plant Biotechnology 20:1–11

Pegler DN, Spooner BM, Young TWK (1993) British truffles: a revision of British hypogeous fungi. Royal Botanic Gardens, Kew, London, UK

Poma A, Limongi T, Pacioni G (2006) Current state and perspectives of truffle genetics and sustainable biotechnology. Appl Microbiol Biotechnol 72(3):437–441

Rolland S, Jobic C, Fevre M, Bruel C (2003) *Agrobacterium*-mediated transformation of *Botrytis cinerea*, simple purification of monokaryotic transformants and rapid conidia-based identification of the transfer-DNA host genomic DNA flanking sequences. Curr Genet 44(3):164–171

Sheng J, Citovsky V (1996) *Agrobacterium*-plant cell DNA transport: have virulence proteins, will travel. Plant Cell 8(10):1699–1710

Singer K, Shiboleth YM, Li J, Tzfira T (2012) Formation of complex extrachromosomal T-DNA structures in *Agrobacterium tumefaciens*-infected plants. Plant Physiol 160(1):511–522

Thomas CM, Jones JD (2007) Molecular analysis of *Agrobacterium* T-DNA integration in tomato reveals a role for left border sequence homology in most integration events. Mol Genet Genomics 278(4):411–420

Wang JY, Li HY (2008) *Agrobacterium tumefaciens*-mediated genetic transformation of the phytopathogenic fungus *Penicillium digitatum*. J Zhejiang Univ Sci 9(10):823–828

Zupan J, Muth TR, Draper O, Zambryski P (2000) The transfer of DNA from *Agrobacterium tumefaciens* into plants: a feast of fundamental insights. Plant J 23(1):11–28

2

Highly conserved salt bridge stabilizes a proteinase K subfamily enzyme, Aqualysin I, from *Thermus aquaticus* YT-1

Masayoshi Sakaguchi[1*†], Kanae Osaku[1†], Susumu Maejima[1], Nao Ohno[1], Yasusato Sugahara[1], Fumitaka Oyama[1] and Masao Kawakita[1,2]

Abstract

The proteinase K subfamily enzymes, thermophilic Aqualysin I (AQN) from *Thermus aquaticus* YT-1 and psychrophilic serine protease (VPR) from *Vibrio* sp. PA-44, have six and seven salt bridges, respectively. To understand the possible significance of salt bridges in the thermal stability of AQN, we prepared mutant proteins in which amino acid residues participating in salt bridges common to proteinase K subfamily members and intrinsic to AQN were replaced to disrupt the bridges one at a time. Disruption of a salt bridge common to proteinase K subfamily enzymes in the D183N mutant resulted in a significant reduction in thermal stability, and a massive change in the content of the secondary structure was observed, even at 70°C, in the circular dichroism (CD) analysis. These results indicate that the common salt bridge Asp183-Arg12 is important in maintaining the conformation of proteinase K subfamily enzymes and suggest the importance of proximity between the regions around Asp183 and the N-terminal region around Arg12. Of the three mutants that lack an AQN intrinsic salt bridge, D212N was more prone to unfolding at 80°C than the wild-type enzyme. Similarly, D17N and E237Q were less thermostable than the wild-type enzyme, although this may be partially due to increased autolysis. The AQN intrinsic salt bridges appear to confer additional thermal stability to this enzyme. These findings will further our understanding of the factors involved in stabilizing protein structure.

Keywords: Serine protease; Subtilase; Proteinase K subfamily; Salt bridge; Thermal stability

Introduction

The molecular bases of protein adaptation to high and low temperatures are interesting from both basic and practical standpoints, as knowledge regarding these factors would enable the construction of genetically engineered proteins that could function under a variety of conditions. Psychrophilic and mesophilic enzymes are used in biotechnological applications requiring high activity at mild temperatures or quick heat-inactivation at moderate temperatures. Thermophilic and hyperthermophilic enzymes have major biotechnological advantages over mesophilic and psychrophilic enzymes because of their high activities at higher temperatures and substrate concentrations as well as their resistance to chemical denaturants. Thus far, various intramolecular interactions, including ionic interactions, hydrogen bonding and hydrophobic interactions, are assumed to make important contributions to the stability and maintenance of enzyme structure as well as the catalytic functions; however, their contributions have not been fully defined in individual cases. In addition, the comparative structural analysis of psychrophilic, mesophilic and thermophilic enzymes indicated that each protein family adopts a different structural strategy to adapt to different temperature ranges (Siezen & Leunissen 1997; Struvay & Feller 2012). To understand the thermal adaptation strategy of proteins, comparative studies among members of a protein family using site-directed mutagenesis as well as laboratory evolution via random mutagenesis using error-prone PCR will provide valuable information.

According to the MEROPS peptidase database (http://merops.sanger.ac.uk/), subtilisin-like protease (subtilase) superfamily members are classified as the S8 subfamily in the serine protease superfamily. These proteins exhibit a

* Correspondence: bt11532@ns.kogakuin.ac.jp
†Equal contributors
[1]Department of Applied Chemistry, Kogakuin University, 2,665-1 Nakano-cho, Hachioji, Tokyo 192-0015, Japan
Full list of author information is available at the end of the article

highly conserved arrangement of amino acids in the active site and have very similar overall structures consisting of an α/β protein scaffold. Nevertheless, their temperature stability profiles differ widely, and they may be psychrophilic, mesophilic, thermophilic or hyperthermophilic depending on the characteristics of the organisms from which they are derived. Because of these characteristics, they appear to be suitable for comparative studies to elucidate the basis of structure-function relationships. Aqualysin I (AQN) is an alkaline serine-protease produced by the Gram-negative thermophilic bacterium *Thermus aquaticus* YT-1 (Matsuzawa et al. 1983; Matsuzawa et al. 1988). Based on an analysis of sequence homology, AQN is classified into the proteinase K subfamily, which consists of a group of Gram-negative bacteria-derived proteinases within the subtilase superfamily (Siezen & Leunissen 1997). In our previous study, we demonstrated that Pro residues in the surface loops of AQN in the N-terminal region contribute significantly to its thermophilicity, and one of two disulfide bonds in AQN is more important for the catalytic activity and conformational stability of AQN than the other (Sakaguchi et al. 2007; Sakaguchi et al. 2008b). These results are consistent with those reported for a subtilisin-like serine protease from *Vibrio* sp., VPR, which is a psychrophilic counterpart of AQN in the proteinase K subfamily (Kristjánsson et al. 1999; Arnórsdóttir et al. 2002). It was found that the introduction of Pro residues into VPR at positions corresponding to those in AQN could improve its thermal stability (Arnórsdóttir et al. 2009).

To enhance the thermal stability of a protein, a common strategy is to introduce more favorable surface charge-charge interactions. However, the role of salt bridges in the stabilization of proteins remains controversial. Thermophilic proteins have an increased number of salt bridges compared with their mesophilic homologues (Kumar et al. 2000; Szilágyi & Závodszky 2000; Vogt et al. 1997). In VPR, a psychrophilic enzyme, seven salt bridges (Arg10-Asp183, Arg14-Asp274, Asp56-Arg95, Asp59-Arg95, Asp138-Arg169, Arg185-Asp260 and Glu236-Arg252) have been identified in the known structure of the enzyme (PDB accession number: 1SH7; (Arnórsdóttir et al. 2005)). However, AQN has six salt bridges (Arg12-Asp183, Asp17-Arg259, Arg31-Asp237, Arg43-Asp212, Asp58-Arg95 and Asp138-Arg169) in its structure (4DZT; (Green et al. 1993)) (Table 1). Both enzymes have a similar number of salt bridges; however, their thermal stabilities are quite different. The VPR N15D mutant, in which an Asp residue is substituted for Asn15 to form a new salt bridge (Asp15-Lys257) at the position corresponding to the Asp17-Arg259 salt bridge in AQN, exhibited increased thermal stability due to the incorporation of a new salt bridge; the thermal stability of the enzyme did not increase further in the double mutant VPR N15D/K257R (Sigurdardóttir et al. 2009). Inversely, the deletion of the Asp17-Arg259 salt bridge in the AQN D17N mutant resulted in reduced thermal stability compared to the wild-type enzyme without exerting a significant effect on the kinetic parameters of the hydrolysis reaction (Arnórsdóttir et al. 2011). These results may imply that salt bridges at appropriate locations play a vital role in the thermal stability of serine proteases, particularly those of the proteinase K subfamily. To help clarify this complicated issue and to further extend our understanding of the molecular basis of proteinase K-related enzyme stabilization, we aimed to examine the role of salt bridges in AQN by site-directed mutagenesis.

Materials and methods

Strains and growth medium

E. coli TG1 was used as the expression host, and *E. coli* DH5α (TOYOBO, Osaka, Japan) and MV1184 (TAKARA BIO INC., Shiga, Japan) were used as the genetic engineering hosts. LB medium (1% Bacto-tryptone, 0.5% Bacto-yeast extract, 1% NaCl, pH 7.0) was used. The solid medium contained Bacto-agar (1.5%). Ampicillin (50 μg/ml) or kanamycin (50 μg/ml) was added to the medium as needed.

Genetic engineering and chemical reagents

Genetic engineering experiments were performed according to the procedure described by Sambrook and Russell (Sambrook & Russell 2012). The enzymes used for genetic engineering were purchased from TAKARA BIO and used according to the manufacturer's instructions. Bacto-tryptone and Bacto-yeast extract were purchased from Becton Dickinson (Franklin Lakes, NJ, USA). Other reagents used were of the highest quality available and were obtained from Wako Pure Chemicals (Tokyo, Japan) and Sigma-Aldrich (St. Louis, MO, USA), unless otherwise specified.

Plasmid construction

The plasmid pMAQΔc2, which was designed to express wild-type AQN as a fusion protein with maltose binding protein (MBP), was constructed based on pAQNΔC105 and pMAL plasmids (New England Biolabs, Ipswich, MA, USA) as described previously (Sakaguchi et al. 2008a).

To construct plasmids with a mutated aqualysin I gene, site-directed mutagenesis was performed following the ODA-PCR method (Mutan®-Super express Km; TAKARA BIO) using pMAQΔc2 as a template. The oligonucleotide primers (Sigma-Aldrich Life Science, Hokkaido, Japan) used for site-directed mutagenesis are shown in Table 2. The fragments containing either of the mutations listed in the table were inserted into the expression vector, pMAQΔc2. The names of the mutant plasmids are provided in the third column of Table 2 (D17N, *etc.*). The

Table 1 Amino acids that form salt bridges in AQN, VPR and SPRK

	AQN (4DZT)	VPR (1SH7)	SPRK (2B6N)
Common	Arg12-Asp183	Arg10-Asp183	Arg12-Asp187
	Asp58-Arg95	Asp56-Arg95	Asp58-Arg97
	Asp138-Arg169	Asp138-Arg169	Asp140-Arg171
Intrinsic to VPR or SPRK		Asp59-Arg95	Asp61-Arg97
		Arg185-Asp260	Arg187-Asp264
		Glu236-Arg252	Asp240-Arg256
		Arg14-Asp274	Asp2-His25
			Asp198-Lys254
Intrinsic to AQN	Asp17-Arg259		
	Arg31-Glu237		
	Arg43-Asp212		

nucleotide sequences around the mutation sites as well as other parts of the gene were confirmed by DNA sequencing using an Applied Biosystems 3130 Genetic Analyzer (Applied Biosystems, Foster City, CA, USA).

Purification and activity measurement of the wild-type enzyme and its mutants

After induction by isopropyl β-D-thiogalactopyranoside (IPTG) at $OD_{660} = 0.8$, the transformants were further cultivated overnight in LB medium. The cells were harvested by centrifugation and subsequently sonicated, and the crude extract was subjected to heat treatment, hydrophobic chromatography (Butyl Sepharose; GE Healthcare, Buckinghamshire, UK) and cation exchange chromatography (Resource S; GE Healthcare) as described previously (Sakaguchi et al. 2008a). The enzymes were purified to homogeneity to yield a single band on SDS-polyacrylamide gel electrophoresis (PAGE) after staining with Coomassie Brilliant Blue R-250 (CBB R-250) (Laemmli 1970). Prior to SDS-PAGE analysis, the enzymes were treated with 25 mM phenylmethane sulfonyl fluoride (PMSF) dissolved in methanol for 30 min to prevent autolytic degradation. To minimize denaturation and autolysis which may occur at higher temperature, the enzyme activity was measured at 40°C with *N*-succinyl-Ala-Ala-Pro-Phe-*p*-nitroanilide (*N*-suc-AAPF-pNA, Sigma-Aldrich) as a substrate in 50 mM 2-[4-(2-hydroxyethyl)-1-piperazynyl]ethanesulfonic acid (HEPES)-NaOH (pH 7.5) buffer containing 1 mM $CaCl_2$. The change in absorbance at 410 nm was continuously monitored, and the activity was estimated using $\varepsilon_{410} = 8{,}680\ M^{-1}\ cm^{-1}$ as a molar absorption coefficient of *p*-nitroaniline (4-nitroaniline). One unit of enzyme was defined as the amount of enzyme that liberates 1 μmole of *p*-nitroaniline from the substrate in 1 minute. The protein concentration was measured using the micro-assay method (Bio-Rad Laboratories, Hercules, CA, USA), which is based on the Bradford method (Bradford 1976), using bovine serum albumin as a standard.

Table 2 Oligonucleotide primers used for site-directed mutagenesis

Name	Oligonucleotide sequence	Mutant name
AQN-D17N	5'-CCAGCGG*A*ACCTTCCC-3'	D17N
AQN-D58N	5'-GGTAGGCTAT*A*ACGCCTTAGGG-3'	D58N
AQN-G61D	5'-ACGCCTTAG*AT*GGGAACG-3'	G61D
AQN-D138N	5'-CTGCCCTG*A*ACAACGCC-3'	D138N
AQN-D183N	5'-ATCTTCC*A*ACGCCCGTG-3'	D183N
AQN-D212N	5'-ACACCTCG*A*ACACGGCC-3'	D212N
AQN-E237Q	5'-CTTTGTATCTACAGCAAAATCTTC-3'	E237Q
AQN-G262D	5'-GCTTTCGGATATCGGATCG-3'	G262D
AQN-S277D	5'-CCTGCTCG*ATT*CGGGGAG-3'	S277D

Mismatched bases are shown in italic.

Determination of the temperature dependence of the proteolytic activity and heat stability of the wild-type enzyme and its mutants

To examine the temperature dependence of the enzyme activity, 480 μl of 1 mM *N*-suc-AAPF-pNA solution in 100 mM HEPES-NaOH (pH 7.5) containing 1 mM $CaCl_2$ was preincubated at an appropriate temperature for 5 min; subsequently, 20 μl of enzyme solution (5 μg/ml) was added. The change in absorbance at 410 nm was continuously monitored, and the activity was estimated as described above based on the results of triplicate experiments.

To examine the heat stability, the enzymes were diluted with 20 mM 2-morpholinoethanesulfonic acid (MES)-NaOH buffer (pH 6.0) containing 1 mM $CaCl_2$ to yield a 10 μg/ml solution. These experiments were performed at pH 6.0 to diminish the massive autolysis that would occur under more alkaline condition. This enabled us to observe the differential decrease of residual activity among mutants due to structural destabilization during the heat treatment processes. The enzyme solution was incubated for the appropriate time period at 70°C or 80°C and was subsequently cooled quickly. The remaining activity was determined based on the results of triplicate experiments using 1 mM *N*-suc-AAPF-pNA as a substrate at 40°C, as described above.

Kinetic analysis

The initial rates of *N*-suc-AAPF-pNA hydrolysis induced by the wild-type enzyme and mutant enzymes were measured at 40°C in 50 mM HEPES-NaOH (pH 7.5) containing 1 mM $CaCl_2$ as described above. The kinetic parameters V_{max} and K_m were estimated assuming a Michaelis-Menten kinetic model, and the graphics software package DeltaGraph version 6 (Nihon Poladigital K.K., Tokyo, Japan) was used with non-linear regression. The apparent values of k_{cat} were estimated using a molecular mass of 28 kDa.

Unfolding study of AQN proteins based on circular dichroism (CD) measurement

CD analysis was carried out to determine the transition temperature (T_m) and to monitor the unfolding of the wild-type and mutant enzymes. Prior to the CD measurements, purified enzymes were treated with 25 mM PMSF dissolved in methanol for 30 min to prevent autolytic degradation during the measurements. After complete inactivation of the protease activity was confirmed, the samples were dialyzed overnight against 20 mM MES-NaOH buffer (pH 6.0) containing 1 mM $CaCl_2$ and filtered through a MILEX®-HV filter (0.45 μm pore size, Durapore (PVDF), Merck Millipore Ltd., Carrigtwohill, Ireland). CD measurements were conducted using a JASCO-725 circular dichroism spectropolarimeter equipped with PTC-348 Peltier type single cell holder, and the change in ellipticity at 220 nm was monitored under a constant heating rate (1°C/min) at temperatures ranging from 40 to 105°C. The melting curves were normalized according to the methods in the literature (Arnórsdóttir et al. 2002), and the melting temperature (T_m) values of the enzymes were estimated using a graphics software package, Delta graph. The experiments were performed in duplicate.

The unfolding of proteins as a function of time was observed at a constant temperature (70°C or 80°C) by CD measurement over a range of 200–250 nm. Measurements were performed with a JASCO-725 circular dichroism spectropolarimeter (JASCO, Tokyo, Japan) equipped with PTC-348 Peltier type single cell holder, and the change in ellipticity was monitored at every 5 min over a 30-min period at a constant temperature. Wavelength scans in the range of 200–250 nm were performed in rectangular quartz cells (JASCO model: T-11-ES-1) with a path length of 0.1 cm.

Results

Mutagenesis of salt bridge-forming residues in AQN and purification of the wild-type enzyme and its mutants

Table 1 lists the salt bridge-forming amino acid residues in AQN, VPR and *Serratia* proteinase K-like enzyme (SPRK; (Larsen et al. 2006)), which were identified based on the known structures of AQN (4DZT; (Green et al. 1993)), VPR (1SH7; (Arnórsdóttir et al. 2005)) and SPRK (2B6N; (Helland et al. 2006)), respectively. There are three conserved salt bridges in both AQN and VPR (Figure 1a and b). In addition, AQN and VPR have three (Asp17-Arg259, Arg31-Glu237 and Arg43-Asp212) and four (Asp59-Arg95, Arg14-Asp274, Arg185-Asp260 and Glu236-Arg252) intrinsic salt bridges, respectively (Figure 1c, d, e and f). To examine the role of salt bridges in AQN, the Asp and Glu residues were replaced with Asn and Gln residues, respectively, to make them incapable of forming salt bridges. To introduce new salt bridges in AQN at the positions at which the VPR-intrinsic salt bridges are located, the residues Gly61, Gly262 and Ser277 were converted to Asp residues to make them capable of forming salt bridges with Arg95, Arg185 and Arg16, respectively, which are conserved in both AQN and VPR. The mutant AQN constructs were expressed, and the protein products were purified to homogeneity using essentially the same methodology used for the wild-type enzyme.

Thermal stability of the wild-type enzyme and its mutants

The remaining activity of the wild-type and mutant enzymes after heat treatment at temperatures of 70°C or 80°C was determined at 40°C using *N*-suc-AAPF-pNA as a substrate; the results are illustrated in Figure 2a-f. Figure 2a and b show the time courses of the residual activity at 70°C and 80°C, respectively, for the D17N, D212N and E237Q mutants lacking a salt bridge specific to AQN. At 70°C, the activity of E237Q and D17N decreased more rapidly than that of the wild-type enzyme, indicating that the mutants are less stable at 70°C. D212N behaved similarly to the wild-type enzyme at 70°C; however, at 80°C, it was inactivated as rapidly as E237Q and D17N (Figure 2a and b). These results indicate that the AQN-intrinsic salt bridges Arg31-Glu237, Asp17-Arg259 and Arg43-Asp212 contribute significantly to the maintenance of active enzyme structures at temperatures above 70-80°C.

Disruption of the salt bridges common to AQN and VPR did not yield a common outcome. For example, the residual activity of D183N declined rapidly, with a half-life of approximately 30 min at both 70°C and 80°C, whereas the activities of D58N and D138N were almost indistinguishable from that of the wild-type enzyme at both temperatures (Figure 2c and d). The above results indicate that the Arg12-Asp183 salt bridge is important for conferring structural stability to proteinase K subfamily enzymes, although the other two common salt bridges

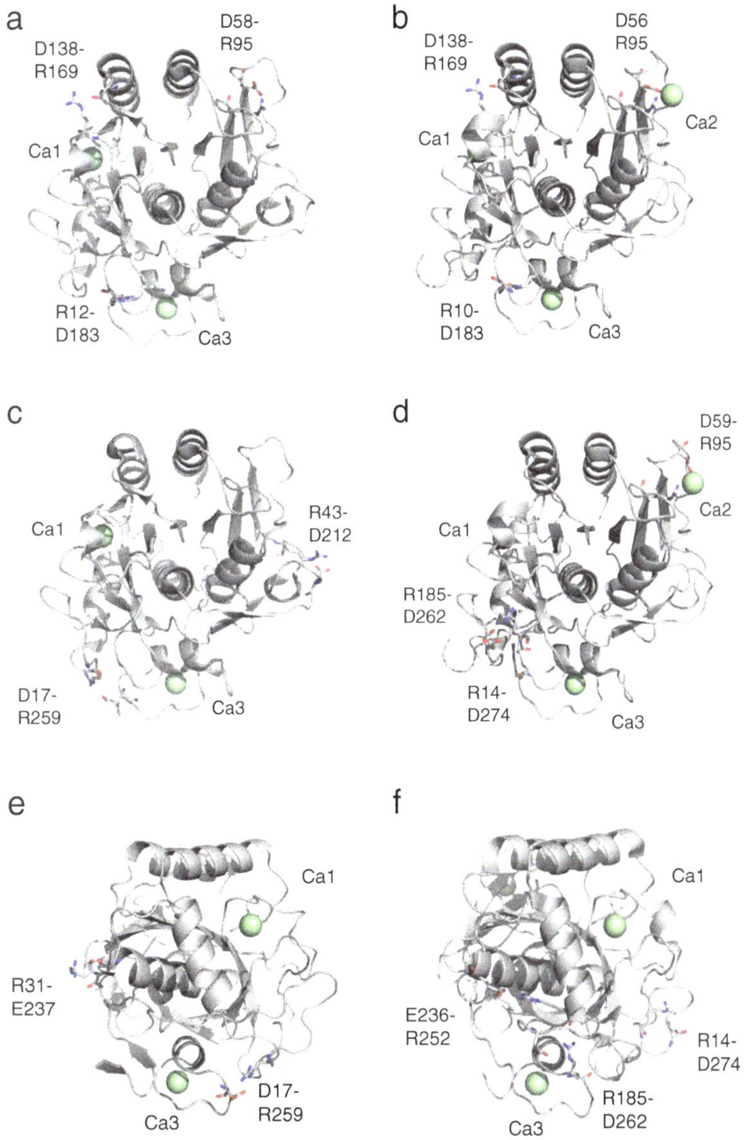

Figure 1 Crystal structures of AQN (a, c, e; 4DZT) and VPR (b, d, f; 1SH7). (a, b) Common salt bridge positions, **(c, d)** intrinsic salt bridge positions, **(e, f)** intrinsic salt bridge positions rotated approximately 90 degrees horizontally from **(c, d)**. The salt bridges of interest are indicated by residue numbers and are represented as stick models. Calcium ions (Ca1, Ca2 and Ca3) are represented as pale green spheres.

are not significantly involved in thermal stabilization. Additional salt bridges were introduced at the positions where VPR-intrinsic salt bridges are located in mutants G61D, G262D and S277D. The inactivation time courses of the three mutants were similar to that of the wild-type enzyme at 70°C and 80°C. In fact, S277D was slightly less stable than the wild-type enzyme at both 70°C and 80°C (Figure 2e and f).

Temperature dependence of wild-type and mutant enzyme activity

Figures 3a-c compares the temperature dependence of wild-type and mutant enzyme activity. All mutants displayed similar temperature-activity profiles over a range of 30-90°C. Disrupting a salt bridge or introducing a potential salt bridge-forming mutation did not lead to an extensive reduction of enzyme activity. In fact, the enzyme activity of the mutants tended to be slightly higher than that of the wild-type enzyme especially at higher temperatures. D17N showed the highest activity among the mutants tested: its activity was about 1.5-fold as high as that of the wild-type enzyme at 90°C. The optimum temperature was not altered significantly by the mutations, and every mutant showed full activity at approximately 90°C, similar to the wild-type enzyme.

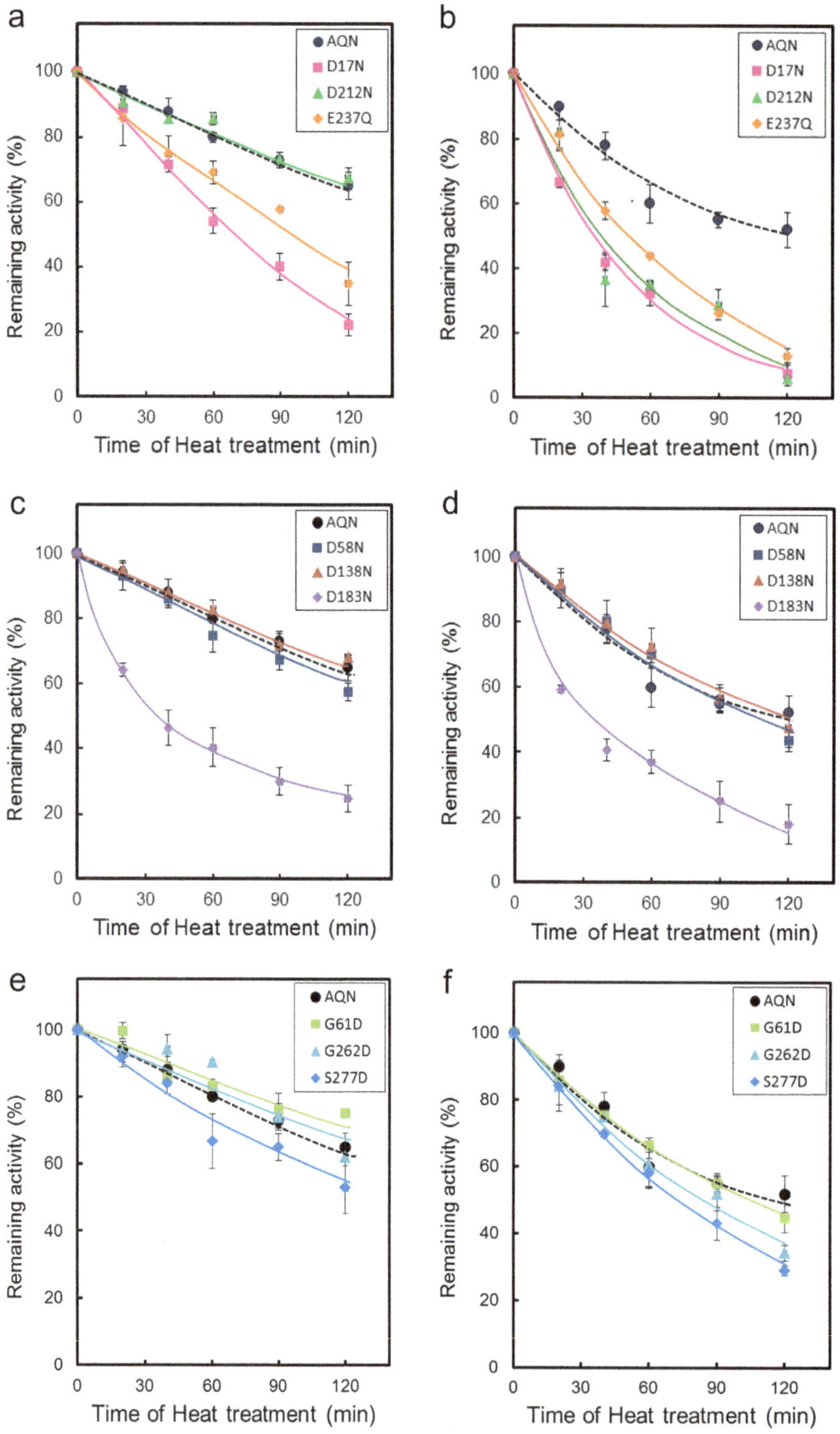

Figure 2 Heat stability of the wild-type and mutant enzymes. Heat stability was determined as described in the Materials and methods section. The remaining activities after heat treatment at 70°C **(a, c and e)** or 80°C **(b, d and f)** for the indicated times relative to the activity before heat treatment are plotted for each mutant. **(a, b)** mutants lacking an intrinsic salt bridge, **(c, d)** mutants lacking a common salt bridge, **(e, f)** mutants in which a salt bridge may have been introduced.

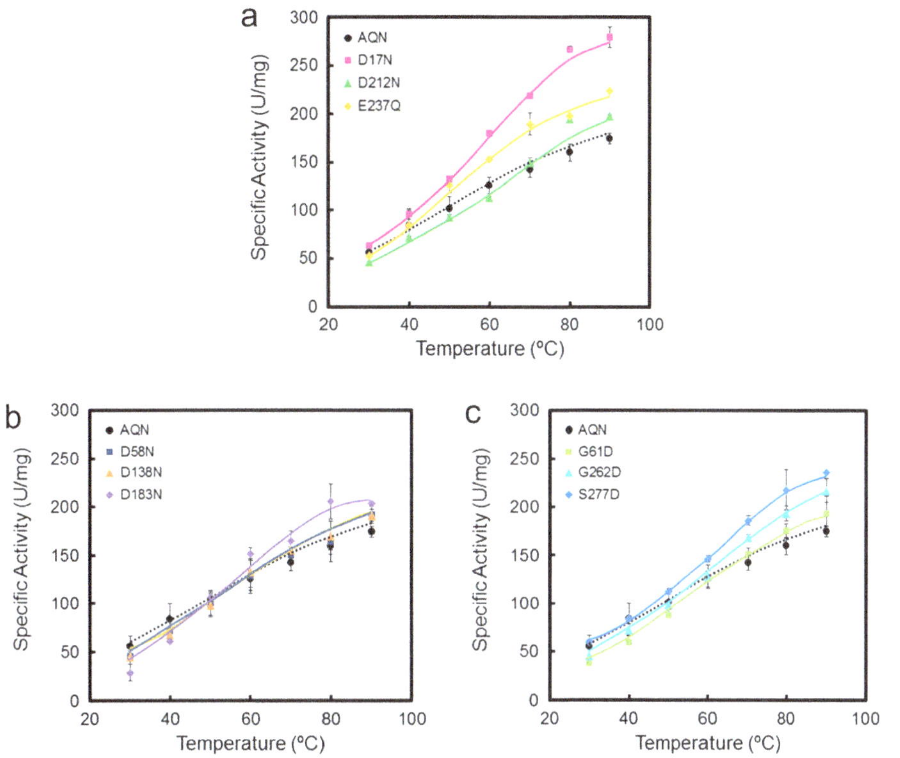

Figure 3 Temperature-dependence of the activity of the wild-type and mutant enzymes. The temperature-dependence of the activity was measured as described in the Materials and methods section. **(a)** mutants lacking an intrinsic salt bridge, **(b)** mutants lacking a common salt bridge, **(c)** mutants in which a salt bridge may have been introduced.

Kinetic analysis of wild-type and mutant enzymes

The kinetic parameters of wild-type and mutant enzymes in the presence of a synthetic substrate, *N*-suc-AAPF-pNA, at 40°C are shown in Table 3. The K_m values of the mutants were similar to that of the wild-type enzyme, except that the K_m value of D212N was slightly higher than that of the wild-type enzyme and the other mutants. The k_{cat} values of the mutants were in the range of 47.2-110 s^{-1}; these values were 52-120% of the wild-type enzyme value. This result suggests that depletion of a salt bridge or introduction of a new salt bridge to these sites does not profoundly affect the integrity of the structure of AQN at 40°C. Structural integrities of mutants were also confirmed below by the similarity between CD spectra of wild type and mutant enzymes (see Figure 4).

Unfolding study on AQN proteins based on CD measurements

To analyze the unfolding process of active AQN mutants at a fixed temperature, changes in the CD spectra of the wild-type enzyme, D17N, E237Q, D212N or D183N as a function of time were recorded at 70°C and 80°C. Figure 4a and b show the CD spectra of the wild-type enzyme during the 30-min incubations at 70°C and 80°C, respectively. The spectrum was not significantly changed during the 30-min incubation at 70°C. However, at 80°C, the ellipticity at 222 nm was slightly but significantly decreased in the first 5 min, and it subsequently remained unchanged for up to 30 min. Figure 4c and d show the CD profiles of D17N during the 30-min incubations at 70°C and 80°C, respectively. The change in ellipticity at 222 nm at 70°C followed a time course that was very similar to that of the wild-type enzyme at 80°C. However,

Table 3 Kinetic parameters of the wild-type and mutant enzymes[1]

Enzyme	k_{cat} (s^{-1})	K_m (mM)	k_{cat}/K_m ($mM^{-1} s^{-1}$)
Wild-type	91.6 ± 2.75	0.79 ± 0.04	116
D17N	110 ± 4.48	0.98 ± 0.05	113
D212N	75.2 ± 3.13	1.10 ± 0.10	68.6
E237Q	96.1 ± 0.98	0.91 ± 0.03	105
D58N	68.5 ± 1.37	0.99 ± 0.07	70.0
D138N	59.5 ± 6.14	0.77 ± 0.01	77.3
D183N	53.0 ± 3.03	0.74 ± 0.05	71.6
G61D	47.2 ± 1.77	0.82 ± 0.13	63.6
G262D	61.3 ± 1.86	0.93 ± 0.15	66.2
S277D	73.6 ± 5.51	0.80 ± 0.12	92.5

[1]Parameters were determined at 40°C with *N*-suc-AAPF-pNA as a substrate. The experiments were performed in either duplicate or triplicate.

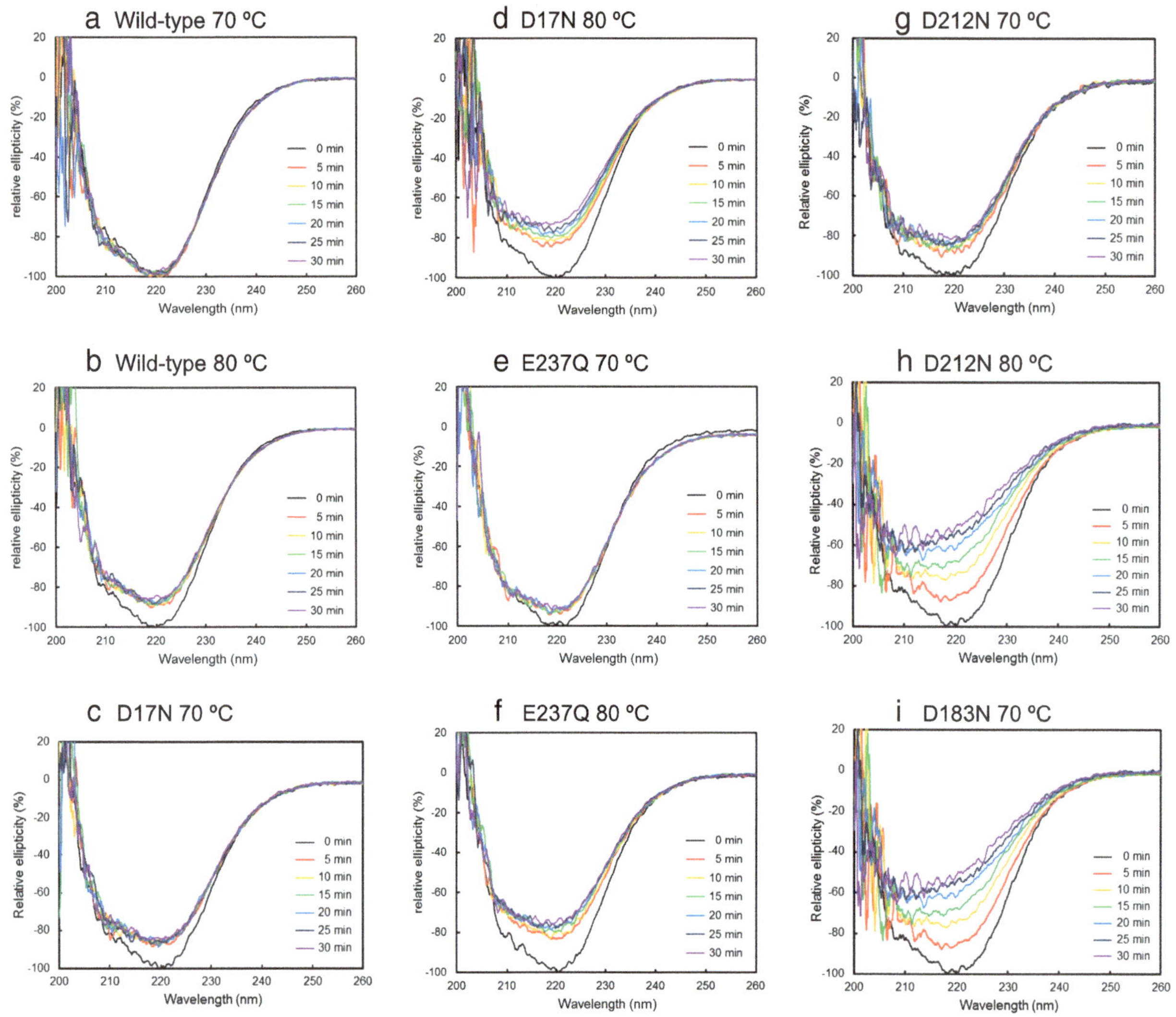

Figure 4 The change in CD spectra of the wild-type and mutant enzymes over a range of 0–30 min at a constant temperature. (a) Wild-type enzyme at 70°C, **(b)** wild-type enzyme at 80°C, **(c)** D17N at 70°C, **(d)** D17N at 80°C, **(e)** E237Q at 70°C, **(f)** E237Q at 80°C, **(g)**, D212N at 70°C, **(h)**, D212N at 80°C, **(i)** D183N at 70°C. The experimental conditions are described in the Materials and methods section.

at 80°C, the initial change at 5 min was slightly more pronounced and there was a gradual decrease in ellipticity that continued for up to 30 min in the D17N mutant. The change in the CD spectrum of E237Q was similar to that of D17N during incubation at both 70°C and 80°C (Figure 4e and f). These results indicated that the secondary structure contents of D17N and E237Q did not show major decreases for 30 min at 70°C. The change in the CD profile of D212N was very similar to that of D17N at 70°C (Figure 4g). However, at 80°C, the ellipticity of D212N in the 200–240 nm range continued to decrease for up to 30 min in parallel with the rapid inactivation of this mutant at 80°C (Figures 4h and 2b). The change in the CD profile of D183N at 70°C is shown in Figure 4i. The peak at 222 nm gradually decreased as a function of time down to 40% of the value before treatment, indicating that more than half of the secondary structures of D183N were destroyed. A comparable change in the ellipticity of D212N occurred only at 80°C. These results suggest that a salt bridge involving Asp183 plays a significant role in maintaining the structure of AQN. This may also be true of other members of the proteinase K subfamily, as the salt bridge involving Asp183 is conserved among these enzymes.

The denaturation curve of the PMSF-treated D183N mutant as monitored by the change in the ellipticity at 220 nm is shown in Additional file1: Figure S1. The T_m value of D183N was estimated to be 74°C. This is consistent with the extensive decrease of 222 nm peak intensity observed in Figure 4i. Denaturation of other mutants as well as the wild type enzyme apparently occurred at much higher temperature range than D183N

Figure 5 Alignment of the primary structures of proteinase K subfamily enzymes and subtilisin BPN'. AQN, aqualysin I; VPR, protease from *Vibrio* sp. PA-44; SPRK, proteinase K-like enzyme from *Serratia* sp.; PK, proteinase K from *Tritirachium album* Limber; BPN, subtilisin BPN' from *Bacillus amyloliquefaciens*. "#" denotes the catalytic residues Asp, His and Ser. The positions of common (closed circles), AQN-intrinsic (open arrowheads) and VPR-intrinsic (closed arrowheads) salt bridges are represented.

mutant, but we could not obtain reliable estimates for their T_m values. It should be noted that the change in CD spectrum gradually proceeded over 30 min at constant temperature as shown in Figure 4, while measurement of T_m was carried out at a heating rate of 1°C/min. It is possible that the rate of protein unfolding did not catch up with the elevation of the temperature, and this might give apparent T_m higher than true T_m.

Discussion

Subtilases, a group of serine proteases in the subtilisin superfamily, are composed of approximately 275 amino acid residues. Mutation studies on more than 50% of the amino acid residues in their primary structure have been described in the literature (Bryan 2000). Various factors appear to make complex contributions to subtilase stability. There are six subfamilies in the subtilisin superfamily, and it is possible that the mechanism by which thermal stability is conferred may differ from one subfamily to another. In this report, we investigated the roles of salt bridges in the thermal stabilization of AQN, a proteinase K subfamily member.

Regarding the salt bridges intrinsic to AQN (Asp17-Arg259, Arg31-Glu237 and Arg43-Asp212), the k_{cat} values of D17N and E237Q at 40°C and the activities at elevated temperatures toward a synthetic substrate were slightly

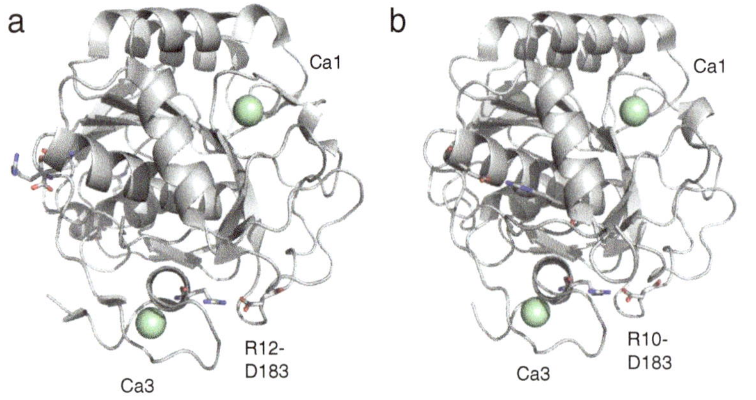

Figure 6 Crystal structures of AQN (a, 4DZT) and VPR (b, 1SH7). The positions of the salt bridges, **(a)** Arg12-Asp183 of AQN and **(b)** Arg10-Asp183 of VPR, are represented as stick models. Calcium ions (Ca1 and Ca3) are represented as pale green spheres.

increased compared to that of the wild-type enzyme; however, these mutants were apparently less thermostable than the wild-type enzyme during heat treatment experiments. These results suggest that the decline of the residual activities of the mutant enzymes during prolonged incubation at high temperatures might be caused in part by extensive autolysis under conditions in which these enzymes exhibit higher activity than the wild-type enzyme. However, it should also be noted that examination by CD spectrometry indicated that D17N and E237Q showed a small change in ellipticity at 222 nm in the first 5 min at 70°C, although the ellipticity remained unchanged over the course of the subsequent incubation for up to 30 min (Figure 4c and e). These results suggest that incubation at 70°C may affect regions of the D17N and E237Q mutants that are not rich in secondary structures, including loop regions, but that the structural perturbation due to the D17N or E237Q mutation may destabilize the active site conformation during a prolonged incubation at 70°C and above. This result agrees with a previous report indicating that the D17N mutant exhibited reduced thermal stability compared to the wild-type enzyme (Arnórsdóttir et al. 2011).

Recently, Jakob et al. reported an intensive analysis of the roles of charged amino acid residues in a *Bacillus gibsonii* subtilisin protease, BgAP, using site-directed mutagenesis. BgAP Q230E showed increased thermal resistance compared to wild-type BgAP (Jakob et al. 2013). This result is consistent with our data on E237Q. Glu237 of AQN corresponds to Gln230 of BgAP, and disruption of a salt bridge in E237Q resulted in a rapid decrease of activity during incubation at 70°C and 80°C.

The stability of D212N was similar to that of the wild-type enzyme at 70°C; however, it was inactivated rapidly at 80°C (Figure 2a and b). This result is consistent with the results of the CD spectrometry analysis showing that the secondary structure content was rapidly decreased as a function of time at 80°C. The inactivation mechanism of D212N at 80°C may be different from that of D17N and E237Q at 70°C.

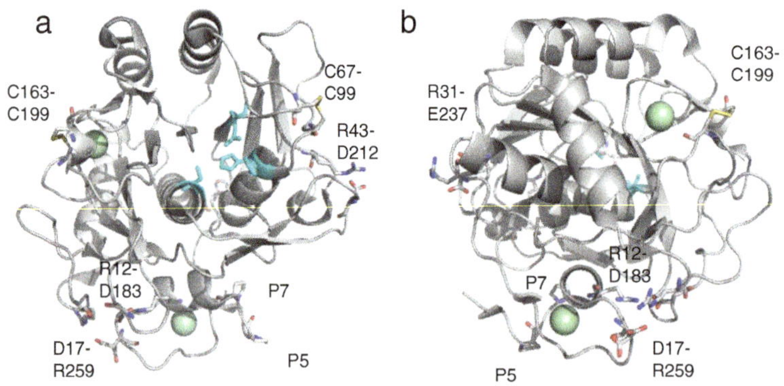

Figure 7 Crystal structure of AQN (4DZT) with residues contributing to enzyme stabilization. (a) The residues contributing to the salt bridges Arg12-Asp183, Asp17-Arg259, Arg31-Glu237 and Arg43-Asp212 and the disulfide bridges Cys67-Cys99 and Cys163-Cys199 are indicated and are represented as stick models. Calcium ions are represented as pale green spheres. The catalytic residues are represented as blue stick models. **(b)** The structure is rotated approximately 90 degrees horizontally from **(a)**.

Introducing additional salt bridges at positions mimicking VPR-intrinsic salt bridges in G61D, G262D and S277D resulted in k_{cat} values lower than that of the wild-type enzyme; however, this did not affect the thermal stability.

D183N, in which one of the three salt bridges common to AQN and VPN was destroyed, was the most heat-labile among all the mutants tested in this study. Asp183 is conserved among the proteinase K subfamily members and is capable of forming a salt bridge toward an Arg residue. The Asp residue is not conserved among other subtilases (e.g., subtilisin BPN') (Figure 5). Monitoring of the unfolding process revealed that the secondary structure content decreased as a function of time at 70°C (Figure 4i). The temperature-activity profile shown in Figure 3b indicates that D183N retained full activity during a brief activity measurement period; however, during prolonged incubation at 70°C, the mutant was progressively denatured.

Arg12-Asp183 may serve as a link between the N-terminal α-helix and the long loop region within the 6th and 7th beta-sheets (as named by Siezen and Leunissen (Siezen & Leunissen 1997)) and is located near the Ca3 site (Figure 6a and b). The gradual decrease in the secondary structure content of D183N at 70°C observed by CD spectrometry is most likely initiated by the detachment of these regions. In our previous studies, we demonstrated that Pro residues in the N-terminal loop structure contribute greatly to the thermal stability of AQN and that the disulfide bond, Cys163-Cys194, is important for conformational stability (Sakaguchi et al. 2007; Sakaguchi et al. 2008b). The present results on D183N also emphasize the importance of the N-terminal loop region as well as the region including the Cys163-Cys194 disulfide bond and Asp183. The Asp17-Arg259 salt bridge may also be contributing to the stabilization of AQN to a lesser extent because Asp17 is also positioned near the N-terminal region. In addition, the Arg31-Glu237 and Arg43-Asp212 salt bridges appear to serve as additional stabilizing factors that are unique to AQN (Figure 7a and b).

In conclusion, mutation of a salt bridge-forming amino acid that is highly conserved among members of the proteinase K subfamily, Asp183 of AQN, destabilized the protein structure, indicating that this salt bridge plays the most important role in the stability of AQN compared with the other salt bridges. Furthermore, AQN-intrinsic salt bridges confer additional thermal stability to AQN.

Additional file

Additional file 1: Figure S1. Normalized denaturation curve of PMSF-treated D183N. The denaturation curve is normalized according to Fu = (y_f-y)/(y_f-y_u), assuming a two-state transition, where y_f and y_u are the CD signals at 220 nm for the folded and unfolded states, respectively, and y is the CD signal at 220 nm at each data point.

Competing interests
The authors declare that the research was conducted in the absence of any commercial or financial relationships that could be construed as potential conflicts of interest.

Authors' contributions
MS conceived of the study, participated in its design and coordination, performed the experiments, interpreted the data and drafted the manuscript. KO participated in its design and coordination, performed the experiments, interpreted the dates and helped to draft the manuscript. SM performed the experiments and the data analysis. NO performed the experiments and the data analysis. YS conceived of the study, participated in its design and coordination, and interpreted the data. FO conceived of the study, participated in its design and coordination, interpreted the data and helped to draft the manuscript. MK conceived of the study, participated in its design and coordination, interpreted the data and drafted the manuscript. All authors read and approved the final manuscript.

Acknowledgments
This study was supported in part by a grant of Strategic Research Foundation Grant-aided Project for Private Universities from Ministry of Education, Culture, Sport, Science, and Technology, Japan (MEXT), 2014-2018 (S1411005).

Author details
[1]Department of Applied Chemistry, Kogakuin University, 2,665-1 Nakano-cho, Hachioji, Tokyo 192-0015, Japan. [2]Center for Medical Research Cooperation, Tokyo Metropolitan Institute of Medical Science, 2-1-6 Kami-kitazawa Setagaya-ku, Tokyo 156-8506, Japan.

References
Arnórsdóttir J, Smáradóttir RB, Magnússon OT, Thorbjarnardóttir SH, Eggertsson G, Kristjánsson MM (2002) Characterization of a cloned subtilisin-like serine proteinase from a psychrotrophic *Vibrio* species. Eur J Biochem 269:5536–5546, doi:10.1046/j.1432-1033.2002.03259.x
Arnórsdóttir J, Kristjánsson MM, Ficner R (2005) Crystal structure of a subtilisin-like serine proteinase from a psychrotrophic *Vibrio* species reveals structural aspects of cold adaptation. FEBS J 272:832–845, doi:10.1111/j.1742-4658.2005.04523.x
Arnórsdóttir J, Sigtryggsdóttir AR, Thorbjarnardóttir SH, Kristjánsson MM (2009) Effect of proline substitutions on stability and kinetic properties of a cold adapted subtilase. J Biochem 145:325–329, doi:10.1093/jb/mvn168
Arnórsdóttir J, Magnúsdóttir M, Friðjónsson OH, Kristjánsson MM (2011) The effect of deleting a putative salt bridge on the properties of the thermostable subtilisin-like proteinase, aqualysin I. Protein Pept Lett 18:545–551, doi:10.2174/092986611795222759
Bradford MM (1976) A rapid and sensitive method for the quantitation of microgram quantities of protein utilizing the principle of protein-dye binding. Anal Biochem 7:248–254
Bryan PN (2000) Protein engineering of subtilisin. Biochim Biophys Acta 1543:203–222, doi:10.1016/S0167-4838(00)00235-1
Green PR, Oliver JD, Strickland LC, Toerner DR, Matsuzawa H, Ohta T (1993) Purification, crystallization and preliminary X-ray investigation of aqualysin I, a heat-stable serine protease. Acta Crystallogr D49:349–352, doi:10.1107/S0907444992012083
Helland R, Larsen AN, Smalås AO, Willassen NP (2006) The 1.8 Å crystal structure of a proteinase K-like enzyme from a psychrotroph *Serratia* species. FEBS J 273:61–71, doi:10.1111/j.1742-4658.2005.05040.x
Jakob F, Martinez R, Mandawe J, Hellmuth H, Siegert P, Maurer KH, Schwaneberg U (2013) Surface charge engineering of a *Bacillus gibsonii* subtilisin protease. Appl Microbiol Biotechnol 97:6793–6802, doi:10.1007/s00253-012-4560-8
Kristjánsson MM, Magnússon OT, Gudmundsson HM, Alfredsson GA, Matsuzawa H (1999) Properties of a subtilisin-like proteinase from a psychrotrophic *Vibrio* species comparison with proteinase K and aqualysin I. Eur J Biochem 260:752–760, doi:10.1046/j.1432-1327.1999.00205.x
Kumar S, Tsai CJ, Nussinov R (2000) Factors enhancing protein thermostability. Protein Eng 13:179–191, doi:10.1093/protein/13.3.179
Laemmli UK (1970) Cleavage of structural proteins during the assembly of the head of bacteriophage T4. Nature 227:680–685, doi:10.1038/227680a0

Larsen AN, Moe E, Helland R, Gjellesvik DR, Willassen NP (2006) Characterization of a recombinantly expressed proteinase K-like enzyme from a psychrotrophic *Serratia* sp. FEBS J 273:47–60, doi:10.1111/j.1742-4658.2005.05044.x
Matsuzawa H, Hamaoki M, Ohta T (1983) Production of thermophilic extracellular proteases (Aqualysin I and II) by *Thermus aquaticus* YT-1, an extreme thermophile. Agri Biol Chem 47:25–28
Matsuzawa H, Tokugawa K, Hamaoki M, Mizoguchi M, Taguchi H, Terada I, Kwon ST, Ohta T (1988) Purification and characterization of aqualysin I (a thermophilic alkaline serine protease) produced by *Thermus aquaticus* YT-1. Eur J Biochem 171:441–447, doi:10.1111/j.1432-1033.1988.tb13809.x
Sakaguchi M, Matsuzaki M, Niimiya K, Seino J, Sugahara Y, Kawakita M (2007) Role of proline residues in conferring thermostability on aqualysin I. J Biochem 141:213–220, doi:10.1093/jb/mvm025
Sakaguchi M, Niimiya K, Takezawa M, Toki T, Sugahara Y, Kawakita M (2008a) Construction of an expression system for aqualysin I in *Escherichia coli* that gives a markedly improved yield of the enzyme protein. Biosci Biotechnol Biochem 72:2012–2018, doi:10.1271/bbb.80132
Sakaguchi M, Takezawa M, Nakazawa R, Nozawa K, Kusakawa T, Nagasawa T, Sugahara Y, Kawakita M (2008b) Role of disulphide bonds in a thermophilic serine protease aqualysin I from *Thermus aquaticus* YT-1. J Biochem 143:625–632, doi:10.1093/jb/mvn007
Sambrook J, Russell DW (2012) Molecular Cloning, a Laboratory Manual, 4th edn. Cold Spring Harbor Laboratory Press, Cold Spring Harbor
Siezen RJ, Leunissen AM (1997) Subtilases: the superfamily of subtilisin-like serine proteases. Protein Sci 6:501–523, doi:10.1002/pro.5560060301
Sigurdardóttir AG, Arnórsdóttir J, Thorbjarnardóttir SH, Eggertsson G, Suhre K, Kristjánsson MM (2009) Characteristics of mutants designed to incorporate a new ion pair into the structure of a cold adapted subtilisin-like serine proteinase. Biochim Biophys Acta 1794:512–518, doi:10.1016/j.bbapap.2008.11.018
Struvay C, Feller G (2012) Optimization to low temperature activity in psychrophilic enzymes. Int J Mol Sci 13:11643–11665, doi:10.3390/ijms130911643
Szilágyi A, Závodszky P (2000) Structural differences between mesophilic, moderately thermophilic and extremely thermophilic protein subunits: results of a comprehensive survey. Structure 8:493–504, doi:10.1016/S0969-2126(00)00133-7
Vogt G, Woell S, Argos P (1997) Protein thermal stability, hydrogen bonds, and ion pairs. J Mol Biol 269:631–643, doi:10.1006/jmbi.1997.1042

3

Overexpression of the *Aspergillus niger* GatA transporter leads to preferential use of D-galacturonic acid over D-xylose

Jasper Sloothaak, Mike Schilders, Peter J Schaap and Leo H de Graaff*

Abstract

Pectin is a structural heteropolysaccharide of the primary cell walls of plants and as such is a significant fraction of agricultural waste residues that is currently insufficiently used. Its main component, D-galacturonic acid, is an attractive substrate for bioconversion. The complete metabolic pathway is present in the genome of *Aspergillus niger*, that is used in this study. The objective was to identify the D-galacturonic acid transporter in *A. niger* and to use this transporter to study D-galacturonic acid metabolism.

We have functionally characterized the gene An14g04280 that encodes the D-galacturonic acid transporter in *A. niger*. In a mixed sugar fermentation it was found that the An14g04280 overexpression strain, in contrast to the parent control strain, has a preference for D-galacturonic acid over D-xylose as substrate. Overexpression of this transporter in *A. niger* resulted in a strong increase of D-galacturonic acid uptake and induction of the D-galacturonic acid reductase activity, suggesting a metabolite controlled regulation of the endogenous D-galacturonic acid catabolic pathway.

Keywords: D-galacturonic acid; Pectin; Sustainable resources; *Aspergillus niger*; Transmembrane transport

Introduction

Global limits in food and energy availability are becoming a major concern. Arable land is needed for production of food and large waste streams come from processing of our largest food resources, like grain, maize, potato and rice. These waste streams are currently insufficiently exploited and mainly used for feed and energy production (Howard et al. 2003). Alternatively, agricultural waste streams can be used as substrate for fermentative production of chemicals by microorganisms.

Historically, *Aspergillus* species are used for the production of food additives and platform chemicals such as citric acid (*A. niger*) and itaconic acid (*A. terreus*) (Willke and Vorlop 2001). The filamentous fungus *Aspergillus niger* is additionally exploited for production of enzymes for food and feed applications, many of which have been rewarded the GRAS (Generally Recognized As Safe) status (Howard et al. 2003; van Dijck et al. 2003).

* Correspondence: leo.degraaff@wur.nl
Microbial Systems & Synthetic Biology, Laboratory of Systems and Synthetic Biology, Wageningen University, Dreijenplein 10, NL-6703 HB, Wageningen, Netherlands

Grain, maize and rice waste stream material contains around 40% cellulose, 35% hemicellulose, 20% lignin and 5% pectin, while waste streams from other plants, such as sugar beet and potato, contain around 20% cellulose and hemicellulose, less than 1% lignin and 30% to 40% pectin (Ángel Siles López et al. 2010; Howard et al. 2003; Micard et al. 1996). *A. niger* has a high capacity for degrading the hemicellulose, pectin and, to a lesser extend, cellulose fractions as the genome of *A. niger* encodes multiple gene variants of the enzymes required for the efficient degradation of these polysaccharides. (van den Brink and de Vries 2011).

D-glucose and D-xylose, released from plant wall material, are carbon sources that will yield high energy upon being metabolized. The genome of *A. niger* is wired to specifically produce and secrete the enzymes involved in degradation of the cellulose and hemicellulose fractions to release D-glucose and D-xylose. These sugars are then taken up and metabolized, before energy is invested in the release of other carbon sources from polysaccharides. This preferential uptake is regulated by the interplay of activating and repressing transcription factors that respond to extracellular concentrations of inducers and sugars.

Expression of genes that code for enzymes involved in cellulose and hemicellulose degradation in *A. niger* is regulated by the transcription factor XlnR (Gielkens et al. 1999). XlnR is activated by inducers derived from xylan, like D-xylose, though at a higher concentration this activation is attenuated by CreA (de Vries et al. 1999; Ruijter et al. 1999; van Peij et al. 1998). Tight regulation of the specific genes encoding proteins involved in the release, uptake and metabolism of sugars gives the fungus the ability to utilize the available sugars in a physiologically most efficient way. However, circumventing this regulatory mechanism would potentially lead to simultaneous uptake of all sugar substrates and improved fermentation yields.

Genes encoding proteins involved in the degradation of pectin and the D-galacturonic acid metabolic pathway are found to be highly upregulated when *A. niger* is growing in the presence of D-galacturonic acid, in comparison to when the fungus is growing in the presence of D-glucose or D-fructose (Table 1). Three genes among the genes that are upregulated are coding for proteins with a strong similarity to transporters (Martens-Uzunova and Schaap 2008; van der Veen et al. 2009).

In this study we have functionally identified the D-galacturonic mono-acid sugar transporter by overexpression in *A. niger*. The effect of the increased uptake of D-galacturonic acid on the preferential uptake and metabolism of substrates has been assessed by mixed sugar fermentations and enzyme activity measurements.

Materials and methods

Strains

All *Aspergillus* strains used are descendants of *A. niger* N400 (CBS 120.49). *A niger* strain NW185 *cspA1, fwnA1, goxC17, prtF28*) was derived from NW131 and has been described by Ruijter et al. 1999. The recipient strain in all transformation experiments NW186 (*ΔargB; pyrA6; prtF28; goxC17; cspA1*) is an *argB* and *pyrA* derivative of NW185.

Vector construction

Primers used in the construction of vectors are summarized in Table 2. Plasmids used for transformations of *A. niger* were constructed according to the general scheme given in Figure 1. Martens-Uzunova and Schaap was the source for the selection of the putative transporters (Martens-Uzunova and Schaap 2008); An14g04280 and An03g01620 were obtained from the *Aspergillus* genome database (Arnaud et al. 2012; Pfaffl 2001). These sequences were used to synthesize the respective coding sequences that were modified to remove conflicting restriction sites (BaseClear, Leiden, Netherlands). The An07g00780 fragment was amplified by PCR using primers JS_gatC_FW and JS_gatC_RV and *A. niger* NW186 genomic DNA as template and was subcloned into a pJET1.2 vector (Thermo scientific). The *BglII* restriction site was removed from the coding sequence by using the Quikchange Lightning Site-directed mutagenesis kit (Agilent Technologies) with the primers JS_gatCmut2_FW and JS_gatCmut2_RV. The PCR fragments were cloned into a pUC19 derived expression vector under control of an modified *xlnD* promoter and the *xlnD* terminator of *A. niger*. Plasmids were propagated in DH5α *E. coli*. The medium used for *E. coli* growth was LB (10 g/L Bacto tryptone, 5 g/L Yeast extract, 10 g/L NaCl) supplemented with 100 mg/L ampicillin when appropriate.

Transformation of *Aspergillus niger*

For the transformation of *A. niger* NW186, protoplasts were generated using Novozyme-234 lysing enzyme cocktail. The An14g04280, An03g01620 and An07g00780 constructs were introduced in *A. niger* by transformation (Kusters-van Someren et al. 1991) using the *pyrA* gene as a primary selection marker, relieving uridine auxotrophy. To generate the parent control reference strain, NW186 was transformed using pGW635, the resulting NW185 PYR A^+ strain was used as a control strain in our studies. Colonies were randomly picked from the primary transformation plates and re-plated on selective medium to purify the single colonies.

Analysis of *A. niger transformants*

Screening of randomly picked transformants was done by PCR on genomic DNA isolates. For that, fresh mycelium was disrupted using MP lysing matrix C tubes and 400 μl DNA extraction buffer (0.1 M Tris HCl pH 8.0, 1.2 M NaCl, 5 mM EDTA). DNA was extracted using phenol-chloroform extraction. The pellet was washed with 70% cold ethanol, air-dried and re-suspended in 50 μl MQ

Table 1 Data used for selection of putative D-galacturonic acid transporters (Martens-Uzunova and Schaap 2008)

Locus tag	Functional annotation	Relative expression t = 4 h after induction on		
		galA	poly-galA	pectin
An07g00780	Major facilitator superfamily strong similarity to monocarboxylate transporter	23.10	9.98	5.27
An14g04280	Major facilitator superfamily strong similarity to hexose transporter	26.53	22.74	27.95
An03g01620	Major facilitator superfamily strong similarity to hexose transporter	6.50	1.345	37.54

Table 2 Primers used in vector construction

Primer	Sequence
JS_*gatC*_FW	5'-GAGAATGCATATGTCTGAGCCTAAGAACCAGC-3'
JS_*gatC*_RV	5'-GAGAGCGGCCGCTCATATTTTGCTCGTATTCC-3'
JS_*gatC*mut2_FW	5'-GTTGCCAAAGTGGGAGACCTGCAAACCCTGGGTC-3'
JS_*gatC*mut2_RV	5'-GACCCAGGGTTTGCAGGTCTCCCACTTTGGCAAC-3'
JS_XlnDp_FW	5'-AGTCAATCGGTCATTCTCCG-3'
JS_*gatA*_RV	5'-CTCTTGGAGGACGCAAAACC-3'
JS_*gatB*_RV	5'-CATGGAAGCGGTGATCTAGG-3'

water. The presence of An14g04280, An03g01620 and An07g00780 expression constructs was confirmed by PCR using the JS_XlnDp_FW primer, binding to the promoter region of the construct, and JS_gatA_RV, JS_gatB_RV or JS_gatC_RV specifically binding to the complementary strand in the corresponding coding region. Transformants with confirmed gene constructs were re-plated on complete medium. Spores are harvested after 4 days of growth at 30°C.

Growth in shake flasks

Strains were grown in duplicate in 200 mL PM medium (1.2 g $NaNO_3$, 0.5 g KH_2PO_4, 0.2 g $MgSO_4 7 H_2O$, 0.5 g Yeast extract and 40 μg Vishniac per liter) (Ruijter et al. 1999) with 91 mM sorbitol and 9 mM D-galacturonic acid or 100 mM of the synthetic cell wall hydrolysate (SCH) (Carpita and Gibeaut 1993) (Table 3) as a carbon source in 1 L flasks. For inoculation, a final concentration of $1 \cdot 10^6$ spores per mL was used. Induction of the *xlnD* promoter with 10 mM D-xylose was done at t = 0, 18 hours after inoculation (van der Veen et al. 2009). Transformants were grown for 5 days at 30°C and 250 rpm. Medium samples were taken by filtration of 2 ml of the culture broth through a 5-micron pore size nylon filter. Biomass samples were obtained by washing of the retentate with demineralized water and flash freezing using liquid nitrogen. The samples were taken at T = −18 h, 0 h, 6 h, 30 h, 54 h for the first growth

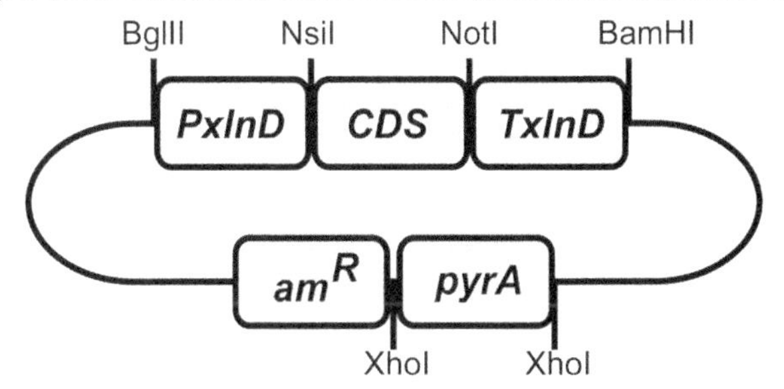

Figure 1 Schematic representation of Funbrick expression vector. Restriction sites flank functional modules. Capital P and T represent promoter and terminator. Bacterial marker: *amR*. Fungal uridine auxotrophy complementation marker: *pyrA*.

Table 3 Composition of synthetic cell wall hydrolysate (SCH)

Compound %	Concentration (g/l)
D-glucose; 65%	13
D-xylose; 20%	4
D-galacturonic acid; 10%	2
L-arabinose; 3%	0.6
D-galactose; 1%	0.2
D-mannose; 1%	0.2

experiment on D-galacturonic acid. For the second D-galacturonic acid growth experiment, samples were taken at T = −18 h, each hour from T = 0 h to T = 10 h, 24 h, 30 h, 54 h and 78 h. For the growth experiment with SCH as a carbon source, samples were taken from T = 0 h with steps of 24 and 6 hours: T = 0 h, T = 24 h, T = 30 h, T = 48 h, T = 54 h.

Growth in fermentors

Strains were grown in duplicate in 500 mL PM medium (Ruijter et al. 1999) with 100 mM sorbitol as a carbon source in 1 L flasks. For inoculation, a final concentration of $1 \cdot 10^6$ spores per mL was used. Cultures were incubated for 18 hours at 30°C with 250 rpm orbital shaking. Mycelium was aseptically harvested by filtering over a sterile nylon cloth, washed with PM without carbon source and transferred to 1 L fermentors. Fermentors contained 750 ml PM with 50 mM D-xylose and 50 mM D-galacturonic acid, adjusted to pH6. Air was flushed at a stirrer speed of 600 rpm. Medium and biomass samples were taken as previously described, with an additional sampling of mycelium from 10 ml broth volume for enzymatic assays. Carbon dioxide and oxygen concentrations in off-gas were measured as well as pH.

HPLC analysis

High-pressure liquid chromatography (HPLC) was used to determine the extracellular concentrations of D-glucose, L-sorbitol, D-xylose, citric acid and D-galacturonic acid in the culture broth samples. A Shodex KC-811 column was used at 30°C that was eluted with 0.01 N H_2SO_4 at a flow rate of 0.8 mL min^{-1}. A refractive index detector (Spectrasystem RI-150, sample frequency 5.00032 Hz) and a UV–VIS detector (Spectrasystem UV1000, λ = 210 nm) were used for detection of the eluting compounds. Crotonate at a concentration of 6 mM was used as an internal standard.

Enzyme assay

To determine the activity of the D-galacturonic acid metabolism, the first step in the enzymatic conversion, D-galacturonic acid to L-galactonic acid, was measured. D-galacturonic acid reductase assay based on conversion of NADPH to $NADP^+$, previously described by Kuorelahti et al., was applied (Kuorelahti et al. 2005). 20 mg of frozen mycelium was added to 400 μl of extraction buffer (100 mM phosphate buffer pH 7.0, 0.1 mM EDTA, 1 mM DTT and fungal protease inhibitor cocktail) in lysing matrix C tubes (MP biomedicals) and disrupted at level 6 for 30 seconds in a bead beater (MP fastprep-24). Cell debris was removed by centrifugation and supernatant was used as cell extract. 25 μl of cell extract was added to 200 μl of assay buffer (100 mM sodium phosphate buffer pH 7.0, 0.25 mM NADPH) and reaction was started by the addition of 25 μl of substrate (1 M D-galacturonic acid pH 6). Decrease of absorbance at 340 nm was measured on a platereader (Biotek Synergy) in parallel with sample controls, for which 25 μl of buffer was added. Background activity of sample control without D-galacturonic acid was substracted from sample measurements with D-galacturonic acid. Protein content was determined by photometric assay (Peterson 1977).

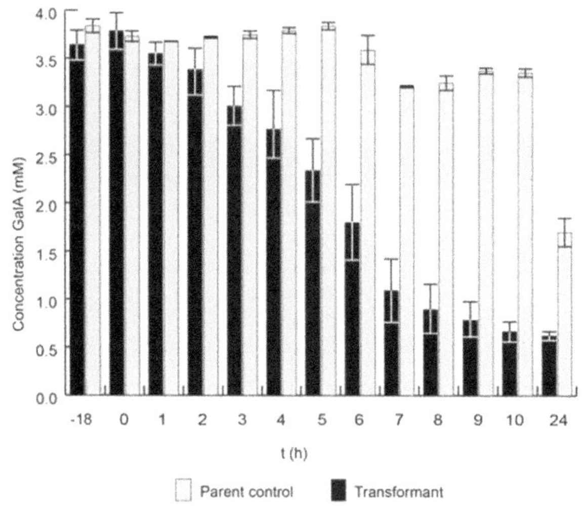

Figure 2 Concentration of GalA in the medium during growth on sorbitol and GalA. Induction of expression system with D-xylose at T = 0. Overexpression strains of *gatA* (number of replicates =4) in comparison to the control NW185 PYR A^+ (number of replicates =2).

Results

Strains that overexpress GatA have increased D-galacturonic acid uptake in shakeflask cultures

To study the effect of the expression of the putative D-galacturonic acid transporter encoding gene, the D-galacturonic acid concentration in the medium was measured during growth. A faster uptake of D-galacturonic acid from the medium was observed for the cultures with the An14g04280 overexpression strains (transformant) in comparison to the NW185 PYR A^+ control strain. At 6 hours after induction, the strains that overexpress An14g04280 had taken up over 50% of D-galacturonic acid present at the start of the cultivation, while no D-galacturonic acid had been taken up by the control strain. The strains containing the other constructs did not show any difference in D-galacturonic acid uptake in comparison to the NW185 PYR A^+ control strain. The effect of the An14g04280 overexpression strain is most pronounced between time points 6 and 30 h.

A second experiment was performed to get a more time resolved determination of the uptake of D-galacturonic acid by the best performing transformant strain (JS013). The difference in D-galacturonic acid uptake starts to become evident from 4 hours after induction, with the highest contrast between 5 and 6 hours after induction (Figure 2).

To study the effect of An14g04280 in the uptake of D-galacturonic acid during growth on more complex mixed sugar substrates, JS013 was grown in the presence of a substrate mix (SCH), mimicking the composition of a plant cell wall hydrolysate (Carpita and Gibeaut 1993). When comparing the uptake of the three main components D-glucose, D-xylose and D-galacturonic acid between the transformants and the control strain, a clear difference is found (Figure 3a and b). Till 24 hours after inoculation, substrate uptake is similar for all strains. After 30 hours, differences in D-galacturonic acid and D-xylose uptake become evident. At time point 48 hours, the control strain has almost depleted the D-xylose and most of the D-galacturonic acid remains in the medium, while the transformant has consumed 80% of the D-galacturonic acid and most D-xylose remains in the medium. While D-glucose and D-xylose are depleted in the control culture after 54 hours, depletion of D-xylose in the transformant cultures can only be seen at time point 72 h.

The biomass formed and the citrate produced were measured for the An14g04280 overexpression strain and the control strain. Recovery of substrate in biomass and citrate produced was 35% for the control strain and 30% for the transformants. It was found that for the overexpression strain 30% of this is citrate and 70% is biomass, while for the control strain, 10% is citrate and 90% is biomass (Figure 4). Cmoles that are not accounted for are assumed to be CO_2 and enzyme production (not measured). For these calculations the generalized molecular formula for fungal biomass $CH_{1.72}O_{0.55}N_{0.17}$ was used (Carlsen and Nielsen 2001).

Growth of transporter overexpression strains in fermentors

When strains were grown in the presence of equimolar concentration of D-xylose and D-galacturonic acid in fermentors, preferential uptake of D-galacturonic acid over D-xylose of JS013 in comparison to NW185 PYR A^+ control strain is observed. This is deduced from measured concentrations of D-galacturonic acid and D-xylose, but also from an increase of the pH of the medium, after D-

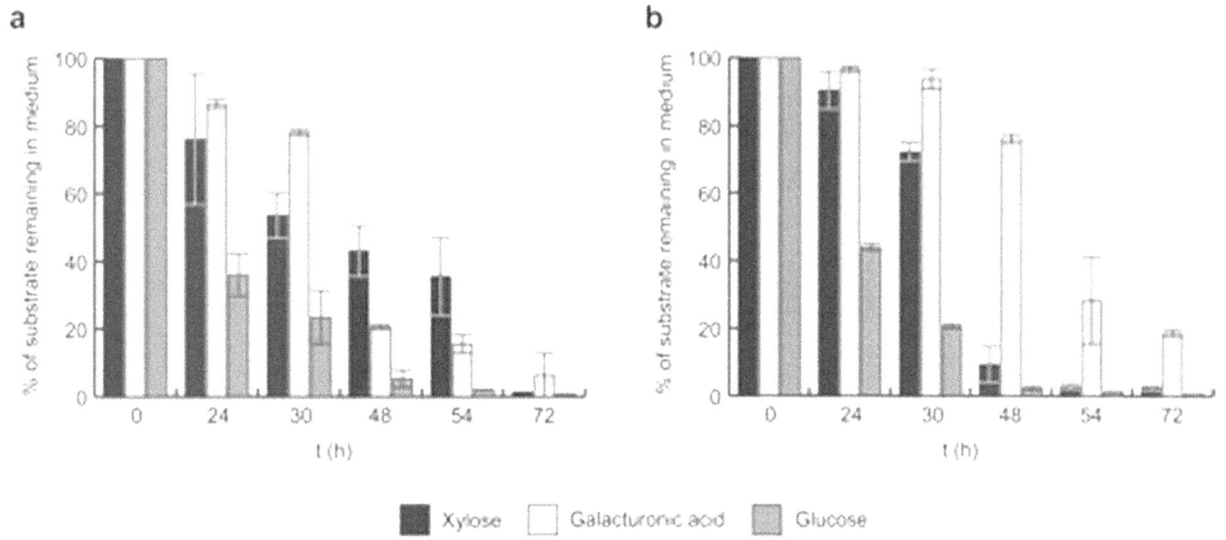

Figure 3 **Percentage remaining in medium of the three most important substrates from the synthetic cell-wall hydrolysate for *gatA* overexpression strain (a) (number of replicates =2) and the parent (NW185 PYR A$^+$ strain) (b) (number of replicates = 2).**

galacturonic acid is taken up (Figure 5). Increased activity of the D-galacturonic acid metabolic pathway is observed for JS013 transformant strain in comparison to the control. For the first 4 hours after transfer to the fermentors, no difference in D-galacturonic acid reductase activity could be seen. 8 hours after transfer, however, specific activity of JS013 is two-fold higher than that of the control strain. The concentration of CO_2 in the offgas went up to 0.35 percent by volume in the first two hours and slowly down to 0.2 percent by volume in the next 6 hours, while ambient concentration was 0.04 percent by volume (Figure 6).

Phylogenetic analysis of GatA

In *Botrytis cinerea* and *Neurospora crassa*, D-galacturonic acid transporter proteins have been functionally identified (Benz et al. 2014; Zhang et al. 2013). After import, D-galacturonic acid is catabolized by three key enzymes: D-galacturonic acid reductase, L-galactonate dehydratase and 2-keto-3-deoxy-galactonate aldolase, in *A. niger*

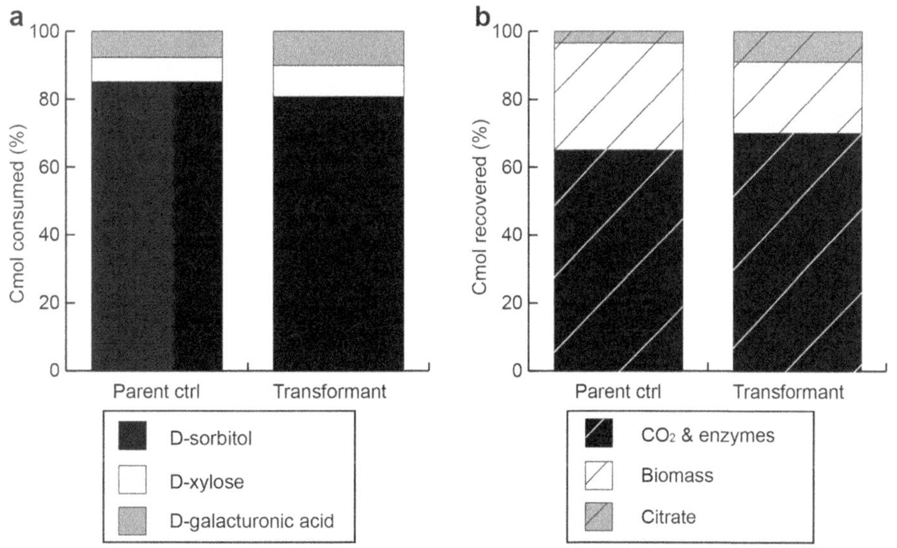

Figure 4 **Carbon consumed (a) and produced (b), expressed in Cmoles, for control strain (NW185 PYR A$^+$ strain) and transformant JS013 after 72 h of growth on SCH medium.** Samples taken from direct cultures, D-xylose in the mixed substrate medium serves as an inducer of the *xlnD* promoter. Totals represent 0.1 Cmoles.

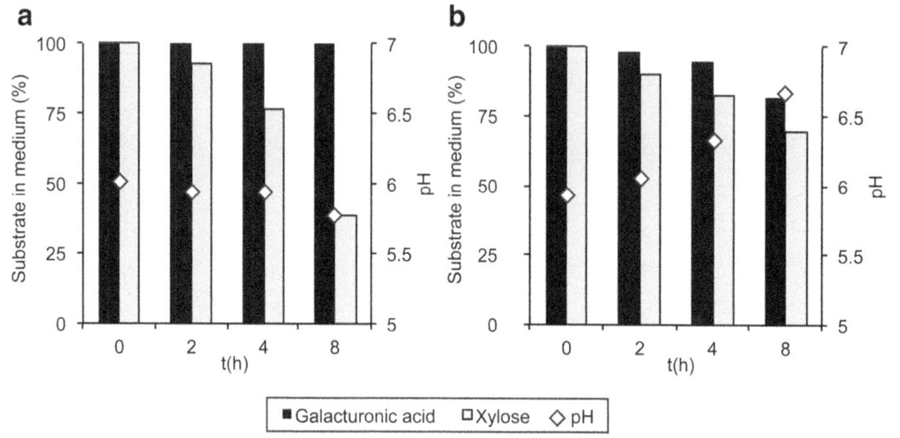

Figure 5 Substrate remaining in culture medium and pH during growth in fermentors of the control, (NW185 PYR A$^+$ strain) (a) and overexpression strain, JS013 (b). The strains were pre-grown and mycelium was transferred at T = 0 to fermentors containing D-xylose and D-galacturonic acid as substrates.

encoded by *gaaA*, *gaaB* and *gaaC* respectively. Orthologs of the *gaaA* and *gaaC* genes have been identified in *B. cinerea* and *N. crasssa* and 16 other fungal species that all have a conserved bidirectional organization of their *gaaA* – *gaaC* promoter region (Martens-Uzunova and Schaap 2008). For each of these 18 species a bidirectional best BLAST hits procedure (Overbeek et al. 1999) yielded a putative GatA ortholog. Orthologous amino acid sequences were aligned with Clustal (Larkin et al. 2007) and a phylogenetic tree was constructed (Figure 7). All orthologs have 12 predicted transmembrane helices (Viklund and Elofsson 2004) separating two interacting domains and the substrate binding domain (Yan 2013) (Figure 8). In these three domains the three characterized D-galacturonic acid transporters share 18 conserved residues not present in the *A. niger* MstA D-glucose transporter protein sequence, which was used as an outgroup. The presence of this motif was used to assess the other 16 putative orthologs. Fifteen putative orthologs shared 8 of these residues, excluding the *Fusarium verticilliodes* putative ortholog which clusters with the D-glucose transporter MstA (Figure 7) and can be a transporter for another sugar.

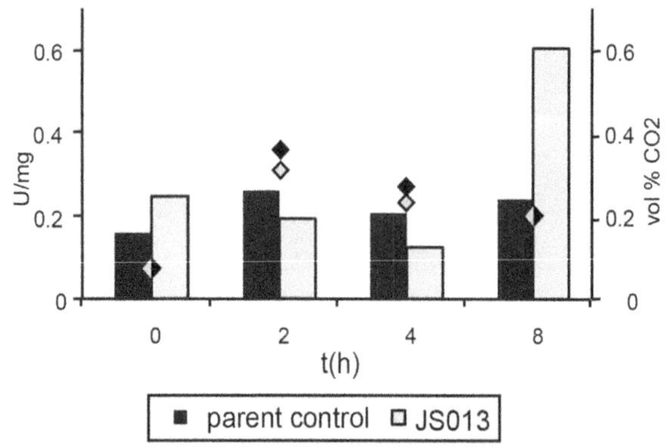

Figure 6 Galacturonic acid reductase activity (Bars; left Y axis) and CO_2 in offgas (Diamonds; right Y axis). Samples were taken during growth on 25mM of D-xylose and 25mM of D-galacturonic acid as a carbon source in fermentors.

Discussion

Three genes that potentially encode the D-galacturonic acid transporter in *A. niger* are investigated. *A. niger* transformants that overexpress the An14g04280 gene under control of the D-xylose induced *xlnD* promoter show a significant increase in D-galacturonic acid uptake. From this we conclude that this gene, *gatA*, encodes for the D-galacturonic acid transporter, GatA.

By analysis of the amino acid sequence using the α-helical transmembrane protein topology prediction software (PRODIV-TNHMM) (Viklund and Elofsson 2004) GatA is predicted to be part of the Major Facilitator Superfamily (MFS). MFS transporter proteins consist of 12 transmembrane helices and have both C- and N-termini located in the cytoplasm.

This study shows that the overexpression of *gatA*, using a D-xylose inducible promoter, leads to differences in the uptake of sugars when the strains are grown on various mixed sugar substrates. GatA overproducing strains preferentially use D-galacturonic acid over D-xylose, while the wild type strains prefer D-xylose. D-xylose is used for overexpression of the GatA transporter in all growth experiments and it is also used as a substrate. However, the main carbon source present in the medium is sorbitol, a non-inducing, non-repressing carbon source. Our extensive previous studies have shown that in this condition induction of expression occurs at

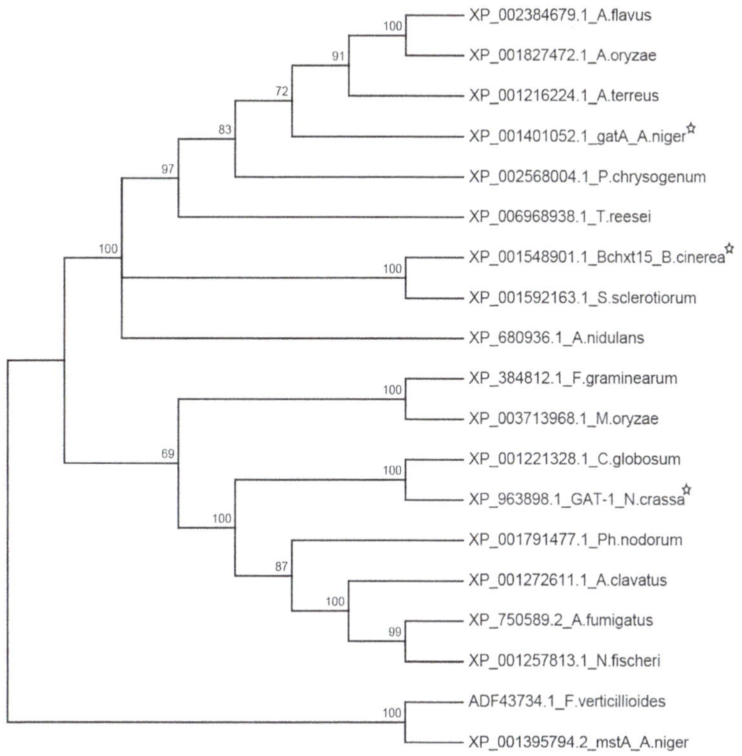

Figure 7 Neighbor-joining tree of GatA and 17 putative orthologs, constructed with 1000 bootstrap replicates using MEGA6.06 (Tamura et al. 2013), and a characterized high-affinity glucose transporter protein (*mstA_A.niger*) as an outgroup (vanKuyk et al. 2004). Stars indicate the three currently characterized D-galacturonic acid transporters. Bootstrap values are indicated at the branch-points. Labels indicate Protein Accession numbers followed by species abbreviations. *Abbreviations: Aspergillus clavatus; A. clavatus, Aspergillus flavus; A. flavus, Aspergillus fumigatus; A. fumigatus, Aspergillus nidulans; A. nidulans, Aspergillus niger; A. niger, Aspergillus oryzae; A. oryzae, Aspergillus terreus; A. terreus, Botrytis cinerea; B. cinerea, Chaetomium globosum; C. globosum, Fusarium verticilliodes; F. verticilliodes, Neosartorya fischeri; N. fischeri, Neurospora crassa; N. crassa, Penicillium chrysogenum; P. chrysogenum, Phaeosphaeria nodorum; Ph. nodorum, Sclerotinia sclerotiorum; S. sclerotiorum, Trichoderma reesei; T. reesei.*

concentrations of 0.1 to 50 mM D-xylose (van der Veen et al. 2009; Mach-Aigner et al. 2012).

The increased D-galacturonic acid uptake of the transformant was not reflected in the basal GaaA enzyme activity during the first hours of growth on D-xylose and D-galacturonic acid. It is, however, reflected in an increased GaaA enzyme activity after 8 hours. This time delay suggests that GaaA enzyme activity is not limiting D-galacturonic acid uptake and metabolism during the first few hours of growth on these substrates and that regulation of expression of GaaA is taking place "at the gate", *via* import of D-galacturonic acid. The repression

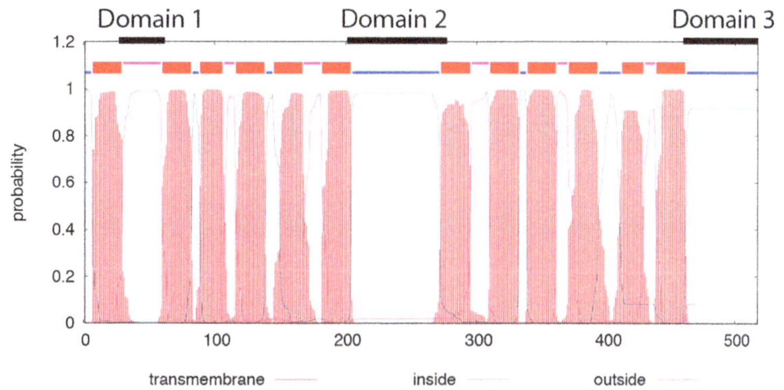

Figure 8 Transmembrane domain prediction and indication of the two interacting domains (1, 3) and the putative substrate binding domain (2) used for comparison of the putative GatA orthologs (Viklund and Elofsson 2004).

of induction of GaaA in the presence of D-xylose in the parent strain, is bypassed by an increased influx of substrate in the GatA overexpressing strain, suggesting that the endogenous D-galacturonic acid catabolic pathway is controlled by a pathway intermediate such as L-galactonate or keto-deoxy-L-galactonate as has been suggested for *T. reesei* (Kuivanen et al. 2012).

Several repressing and inducing regulatory systems are known to be functioning in *Aspergillus niger*. The glucose carbon catabolite repressor *creA* is the most studied and is known to be dominant in most cases (de Vries et al. 1999). The promoter region of the *gatA* gene shows two consensus sequences for binding of this protein, which suggests that D-glucose is repressing the expression of the endogenous gene. The *xlnR* xylanolytic induction system is also well studied and an inducer consensus binding sequence is known (van Peij et al. 1998). This consensus sequence can not be found in the upstream region of the endogenous *gatA* gene and it is therefore not induced by the presence of D-xylose. Additionally, GatA expression has been found to be strickly coregulated with D-galacturonic acid catabolic enzymes and a number of extracellular galacturonases (Martens-Uzunova and Schaap 2008). The regulator of the D-galacturonic acid metabolism, however, remains unknown for now.

We have investigated the application of our findings by growing our strains on a synthetic medium mimicking a plant cell wall hydrolysate. Under these conditions the *gatA* overexpressing strains preferentially use D-galacturonic acid over D-xylose. In these experiments it was found that GatA overexpression strains had a higher citric acid yield. While the effect seems pronounced on a simulated hydrolysate substrate, further experiments are needed to study the benefit on second generation feedstocks with high D-galacturonic acid content like sugar beet pulp to investigate the potential of GatA in biotechnological applications.

Competing interests

The authors declare that they have no competing interests.

Authors' contributions

JS and LG designed the study. JS and MS designed and performed the experimental work. PS contributed to the design and execution of the phylogenetic analysis. JS wrote the manuscript PS and LG participated herein. All authors read and approved the submission of the manuscript.

Acknowledgements

This work has been carried out on the basis of a grant in the framework of the BE- BASIC program F01.011 Transport processes in the production of organic acids by *Aspergillus niger*. We would like to thank Tom Schonewille for his assistance with the analysis equipment.

References

Ángel Siles López J, Li Q, Thompson IP (2010) Biorefinery of waste orange peel. CRC Cr Rev Biotechn 30:63–69, doi:10.3109/07388550903425201

Arnaud MB, Cerqueira GC, Inglis DO, Skrzypek MS, Binkley J, Chibucos MC, Crabtree J, Howarth C, Orvis J, Shah P, Wymore F, Binkley G, Miyasato SR, Simison M, Sherlock G, Wortman JR (2012) The *Aspergillus* Genome Database (AspGD): recent developments in comprehensive multispecies curation, comparative genomics and community resources. Nucleic Acids Res 40:D653–D659

Benz J, Protzko RJ, Andrich JM, Bauer S, Dueber JE, Somerville CR (2014) Identification and characterization of a galacturonic acid transporter from *Neurospora crassa* and its application for *Saccharomyces cerevisiae* fermentation processes. Biotechnol Biofuels 7:20–14, doi:10.1111/j.1567-1364.2009.00523.x

Carlsen M, Nielsen J (2001) Influence of carbon source on alpha-amylase production by *Aspergillus oryzae*. Appl Microbiol Biotechnol 57:346–349

Carpita NC, Gibeaut DM (1993) Structural models of primary cell walls in flowering plants: consistency of molecular structure with the physical properties of the walls during growth. Plant J 3:1–30

de Vries RP, Visser J, de Graaff LH (1999) CreA modulates the XlnR-induced expression on xylose of *Aspergillus niger* genes involved in xylan degradation. Res Microbiol 150(4):281–285

Gielkens MMC, Dekkers E, Visser J, de Graaff LH (1999) Two cellobiohydrolase-encoding genes from *Aspergillus niger* require D-xylose and the xylanolytic transcriptional activator XlnR for their expression. Appl Environ Microbiol 65:4340–4345

Howard RL, Abotsi E, van Rensburg EJ, Howard S (2003) Lignocellulose biotechnology: issues of bioconversion and enzyme production. Afr J Biotechnol 2:602–619, doi:10.4314/ajb.v2i12.14892

Kuivanen J, Mojzita D, Wang Y, Hilditch S, Penttila M, Richard P, Wiebe MG (2012) Engineering filamentous fungi for conversion of D-galacturonic acid to L-galactonic acid. Appl Environ Microbiol 78:8676–8683, doi:10.1128/AEM.02171-12

Kuorelahti S, Kalkkinen N, Penttil M, Londesborough J, Richard P (2005) Identification in the mold *Hypocrea jecorina* of the first fungal D-galacturonic acid reductase. Biochemistry 44:11234–11240, doi:10.1021/bi050792f

Kusters-van Someren MA, Harmsen JA, Kester HC, Visser J (1991) Structure of the *Aspergillus niger pelA* gene and its expression in *Aspergillus niger* and *Aspergillus nidulans*. Curr Genet 20:293–299

Larkin MA, Blackshields G, Brown NP, Chenna R, McGettigan PA, McWilliam H, Valentin F, Wallace IM, Wilm A, Lopez R, Thompson JD, Gibson TJ, Higgins DG (2007) Clustal W and Clustal X version 2.0. Bioinformatics 23:2947–2948, doi:10.1093/bioinformatics/btm404

Mach-Aigner AR, Omony J, Jovanovic B, van Boxtel AJB, de Graaff LH (2012) D-Xylose concentration-dependent hydrolase expression profiles and the function of CreA and XlnR in *Aspergillus niger*. Appl Environ Microbiol 78(9):3145–3155, doi:10.1128/AEM.07772-11

Martens-Uzunova ES, Schaap PJ (2008) An evolutionary conserved D-galacturonic acid metabolic pathway operates across filamentous fungi capable of pectin degradation. Fungal Genet Biol 45:1449–1457, doi:10.1016/j.fgb.2008.08.002

Micard V, Renard CMGC, Thibault JF (1996) Enzymatic saccharification of sugar-beet pulp. Enzyme Microb Technol 19:162–170, doi:10.1016/0141-0229(95)00224-3

Overbeek R, Fonstein M, D'Souza M, Pusch GD, Maltsev N (1999) The use of gene clusters to infer functional coupling. Proc Natl Acad Sci U S A 96:2896–2901

Peterson GL (1977) A simplification of the protein assay method of Lowry et al. which is more generally applicable. Anal Biochem 83:346–356, doi:10.1016/0003-2697(77)90043-4

Pfaffl MW (2001) A new mathematical model for relative quantification in real-time RT-PCR. Nucleic Acids Res 29:45e–45, doi:10.1093/nar/29.9.e45

Ruijter GJG, van de Vondervoort PJI, Visser J (1999) Oxalic acid production by *Aspergillus niger*: an oxalate-non-producing mutant produces citric acid at pH 5 and in the presence of manganese. Microbiology 145(Pt 9):2569–2576

Tamura K, Stecher G, Peterson D, Filipski A, Kumar S (2013) MEGA6: molecular evolutionary genetics analysis version 6.0. Mol Biol Evol 30:2725–2729, doi:10.1093/molbev/mst197

van den Brink J, de Vries RP (2011) Fungal enzyme sets for plant polysaccharide degradation. Appl Microbiol Biotechnol 91:1477–1492, doi:10.1007/s00253-011-3473-2

van der Veen D, Oliveira JM, van den Berg WAM, de Graaff LH (2009) Analysis of variance components reveals the contribution of sample processing to transcript variation. Appl Environ Microbiol 75:2414–2422

van Dijck PWM, Selten GCM, Hempenius RA (2003) On the safety of a new generation of DSM *Aspergillus niger* enzyme production strains. Regul Toxicol Pharmacol 38:27–35, doi:10.1016/S0273-2300(03)00049-7 van Peij NNME, Visser J, de Graaff LH (1998) Isolation and analysis of *xlnR*, encoding a transcriptional activator co-ordinating xylanolytic expression in *Aspergillus niger*. Mol Microbiol 27:131–142, doi:10.1046/j.1365-2958.1998.00666.x

vanKuyk PA, Diderich JA, MacCabe AP, Hererro O, Ruijter G, Visser J (2004) *Aspergillus niger mstA* encodes a high-affinity sugar/H + symporter which is regulated in response to extracellular pH. Biochemical J 379:375–383

Viklund H, Elofsson A (2004) Best α-helical transmembrane protein topology predictions are achieved using hidden Markov models and evolutionary information. Protein Sci 13:1908–1917, doi:10.1110/ps.04625404

Willke T, Vorlop KD (2001) Biotechnological production of itaconic acid. Appl Microbiol Biotechnol 56:289–295

Yan N (2013) Structural advances for the major facilitator superfamily (MFS) transporters. Trends Biochem Sci 38:151–159, doi:10.1016/j.tibs.2013.01.003

Zhang L, Hua C, Stassen JHM, Chatterjee S, Cornelissen M, van Kan JAL (2013) Genome-wide analysis of pectate-induced gene expression in *Botrytis cinerea*: Identification and functional analysis of putative D-galacturonate transporters. Fungal Genet Biol, doi:10.1016/j.fgb.2013.10.002

4

Production of therapeutic proteins in the chloroplast of *Chlamydomonas reinhardtii*

Alma Lorena Almaraz-Delgado[†], José Flores-Uribe[†], Víctor Hugo Pérez-España, Edgar Salgado-Manjarrez and Jesús Agustín Badillo-Corona[*]

Abstract

Chloroplast transformation in the photosynthetic alga *Chlamydomonas reinhardtii* has been used to explore the potential to use it as an inexpensive and easily scalable system for the production of therapeutic recombinant proteins. Diverse proteins, such as bacterial and viral antigens, antibodies and, immunotoxins have been successfully expressed in the chloroplast using endogenous and chimeric promoter sequences. In some cases, proteins have accumulated to high level, demonstrating that this technology could compete with current production platforms. This review focuses on the works that have engineered the chloroplast of *C. reinhardtii* with the aim of producing recombinant proteins intended for therapeutical use in humans or animals.

Keywords: *Chlamydomonas*; Chloroplast transformation; Recombinant therapeutic proteins

Introduction

For centuries, algae have been used for human and animal consumption (Spolaore et al. 2006). More recently algae have been cultivated for the production of fatty acids, secondary metabolites such as pigments, colorants, and bioactive compounds with antitumor, antiviral, and antifungal activities (Borowitzka 1999; Singh et al. 2005; Spolaore et al. 2006). Despite the high number of algae species estimated to populate the planet (Guiry 2012), only a handful of them are currently used with commercial success. The main species cultivated for human or animal consumption are *Chlorella* and *Spirulina*, while *Dunaliella salina* and *Haematococcus pluvialis* have been cultured for the extraction of β-carotene and astaxanthin, respectively (Borowitzka 1999).

High demand and low productivities are driving the efforts in algal technology towards the development of genetically engineered strains. At the top of the list of the algae species that have been genetically modified comes *Chlamydomonas reinhardtii*, an alga that has been used for a long time as a model for the study of photosynthesis, cell motility, phototaxis, and cell wall biogenesis among other genetic and metabolic processes (Harris 2001). Transformation protocols exist for the genetic modification of the nucleus (Kindle 1990), the mitochondria (Remacle et al. 2006), and the chloroplast (Boynton et al. 1988). All three compartments have been genetically engineered with different purposes but it has been the chloroplast which has received considerable attention, mainly to demonstrate that the accumulation of recombinant proteins in this organelle is feasible. Chloroplast transformation is carried out using a particle bombardment device as schematised in Figure 1. Chloroplast transformation technology has demonstrated that not only could it compete with current production platforms but it would be better suited for the production of some proteins such as immunotoxins (Tran et al. 2013a; Tran et al. 2013b), non glycosilated antigens (Gregory et al. 2012; Gregory et al. 2013), and antibodies (Mayfield et al. 2003; Tran et al. 2009) which have been difficult to produce in other systems. This review will comment on the most representative cases that have made clear that chloroplast transformation technology to produce therapeutical proteins is beyond the proof-of-concept stage.

Biopharming in the chloroplast of *C. reinhardtii*

Biopharming refers to the technology that aims, through the use of genetic engineering or, more recently, synthetic biology, to produce health-related products such

* Correspondence: jbadilloc@ipn.mx
[†]Equal contributors
Unidad Profesional Interdisciplinaria de Biotecnología, Instituto Politécnico Nacional, Av. Acueducto SN Col. Barrio la Laguna Ticoman, 07340 Mexico City, México

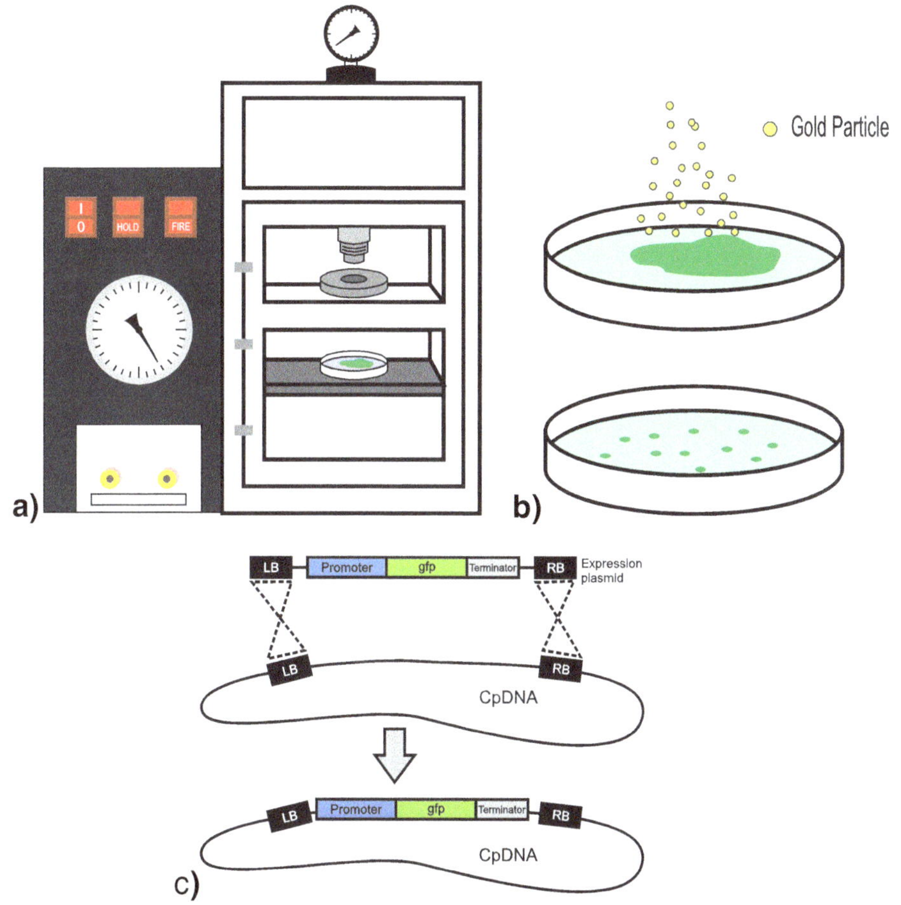

Figure 1 Chloroplast transformation in *Chlamydomonas reinhardtii*. a) Introduction of foreign DNA material into the chloroplast of *C. reinhardtii* is carried out using a particle bombardment device. The device uses helium gas to accelerate particles towards algae placed at the interior of a vacuum chamber. **b)** Gold or tungsten particles are coated with a plasmid carrying the genes of interest (in this case the green fluorescence protein GFP) and when accelerated penetrate the cells placed on top of selection medium. After a few weeks, transformed cells proliferate in the presence of a selection antibiotic. **c)** When the plasmid carrying the genes reaches de chloroplast, genes integrate into the plastid genome by homologous recombination between regions present in the plasmid (LB and RB) and in the chloroplast genome (CpDNA).

as antibodies, antigens, human blood components, growth factors, etc. from plants and animals, as opposed to the traditional chemical synthesis in the pharmaceutical industry, the extraction from the natural source or production in bacteria, yeast or mammalian cells (Gimpel et al. 2013; Specht et al. 2010). Biopharming in plants and algae has several advantages over bacteria and yeast; three of the most import are: a) a product that is free of human pathogens and more quality-consistent; b) larger quantities can be obtained in a reduced or confined area and; c) low production cost when organisms are grown in economic media.

Microbes, yeast, and mammalian cells have been explored to produce therapeutic proteins (Demain and Vaishnav 2009), in fact, there are products derived from genetic engineering currently in the market (Leader et al. 2008). Because *C. reinhardtii* has many advantages over other species, it is not surprising that it has also been proposed as a platform for the production of high-value proteins with therapeutical applications (Specht et al. 2010). Some of the advantages *C. reinhardtii* has include: a fast growth rate compared to plants and animal cells, it is a Generally Recognized as Safe (GRAS) organism by the Food and Drug Administration from the United States of America, cell cultures can be grown in bioreactors, thus reducing the risk of transgene dispersion, or in open raceways without competing with arable land.

Genetic tools and transformation protocols have been developed for the chloroplast of *C. reinhardtii* for the production of high value molecules for the biotechnological industry (Rosales-Mendoza et al. 2012; Specht et al. 2010). However, unlike plants, where chloroplast genetic engineering has been used more widely, for instance, to produce oils, pigments, vitamins, enzymes, or to make drought and salt tolerant plants (Bock 2014;

Maliga and Bock 2011), chloroplast genetic engineering *C. reinhardtii* has mainly been limited to the production of proteins with therapeutical applications. In Table 1, a comprehensive list of the therapeutic proteins produced in the chloroplast of *C. reinhardtii*, showing the genetic elements used for expression and relevant information can be consulted. Such proteins include: viral and bacterial antigens, antibodies, immunotoxins, and enzymes. In the following section representative cases that have dealt with the production of these proteins will be discussed. For a more complete review of chloroplast genetic tools (Bateman and Purton 2000; Day and Goldschmidt-Clermont 2011), synthetic biology in algae (Georgianna and Mayfield 2012; Gimpel et al. 2013), and application perspectives (Beer et al. 2009; Gimpel et al. 2013) the reader is invited to consult the appropriate references.

Antibodies

Demand for antibodies to treat diverse diseases is constantly increasing and is likely to be difficult to meet in the near future. Antibodies have been difficult to produce in prokaryotic systems and thus have to be produced in eukaryotic systems (Makino et al. 2011). Most of the commercially available antibodies are currently produced in mammalian cells making the final product expensive and sometimes scarce (Leader et al. 2008). If antibodies could be efficiently produced in a photosynthetic organism, price would drop and the product would be more accessible to developing countries.

Antibodies have been successfully expressed in the chloroplast of *C. reinhardtii*. Mayfield et al. (2003) have expressed a monoclonal antibody directed against a glycoprotein of the herpes simplex virus D (HSV8). The antibody was expressed from a large single chain (lsc) coding sequence, which consisted on the entire IgA heavy chain protein fused to the variable region of the light chain by a flexible linker peptide. The expression of the HSV8-lsc was under the control of the chloroplast *atpA* and *rbcL* promoters. The lsc antibody accumulated as a soluble protein in transplastomic strains (that is, strains with a transgenic chloroplast genome), and was capable of binding to herpes virus protein D *in vitro*. The lsc antibodies formed disulphide bonds and assembled as dimers in the chloroplast.

In another successful case of an antibody produced in *C. reinhardtii* chloroplast, (Tran et al. 2009) have demonstrated that a monoclonal antibody against a protein from *Bacillus anthracis* could be assembled into a functional tetrameric structure consisting on 2 heavy-chain and 2 light-chain subunits, capable of binding the PA83 antigen from the pathogen *in vitro*. Affinity was determined for the antibodies produced by the alga and an antibody expressed in a mammalian system and were shown to be similar in function. A comparison with antibody 83K7C, expressed in CHO cells, which had been demonstrated to provide protection from anthrax toxicity, in both cell-based assays and animal models, (Tran et al. 2009) showed that, apart from the lack of glycosylation (inherent to chloroplasts), the chloroplast-produced antibody had the same PA83 binding properties than the antibody produced in the mammalian cell-line. The antibody producing strain was reported to be stable and continued to produce the antibody for more than one year.

Immunotoxins

Immunotoxins, antibodies linked to a proteic toxin, have found application in cancer treatment as eukaryotic toxins can be coupled to antibodies with cancer-cell specific binding affinity, thus targeting the apoptosis-inducing toxin. Immunotoxins have been difficult to produce in eukaryotic systems, because due to their nature they kill eukaryotic cells, and production in bacteria has had limited success because prokaryotes are unable to fold multiple domain proteins.

The Stephen Mayfield group at the University of California San Diego has shown that immunotoxins can be produced and confined in the chloroplast of *C. reinhardtii*. A single chain antibody recognizing the CD22 surface receptor from B-cells was fused either to domain II and III of Exotoxin A from *Pseudomonas aeruginosa* (Tran et al. 2013b) or to the ribosome inactivating protein, gelonin, from *Gelonium multiflorm* (Tran et al. 2013a). In both cases it was demonstrated that the immunotoxins were capable of specifically binding to B-cell *in vitro* and that in the case of the immunotoxin Exotoxin A, survival of mice implanted with a human B-cell tumor, the life-span was extended. The two reports are the first cases in which entire immunotoxins have been produced in eukaryotic cells. This has been possible because the chloroplast served as a confined compartment sequestering the eukaryotic toxin from the cytosol, where it is toxic.

Antigens for vaccination

Of the total deaths in the world, 25% result from infectious diseases; that number increases to 45% in developing countries (Arntzen et al. 2005). In humans, vaccination is the most effective mean of inducing resistance to a certain pathogen. The World Health Organization (WHO) and the United Nations Children's Fund (UNICEF) have emphasised the need for new production technologies that can satisfy the demand for global immunisation (UNICEF; WHO). They have also made clear that needle-free, heat-stable, and ease of distribution are important elements for these vaccines to be cost-benefit acceptable.

If antigens intended for vaccination (subunit vaccination) could be produced in GRAS organisms, such as *C. reinhardtii*, the purification process would not be concerned with the removal of toxins or rests of a pathogen, making the process less expensive and the product safer than when it is obtained from microbial cultures. Furthermore, for certain applications, whole GRAS organism (lysed or lyophilized) could even be used as vehicles for administration (Yoon et al. 2011). A few attempts have been made to express viral or bacterial antigens in the chloroplast of *C. reinhardtii*, for which some examples are given below.

The first subunit viral vaccine produced in the chloroplast of *C. reinhardtii* was the translational fusion of the structural protein VP1, from the Foot and Mouth Disease Virus (FMDV), and the cholera toxin B subunit (CtxB) (Sun et al. 2003). FMD in livestock can be prevented by subunit vaccination. The structural protein VP1 of FMDV carries critical epitopes, which can induce neutralizing antibodies while the CtxB is a potent mucosal adjuvant that can bind to intestinal epithelial surfaces via GM1 ganglioside receptors. The CtxB-VP1 fusion protein produced in the chloroplast retained both GM1-ganglioside binding affinity and antigenicity. The CtxB-VP1 fusion protein accumulated up to 3% of the total soluble protein (TSP), a 30-fold increase in comparison with the accumulation level obtained for the same antigen when produced from the nuclear genome of alfalfa and *Arabidopsis thaliana* (Carrillo et al. 1998; Wigdorovitz et al. 1999).

Bacterial antigens have also been shown to accumulate in the chloroplast of *C. reinhardtii* (Michelet et al. 2011). Genes *acrV* and *vapA*, from *Aeromonas salmonicida*, an important pathogen causing fish furunculosis were expressed using the *atpA*, *psbA*, and *psbD* promoters, previously reported to produce high levels of expression for various transgenes (Rasala et al. 2011). The genes were also expressed from a newly proposed promoter, the promoter from gene *psaA* along with exon 1 from the same gene (P*psaA*-exon1). The authors postulated that, since in wild-type strains the *psaA* mRNA is spliced in trans from three separate precursors, the precursor *psaA*-exon1 would over-accumulate in mutants defective in *psaA* transsplicing. The same should happen to a chimeric mRNA with the P*psaA*-exon1. Furthermore, as the unassembled product of *psaA* inhibits the translation of their own mRNA, in a trans-splicing-deficient mutant where *psaA* product cannot be produced, feedback inhibition would be circumvented, and the translation of a recombinant protein expressed under the P*psaA*-exon1 would be higher. It was found that by expressing *acrV* and *VapA* from the P*psaA*-exon1 promoter in a nuclear mutant-deficient on the *psaA* trans-splicing factor, *VapA* accumulated to 0.3% and *AcrV* to 0.8% total protein (TP). In the case of *VapA*, this level was approximately 15 to 30-fold higher than the level of *VapA* present in cells where the nuclear mutation was absent. Even though this increase in the level of accumulation looks attractive, the fact that the transgenic lines obtained are non photosynthetic is a drawback, because, even though lines could be cultivated with acetate as the only carbon source, costs could be reduced if lines were simply grown on CO_2 and light.

The fact that chloroplasts do not glycosilate proteins might be seen as a drawback. However, this characteristic has been exploited positively for certain proteins, (Gregory et al. 2013) have successfully expressed the structurally complex non-glycosilated proteins Pfs25 and Pfs28 from *Plasmodium falciparum*, the parasite responsible of malaria. Proteins accumulated to levels of 0.2-0.5% TSP when expressed from the *psbA* promoter in photosynthetic deficient strains. It has been suggested that Pfs25 and Pfs28 could be used to generate antibodies that bind, and neutralize them *in vivo*, thus preventing the formation of *P. falciparum* sporozoites in the mosquitos, which are ultimately responsible for the transmission of the parasite to humans. Producing these proteins in other systems has been difficult as diverse conformations occur and the proteins are glycosylated when produced in eukaryotic systems whilst native Pfs25 and Pfs28 from the parasite are not. When produced in the chloroplast, proteins folded correctly, lacked glycosylation, and generated antibodies that blocked the transmission of the parasite by preventing the formation of the sporozoites in the mosquito *in vitro*. This shows the fitness of the *C. reinhardtii* chloroplast for the expression of properly folded complex proteins.

Oral delivery of chloroplast-produced antigens

The use of whole lyophilized algae as a vehicle to orally administer biologically active proteins has also been demonstrated, even if only for a limited number of cases. The D2 fribronectin-binding domain from *Staphylococcus aureus* (Georgianna and Mayfield 2012) and the transmission blocking Pfs25 antigen from *Plasmodium falciparum*, the causative agent of malaria (Gregory et al. 2013), have been shown to elicit systemic and mucosal responses when fed to mice in a lyophilized powder of algae. The idea that algae expressing antigens could be administered orally seems very attractive because presumably the production of IgG and IgA antibodies could be induced and hence confer protection to the animal or person vaccinated.

In the case of the D2 fribronectin-binding domain from *S. aureus*, specific mucosal (measured by the presence of IgA antibodies) and systemic (measured by the presence of IgG antibodies) immune responses were demonstrated following oral delivery of lyophilized algae expressing the D2 protein. Furthermore, immunized mice were able to reduce pathogen load when *S. aureus* was inoculated in the intestine and the spleen.

In the case of Pfs25 from *P. falciparum*, when it was translationally fused to CtxB and administered orally

Table 1 Overview of recombinant proteins produces in the chloroplast of *Chlamydomonas reinhardtii*

Recombinant therapeutic protein	Yield	Relevant information	Reference
VP1-CTB; Protein VP1 from foot and mouth disease virus (FMDV) fused to cholera toxin B (CTB)	3-4% Total Soluble Protein (TSP)	Demonstrated that the *C. reinhardtii* chloroplast derived VP1-CTB could bind to GM1-ganglioside receptor *in vitro*	(Sun et al. 2003)
HSV-lsc; Large single chain (lsc) antibody directed against glycoprotein D protein from Herpes simplex virus (HSV)	Not reported	First report to show that the *C. reinhardtii* chloroplast can efficiently fold antibodies and form disulfide bonds	(Mayfield et al. 2003)
TRAIL; Tumor necrosis factor-related apoptosis-inducing ligand	0.43%-0.67% TSP		(Yang et al. 2006)
M-SAA; Mammary-associated serum amyloid	3%-5% TSP	M-SAA was shown to generate mucin induction in a human intestinal epithelial cell line. Demonstrated that the *psbA* promoter yields high level or recombinant protein accumulation when the endogenous *psbA* gene is absent	(Manuell et al. 2007)
CSFV-E2; Classical swine fever virus (CSFV) structural protein E2	1.5-2% TSP	Subcutaneous immunization of mice with E2 was shown to induced IgG antibodies	(He et al. 2007)
Human glutamic acid decarboxylase (hGAD65)	0.25-0.3% TSP	The protein was shown to immunoreact with sera from diabetic mice	(Wang et al. 2008)
IBDV-VP2; Infectious burial disease virus VP2 protein	4-0.8% Total cell protein (TCP)	This report looked at the expression of 11 proteins. Nine proteins showed some level of accumulation, while the rest could not be detected. It showed that there are variations in the level of expression even amongst lines obtained with the transformation construct. Authors postulated the existence of the transformosome, a state in which particular genomic characteristics, induced incidentally with transformation, affect, negatively or positively, the expression of the transgene	(Surzycki et al. 2009)
IHNV-G; Infectious haematopoietic necrosis virus	< 0.5% TCP		
IPNV-VP2; Infectious pancreatic necrosis virus	< 0.3% TCP		
VP2 protein	1-0.1% TCP		
IPNV-VP2 SBC; Infectious pancreatic necrosis virus	1-0.2% TCP		
Quorum sensing-regulated gene (LecA) p57	< 0.5 TCP		
PCV2; Porcine circovirus type 2	0.9-0.2% TCP		
VP-2C	< 0.5% TCP		
VP28	21-0.2% TCP		
HC-83K7C; Heavy chain human monoclonal antibody against anthrax protective antigen 83 (PA83)	0.01% dwt	It was shown that the heavy and light chains expressed in trans could assembled into a fully-functional monoclonal antibody against PA83	(Tran et al. 2009)
LC-83K7C; Light chain human monoclonal antibody against anthrax PA83			
CTB-D2; D2 fibronectin-binding domain of *Staphylococcus aureus* fused to the cholera toxin B subunit	0.7% TSP	First report to show that an orally-administered alga expressing an antigen in the chloroplast triggers a mucosal and systemic immune response in mice	(Dreesen et al. 2010)
14FN3; Domain 14 of human fibronectin	3%-0.15% TSP	This report looked at the expression of seven therapeutic proteins. For three of the proteins, a level of accumulation above 1% was observed, whereas for the rest of the proteins, erythropoietin, interferon β, and proinsulin no protein was detected. Biological activity was evaluated for VEGF and HMGB1	(Rasala et al. 2010)
VEGF; Human vascular endothelial growth factor	2%-0.1% TSP		
HMGB1; High mobility group protein B1	2.5%-1% TSP		
acrV2 and vapA2; antigens from the fish pathogen *Aeromonas salmonicida*	0.8% and 0.3% TP respectively	Showed that the *psaA* promoter-exon1 element can be used to drive the expression of foreign genes in non-photosynthetic strains	(Michelet et al. 2011)
***Escherichia coli* phytase gene (*appA*)**	N.D.	This study showed that algae expressing a bacterial phytase gene in the chloroplast could be lyophilized and administered orally to broiler chicks. The enzyme was active in the gut and reduce the fecal excretion of phytate.	(Yoon et al. 2011)
Pfs25 and Pfs28; surface proteins from *Plasmodium falciparum*	0.5% and 0.2% TSP respectively	First report to show that Pfs25 and Pfs28 can be produced without glycosilation and in a correct conformation recognized by monoclonal antibodies specific to conformational epitopes	(Gregory et al. 2012)

Table 1 Overview of recombinant proteins produces in the chloroplast of *Chlamydomonas reinhardtii* *(Continued)*

αCD22PE40; monomeric immunotoxin consisting on the single chain antibody that recognizes the CD22 surface protein from B-cells, fused to domains II and III of exotoxin A (PE40) from *Pseudomonas aeruginosa*	0.3%-0.4% TSP	First report to show that immunotoxins can be produced in an eukaryotic system without being toxic to the cell.	(Tran et al. 2013b)
αCD22HCH23PE40; dimeric version of αCD22PE40	0.2%-0.3% TSP		
CtxB-Pfs25; *Plasmodium falciparum* surface protein 25 fused to the β subunit of the cholera toxin from *Vibrio cholera*	0.09% TSP	Demonstrated that the fusion protein can induced IgA antibodies when administered orally as part of a lyophilized powered. However, IgG antibodies could not be elicited with this route of administration	(Gregory et al. 2013)
αCD22Gel; single chain antibody targeting the CD22 receptor from B-cells, fused to the eukaryotic ribosome inactivating protein, gelonin, from *Gelonium multiflorm*	0.2%-0.3% TSP	Demonstrated that immunotoxin can efficiently bind to cancerous B-cells in vitro and kill them without affecting non B-cells	(Tran et al. 2013a)
αCD22CH23Gel; dimeric version of αCD22Gel	0.1% - 0.2% TSP		

using a whole-cell lyophilized preparation, IgA antibodies against CtxB and Pfs25 were detected but no IgG antibodies against Pfs25 (Gregory et al. 2013). This could have been the result of CtxB not being the most appropriate adjuvant for oral vaccination with Pfs25. In contrast, it had been previously demonstrated that when Psf25 was purified and injected into mice an IgG response was observed (Gregory et al. 2012). If other adjuvants can perform better than CtxB in conjunction with Pfs25 or if some antigen will simply not elicit an immune response when administered orally remains to be investigated.

Even though, these two cases clearly demonstrate that it could be possible to lyophilize algae expressing antigens and use the powder as a vehicle to administer antigens orally, further studies are needed to demonstrate the safety and efficacy of a broader number of antigens. It has been suggested that vaccines produced in edible organisms such as plants, algae or yeast could potentially be administered orally, however, it is not clear yet how dosage could be controlled consistently. Freeze-drying the product could help solve this problem, as the concentration would be easier to determine in a powder (Gregory et al. 2013). Antigens in lyophilized algae have been shown to be stable at room temperature over a period of months (Dreesen et al. 2010; Gregory et al. 2013). A powder that is stable at room temperature may help overcome the problems associated with the cold chain in the distribution of vaccines in developing countries.

Other therapeutic proteins

In the first report of a recombinant mammalian protein produced in algae, and one of the highest level of protein accumulation yet reported for the chloroplast of *C. reinhardtii*, (Manuell et al. 2007) have demonstrated that it is possible to produce a bovine mammary-associated serum amyloid (M-SAA) protein at levels above 5% of TSP in the chloroplasts of a photosynthetic strain. M-SAA is found in the colostrum of mammals and can induce mucin synthesis in mammalian gut epithelial cells, which could offer protection against intestinal bacterial and viral infections in newborns. M-SAA accumulated predominantly as a soluble protein and was shown to be bioactive when assessed by the induction of mucin in a human intestinal epithelial cell line (HT29) *in vitro.*

Manuell et al. (2007) have also demonstrated the influence of the product of the *psbA* gene in the expression of genes driven by the *psbA* promoter. When M-SAA was expressed from the *psbA* promoter in a non-photosynthetic strain (that is a line lacking the *psbA* gene) the protein accumulated to 3% TSP. When photosynthesis was restored by introduction of a *psbA* gene under the control of a *psbA* promoter, M-SAA did not accumulate. Interestingly, M-SAA accumulated up to 5% of TSP (12% total protein) when photosynthesis was restored by using the *psbA* gene under the control of the *psbD* promoter. As in both cases the D1 protein was present, there seemed to be a competition between the *psbA* mRNA and the M-SAA mRNA for translation (in which the *psbA* mRNA was favoured). Other reports have also shown that high levels of protein can accumulate from the *psbA* promoter when the endogenous *psbA* gene has been deleted (Rasala et al. 2010).

The second report for a human protein produced in the chloroplast of *C. reinhardtii* was for Glutamic Acid Decarboxylase 65 (hGAD65), which is a key autoantigen in insulin dependent diabetes mellitus (IDDM) (Wang et al. 2008). hGAD65 has been identified as an important marker for the prediction and diagnosis of type 1 diabetes, as antibodies against it are present before the onset of the disease and it is believed that tolerance to this autoantigen can be induced using a vaccination-like scheme. Wang et al. (2008) transformed the chloroplast of *C. reinhardtii* with the native non codon-optimized hGAD65 gene under control of the chloroplast *rbcL* promoter. The presence of hGAD65 was confirmed by western blotting using anti-GAD antibodies and the level of accumulation accounted for up to 0.3% TSP, a significantly higher value when compared to the 0.04% TSP obtained for a nuclear transgenic plant (Avesani et al. 2010). The antigenicity of algal-derived hGAD65 was determined by its immunoreactivity with sera from diabetic patients.

In an extensive study, Rasala et al. (2010) evaluated the expression of a diverse group of therapeutic human proteins in *C. reinhardtii* chloroplast. Four out of the seven proteins reported in the study accumulated to levels above 1% TSP. All four proteins 10FN3, 14FN3, HMGB1, and VEGF were expressed using the *psbA* and *atpA* promoters and were shown to be biologically active. In a follow up report, (Rasala et al. 2011) reported the expression of 14FN3, a potential antibody mimic, using a line that lacked the *psbA* gene, and hence, was non-photosynthetic. Lines expressing the 14FN3 from the wild-type *psbA* promoter and from a chimeric promoter (obtained by fusing the 16S rDNA promoter and the *psbA* 5'UTR) were generated. In parallel, using a photosynthetically active strain, the *psbD*, *atpA*, and the chimeric16S rDNA + *atpA* 5′UTR promoters driving the expression of 14FN3 were used to generate transplastomic lines. The wild type *psbA* promoter allowed for the accumulation of 14FN3 to 0.5% TSP but only in the non-photosynthetic strain. The chimeric 16S + *atpA* and 16S + *psbA* promoters also yielded significant amounts of the heterologous protein in photosynthetic strains, 0.23% and 0.21% respectively. Finally, the wild type *atpA* and *psbD* promoters supported only low levels of expression for 14FN3.

Proteins with enzymatic activity intended for animal use have also been produced in the chloroplast of *C. reinhardtii* (Yoon et al. 2011). Nine versions of a phytase gene of bacterial or fungal origin were used to generate transplastomic lines. Microbial phytases are used as food additives to increase phosphorus utilization from phytate in the livestock industry with the aim of reducing fecal phytates and inorganic phosphate (iP) outputs. (Yoon et al. 2011) recovered transplastomic algae but only in the case of strain where a full-length, codon-optimized version of the *E. coli appA* gene (E-OPT) phytase was used, enzymatic activity could be detected (10 U/g of algae).

Strains of algae expressing the enzyme phytase were lyophilized and directly added as a food additive, termed Chlasate, to deliver the enzyme with no need of purification. Using this transgenic phytase at 500 U/kg in the diet of young broiler chicks, the fecal phytate excretion was reduced by 43% while the assimilation of iP increased to 41%, as compared to those of chicks fed with only the basal diet. The effectiveness of Chlasate to break down the dietary phytates was comparable with that of a commercial fungal phytase.

To make the final product, cell pastes were directly frozen once with liquid nitrogen, thawed, frozen again and dried with a lyophilizer to obtain a dry cell powder. The one-step frozen-thawing procedure seemed to be sufficient to rupture the *C. reinhardtii* cells and release a functional phytase. The major advantage of using transgenic microalgae for oral delivery is that several processing steps from raw material to the edible final product can be omitted.

Final remarks and conclusions

It is now clear that chloroplast transformation in *C. reinhardtii* can be consistently accomplished and used to generate transplastomic strains that accumulate recombinant heterologous proteins with therapeutical applications. Although it is still behind the achievements chloroplast transformation has had in land plants has had (Daniell et al. 2009), the technology has successfully been used in laboratories around the world to produce diverse bacterial, viral, and human proteins.

To achieve high levels of foreign protein accumulation, diverse endogenous and chimeric promoters have been developed. In some cases, recombinant proteins have accumulated up to 20% of the TSP (Surzycki et al. 2009), although 1-5% TSP is the most common accumulation range for several proteins (Manuell et al. 2007; Rasala et al. 2010). In some other cases low to undetectable levels of protein accumulation have been encountered (Rasala et al. 2010; Surzycki et al. 2009), stressing the importance of diverse factors in the regulation of gene expression. At present time there is no way of predicting which genetic elements would be better suited for the expression of a particular foreign gene. Some promoters seem to work and yield reasonable levels of expression for some genes whilst for others they seem to be less appropriate (Rasala et al. 2010).

For some of the proteins revised in this paper, biological activity has been evaluated both *in vitro* and *in vivo*. In the case of antigens intended for vaccination, it has been demonstrated that a systemic and mucosal response can be induced after the administration of a lyophilized powder of the algae expressing the antigen in mice. However, after the purity and efficacy of the antigens has been demonstrated, preclinical trials are needed to demonstrate the safety of the chloroplast derived antigens. Next step would be to carry out clinical trials, though, this may still be a long way from current status; chloroplast-derived vaccines from plants have had a tortuous path from the laboratory to clinical trials (Rybicki 2009). It is clear that the ultimate goal of producing proteins in the chloroplast of *C. reinhardtii* cannot be considered accomplished until there is an algal produced protein in the market.

Besides being used as bioreactors for the production of recombinant proteins, *C. reinhardtii* and algae in general, have the potential of serving as tools to solve diverse problems. The spectrum of metabolites that could potentially be produced in the chloroplast of *C. reinhardtii* by genetic modification goes from secondary metabolites such as vitamins, carotenoids, and oils to bioplastics and biofuels (such as hydrogen, fatty acids, or solvents), etc. This would alleviate two problems, the ever-increasing worry of a decline in oil production and a reduction in atmospheric CO_2. Use of algae in bioremediation, removal of toxic compounds from land, water, and air has also been contemplated but poorly studied.

Also, there is a need to expand chloroplast transformation technology to other algae species. Such species should be: 1) easily scalable, with a better ability to grow in open ponds or photobioreactors and, 2) grow faster and have little susceptibility to contamination, but most importantly, 3) strains that depending on the final use are valuable for the production of biopharmaceuticals, secondary metabolites or high energy molecules, such as lipids and sugars. A wide interest in such strains will detonate the development of genetic tools and transformation methods, which has been so far limited to *C. reinhardtii*.

Competing interests

The authors declare that they have no competing interests.

Author's contributions

All authors participated in drafting the manuscript and ALAD and JFU wrote all sections except for the introduction and perspectives, which were written by JABC and ESM. Table 1 was produced by VHPE. All authors have given final approval of the version to be published.

Acknowledgements

JABC and ESM are recipients of funding for research from Instituto Politécnico Nacional (SIP-20121514). Research in the laboratory of JABC is sponsored by ICYT-DF (PIUTE 10-74, PICSA 10-174). ALAD and JFU are recipients of studentships from PIFI-IPN and from the National Council of Science and Technology (CONACYT). JABC is recipient of fellowships from EDI-IPN and COFAA-IPN. Postdoctoral visit of VHPE in the laboratory of JABC was sponsored by ICYT-DF (PIUTE 10-74). We thank Dr. José L. Castrejón Flores and Dr. Juan S. Aranda Barradas for critically reading the manuscript.

References

Arntzen C, Plotkin S, Dodet B (2005) Plant-derived vaccines and antibodies: potential and limitations. Vaccine 23:1753–1756

Avesani L, Vitale A, Pedrazzini E, DeVirgilio M, Pompa A, Barbante A, Gecchele E, Dominici P, Morandini F, Brozzetti A, Falorni PM (2010) Recombinant human GAD65 accumulates to high levels in transgenic tobacco plants when expressed as an enzymatically inactive mutant. Plant Biotechnol J 8:862–872

Bateman JM, Purton S (2000) Tools for chloroplast transformation in *Chlamydomonas*: expression vectors and a new dominant selectable marker. Mol Gen Genet 263:404–410

Beer LL, Boyd ES, Peters JW, Posewitz MC (2009) Engineering algae for biohydrogen and biofuel production. Curr Opin Biotech 20:264–271

Bock R (2014) Genetic engineering of the chloroplast: novel tools and new applications. Curr Opin Biotech 26:7–13

Borowitzka MA (1999) Commercial production of microalgae: ponds, tanks, and fermenters. Prog Ind M 35:313–321

Boynton JE, Gillham NW, Harris EH, Hosler JP, Johnson AM, Jones AR, Randolph-Anderson BL, Robertson D, Klein TM, Shark KB (1988) Chloroplast transformation in *Chlamydomonas* with high velocity microprojectiles. Science 240:1534–1538

Carrillo C, Wigdorovitz A, Oliveros JC, Zamorano PI, Sadir AM, Gomez N, Salinas J, Escribano JM, Borca MV (1998) Protective immune response to foot-and-mouth disease virus with VP1 expressed in transgenic plants. J Virol 72:1688–1690

Daniell H, Singh ND, Mason H, Streatfield SJ (2009) Plant-made vaccine antigens and biopharmaceuticals. Trends Plant Sci 14:669–679

Day A, Goldschmidt-Clermont M (2011) The chloroplast transformation toolbox: selectable markers and marker removal. Plant Biotechnol J 9:540–553

Demain AL, Vaishnav P (2009) Production of recombinant proteins by microbes and higher organisms. Biotechnol Adv 27:297–306

Dreesen IA, Hamri GC-E, Fussenegger M (2010) Heat-stable oral alga-based vaccine protects mice from *Staphylococcus aureus* infection. J Biotechnol 145:273–280

Georgianna DR, Mayfield SP (2012) Exploiting diversity and synthetic biology for the production of algal biofuels. Nature 488:329–335

Gimpel JA, Specht EA, Georgianna DR, Mayfield SP (2013) Advances in microalgae engineering and synthetic biology applications for biofuel production. Curr Opin Chem Biol 17:489–495

Gregory JA, Li F, Tomosada LM, Cox CJ, Topol AB, Vinetz JM, Mayfield S (2012) Algae-produced Pfs25 elicits antibodies that inhibit malaria transmission. PLoS One 7:e37179

Gregory JA, Topol AB, Doerner DZ, Mayfield S (2013) Alga-produced cholera toxin-Pfs25 fusion proteins as oral vaccines. Appl Environ Microb 79:3917–3925

Guiry MD (2012) How many species of algae are there? J Phycol 48:1057–1063

Harris EH (2001) *Chlamydomonas* as a model organism. Annu Rev Plant Biol 52:363–406

He D-M, Qian K-X, G-F SHEN, Zhang Z-F, Li Y-N, Su Z-L, Shao H-B (2007) Recombination and expression of classical swine fever virus (CSFV) structural protein E2 gene in *Chlamydomonas reinhardtii* chroloplasts. Colloids Surf B Biointerfaces 55:26–30

Kindle KL (1990) High-frequency nuclear transformation of *Chlamydomonas reinhardtii*. Proc Natl Acad Sci U S A 87:1228–1232

Leader B, Baca QJ, Golan DE (2008) Protein therapeutics: a summary and pharmacological classification. Nat Rev Drug Discov 7:21–39

Makino T, Skretas G, Kang TH, Georgiou G (2011) Comprehensive engineering of *Escherichia coli* for enhanced expression of IgG antibodies. Met Eng 13:241–251

Maliga P, Bock R (2011) Plastid biotechnology: food, fuel, and medicine for the 21st century. Plant Physiol 155:1501–1510

Manuell AL, Beligni MV, Elder JH, Siefker DT, Tran M, Weber A, McDonald TL, Mayfield SP (2007) Robust expression of a bioactive mammalian protein in *Chlamydomonas* chloroplast. Plant Biotechnol J 5:402–412

Mayfield SP, Franklin SE, Lerner RA (2003) Expression and assembly of a fully active antibody in algae. Proc Natl Acad Sci U S A 100:438–442

Michelet L, Lefebvre-Legendre L, Burr SE, Rochaix J-D, Goldschmidt-Clermont M (2011) Enhanced chloroplast transgene expression in a nuclear mutant of *Chlamydomonas*. Plant Biotechnol J 9:565–574

Rasala BA, Muto M, Lee PA, Jager M, Cardoso RM, Behnke CA, Kirk P, Hokanson CA, Crea R, Mendez M, Mayfield SP (2010) Production of therapeutic proteins in algae, analysis of expression of seven human proteins in the chloroplast of *Chlamydomonas reinhardtii*. Plant Biotechnol J 8:719–733

Rasala BA, Muto M, Sullivan J, Mayfield SP (2011) Improved heterologous protein expression in the chloroplast of *Chlamydomonas reinhardtii* through

promoter and 5′ untranslated region optimization. Plant Biotechnol J 9:674–683
Remacle C, Cardol P, Coosemans N, Gaisne M, Bonnefoy N (2006) High-efficiency biolistic transformation of *Chlamydomonas* mitochondria can be used to insert mutations in complex I genes. Proc Natl Acad Sci U S A 103:4771–4776
Rosales-Mendoza S, Paz-Maldonado LMT, Soria-Guerra RE (2012) *Chlamydomonas reinhardtii* as a viable platform for the production of recombinant proteins: current status and perspectives. Plant Cell Rep 31:479–494
Rybicki EP (2009) Plant-produced vaccines: promise and reality. Drug Discov Today 14:16–24
Singh S, Kate BN, Banerjee UC (2005) Bioactive compounds from cyanobacteria and microalgae: an overview. Crit Rev Biotechnol 25:73–95
Specht E, Miyake-Stoner S, Mayfield S (2010) Micro-algae come of age as a platform for recombinant protein production. Biotechnol Lett 32:1373–1383
Spolaore P, Joannis-Cassan C, Duran E, Isambert A (2006) Commercial applications of microalgae. J Biosci Bioeng 101:87–96
Sun M, Qian K, Su N, Chang H, Liu J, Shen G (2003) Foot-and-mouth disease virus VP1 protein fused with cholera toxin B subunit expressed in *Chlamydomonas reinhardtii* chloroplast. Biotechnol Lett 25:1087–1092
Surzycki R, Greenham K, Kitayama K, Dibal F, Wagner R, Rochaix J-D, Ajam T, Surzycki S (2009) Factors effecting expression of vaccines in microalgae. Biologicals 37:133–138
Tran M, Zhou B, Pettersson PL, Gonzalez MJ, Mayfield SP (2009) Synthesis and assembly of a full-length human monoclonal antibody in algal chloroplasts. Biotechnol Bioeng 104:663–673
Tran M, Henry RE, Siefker D, Van C, Newkirk G, Kim J, Bui J, Mayfield SP (2013a) Production of anti-cancer immunotoxins in algae: Ribosome inactivating proteins as fusion partners. Biotechnol Bioeng 110:2826–2835
Tran M, Van C, Barrera DJ, Pettersson PL, Peinado CD, Bui J, Mayfield SP (2013b) Production of unique immunotoxin cancer therapeutics in algal chloroplasts. Proc Natl Acad Sci U S A 110:E15–E22
UNICEF The United Nations Children's Fund. http://www.unicef.org/. Accessed 24 Apr 2014
Wang X, Brandsma M, Tremblay R, Maxwell D, Jevnikar AM, Huner N, Ma S (2008) A novel expression platform for the production of diabetes-associated autoantigen human glutamic acid decarboxylase (hGAD65). BMC Biotechnol 8:87
WHO The World Health Organization. http://www.who.int/en/. Accessed 24 Apr 2014
Wigdorovitz A, Carrillo C, Dus Santos MJ, Trono K, Peralta A, Gomez MC, Rios RD, Franzone PM, Sadir AM, Escribano JM, Borca MV (1999) Induction of a protective antibody response to foot and mouth disease virus in mice following oral or parenteral immunization with alfalfa transgenic plants expressing the viral structural protein VP1. Virology 255:347–353
Yang Z, Chen F, Li D, Zhang Z, Liu Y, Zheng D, Wang Y, Shen G (2006) Expression of human soluble TRAIL in *Chlamydomonas reinhardtii* chloroplast. Chin Sci Bull 51:1703–1709
Yoon S-M, Kim SY, Li KF, Yoon BH, Choe S, Kuo MM-C (2011) Transgenic microalgae expressing *Escherichia coli* AppA phytase as feed additive to reduce phytate excretion in the manure of young broiler chicks. Appl Microbiol Biot 91:553–563

5

Stepwise enhancement of catalytic performance of haloalkane dehalogenase LinB towards β-hexachlorocyclohexane

Ryota Moriuchi[1,3], Hiroki Tanaka[1], Yuki Nikawadori[1], Mayuko Ishitsuka[1], Michihiro Ito[1,4], Yoshiyuki Ohtsubo[1], Masataka Tsuda[1], Jiri Damborsky[2], Zbynek Prokop[2] and Yuji Nagata[1*]

Abstract

Two haloalkane dehalogenases, $LinB_{UT}$ and $LinB_{MI}$, each with 296 amino acid residues, exhibit only seven amino acid residue differences between them, but $LinB_{MI}$'s catalytic performance towards β-hexachlorocyclohexane (β-HCH) is considerably higher than $LinB_{UT}$'s. To elucidate the molecular basis governing this difference, intermediate mutants between $LinB_{UT}$ and $LinB_{MI}$ were constructed and kinetically characterized. The activities of $LinB_{UT}$-based mutants gradually increased by cumulative mutations into $LinB_{UT}$, and the effects of the individual amino acid substitutions depended on combination with other mutations. These results indicated that $LinB_{UT}$'s β-HCH degradation activity can be enhanced in a stepwise manner by the accumulation of point mutations.

Keywords: β-Hexachlorocyclohexane; Xenobiotics; Biodegradation; Haloalkane dehalogenase; Protein evolution

Introduction

γ-Hexachlorocyclohexane (γ-HCH; also known as γ-BHC or lindane) is a manmade and xenobiotic halogenated insecticide that was once used worldwide on a large scale. A number of soil bacterial strains that can aerobically degrade γ-HCH have been isolated from geographically distant locations (Lal et al. 2010; Ito et al. 2007; Lal et al. 2006; Mohn et al. 2006; Phillips et al. 2005). As this novel compound was first released into the environment in the 1940s, they must have evolved quickly to utilize it.

An industrial chemical process of benzene photochlorination generates so-called technical-HCH (t-HCH), which consists mainly of five isomers, α- (60-70%), γ- (12-16%), β- (10-12%), δ- (6-10%), and ε-HCH (3-4%) (Vijgen et al. 2011). Among these isomers, only γ-HCH has insecticidal activity; this isomer was therefore purified. The remaining isomers were in many cases improperly deposited, causing serious environmental problems. α- and β-HCH isomers as well as γ-HCH were categorized as persistent organic pollutants (POPs) at the Stockholm Convention (Vijgen et al. 2011). Among the HCH isomers, β-HCH is the most recalcitrant; it is usually the predominant isomer remaining in contaminated soils and in animal tissues and fluids (Willett, et al. 1998). All six chlorines of β-HCH in equatorial positions seem to contribute to its having the greatest chemical stability among the isomers. Several β-HCH-degrading bacterial strains have also been identified (Johri et al. 1998; Gupta et al. 2000, 2001). Haloalkane dehalogenase (HLD) LinB, which was originally described as an enzyme involved in γ-HCH degradation in *Sphingobium japonicum* UT26 ($LinB_{UT}$) (Nagata et al. 1993), was more recently identified as an enzyme possessing β-HCH degradation activity (Nagata et al. 2005; Sharma et al. 2006) (Figure 1).

HLDs belong to the α/β-hydrolase family, and their catalytic mechanism consists of the following steps: substrate binding, cleavage of the carbon-halogen bond in the substrate and simultaneous formation of an intermediate covalently bound to a nucleophile, hydrolysis of the alkyl-enzyme intermediate, and release of halide ion and alcohol (Damborsky and Koca 1999; Janssen 2004; Prokop et al. 2003). $LinB_{MI}$ isolated from *Sphingobium* sp. MI1205 (Ito et al. 2007) and $LinB_{UT}$ each consist of 296 amino acid residues and share 98% sequence identity, with only seven different amino acid residues between them, at the positions 81, 112, 134, 135, 138, 247, and 253 (Figure 2).

* Correspondence: aynaga@ige.tohoku.ac.jp
[1]Department of Environmental Life Sciences, Graduate School of Life Sciences, Tohoku University, Sendai 980-8577, Japan
Full list of author information is available at the end of the article

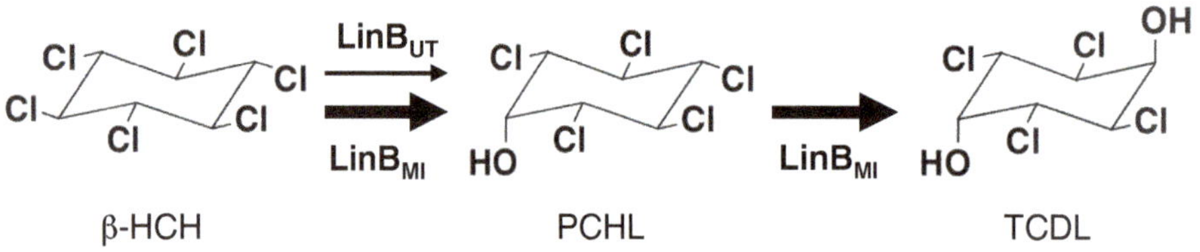

Figure 1 β-HCH degradation reactions catalyzed by LinB$_{UT}$ and LinB$_{MI}$. LinB$_{MI}$ converts β-HCH to PCHL and further to TCDL, while LinB$_{UT}$ catalyzes only the first conversion step of β-HCH to PCHL.

However, these two enzymes exhibit significantly different enzymatic behaviors in β-HCH degradation (Figure 1). LinB$_{MI}$ catalyzes the two-step dehalogenation and converts β-HCH to 2,3,4,5,6-pentachlorocyclohexanol (PCHL) and then to 2,3,5,6-tetrachlorocyclohexane-1,4-diol (TCDL) (LinB$_{MI}$-type activity) (Ito et al. 2007), whereas LinB$_{UT}$ catalyzes only the former step (Nagata et al. 2005) (Figure 1). Furthermore, LinB$_{MI}$ can catalyze the first conversion step an order of magnitude more rapidly than LinB$_{UT}$ (Ito et al. 2007). Substitution of the LinB$_{UT}$ I134 and A247 residues, which form the catalytic pocket, to the LinB$_{MI}$-type V and H residues, respectively, resulted in only a weak effect on LinB$_{MI}$-type activity (Ito et al. 2007). Additionally, the reciprocal double mutant of LinB$_{MI}$ (V134I/H247A) still retained relatively high LinB$_{MI}$-type activity (Ito et al. 2007). These results indicated that one or more of the five other residues are also important for LinB$_{MI}$-type activity. Our previous site-directed mutagenesis and X-ray crystallographic studies of LinB$_{MI}$ (Okai et al. 2013) indicated that (i) these five

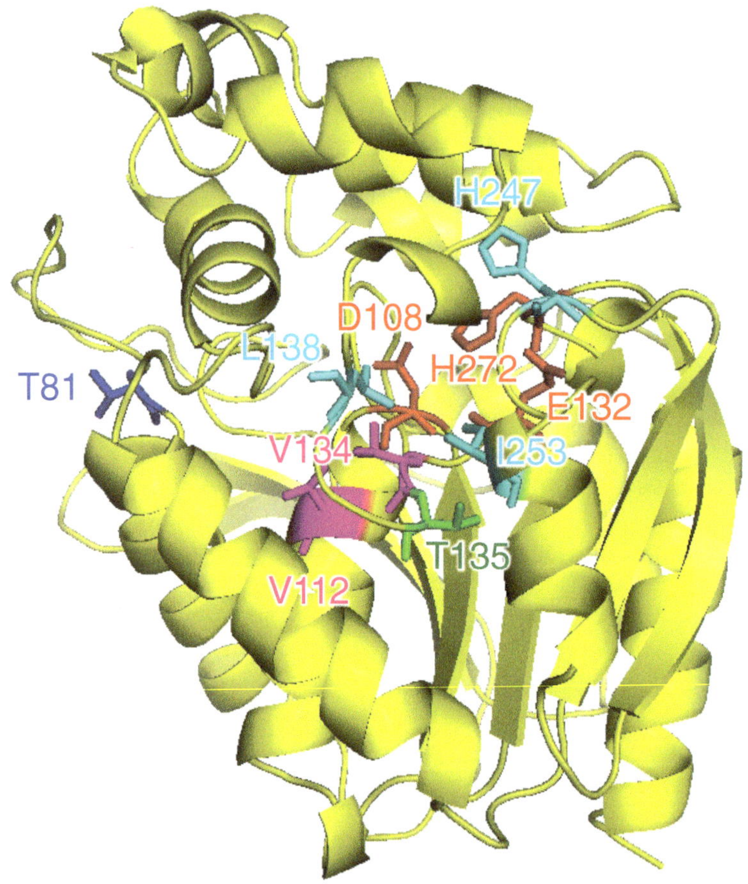

Figure 2 Structure of LinB$_{MI}$ (PDB code 4H77) (Okai et al. 2013) and location of catalytic triad (D108, E132, and H272; shown in red) and the seven dissimilar amino acid residues between LinB$_{MI}$ and LinB$_{UT}$: V134 and V112 (in magenta), L138, H247, and I253 (in cyan), T135 (in green), and T81 (in blue). See text for detail.

residues are not essential to the $LinB_{MI}$-type activity, but they all significantly contribute to this activity, and (ii) three of the five residues, V112, L138, and I253, are more important than T81 and T135 for the conversion of PCHL. The structural basis for the importance of the seven amino acid residues of $LinB_{MI}$ can be partially explained by analysis of its tertiary structure (Figure 2). V134 and V112 are located at the catalytic pocket near the nucleophile residue (D108) and at the bottom of the substrate binding pocket, respectively, while L138, H247, and I253 are located at the access tunnels to the catalytic pocket. Therefore, these five amino acid residues may be directly involved in $LinB_{MI}$-type activity (Okai et al. 2013). The effect of T135 on $LinB_{MI}$-type activity may be due to its interaction with I253. However, it is unclear how T81, which is located at the protein surface and far from the active site, affect the activity.

In this study, cumulative mutations were introduced into $LinB_{UT}$, and the resulting intermediate mutant enzymes between $LinB_{UT}$ and $LinB_{MI}$ were characterized in order to gain more insight into the molecular evolution of LinB towards β-HCH degradation activity. Since the $LinB_{UT}$ I134V/A247H (=M2-1) mutant showed only weak $LinB_{MI}$-type activity in our previous study (Ito et al. 2007), cumulative mutations at the positions A112, I138, and M253 were introduced into the M2-1 mutant. On the basis of kinetic analyses of the resulting three-, four-, and five-point $LinB_{UT}$ mutants (Figures 3, 4 and Table 1), enhancement of the catalytic performance of $LinB_{UT}$ towards β-HCH is discussed.

Materials and methods

Expression and purification of enzymes

The nomenclature of LinB mutants used in this study is shown in Table 1. An established method was used for site-directed mutagenesis (Ito et al. 2007). The expression plasmids for the 10 $LinB_{UT}$ multiple mutants, M3-1 (A112V/I134V/A247H), M3-2 (I134V/I138L/A247H), M3-3 (I134V/A247H/M253I), M4-1 (A112V/I134V/I138L/A247H), M4-2 (I134V/I138L/A247H/M253I), M5-1 (A112V/I134V/I138L/A247H/M253I), M5-2 (A81T/I134V/I138L/A247H/M253I), M5-3 (I134V/A135T/I138L/A247H/M253I), M5-4 (A81T/A112V/I134V/I138L/A247H), and M5-5 (A112V/I134V/A135T/I138L/A247H), were constructed as described previously using the vector pAQNM (Ito et al. 2007). The His-tagged target proteins were expressed under the control of the *tac* promoter and *lacI^q^*. *E. coli* BL21 Star (DE3) cells expressing $LinB_{UT}$ mutants were disrupted by bacteriolysis using a CelLytic B Reagent (Sigma), and His-tagged enzymes were purified by using BD TALON Metal Affinity Resins (BD Biosciences). Only one protein band corresponding to about 33 kDa was observed on sodium dodecyl sulfate (SDS)-polyacrylamide gel electrophoresis after the purification (data not shown).

Enzymatic assays

The purified enzyme was incubated with 17 μM of β-HCH in 50 mM potassium phosphate buffer (pH 7.5) containing 10% (v/v) glycerol at 30°C. The enzyme concentration in the reaction mixture was 150 μg/ml. The mixture (100 μl) was extracted with an equal volume of ethyl acetate and then analyzed by a Shimadzu GC-17A gas chromatograph equipped with a ^{63}Ni electron capture detector (ECD) and Rtx-1 capillary column (30 m × 0.25 mm × 0.25 μm; Restek). The column temperature was increased from 160°C to 200°C at a rate of 4°C/min, and then from 200°C to 260°C at a rate of 20°C/min. The gas flow rate was a constant 30 ml/min. As the internal standard, 10 μM 2,4,5-trichlorophenol was used. Due to the low solubility of β-HCH in water, the k_{cat} and K_m values of the mutants could not be calculated. Kinetic data were fitted to the irreversible two-step reaction scheme of β-HCH conversion to TCDL via PCHL (Scheme 1). Nonlinear regression provided estimates of the specificity constants and standard errors for both reaction steps (k_1 and k_2) by using GEPASI 3.2 software (Mendes 1997).

$$E + HCH \xrightarrow{k1} E + PCHL \xrightarrow{k2} E + TCDL$$

Scheme 1

Results

Characterization of single- and double-point mutants of $LinB_{UT}$

Since M2-1 showed only a weak $LinB_{MI}$-type activity in the previous study (Ito et al. 2007), we introduced further mutations into M2-1 in this study. However, for the critical comparison with other mutants, the seven single-point mutants [A81T (M1-1), A112V (M1-2), I134V (M1-3), A135T (M1-4), I138L (M1-5), A247H (M1-6), and M253I (M1-7)] of $LinB_{UT}$ and M2-1 were also kinetically characterized in this study, and the importance of individual mutations for $LinB_{MI}$ activity towards β-HCH was assessed (Figure 3 and Table 1). Among the seven single-point mutations, only the A112V mutation had a negative effect on β-HCH degradation activity, while the other six mutations showed a slightly positive effect on enzymatic activity towards β-HCH (Figure 3b-h and Figure 5). Interestingly, the A81T mutation had a relatively strong effect on the first conversion (β-HCH to PCHL) step (Figure 3b and Figure 5e). The involvement of T81 in the first step was consistent with the decrease in this step by the reciprocal T81A mutation into $LinB_{MI}$ (M6-1) (Figure 5e and Additional file 1: Figure S1b) (Okai et al. 2013). M2-1 showed higher activity for the second conversion (PCHL to TCDL) step (Table 1) than all the single mutants, but its activity was still weak (Figure 3i and Figure 5a).

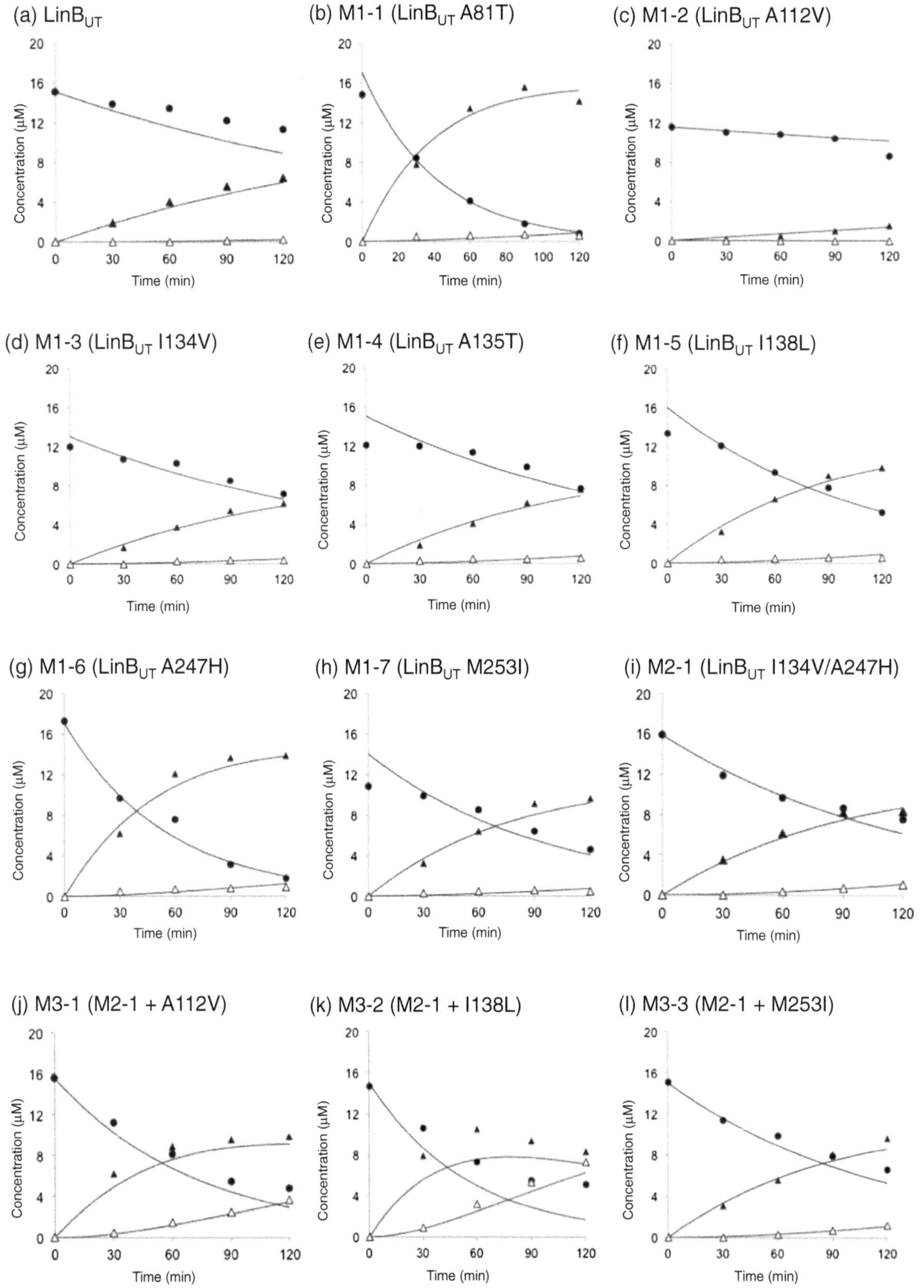

Figure 3 (See legend on next page.)

(See figure on previous page.)

Figure 3 Degradation of β-HCH in reaction mixtures containing $LinB_{UT}$ and its mutant derivatives. $LinB_{UT}$ wild-type **(a)** and its mutants, M1-1 **(b)**, M1-2 **(c)**, M1-3 **(d)**, M1-4 **(e)**, M1-5 **(f)**, M1-6 **(g)**, M1-7 **(h)**, M2-1 **(i)**, M3-1 **(j)**, M3-2 **(k)**, M3-3 **(l)**. The closed circle and closed and open triangles represent β-HCH, PCHL, and TCDL, respectively. Each value given is the mean of triplicates. Kinetic data were fitted to the irreversible two-step reaction scheme of β-HCH conversion to TCDL via PCHL (Scheme 1 in Materials and methods) by using GEPASI 3.2 software (Mendes 1997) and shown in solid lines. The specificity constants and their standard errors for both reaction steps (k_1 and k_2) were obtained from the calculation (Table 1).

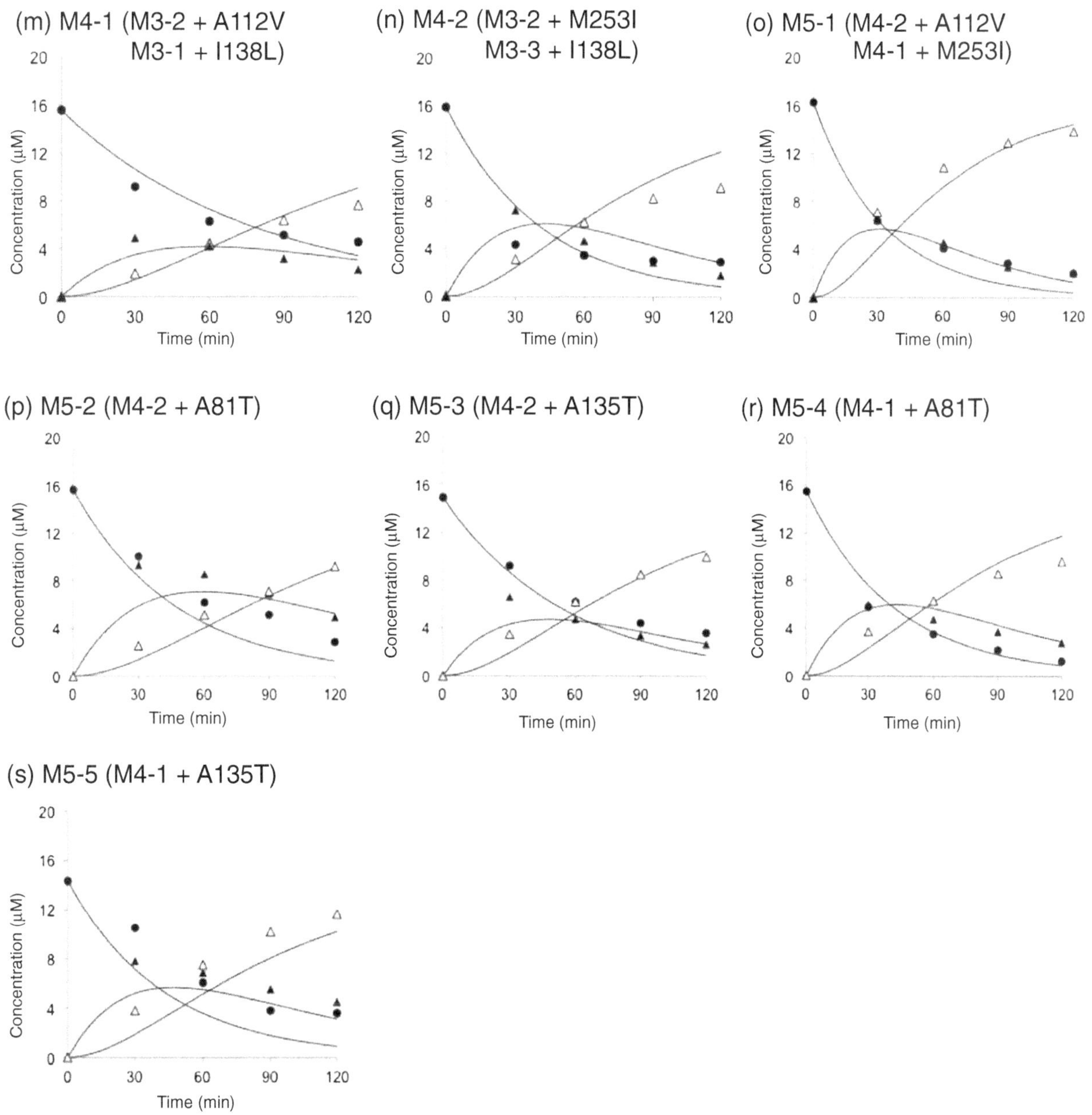

Figure 4 Degradation of β-HCH in reaction mixtures containing $LinB_{UT}$ mutant derivatives. M4-1 **(m)**, M4-2 **(n)**, M5-1 **(o)**, M5-2 **(p)**, M5-3 **(q)**, M5-4 **(r)**, and M5-5 **(s)**. See legend of Figure 3.

Table 1 Specificity constants of $LinB_{UT}$, $LinB_{MI}$, and their intermediate mutants

Enzyme			Position of the different amino acid residues							Specificity constant k_{cat}/K_m ($M^{-1}s^{-1}$)				Data source
Fold[a]	Mutant name		81	112	134	135	138	247	253	HCH → PCHL	% to $LinB_{MI}$	PCHL → TCDL	% to $LinB_{MI}$	
0	$LinB_{UT}$		A	A	I	A	I	A	M	17 ± 0	9	2.0 ± 1	0.2	This study[b]
1	$LinB_{UT}$ A81T	M1-1	T	A	I	A	I	A	M	92 ± 3	50	2.0 ± 1	0.2	This study
1	$LinB_{UT}$ A112V	M1-2	A	V	I	A	I	A	M	4 ± 0	2.1	0	0	This study
1	$LinB_{UT}$ I134V	M1-3	A	A	V	A	I	A	M	21 ± 0	11	5 ± 1	0.5	This study[b]
1	$LinB_{UT}$ A135T	M1-4	A	A	I	T	I	A	M	23 ± 1	12	6 ± 2	0.6	This study
1	$LinB_{UT}$ I138L	M1-5	A	A	I	A	L	A	M	35 ± 1	18	5 ± 1	0.5	This study
1	$LinB_{UT}$ A247H	M1-6	A	A	I	A	I	H	M	68 ± 2	36	4 ± 1	0.4	This study[b]
1	$LinB_{UT}$ M253I	M1-7	A	A	I	A	I	A	I	39 ± 1	21	4 ± 1	0.4	This study
2	$LinB_{UT}$ I134V/A247H	M2-1	A	A	V	A	I	H	M	30 ± 0	16	7 ± 1	0.7	This study[b]
3	$LinB_{UT}$ A112V/ I134V/A247H	M3-1	A	V	V	A	I	H	M	53 ± 3	28	17 ± 2	1.7	This study
3	$LinB_{UT}$ I134V/I138L/A247H	M3-2	A	A	V	A	L	H	M	69 ± 10	36	31 ± 7	3.1	This study
3	$LinB_{UT}$ I134V/A247H/M253I	M3-3	A	A	V	A	I	H	I	33 ± 1	17	7 ± 1	0.7	This study
4	$LinB_{UT}$ A112V/I134V/I138L/A247H	M4-1	A	V	V	A	L	H	M	48 ± 4	25	85 ± 10	8.5	This study
4	$LinB_{UT}$ I134V/I138L/A247H/M253I	M4-2	A	A	V	A	L	H	I	$1.0 \pm 0.2 \times 10^2$	53	83 ± 20	8.3	This study
5	$LinB_{UT}$ A112V/I134V/I138L/A247H/M253I	M5-1	A	V	V	A	L	H	I	$1.2 \pm 0.1 \times 10^2$	63	$1.3 \pm 0.2 \times 10^2$	13	This study
5	$LinB_{UT}$ A81T/I134V/I138L/A247H/M253I	M5-2	T	A	V	A	L	H	I	79 ± 10	42	50 ± 10	5	This study
5	$LinB_{UT}$ I134V/A135T/I138L/A247H/M253I	M5-3	A	A	V	T	L	H	I	68 ± 8	36	90 ± 20	9	This study
5	$LinB_{UT}$ A81T/A112V/I134V/I138L/A247H	M5-4	T	V	V	A	L	H	M	91 ± 9	48	83 ± 10	8.3	This study
5	$LinB_{UT}$ A112V/I134V/A135T/I138L/A247H	M5-5	A	V	V	T	L	H	M	84 ± 20	44	73 ± 20	7.3	This study
6	$LinB_{MI}$ T81A	M6-1	A	V	V	T	L	H	I	68 ± 3	36	$9.5 \pm 5.0 \times 10^2$	95	Okai et al. 2013
6	$LinB_{MI}$ V112A	M6-2	T	A	V	T	L	H	I	$1.0 \pm 0.09 \times 10^2$	53	$2.3 \pm 0.4 \times 10^2$	23	Okai et al. 2013
6	$LinB_{MI}$ V134I	M6-3	T	V	I	T	L	H	I	$1.2 \pm 0.05 \times 10^2$	63	80 ± 3	8	Ito et al. 2007
6	$LinB_{MI}$ T135A	M6-4	T	V	V	A	L	H	I	83 ± 5	44	$4.2 \pm 1 \times 10^2$	42	Okai et al. 2013
6	$LinB_{MI}$ L138I	M6-5	T	V	V	T	I	H	I	$1.3 \pm 0.1 \times 10^2$	68	$2.2 \pm 0.4 \times 10^2$	22	Okai et al. 2013
6	$LinB_{MI}$ H247A	M6-6	T	V	V	T	L	A	I	$2.1 \pm 0.2 \times 10^2$	111	$2.4 \pm 0.2 \times 10^2$	24	Ito et al. 2007
6	$LinB_{MI}$ I253M	M6-7	T	V	V	T	L	H	M	$2.1 \pm 0.3 \times 10^2$	111	$1.4 \pm 0.2 \times 10^2$	14	Okai et al. 2013
7	$LinB_{MI}$		T	V	V	T	L	H	I	$1.9 \pm 0.08 \times 10^2$	100	$1.0 \pm 0.3 \times 10^3$	100	Okai et al. 2013

[a]Number of mutation into $LinB_{UT}$.
[b]The activities of $LinB_{UT}$, M1-3, M1-6, and M2-1 have already been measured in our previous study (Ito et al. 2007), but they were reanalyzed in this study for the critical comparison with other mutants.

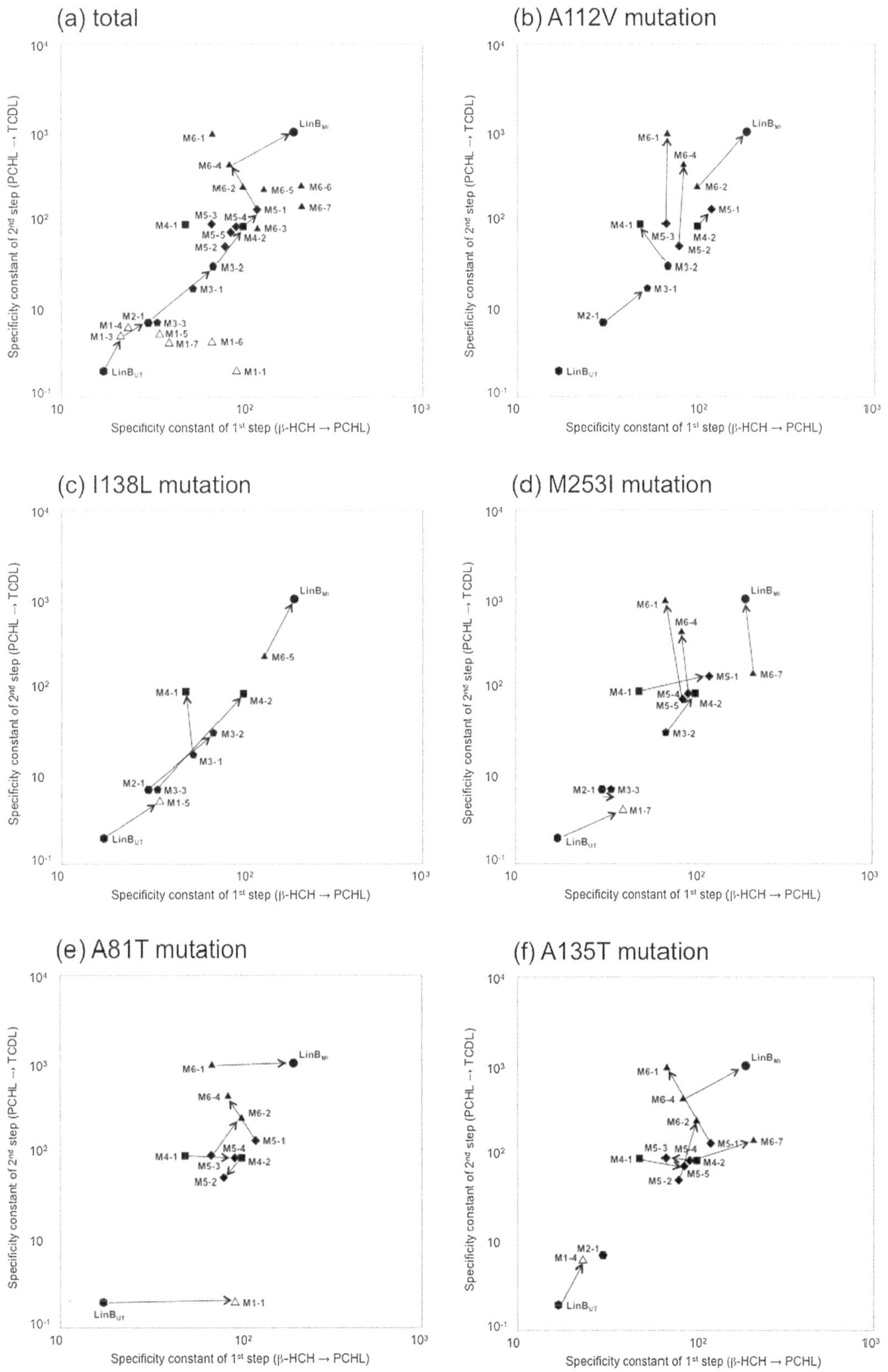

Figure 5 (See legend on next page.)

(See figure on previous page.)
Figure 5 The β-HCH degradation activities of $LinB_{UT}$, $LinB_{MI}$, and their intermediate mutants. Specificity constants of $LinB_{UT}$ (vertical hexagon), $LinB_{MI}$ (circle), and their intermediate mutants (single, open triangle; double, horizontal hexagon; 3-point, pentagon; 4-point, square; 5-point, diamond; and 6-point, closed triangle) for the first conversion (from β-HCH to PCHL: X axis) and the second conversion (from PCHL to TCDL: Y axis) steps (Table 1) were plotted in logarithmical values. The effects of A112V **(b)**, I138L **(c)**, M253I **(d)**, A81T **(e)**, and A135T **(f)** mutations were extracted from the total plot **(a)** and shown by arrows. One potential evolutionary route from $LinB_{UT}$ to $LinB_{MI}$ by the accumulation of seven point mutations is shown by arrows **(a)**.

Characterization of three-point mutants of $LinB_{UT}$

The importance of V112, L138, and I253 for $LinB_{MI}$-type activity was suggested in a previous study (Okai et al. 2013). Therefore, the A112V, I138L, and M253I mutations were independently introduced into M2-1, resulting in three-point mutants: M3-1 (Figure 3j), M3-2 (Figure 3k), and M3-3 (Figure 3l), respectively. These three mutations showed different effects on β-HCH degradation activity (Figure 5a and Table 1). The I138L mutation positively influenced both conversion steps (Figure 5c: M2-1 to M3-2), while the M253I mutation had only a weak positive effect on the first conversion step (Figure 5d: M2-1 to M3-3). The A112V mutation showed positive effects on both conversion steps (Figure 5b: M2-1 to M3-1), although the effects were lower than those of I138L (Figure 5a).

Characterization of four-point mutants of $LinB_{UT}$

Since M3-2 showed the highest β-HCH degradation activity among the evaluated three-point mutants (Figure 5a and Table 1), the A112V and M253I mutations were independently introduced into M3-2, giving rise to four-point mutants, M4-1 (Figure 4m) and M4-2 (Figure 4n), respectively. These mutations resulted in similar levels of positive effects on the second conversion step, but the M253I (M3-2 to M4-2) and A112V (M3-2 to M4-1) mutations had positive and negative effects, respectively, on the rates of the first conversion step (Figure 5a, b, d and Table 1). M4-1 and M4-2 are equivalent to the I138L mutants of M3-1 and M3-3, respectively. Therefore, the I138L mutation had a relatively high level of positive effects on the second conversion step in the cases of M3-1 to M4-1 and M3-3 to M4-2, and also had a positive effect on the first conversion step in the latter case (Figure 5c and Table 1).

Characterization of five-point mutants of $LinB_{UT}$

We further constructed a five-point mutant of $LinB_{UT}$ (A112V/I134V/I138L/A247H/M253I: M5-1) (Figure 4o), which has the mutations at all five amino acid residues that were suggested to be important for β-HCH degradation activity. This mutant can be constructed by the introduction of M253I and A112V mutations into M4-1 and M4-2, respectively. The enzymatic activities of M5-1 indeed increased in both conversion steps compared with those by the parental M4-1 and M4-2 mutants (Figure 5a and Table 1).

All of the other four possible five-point mutants from M4-2 or M4-1 were also constructed: M5-2 (Figure 4p) and M5-3 (Figure 4q) from M4-2, and M5-4 (Figure 4r) and M5-5 (Figure 4s) from M4-1. None of them surpassed the activity of M5-1 (Figure 5a). The rates of both conversion steps using M5-2 and M5-3 decreased or remained at levels similar to those of the M4-2 mutant (Figure 5a). On the other hand, the rates of the first conversion step by M5-4 and M5-5 were higher than that by the parental enzyme, M4-1, but the rates of the second conversion step of these mutants remained indistinguishable from that of M4-1 (Figure 5a).

Discussion

In this study, cumulative substitution mutations were introduced into $LinB_{UT}$, and various intermediate mutant enzymes between $LinB_{UT}$ and $LinB_{MI}$ were kinetically characterized (Figures 3, 4 and Table 1). Since LinB has promiscuous enzymatic activities towards various compounds, including other HCH isomers and various haloalkanes (Lal et al. 2010), the functional evolution of LinB seems too complicated to analyze. However, we focused in this study on its β-HCH degradation activity as a representative model. Overall, the β-HCH degradation activities of the mutants gradually changed to those of a $LinB_{MI}$-type enzyme according to the number of introduced mutations (Figure 5a), indicating that the function of LinB towards this activity can evolve in a stepwise manner. However, the effects of the particular amino acid substitutions depended on the order of the introduced mutations or on the combination with other mutations, and every substitution influenced both the first and second conversion steps (Figure 5b-f). Especially, the I138L mutation showed relatively strong positive effects on both conversion steps in almost all mutant enzymes examined (Figure 5c), suggesting this mutation plays a key role in $LinB_{MI}$-type activity. At the beginning of this study, we mainly focused on three mutations, A112V, I138L, and M253I, on the basis of the results of our previous mutational and structural analyses of $LinB_{MI}$ (Okai et al. 2013), but our additional analysis in this study also confirmed the involvement of A81T and A135T mutations in β-HCH degradation activity (Figure 5e, f). It is of interest that the T81 residue of $LinB_{MI}$ was mainly involved in the first conversion step (Figure 5e). However, it is at present unclear how this residue contributed so substantially to such a step, because T81 is located on the protein surface and far away from the active site and access tunnels (Figure 2) (Okai et al. 2013).

Two examples of plausible evolutionary routes of different protein functions between two highly similar proteins have recently been reported. One example is the route to the formation of atrazine chlorohydrolase (AtzA) and melamine deaminase (TriA) (Noor et al. 2012), which are 98% identical (nine amino acid differences in the 475 amino acid proteins). AtzA catalyzes the dehalogenation of halo-substituted triazine ring compounds but shows no activity towards melamine or ammeline (Seffernick et al. 2001), whereas TriA has no detectable activity toward the halo-triazine substrates (Seffernick et al. 2001). The nine amino acid substitutions for generating the different enzymatic activities could have occurred in either enzyme (Noor et al. 2012). The other example is the route of the evolution of NtdR (a regulator of the nitrotoluene degradation pathway) from NagR (a regulator of the naphthalene degradation pathway) (Ju et al. 2009). Although these two regulators are 98% identical (five differences among 301 amino acids), NtdR, but not NagR, can recognize a wide spectrum of nitroaromatic compounds. It has been proposed that NtdR evolved from NagR by stepwise broadening of the effector range without loss of the original function (Ju et al. 2009). We also demonstrated in this study that $LinB_{UT}$ could be changed to $LinB_{MI}$ by the accumulation of seven point mutations, and that the plausible evolutionary routes could be predicted. For example, if the β-HCH degradation activity of $LinB_{UT}$ increased under relevant selection pressure, the following order of mutations would be most likely: I134V ($LinB_{UT}$ to M1-3) - A247H (M1-3 to M2-1) - I138L (M2-1 to M3-2) - M253I (M3-2 to M4-2) - A112V (M4-2 to M5-1) - A81T (M5-1 to M6-4) - A135T (M6-4 to $LinB_{MI}$) (Figure 5a). In this order, the activity for the second conversion increases gradually by every mutation step. Although the activity for the first conversion decreases at the 6th step, the second conversion seems to be more important for the $LinB_{MI}$-type activity, since $LinB_{UT}$ has nearly no activity for the second conversion.

However, we have to keep in mind that it is impossible at present to predict the diverging processes of $LinB_{MI}$ and $LinB_{UT}$ in the environment, since (i) we have not constructed all possible intermediate mutants between $LinB_{MI}$ and $LinB_{UT}$, and (ii) their common ancestral enzyme remains unknown. All eight nucleotide substitutions (in the seven codons) between *linB*$_{MI}$ and *linB*$_{UT}$ are nonsynonymous (Ito et al. 2007), suggesting that $LinB_{MI}$ and $LinB_{UT}$ diverged relatively recently from a common ancestral LinB protein under strong selection pressure. However, the benefit of the $LinB_{MI}$-type activity to the host cells is still unknown. TCDL is a dead-end product in strain MI1205 (Ito et al. 2007), indicating the inability of this strain to use β-HCH as a carbon and energy source. Furthermore, TCDL seems to be more toxic than β-HCH, because a UT26 derivative whose *linB*$_{UT}$ gene is replaced by *linB*$_{MI}$ showed a growth defect in the presence of β-HCH (unpublished data). In other words, the β-HCH degradation activity itself is apparently unfavorable even for the host cells in the presence of β-HCH. β-HCH degradation activity may be beneficial when cells with this activity coexist with other types of cells having enzymes for the metabolism of TCDL. In the γ-HCH metabolism, LinB converts 1,3,4,6-tetrachloro-1,4-cyclohexadiene, which is produced from γ-HCH by LinA (Nagata et al. 1993). Our preliminary analysis indicated that there was no difference between the $LinB_{UT}$- and $LinB_{MI}$-catalyzed transformation activities towards the intermediate (unpublished data). It has been proposed that enzymatic promiscuity is important for protein evolution (Aharoni et al. 2005; Khersonsky et al. 2006), and $LinB_{MI}$ may be in a more promiscuous state than $LinB_{UT}$. More detailed studies are needed to elucidate the physiological significance of the activity unique to $LinB_{MI}$.

All seven dissimilar amino acid residues are involved in the β-HCH degradation activity unique to $LinB_{MI}$, and their positions in the structure of $LinB_{MI}$ can partially explain their functions in this activity, as described above (Figure 2) (Okai et al. 2013). However, the detailed mechanism by which they contribute to the catalytic activity is not fully understood, because the effect of each successive amino acid substitution depends on the combination of other mutations. Furthermore, all seven mutations showed substantial positive effects at the final stage of their introduction (six-point mutants of $LinB_{UT}$, Figure 5a). These results suggested that the synergetic effects are important for the activity. Numerous naturally occurring LinB variants have recently been reported (Additional file 1: Table S1). Although the seven amino acid residues described herein indeed seem to be the hot spots for mutations in the variants, variations in other amino acid residues, such as 134L, 247S, and 253L, were also found (Additional file 1: Table S1). To discuss the evolution of LinB critically, the effects of such novel substitutions on the β-HCH degradation activity should be addressed in future studies. Furthermore, this study also describes the influence of several mutations on the enzymatic activity of LinB and helps in understanding the structure-function relationship. This information might be useful in future for rational design of LinBs with improved activity.

Additional file

Additional file 1: Figure S1. Degradation of β-HCH (closed circle) and appearance of its metabolites, PCHL (closed triangle) and TCDL (open triangle), in reaction mixtures containing $LinB_{MI}$ wild-type (a), and seven

point mutants of $LinB_{MI}$ (b-h). Values given are the mean of triplicates. Kinetic data were fitted to the irreversible two-step reaction structure of β-HCH conversion to TCDL via PCHL (Scheme 1 in Materials and Methods) by using GEPASI 3.2 software (Mendes 1997) and shown in solid lines. The specificity constants and their standard errors for both reaction steps (k_1 and k_2) were obtained from the calculation (Table 1). The same data were used that have already been published by Ito et al (2007) (panels d and g) and Okai et al (2013) (panels a, b, c, e, f, and h).
Table S1. Naturally occurring LinB variants.

Competing interests
The authors declared that they have no competing interests.

Authors' contributions
RM, HT, YNit, MIsh, and MIto designed and performed experiments. YO participated in the design of the study. JD and ZP analyzed data. MT, JD, ZP, and YNag participated in the design and coordination of this study and drafted the manuscript. YNag conceived the study and is the responsible of the entire project. All authors read and approved the final manuscript.

Acknowledgements
This work was supported by Grant-in-Aids for Scientific Research from Ministry of Education, Culture, Sports, Science, and Technology, and Ministry of Agriculture, Forestry, and Fisheries of Japan, the Grant Agency of the Czech Republic (P503/12/0572) and the Czech Ministry of Education (LO1214).

Author details
[1]Department of Environmental Life Sciences, Graduate School of Life Sciences, Tohoku University, Sendai 980-8577, Japan. [2]Loschmidt Laboratories, Department of Experimental Biology and Research Centre for Toxic Compounds in the Environment RECETOX, Faculty of Science, Masaryk University, Kamenice 5/A13, 625 00 Brno, Czech Republic. [3]The United Graduate School of Agricultural Science, Gifu University 1-1 Yanagido, Gifu 501-1193, Japan. [4]Consolidated Research Institute for Advanced Science and Medical Care, Waseda University, 2-2 Wakamatsu-cho, Shinjuku, Tokyo 162-8480, Japan.

References
Aharoni A, Gaidukov L, Khersonsky O, Mc QGS, Roodveldt C, Tawfik DS (2005) The 'evolvability' of promiscuous protein functions. Nature Genet 37:73–76
Damborsky J, Koca J (1999) Analysis of the reaction mechanism and substrate specificity of haloalkane dehalogenases by sequential and structural comparisons. Protein Eng 12:989–998
Gupta A, Kaushik C, Kaushik A (2000) Degradation of hexachlorocyclohexane (HCH; α, β, γ and δ) by *Bacillus circulans* and *Bacillus brevis* isolated from soil contaminated with HCH. Soil Biol Biochem 32:1803–1805
Gupta A, Kaushik CP, Kaushik A (2001) Degradation of hexachlorocyclohexane isomers by two strains of *Alcaligenes faecalis* isolated from a contaminated site. Bull Environ Contam Toxicol 66:794–800
Ito M, Prokop Z, Klvana M, Otsubo Y, Tsuda M, Damborsky J, Nagata Y (2007) Degradation of β-hexachlorocyclohexane by haloalkane dehalogenase LinB from γ-hexachlorocyclohexane-utilizing bacterium *Sphingobium* sp. MI1205. Arch Microbiol 188:313–325
Janssen DB (2004) Evolving haloalkane dehalogenases. Curr Opin Chem Biol 8:150–159
Johri A, Dua M, Tuteja D, Saxena R, Saxena D, Lal R (1998) Degradation of α, β, γ and δ-hexachlorocyclohexanes by *Sphingomonas paucimobilis*. Biotechnol Lett 20:885–887
Ju KS, Parales JV, Parales RE (2009) Reconstructing the evolutionary history of nitrotoluene detection in the transcriptional regulator NtdR. Mol Microbiol 74:826–843
Khersonsky O, Roodveldt C, Tawfik DS (2006) Enzyme promiscuity: evolutionary and mechanistic aspects. Curr Opin Chem Biol 10:498–508
Lal R, Dogra C, Malhotra S, Sharma P, Pal R (2006) Diversity, distribution and divergence of *lin* genes in hexachlorocyclohexane-degrading sphingomonads. Trends Biotechnol 24:121–130
Lal R, Pandey G, Sharma P, Kumari K, Malhotra S, Pandey R, Raina V, Kohler HP, Holliger C, Jackson C, Oakeshott JG (2010) Biochemistry of microbial degradation of hexachlorocyclohexane and prospects for bioremediation. Microbiol Mol Biol Rev 74:58–80
Mendes P (1997) Biochemistry by numbers: simulation of biochemical pathways with Gepasi 3. Trends Biochem Sci 22:361–363
Mohn WW, Mertens B, Neufeld JD, Verstraete W, de Lorenzo V (2006) Distribution and phylogeny of hexachlorocyclohexane-degrading bacteria in soils from Spain. Environ Microbiol 8:60–68
Nagata Y, Nariya T, Ohtomo R, Fukuda M, Yano K, Takagi M (1993) Cloning and sequencing of a dehalogenase gene encoding an enzyme with hydrolase activity involved in the degradation of γ-hexachlorocyclohexane in *Pseudomonas paucimobilis*. J Bacteriol 175:6403–6410
Nagata Y, Prokop Z, Sato Y, Jerabek P, Kumar A, Ohtsubo Y, Tsuda M, Damborsky J (2005) Degradation of β-hxachlorocyclohexane by haloalkane dehalogenase LinB from *Sphingomonas paucimobilis* UT26. Appl Environ Microbiol 71:2183–2185
Noor S, Taylor MC, Russell RJ, Jermiin LS, Jackson CJ, Oakeshott JG, Scott C (2012) Intramolecular epistasis and the evolution of a new enzymatic function. PLoS One 7(6):e39822
Okai M, Ohtsuka J, Imai LF, Mase T, Moriuchi R, Tsuda M, Nagata K, Nagata Y, Tanokura M (2013) Crystal structure and site-directed mutagenesis analyses of haloalkane dehalogenase LinB from *Sphingobium* sp. MI1205. J Bacteriol 195:2642–2651
Phillips TM, Seech AG, Lee H, Trevors JT (2005) Biodegradation of hexachlorocyclohexane (HCH) by microorganisms. Biodegradation 16:363–392
Prokop Z, Monincova M, Chaloupkova R, Klvana M, Nagata Y, Janssen DB, Damborsky J (2003) Catalytic mechanism of the haloalkane dehalogenase LinB from *Sphingomonas paucimobilis* UT26. J Biol Chem 278:45094–45100
Seffernick JL, de Souza ML, Sadowsky MJ, Wackett LP (2001) Melamine deaminase and atrazine chlorohydrolase: 98 percent identical but functionally different. J Bacteriol 183:2405–2410
Sharma P, Raina V, Kumari R, Malhotra S, Dogra C, Kumari H, Kohler HP, Buser HR, Holliger C, Lal R (2006) Haloalkane dehalogenase LinB is responsible for β- and δ-hexachlorocyclohexane transformation in *Sphingobium indicum* B90A. Appl Environ Microbiol 72:5720–5727
Vijgen J, Abhilash PC, Li YF, Lal R, Forter M, Torres J, Singh N, Yunus M, Tian C, Schaffer A, Weber R (2011) Hexachlorocyclohexane (HCH) as new Stockholm Convention POPs-a global perspective on the management of Lindane and its waste isomers. Environ Sci Pollut Res Int 18:152–162
Willett KL, Ulrich EM, Hites RA (1998) Differential toxicity and environmental fates of hexachlorocyclohexane isomers. Environ Sci Technol 32:2197–2207

6

(S)-3-hydroxyacyl-CoA dehydrogenase/enoyl-CoA hydratase (FadB') from fatty acid degradation operon of *Ralstonia eutropha* H16

Elena Volodina[1] and Alexander Steinbüchel[1,2*]

Abstract

In this study (*S*)-3-hydroxyacyl-CoA dehydrogenase/enoyl-CoA hydratase (H16_A0461/FadB', gene ID: 4247876) from one of two active fatty acid degradation operons of *Ralstonia eutropha* H16 has been heterologously expressed in *Escherichia coli*, purified as protein possessing a His-Tag and initially characterized. FadB' is an enzyme with two catalytic domains exhibiting a single monomeric structure and possessing a molecular weight of 86 kDa. The C-terminal part of the enzyme harbors enoyl-CoA hydratase activity and is able to convert *trans*-crotonyl-CoA to 3-hydroxybutyryl-CoA. The N-terminal part of FadB' comprises an NAD^+ binding site and is responsible for 3-hydroxyacyl-CoA dehydrogenase activity converting (*S*)-3-hydroxybutyryl-CoA to acetoacetyl-CoA. Enoyl-CoA hydratase activity was detected spectrophotometrically with *trans*-crotonyl-CoA. (*S*)-3-Hydroxyacyl-CoA dehydrogenase activity was measured in both directions with acetoacetyl-CoA and 3-hydroxybutyryl-CoA. FadB' was found to be strictly stereospecific to (*S*)-3-hydroxybutyryl-CoA and to prefer NAD^+. The K_m value for acetoacetyl-CoA was 48 μM and V_{max} 149 μmol mg^{-1} min^{-1}. NADP(H) was utilized at a rate of less than 10% in comparison to activity with NAD(H). FadB' exhibited optimal activity at pH 6–7 and the activity decreased at alkaline and acidic pH values. Acetyl-CoA, propionyl-CoA and CoA were found to have an inhibitory effect on FadB'. This study is a first report on biochemical properties of purified (*S*)-stereospecific 3-hydroxyacyl-CoA dehydrogenase/enoyl-CoA hydratase with the inverted domain order from *R. eutropha* H16. In addition to fundamental information about FadB' and fatty acid metabolism, FadB' might be also interesting for biotechnological applications.

Keywords: Fatty acid metabolism; 3-hydroxyacyl-CoA dehydrogenase/enoyl-CoA hydratase, *Ralstonia eutropha* H16

Introduction

Ralstonia eutropha H16 is a Gram-negative β-proteobacterium. Recently, this bacterium was reclassified to *Cupriavidus necator* H16 (Vandamme and Coenye 2004); however, *R. eutropha* H16 is the most used designation in the recent scientific literature. *R. eutropha* is able to utilize various fatty acids (FAs) as sole carbon source and to synthesize poly(3-hydroxybutyrate) (PHB) up to 80% or even more of its dry cell weight (Anderson and Dawes 1990). Two (*S*)-3-hydroxyacyl-CoA dehydrogenase/enoyl-CoA hydratases, H16_A0461/FadB' and H16_A1526/FadB1, are involved in the FA degradation in *R. eutropha* H16 (Brigham et al. 2010). According to their annotation, FadB' and FadB1 possess an enoyl-CoA hydratase activity, catalyzing hydrogenation of the unsaturated enoyl Coenzyme A (CoA); and a 3-hydroxyacyl-CoA dehydrogenase activity, i.e. oxidation of the hydroxyl group into a keto group using one NAD^+ molecule.

The genome sequence of *R. eutropha* H16 unravelled a variety of genes and operons, which could be responsible for the degradation of FAs in this organism (Pohlmann et al. 2006; Reinecke and Steinbüchel 2009). However, little is still known about how the fatty acid metabolism of *R. eutropha*. It is assumed that β-oxidation in *R. eutropha* H16 is similar to the well-studied pathway of *Escherichia coli* (Binstock and Schulz 1981; Black and DiRusso 1994; Clark and Cronan 1996; Riedel et al. 2014). *E. coli* contains two *f*atty *a*cid *d*egradation (*fad*) operons. Their expression is induced when cells are grown in presence of FAs containing 12 or more carbon atoms (Klein et al. 1971).

* Correspondence: steinbu@uni-muenster.de
[1]Institut für Molekulare Mikrobiologie und Biotechnologie, Westfälische Wilhelms-Universität Münster, Corrensstraße 3, D-48149 Münster, Germany
[2]King Abdulaziz University, Jeddah, Saudi Arabia

Recently, Brigham et al. (2010) showed that two *fad* operons (H16_A0459-A0464 and H16_A1526-A1531) were also up-regulated when *R. eutropha* was grown in presence of trioleate. However, their moderate expression is also observed in the absence of FAs (Schimizu et al. 2013). Initially, FAs are activated to acyl-CoA by an ATP-depended acyl-CoA synthase (FadD) resulting in an acyl-CoA (Black et al. 1992). Then an acyl-CoA dehydrogenase (FadE) introduces a double bound between the FA's 2nd and the 3rd carbon atoms yielding *trans*-2-enoyl-CoA. Next, two reactions catalyzed by FadB and yielding 3-ketoacyl-CoA (see above) are following. $FadB_{Ec}$ is a single trifunctional fused polypeptide exhibiting enoyl-CoA hydratase (EC 4.2.1.17), epimerase (EC 5.1.2.3) and isomerase/3-hydroxyacyl-CoA dehydrogenase (EC 1.1.1.35; EC 5.3.3.8) activities. $FadB_{Ec}$ forms, together with a thiolase (FadA), a multienzyme complex (Black and DiRusso 1994; Pramanik et al. 1979; Pawar and Schulz 1981; Yang et al. 1988). The final step of the FA degradation cycle, catalyzed by FadA, is a cleavage of 3-ketoacyl-CoA to an acetyl-CoA and an acyl-CoA molecule shortened for two carbon atoms. The shortened acyl-CoA enters again the β-oxidation, whereas acetyl-CoA enters the Krebs cycle (Black and DiRusso 1994; Clark and Cronan 1996) or, as it is shown for *R. eutropha*, the PHB synthesis.

Synthesis of PHB in *R. eutropha* comprises three steps. The first two steps of this pathway are similar with the last reactions of the β-oxidation, though in a reverse direction. A condensation of two acetyl-CoA molecules by a thiolase (PhaA) to acetoacetyl-CoA (Haywood et al. 1988a; Oeding and Schlegel 1973) and a reduction of acetoacetyl-CoA to (*R*)-3-hydroxybutyryl-CoA by a NADPH-dependent acetoacetyl-CoA reductase, PhaB (EC.1.1.1.36; Haywood et al. 1988a). The third step of PHB synthesis is a stereospecific polymerization of (*R*)-3-hydroxybutyrate (3HB) by PHB synthase (PhaC) (Peoples and Sinskey 1989; Schubert et al. 1991). Beside (*R*)-3HB, (*R*)-3-hydroxyvalerate and (*R*)-3-hydroxyhexanoate can be incorporated into polyhydroxyalkanoates (PHAs) (Dennis et al. 1998; Slater et al. 1998).

Thus, 3-ketoacyl-CoAs and 3-hydroxyacyl-CoAs are the intermediates of either β-oxidation or PHA metabolism. Hence, (*S*)-3-hydroxyacyl-CoA dehydrogenase/enoyl-CoA hydratases, beside their fundamental role in FA degradation (Brigham et al. 2010), might mediate between FA metabolism and PHA synthesis. The studies on a *fadB'* and *fadB1* deletion mutants of *R. eutropha* revealed a special role of FadB' in provision of precursors for PHA synthesis (Insomphun et al. 2014). The *fadB'* deletion led to the reduction of the 3HB content in the copolymer on FAs as sole carbon source. However, if only FadB' was active in the *fadB1* deletion mutant grown on fatty acids, the 3HB content was similar with the wild type. On the other hand, the *fadB1* deletion mutant exhibited an increased 3-hydroxyhexanoate (3HHx) fraction in the copolymer, compared to the wild type and *fadB'* gene deletion mutant (Insomphun et al. 2014). The different role of FadB', in comparison to FadB1, in the PHA precursor supply is not clear. Moreover, FadB' exhibits a different structure as its isoenzyme FadB1 and well studied enzyme of *E. coli* (Insomphun et al. 2014). FadB' is a polypeptide comprising a N-terminal 3HCDH_N-domain, responsible for NAD binding, and a C-terminal enoyl-CoA hydratase domain and little is known about the enzymatic properties of the novel type of enzyme with the inverted domain order.

Amongst the provision of the intermediates for the PHA synthesis, a different field where FadB' could be exploited is the production of enantiomeric pure chemicals, as it was shown for (*S*)- or (*R*)-3-hydroxybutyrate in *E. coli* (Lee et al. 2008; Tseng et al. 2009). Another example is an engineered microbial pathway for synthesis of linear alcohols, which is based on β-oxidation and requires both 3-hydroxyacyl-CoA dehydrogenase and enoyl-CoA hydratase activities (Machado et al. 2012). For this purpose another 3-hydroxyacyl-CoA dehydrogenase (PaaH1) from *R. eutropha* and a crotonase were studied intensively. PaaH1 was heterlogously expressed and crystallized (Machado et al. 2012; Kim et al. 2014). At last, since the purified FadB' is stable over a relatively long time period, its *in vitro* use for coupled enzyme assays may have a potential application (Haywood et al. 1988a; Volodina et al. 2014). Here, we report about the purification and initial characterization of a heterologously expressed His-tagged H16_A0461/FadB' (Gene ID: 4247876) from one of the active *fad* operons of *R. eutropha*.

Materials and methods

Bacterial strains, oligonucleotides and plasmids

Bacterial strains, oligonucleotides and plasmids used in this study are listed in the Table 1. *Escherichia coli* BL21 was used for inducible expression of the enzymes of interest.

Chemicals and materials

Trans-crotonyl-CoA, (*R*)-3-hydroxybutyrate, (*S*)-3-hydroxybutyric acid, lactate dehydrogenase were purchased from Sigma-Aldrich (Steinheim, Germany). Coenzyme A, NAD^+, NADH, $NADP^+$ and NADPH were available from Gerbu Biochemicals GmbH (Gaiberg, Germany). Acetic acid anhydride was form Acros Organics (Geel, Belgium) and propionic and crotonic acid anhydrides were from Merck Millipore (Darmstadt, Germany) and His SpinTrap™ chelating metal affinity columns were purchased from GE Healthcare (Munich, Germany).

Synthesis of acyl-CoA thioesters

Acetyl-CoA, propionyl-CoA and crotonyl-CoA were prepared chemically from corresponding anhydrides and

Table 1 Bacterial strains, plasmids and oligonucleotides used in this study

	Description	Reference or source
Strains		
Ralstonia eutropha H16	Wild type	DSM 428
Escherichia coli		
BL21 (DE3)	F^- *ompT* *hsdS*$_B$ (r_B^-, m_B^-) *gal dcm* (DE3)	Novagen
Plasmids		
pCOLADuet-1::*3HAD*	pCOLADuet-1 with *fadB'* as *Bam*HI/*Hin*dIII fragment	Volodina et al. 2014
pET-19b::*pct*	pET-19b with *pct* as *Nde*I/*Xho*I fragment	Lindenkamp et al. 2013
pET-23a	*E. coli* expression vector (Ap^r, T7 promoter)	Novagen
pET-23a::*bktB*	pET-23a with *bktB* as *Nde*I/*Hin*dIII fragment	This study
Oligonucleotides		
Forward primer BktB *Nde*I*	AAA*CATATG*ACGCGTGAAGTGGTAGTGGTAAGC	This study
Reverse primer BktB *Hin*dIII*	AAA*AAGCTT*GATACGCTCGAAGATGGCGGC	This study

*Restriction sites are highlighted in italics.

CoA according to the method of Simon and Shemin (1953). (*S*)-3-hydroxybutyryl-CoA was synthesized enzymatically by propionate-CoA transferase for each assay *in situ* as described below.

DNA isolation and recombinant DNA techniques

Chromosomal DNA of *R. eutropha* H16 was isolated according to the method of Marmur (1961). Plasmid DNA was isolated by using the peqGOLD Plasmid Miniprep Kit (Peqlab) according to the manufacturer's manual. PCRs were carried out in an Omnigene HBTR3CM DNA thermal cycler (Hybaid, Heidelberg Germany) using genomic DNA of *R. eutropha* H16 as a template, oligonucleotides (Table 1) and *Taq*-DNA polymerase according to the manufacturer's instructions (Thermo Scientific, Dreieich, Germany). The oligonucleotide annealing temperature was 56°C, and the elongation time at 72°C was 2.5 min. PCR products were isolated from an agarose gel and were purified by using a peqGOLD Gel Extraction Kit (Peqlab) according to the manufacturer's instructions. T4 DNA Ligase was purchased from Thermo Scientific (Dreieich, Germany). Oligonucleotides were synthesized by Eurofins MWG (Ebersberg, Germany). Sequencing reactions of DNA fragments were carried out according to a standard procedure at the Sequence Laboratories Göttingen GmbH (Göttingen, Germany). Competent cells of *E. coli* were prepared and transformed with plasmids by the $CaCl_2$ procedure as described by Hanahan (Hanahan 1983).

Sequence data analysis

The nucleotide and amino acid sequences were extracted from the National Center for Biotechnology Information (NCBI) nucleotide sequence and amino acid database (http://www.ncbi.nlm.nih.gov/Genbank/index.html). Gene accession numbers: *bktB* (AM260479) and *fadB'* (AM260479). For determination of amino acid identity the program BlastP from the National Center for Biotechnology Information (http://www.ncbi.nlm.nih.gov/) was used (Altschul et al. 1997). Phylogenetic tree and amino acid sequence alignment were generated using CLUSTAL X program with PhaA as a reference (Thompson et al. 1997).

Heterologous expression and purification of FadB' and coupling enzymes

To obtain overexpressed enzymes, strains of *E. coli* BL21 harboring pCOLADuet-1::*3HAD* (Volodina et al. 2014), pET19b::*pct* (Lindenkamp et al. 2013) or pET23a::*bktB* were used. β-Ketothiolase (BktB) and propionate CoA-transferase (Pct) (Lindenkamp et al. 2013) from *R. eutropha* H16 were used as auxiliary enzymes for the enzyme assays. FadB', Pct and β-ketothiolase were heterologously expressed in *E. coli* BL21. Isopropyl-β-D-1-thiogalactopyranoside (IPTG) was used for induction of the expression (0.5 mM IPTG, 30°C). Purification of the enzymes was carried out with a His SpinTrap™ column according to the manufacturer's instructions via N-terminal His-tag and stored as described before (Lindenkamp et al. 2013; Volodina et al. 2014). Binding buffer (100 mM Tris–HCl, 500 mM NaCl, 20 mM imidazole, pH 7.5) was used for cells resuspension and His SpinTrap™ column equilibration. The bound enzyme was washed with 3 volumes of 100 mM Tris–HCl containing 500 mM NaCl plus 100 mM imidazole (pH 7.5) and was then eluted with the same buffer containing 500 mM imidazole (Volodina et al. 2014). FadB' was stabilized by addition of glycerol to a final concentration of 50% (v/v) and was then stored at −20°C for over one month without loss of activity.

The purity of enzymes was determined by sodium dodecyl sulfate (SDS) polyacrylamide gel electrophoresis

(PAGE) (Suppl. 3; Laemmli 1970). Protein concentrations were measured by using the method of Bradford (Bradford 1976). The protein samples (5–10 μg protein) were resuspended in gel loading buffer (0.6% (w/v) SDS, 1.25% (v/v) β-mercaptoethanol, 0.25 mmol l^{-1} EDTA, 10% (v/v) glycerol, 0.001% (w/v) bromophenol blue, 12.5 mM Tris–HCl [pH 6.8]) and were separated in 12.5% (w/v) SDS-polyacrylamide gels as described by Laemmli (1970). The proteins were stained with Coomassie brilliant blue R-250 (Weber and Osborn 1969).

Determination of the molecular mass

The molecular mass of FadB' was determined on a Superdex 200 HR 10/30 column (Amersham Pharmacia Biotech) at a flow rate of 1 ml/min in 50 mM sodium phosphate buffer (pH7.4) containing 150 mM NaCl. Molecular mass standards used for calibration were from the Gel Filtration Calibration Kit HMW (GE Healthcare, UK).

High performance liquid chromatography/mass spectrometry (HPLC/MS) assay

The purity of the synthesized CoA-thioesters and enzyme reaction products was determined with reverse phase high performance liquid chromatography-mass spectrometry (RP-HPLC-MS) by employing an UltiMate® 3000 HPLC apparatus (Dionex GmbH, Idstein, Germany) connected directly to an LXQ™ Finnigan™ (ThermoScientific, Dreieich, Germany) mass spectrometer. An Acclaim 120 C18 Reversed-Phase LC Column (4.6 × 250 mm, 5 μm, 120 Å pores; Dionex GmbH) was used at 30°C based on a method described earlier with slight modifications (Lindenkamp et al. 2013; Schürmann et al. 2011). A gradient system was used, with 50 mM ammonium acetate, pH 5.0 adjusted with acetic acid (A), and 100 % (v/v) methanol (B) as eluents. Elution occurred at a flow rate of 0.5 ml/min. Ramping was performed as follows: equilibration with 90 % A for 2 min before injection and afterwards a change from 90% to 10% eluent A in 25 min, followed by holding for 10 min and then returning to 90% eluent A within 10 min. CoA-thioesters were detected at 259 nm by a photodiode array detector. Tuning of the instrument was done by direct infusion of a solution of 0.4 mM CoA as shown before (Schürmann et al. 2011).

Enzyme assays

All the enzyme activities were assayed by continuous spectrophotometric assays as described below. The substrates were prepared fresh directly before each enzyme assay and the coupling enzymes were applied in purified form. All values are presented as mean values of at least two measurements with standard deviations.

(i) Enoyl-CoA hydratase activity of the FadB' was measured with *trans*-2-crotonyl-CoA (Sigma-Aldrich, Germany) as a substrate. The reaction mixture contained 0.1 mM *trans*-crotonyl-CoA, 1.5 mM NAD^+, 0.3 μg $FadB'_{Re}$, 1 μg β-ketothiolase, 0.1 mM CoA in 100 mM Tris–HCl buffer, pH 7. The increase of absorption was monitored at 340 nm and 30°C (Binstock and Schulz 1981).

For determination of inhibition of enoyl-CoA hydratase activity of FadB' 0.1 mM crotonyl-CoA was prepared from crotonic acid anhydride and CoA. NAD^+ regeneration with lactate dehydrogenase was applied to ensure the processing of the reaction in desired direction as described before (Haywood et al. 1988a; Volodina et al. 2014). The reaction mixture contained 100 mM Tris–HCl (pH 8.1), 0.1 mM crotonyl-CoA, 2 mM pyruvate, lactate dehydrogenase (9 U), 1.5 mM NAD^+, 25 mM $MgCl_2$ and different concentrations of CoA, acetyl-CoA or propionyl-CoA (0.1 mM, 0.5 mM or 1 mM) were additionally applied. The reaction was started by addition of 0.3 μg of FadB' and the formation of acetoacetyl-CoA-Mg^{2+} ($\varepsilon_{303} = 16500\ M^{-1}cm^{-1}$) complexes was monitored for 2 min at 303 nm at 30°C. One unit of enzyme activity was defined as the formation of 1 μmol acetoacetyl-CoA per min.

(ii) 3-Hydroxyacyl-CoA dehydrogenase activity of FadB' was measured at 340 nm and 30°C with acetoacetyl-CoA in direction of 3-hydroxybutyryl-CoA formation. The reaction mixture contained 0.1 mM acetoacetyl-CoA, 0.2 mM NADH, 0.3 μg $FadB'_{Re}$ in 100 mM Tris–HCl buffer, pH 7 (Haywood et al. 1988b). For determination of inhibition of the 3-hydroxyacyl-CoA dehydrogenase activity of FadB' by the potential inhibitors CoA or acetyl-CoA were added at a final concentration of 0.1 mM, 0.5mM or 1 mM. The kinetic parameters K_m and V_{max} for acetoacetyl-CoA were ascertained by varying its concentration (0.01 mM, 0.05 mM, 0.1 mM, 0.3 mM, 0.5 mM and 0.75 mM) in presence and absence of 0.1 mM acetyl-CoA and fitting the data to the Michaelis–Menten equation. Temperature optimum was determined by varying the temperature: 25°C, 30°C, 35°C, 40°C, 45°C, and 50°C. For determination of optimal pH at room temperature 100 mM Tris–HCl buffer at a range of 6–10 was applied, whereas the range 4–6 was measured in 10 mM acetic acid/acetate buffer.

(iii) Stereospecificity of 3-hydroxybutyryl-CoA dehydrogenase was determined by another coupled enzyme assay. Here, the ability of Pct to synthesize 3-hydroxybutyryl-CoA from acetyl-CoA and 3-hydroxybutyrate was used for *in situ* substrate synthesis as described before (Haywood et al. 1988a; Volodina et al. 2014). The reaction mixture contained: 100 mM Tris–HCl (pH 8.1), 0.1 mM acetyl-CoA, 25 mM (*S*)- or (*R*)-3-hydroxybutyric acid, 2 mM pyruvate, lactate dehydrogenase (9 U), 1.5 mM NAD^+, Pct (10 μg) and 25 mM $MgCl_2$. After pre-incubation of the enzyme mixture for 10 min at 30°C the reaction was started with addition of 0.3 μg of purified FadB'. The generated 3-hydroxybutyryl-CoA was converted to acetoacetyl-CoA

by FadB', and the formation of acetoacetyl-CoA-Mg^{2+} (ε_{303} = 16,500 $M^{-1}cm^{-1}$) complexes was monitored for 2 min at 303 nm at 30°C. One unit of enzyme activity was defined as the formation of 1 μmol acetoacetyl-CoA per min.

(iv) NAD(H) and NADP(H) cofactor utilization capability of FadB' was ascertained in two directions. First, acetoacetyl-CoA was applied as a substrate for the enzyme assay, and the oxidation of 0.1 mM NADH or 0.1 mM NADPH was measured at 340 nm as described above. Second, (*S*)-3-hydroxybutyryl-CoA was synthesized by Pct as described for 3-hydroxybutyryl-CoA dehydrogenase assay, however, BktB was applied as coupling enzyme, and the assay mixture contained 0.1 mM acetyl-CoA, 25 mM 3-hydroxybutyrate, 10 μg Pct, 0.1 mM CoA, 1 μg BktB, 0.3 μg FadB. The reaction mixture was pre-incubated for 10 min at 30°C, and the change of absorption at 340 nm was measured for 5 minutes after addition of NAD^+ or $NADP^+$ to the final concentration 1 mM.

Results

In silico assay

FadB' consists of 806 amino acids (AA) with a theoretical molecular weight of 85.9 kDa and a pI of 7.66. The C terminus is annotated as a crotonase-like family/enoyl-CoA hydratase (ECH) domain and reveals similarity to the N-terminus of FadB1 (34% identity), a variety of enoyl-CoA hydratases like H16_B0389 (261 AA, 34% identity) and H16_B0987 (255 AA, 30% identity) from *R. eutropha* H16 and other organisms. The enoyl-CoA hydratase reaction is shown in Figure 1. The N-terminus of FadB'$_{Re}$ reveals similarities to genes of other 3-hydroxyacyl-CoA dehydrogenases of *R. eutropha* H16 like H16_A1102 (507 AA, 31% identity), H16_A0282 (284 AA, 33% identity), FadB1 (693 AA, 31% identity), and H16_B1652 (322 AA, 30% identity). This terminus comprises NAD binding (3HCDH_N) and 3HCDH domains and catalyzes the second step of the reaction shown in Figure 1. FadB1 is analogous to FadB', although it possesses a reverse domain order (Insomphun et al. 2014). Interestingly, a separate enoyl-CoA hydratase (H16_A0464, 11% identity) is located downstream of the H16_A0459-A0464 operon and reveals no significant similarity to the ECH-domain of FadB'$_{Re}$. Amongst other bacteria most homologous to FadB'$_{Re}$ enzymes were found in many of species of the genus *Burkholderia* and other bacteria. Similarly organized 3-hydroxyacyl-CoA dehydrogenases were also found in *Bacillus subtilis* and *Chromobacterium violaceum* (Figure 2). According to the amino acid sequence and the order of domains two different groups of 3-hydroxyacyl-CoA dehydrogenases could be distinguished (Clustal X, Thompson et al. 1997). The first group comprises two well-studied enzymes from *E. coli*, *Pseudomonas mendocina* and mammalians (as an example, *Rattus norvegicus*) with the so-called *E. coli*-like structure of 3-hydroxyacyl-CoA dehydrogenases. The representatives of the second group with an inverted order of the domains are less studied and are referred here to as *Burkholderia*-like enzymes.

Determination of the molecular mass

The apparent molecular weight of FadB' with N-terminal His-tag defined by SDS-PAGE was 86 kDa, which corresponds to the theoretically calculated molecular weight of one subunit (see Additional file 1) (Laemmli 1970; Weber and Osborn 1969). The molecular weight of the native enzyme as determined by gel permeation chromatography (Superdex 200 HP) indicated a monomeric structure (87 kDa).

Enzyme assays

(i) *Enoyl-CoA hydratase activity.* Specific activity of FadB' with *trans*-crotonyl-CoA reached 98 ± 3 μmol mg^{-1} min^{-1}. The intermediate and products of the reaction, 3-hydroxybutyryl-CoA and acetyl-CoA were detected via RP-HPLC/MS. Since no *cis*-crotonic acid was commercially available, it was not possible to measure the FadB' activity with the CoA-thioester of this substrate.

Enoyl-CoA hydratase activity of FadB' was inhibited by 70% in presence of 0.1 mM acetyl-CoA and by 30% in presence of 0.1 mM CoA (Figure 3). Propionyl-CoA was tested additionally to verify the inhibitory effect. Similarly, as with acetyl-CoA about 60% of the FadB' activity was lost in presence of 0.1 mM propionyl-CoA. Higher concentrations of the mentioned CoA-thioesters and CoA inhibited the FadB' activity stronger. 0.5 mM propionyl-CoA reduced the activity of FadB' to 28% and 0.5 mM acetyl-CoA resulted in only 7% of residual activity. However, FadB' retained 7% of the initial activity in presence of 1 mM acetyl-CoA and 11% - in presence of propionyl-CoA.

(ii) 3-Hydroxyacyl-CoA dehydrogenase activity. FadB'$_{Re}$ showed maximal activity with NADH (92.7 ± 3.8 μmol

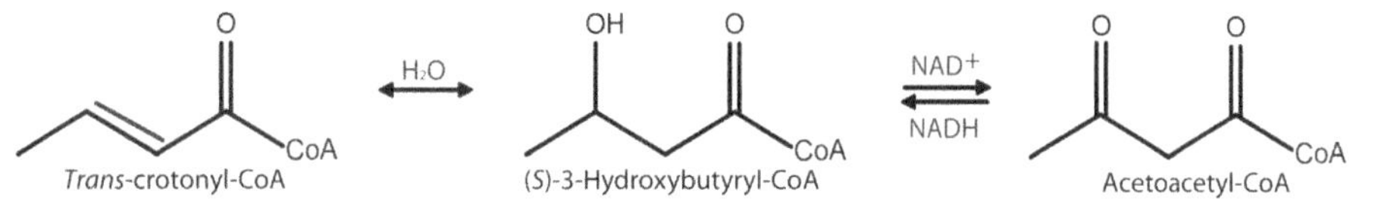

Figure 1 Enoyl-CoA hydratase and (S)-3-hydroxyacyl-CoA dehydrogenase reactions of FA degradation pathway, catalyzed by FadB'.

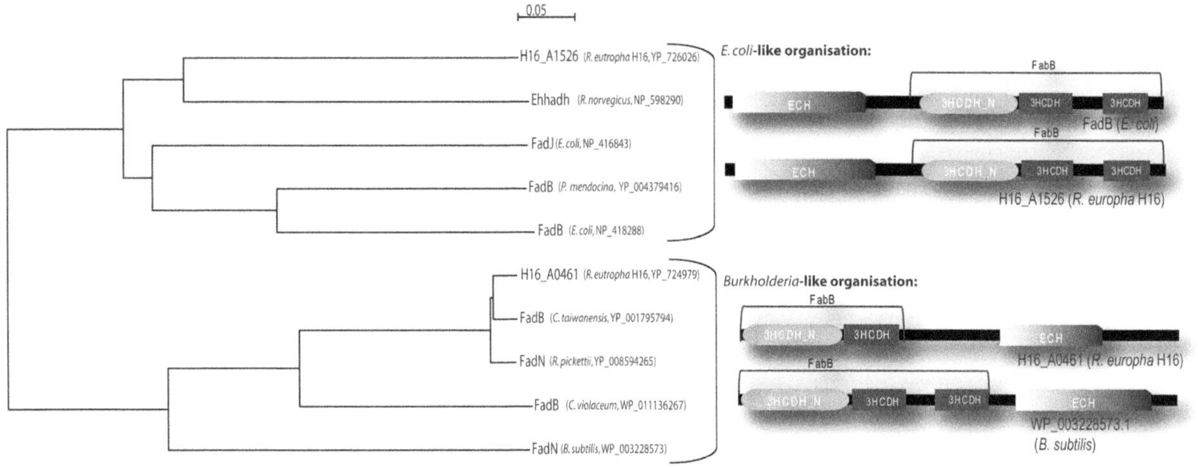

Figure 2 Neighbor-joining phylogenetic tree (Clustal X; Thompson et al. 1997) based on amino acid sequence of different 3-hydroxyacyl-CoA dehydrogenases and the organization of FadB enzymes with different domain order (modified from Insomphun et al. 2014): I- enzymes with *E. coli*-like structure; II- enzymes with *Burkholderia*-like structure. Accession numbers are given at the braces. Bar, 0.05 amino acid substitution per site.

mg^{-1} min^{-1}) and was less active with NADPH (about 5 %, 4.9 ± 1.5 μmol mg^{-1} min^{-1}) (Table 2). Kinetic values for acetoacetyl-CoA were determined. FadB' exhibited a high affinity to acetoacetyl-CoA (K_m 48 μM) and a V_{max} of 149 μmol m $^{-1}min^{-1}$. Presence of 0.1 mM acetyl-CoA resulted in the same V_{max} (149 μmol mg^{-1} min^{-1}) values but a higher K_m (76 μM). The optimum pH was between 6 and 7, and the activity decreased with acidic and basic pH; at pH 5 and 8 the enzyme showed 65% and 75%, respectively, and at pH 10 only 46% of the maximal activity. An acidic pH of 4 gave only 26% of maximal activity. Dependence of FadB' on the temperature revealed that no increase of specific activity was determined after 35°C and the highest activity was observed at 35 and 40°C. FadB' lost 20% and almost 50% of its activity with acetoacetyl-CoA in presence of 0.1 mM CoA or 0.1 mM acetyl-CoA, respectively (Figure 3). Further increases of the inhibitor concentrations reduced the activity of FadB' drastically.

(iii) Stereospecificity of 3-hydroxyacyl-CoA dehydrogenase. To analyze the stereospecificity of FadB' the formation of (*S*)- or (*R*)-3-hydroxybutyryl-CoA from (*S*)- or (*R*)-3HB and acetyl-CoA was carried out by an auxiliary enzyme Pct *in situ* in a similar way as described before (Volodina et al. 2014). $FadB'_{Re}$ converted (*S*)-3-hydroxybutyryl-CoA to acetoacetyl-CoA (16.5 ± 0.4 μmol mg^{-1} min^{-1}), while no

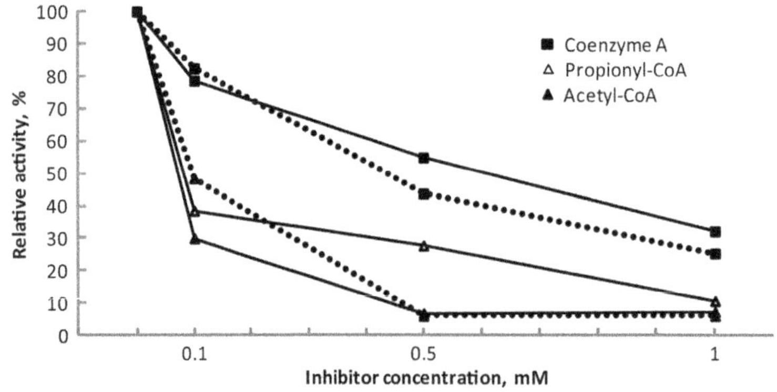

Figure 3 Dependence of the inhibition of FadB' activity on the concentration of acetyl-CoA, propionyl-CoA and CoA. The dotted lines demonstrate the inhibitory effect on FadB' activity measured with acetoacetyl-CoA as substrate. The reaction mixture contained in 100 mM Tris–HCl buffer (pH 7.0) 0.1 mM acetoacetyl-CoA, 0.2 mM NADH, 0.3 μg $FadB'_{Re}$ and the inhibitor CoA or acetyl-CoA at the concentrations as indicated. The straight lines demonstrate the inhibitory effect on FadB' measured with crotonyl-CoA as substrate. The reaction mixture contained in 100 mM Tris–HCl (pH 8.1) 0.1 mM crotonyl-CoA, 2 mM pyruvate, lactate dehydrogenase (9 U), 1.5 mM NAD^+, 25 mM $MgCl_2$ and the inhibitor CoA, acetyl-CoA or propionyl-CoA at the concentrations as indicated. All values are presented as mean values of at least two measurements, standard deviations varied between 0.01% and 3.58%.

Table 2 Enzyme characteristic of FadB' in comparison with other enzymes

	Cofactor		Trifunctional activity	Specificity to (*S*)- or (*R*)-3-hydroxyacyl-CoA	Reference
	NAD^+	$NADP^+$			
FadB'$_{Re}$	100*	5*	+	S**	this work
PhaB1$_{Re}$	20	100	-	R	Haywood et al. 1988b
Unidentified acetoacetyl-CoA reductase (*R. eutropha* H16)	100	2	n.a.	S/R	Haywood et al. 1988b
FadB$_{Ec}$	+	n.a.	+	S	Binstock and Schulz 1981

*-The activity of FadB' was measured at 340 nm and 30°C with acetoacetyl-CoA in direction of 3-hydroxybutyryl-CoA formation as described in "Materials and Methods".
**-The stereospecificity of FadB' was measured at 303 nm and 30°C with (*S*)- and (*R*)-3hydroxybutyryl-CoA as described in "Materials and Methods".

conversion of (*R*)-3-hydroxybutyryl-CoA was detected. Formation of (*R*)-3-hydroxybutytryl-CoA by Pct was confirmed by employing RP-HPLC/MS. Relative activity of FadB' measured with $NADP^+$ was less than 10% in comparison to the activity measured with NAD^+ (1.5 ± 0.3 μmol mg^{-1} min^{-1}).

Discussion

The expression of the β-oxidation genes in *R. eutropha* H16 is up-regulated, when the cells are grown in presence of FAs (Brigham et al. 2010). Two operons located on chromosome #1 are responsible for the degradation of FAs in *R. eutropha* H16: H16_A0459-A0464 and H16_A1526-A1531. The latter operon contains among others the gene for a 3-hydroxyacyl-CoA dehydrogenase/enoyl-CoA hydratase (FadB1), which is similar to the enzymes from *E. coli* (Insomphun et al. 2014). On the other hand, FadB'$_{Re}$ from H16_A0459-A0464 due to its primary structure belongs to the *Burkholderia*-like subgroup of 3-hydroxyacyl-CoA dehydrogenases (Figure 2). It exhibits an inverted order of catalytic domains in comparison to FadB1, well-studied enzymes from mammalians and *E. coli*. FadB' is more related to the enzymes from *B. subtilis* and *C. vinosum*. Apparently FadB' and FadB1 have different ancestors and fusion of both 3HCDH_N and ECH domains in FadB' and FadB1 has occurred independently and might be the result of a convergent evolution. In addition, the separate enoyl-CoA hydratase is located on the same operon downstream of the *fadB'*. Nevertheless, it has been shown that the enoyl-CoA hydratase domain of FadB' retains its activity. Since both genes (H16_A0464, enoyl-CoA hydratase and H16_A0461, FadB') were found to be expressed in presence of fatty acids (Brigham et al. 2010), they seem to function simultaneously and might exhibit different substrate specificities.

Thus, FadB' is the first enzyme of the fatty acid metabolism of *R. eutropha* characterized *in vitro*. Heterologously expressed His-tagged FadB' exhibited a single monomeric structure. However, an oligomeric structure of the enzyme *in vivo* cannot be excluded, as the enzyme might function as part of a multienzyme complex, similarly as FadB in *E. coli*. Nevertheless the solely expressed and folded FadB'$_{Re}$ was active, both in the absence of potential co-monomer(s), such as β-ketothiolase, and in presence of N-terminal His-tag. Since, the activities for both, the enoyl-CoA hydratase and 3-hydroxyacyl-CoA dehydrogenase reactions, were determined, the mechanism of the reaction is presumably as follows: enoyl-CoA, once being captured by FadB', undergoes two consequent conversions finally yielding 3-ketoacyl-CoA (Figure 1).

FadB' was found to be an NADH-dependent enzyme. 3-Hydroxyacyl-CoA dehydrogenases from β-oxidation are presupposed to utilize NADH as a cofactor, due to the organisation of the central metabolism. One cycle of β-oxidation reduces one NADH molecule and produces one molecule of acetyl-CoA. The latter is directed to the Krebs cycle, which supplies three NADH molecules pro cycle. On the other hand, NADP(H) could also be used as a cofactor, although the enzyme showed only 5-10% of the activity with NADP(H) in comparison to NAD (H). However, NADPH is the main cofactor of PHB synthesis and FA *de novo* synthesis and, consequently, FadB' from FA degradation favours NAD(H), avoiding a competition for the cofactor with other enzymes. FadB' was found to be strictly (*S*)-stereospecific with 3-hydroxybutyryl-CoA, which is also characteristic for enzymes from β-oxidation. This corroborates with the theory, that the fatty acid metabolism does not intersect the PHB synthesis and (*R*)-isomers are the only precursors of PHB synthase, and (*S*)-isomers are engaged in the degradation of fatty acids. Thus, as expected, FadB' does not compete with PhaB from PHB metabolism for the substrates. Nevertheless, the deletion of the FadB' encoding gene led to a decrease of PHA content, suggesting an indirect role of FadB' in PHA precursor supply (Insomphun et al. 2014).

Interestingly, acetyl-CoA and propionyl-CoA, which are, on one hand, the end products of FA degradation and, on the other hand, the precursors for the PHA synthesis, are inhibiting FadB'. The activity of FadB' in presence of free CoA was also lower (Figure 3). The inhibitory effect of high concentrations of these substrates might demonstrate a modulatory character of the FA degradation pathway. The end products of β-oxidation seem to negatively affect the rate of FA oxidation in *R. eutropha*. Additionally, free CoA, elaborated in Krebs cycle, is known to inhibit the β-ketothiolase from PHB synthesis (Oeding and Schlegel 1973). So, during the growth phase abundant acetyl-CoA is oxidized releasing CoA, and PHB synthesis is slowed down. Apparently the activity of at least one enzyme from FA degradation, FadB', is inhibited by the increased concentration of CoA, acetyl-CoA and propionyl-CoA. Concerning the mechanism of the inhibition, it was shown, that with increasing concentrations of acetoacetyl-CoA the inhibitory effect of acetyl-CoA was weakened. A higher apparent K_m value and unchanged V_{max} for acetoacetyl-CoA in presence of acetyl-CoA suggest competitive mechanism of acetyl-CoA's inhibition; however, double reciprocal Lineweaver-Burk analysis (data not shown) revealed a mixed type of inhibition. Recently, it was demonstrated that if only FadB' was active in *R. eutropha* lacking *fadB1*, the mutant stored comparable amounts of PHA as the wild type but with higher 3HHx content (Insomphun et al. 2014). Taken into a consideration, that during degradation of long chain fatty acids, the concentration of acetyl-CoA/propionyl-CoA increases and the relative concentration of long chain fatty acids decreases, this might cause the inhibitory effect on FadB'. Under these conditions in the strain lacking *fadB1* the FAs are, putatively, not degraded completely and more 3HHx is available for the PHA synthesis. The regulation of FA degradation seems to be linked between the PHA synthesis and the central metabolism. First, the deletion of FadB' has an indirect influence on PHA synthesis, when the strain is grown on fatty acids (Insomphun et al. 2014). Second, the activity of FadB' is inhibited by high concentrations of the end products of FA degradation and Krebs cycle.

The existence of two analogous enzymes, FadB' and FadB1, demonstrates genetic diversity of *R. eutropha*. The presence of multiple isoenzymes is beneficial for the organism for an adaptation to changing environments and survival. As it was shown, although both *fad* operons function simultaneously, deletion of only one of them does not impair the fatty acid utilization capacity of *R. eutropha* (Brigham et al. 2010). Multiple copies of β-ketothiolases and acetoacetyl-CoA reductases assist in PHB metabolism and compensate the lack of one or several genes (Budde et al. 2010; Lindenkamp et al. 2012).

The determination of the substrate specificity of FadB' with medium and long chain length acyl-CoAs especially in comparison to FadB1 and additional enoyl-CoA hydratase (H16_A0464) could provide in the future more information about the role of FadB' in β-oxidation. 3-Hydroxyacyl-CoA dehydrogenase/enoyl-CoA hydratase from *E. coli*, for example, possesses also Δ^3-*cis*,Δ^2-*trans*-enoyl-CoA isomerase and (*S*)/(*R*)-epimerase activities. However, further studies have to be done to clarify, if FadB' from *R. eutropha* despite its different domain organisation is also capable of converting *cis*- and (*R*)-isomers of CoA thioester to the *trans*- and (*S*)-stereoisomers, respectively.

Thus, FadB' as well as 3-hydroxyacyl-CoA dehydrogenases (*paaH1* and *paaH2*) and acetoacetyl-CoA reductases (*phaB1* and *phaB3*) are able to convert 3-ketoacyl-CoAs to 3-hydroxyacyl-CoAs (Table 2; Budde et al. 2010) and, therefore, are interesting for biotechnical applications. It is assumed that (*S*)-3-hydroxyacyl-CoA dehydrogenase (PaaH1) might refer to the non-stereospecific NADH-dependent acetoacetyl-CoA reductase with broad substrate range purified by Haywood et al. (1988b) (Machado et al. 2012), however, a mixture of the above mentioned enzymes could have been purified. PhaB1 and PhaB3 are important for PHA synthesis. Though, additional application of (*S*)-specific FadB' together with PhaB can enhance the flux of the β-oxidation intermediates to PHA synthesis during growth on FAs. PaaH1 has been applied in an artificial alcohol synthesis pathway in *E. coli* (Machado et al. 2012). This pathway comprises, amongst others, three reverse steps of β-oxidation, catalyzed by three separate enzymes: β-ketothiolase, 3-hydroxyacyl-CoA dehydrogenase and enoyl-CoA hydratase reactions. Concerning that FadB' assisted in the incorporation of 3HHx into a copolymer (Insomphun et al. 2014) and exhibited the enoyl-CoA hydratase activity, it might be promising to exchange the PaaH1 and a crotonase with FadB' in linear alcohol synthesis pathway. (*S*)-Stereospecific 3-hydroxyacyl-CoA dehydrogenases are also interesting for biotechnological production of enantiomeric pure compounds *in vivo* (Lee et al. 2008). Since FadB' can be purified in one step and, if stored in glycerol, it does not loose activity for months, making the FadB' applicable as a coupling enzyme for *in vitro* assays (Volodina et al. 2014).

Additional file

Additional file 1: SDS-PAGE of FadB' after His-Tag purification.

Competing interests

The authors declare that they have no competing interests.

Authors' contributions

EV conceived, designed and performed the experiments. Acquisition of data, analysis and interpretation of data: EV. AS supervised the work and has been involved in drafting the manuscript, revising it critically for important intellectual content; and has given final approval of the version to be published. Both authors read and approved the final manuscript.

Acknowledgements

Elena Volodina is indebted to the Deutscher Akademischer Austauschdienst (DAAD, Germany) for the award of a doctoral scholarship. We acknowledge support by Deutsche Forschungsgemeinschaft and Open Access Publication Fund of University of Muenster.

References

Altschul FA, Madden TL, Schäffer AA, Zhang J, Zhang Z, Miller W, Lipman DJ (1997) Gapped BLAST and PSI-BLAST: a new generation of protein database search programs. Nucleic Acids Res 25:3389–3402

Anderson AJ, Dawes EA (1990) Occurrence, metabolism, metabolic role, and industrial uses of bacterial polyhydroxyalkanoates. Microbiol Rev 54:450–472

Binstock JF, Schulz H (1981) Fatty acid oxidation complex from *Escherichia coli*. Methods Enzymol 71:403–411

Black PN, DiRusso CC, Metzger AK, Heimert TL (1992) Cloning, sequencing, and expression of the *fadD* gene of *Escherichia coli* encoding acyl coenzyme A synthetase. J Biol Chem 267:25513–25520

Black PN, DiRusso CC (1994) Molecular and biochemical analysis of fatty acid transport, metabolism, and gene regulation in *Escherichia coli*. Biochim Biophys Acta 1210:123–145

Bradford MM (1976) Rapid and sensitive method for the quantitation of microgram quantities of protein utilizing the principle of protein-dye binding. Anal Biochem 72:248–254

Brigham CJ, Budde CF, Holder JW, Zeng Q, Mahan AE, Rha C, Sinskey AJ (2010) Elucidation of β-oxidation pathways in *Ralstonia eutropha* H16 by examination of global gene expression. J Bacteriol 192:5454–5464

Budde CF, Mahan AE, Lu J, Rha CK, Sinskey AJ (2010) Roles of multiple acetoacetyl coenzyme a reductases in polyhydroxybutyrate biosynthesis in *Ralstonia eutropha* H16. J Bacteriol 192:5319–5328

Clark DP, Cronan JE (1996) Two-carbon compounds and fatty acids as carbon sources. In: Neidhardt FC, Curtiss R III, Ingraham JL, Lin ECC, Low KB, Magasanik B, Reznikoff WS, Riley M, Schaechter M, Umbarger HE (eds) *Escherichia coli* and *Salmonella*: cellular and molecular biology, 2nd edn. ASM Press, Washington, D.C, pp 343–357

Dennis D, McCoy M, Stangl A, Valentin HE, Wu Z (1998) Formation of poly (3-hydroxybutyrate-co-3-hydroxyhexanoate) by PHA synthase from *Ralstonia eutropha*. J Biotechnol 64:177–186

Hanahan D (1983) Studies on transformation of *Escherichia coli* with plasmids. J Mol Biol 166:557–580

Haywood GW, Anderson AJ, Chu L, Dawes EA (1988a) Characterization of two 3-ketothiolases possessing differing substrate specificities in the polyhydroxyalkanoate synthesizing organism *Alcaligenes eutrophus*. FEMS Microbiol Lett 52:91–96

Haywood GW, Anderson AJ, Chu L, Dawes EA (1988b) The role of NADH- and NADPH-linked acetoacetyl-CoA reductases in the poly-3-hydroxybutyrate synthesizing organism *Alcaligenes eutrophus*. FEMS Microbiol Lett 52:259–264

Insomphun C, Mifune J, Orita I, Numata K, Nakamura S, Fukui T (2014) Modification of β-oxidation pathway in *Ralstonia eutropha* H16 for production of poly (–3-hydroxybutyrate-*co*-3-hydroxyhexanoate) from soybean oil. J Biosci Bioeng 117:184–190

Kim J, Chang JH, Kim KJ (2014) Crystal structure and biochemical properties of the (*S*)-3-hydroxybutyryl-CoA dehydrogenase PaaH1 from *Ralstonia eutropha*. Biochem Biophys Res Commun 448:163–168

Klein K, Steinberg R, Fiethen B, Overath P (1971) Fatty acid degradation in *Escherichia coli*. An inducible system for the uptake of fatty acids and further characterization of old mutants. Eur J Biochem 19:442–450

Laemmli UK (1970) Cleavage of structural proteins during the assembly of the head of bacteriophage T4. Nature 227:680–685

Lee SH, Park SJ, Lee SY, Hong SH (2008) Biosynthesis of enantiopure (*S*)-3-hydroxybutyric acid in metabolically engineered *Escherichia coli*. Appl Microbiol Biotechnol 79:633–641

Lindenkamp N, Volodina E, Steinbüchel A (2012) Genetically modified strains of *Ralstonia eutropha* H16 with β-ketothiolase gene deletion for production of copolyesters with defined 3-hydroxyvaleric acid contents. Appl Environ Microbiol 78:5375–5383

Lindenkamp N, Schürmann M, Steinbüchel A (2013) A propionate CoA-transferase of *Ralstonia eutropha* H16 with broad substrate specificity catalyzing the CoA thioester formation of various carboxylic acids. Appl Microbiol Biotechnol 97:7699–7709

Machado HB, Dekishima Y, Luo H, Lan EI, Liao JC (2012) A selection platform for carbon chain elongation using the CoA-dependent pathway to produce linear higher alcohols. Metab Eng 14:504–511

Marmur J (1961) A procedure for the isolation of desoxyribonucleic acids from microorganisms. J Mol Biol 3:208–218

Oeding V, Schlegel HG (1973) β-ketothiolase from *Hydrogenomonas eutropha* H16 and its significance in the regulation of poly-p-hydroxybutyrate metabolism. J Biochem 134:239–248

Pawar S, Schulz H (1981) The structure of the multienzyme complex of fatty acid oxidation from *Escherichia coli*. J Biol Chem 256:3894–3899

Peoples OP, Sinskey AJ (1989) Poly-beta-hydroxybutyrate (PHB) biosynthesis in *Alcaligenes eutrophus* H16. Identification and characterization of the PHB polymerase gene (*phbC*). J Biol Chem 264:15298–15303

Pohlmann A, Fricke WF, Reinecke F, Kusian B, Liesegang H, Cramm R, Eitinger T, Ewering C, Pötter M, Schwartz E, Strittmatter A, Voss I, Gottschalk G, Steinbüchel A, Friedrich B, Bowien B (2006) Genome sequence of the bioplastic-producing "Knallgas" bacterium *Ralstonia eutropha* H16. Nat Biotechnol 24:1257–1262

Pramanik A, Pawar S, Antonian E, Schulz H (1979) Five different enzymatic activities are associated with the multienzyme complex of fatty acid oxidation from *Escherichia coli*. J Bacteriol 137:469–473

Reinecke F, Steinbüchel A (2009) *Ralstonia eutropha* strain H16 as model organism for PHA metabolism and for biotechnological production of technically interesting biopolymers. J Mol Microbiol Biotechnol 16:91–108

Riedel SL, Lu J, Stahl U, Brigham CJ (2014) Lipid and fatty acid metabolism in *Ralstonia eutropha* H16: relevance for the biotechnological production of value-added products. Appl Microbiol Biotechnol 98:1469–1483

Schimizu R, Chou K, Orita I, Suzuki Y, Nakamura S, Fukui T (2013) Detection of phase-dependent transcriptomic changes and Rubisco-mediated CO_2 fixation into poly (3-hydroxybutyrate) under heterotrophic condition in *Ralstonia eutropha* H16 based on RNA-seq and gene deletion analysis. BMC Microbiol 13:1–14

Schubert P, Krüger N, Steinbüchel A (1991) Molecular analysis of the *Alcaligenes eutrophus* poly (3-Hydroxybutyrate) biosynthetic operon: identification of the N terminus of poly (3-hydroxybutyrate) synthase and identification of the promoter. J Bacteriol 173:168–175

Schürmann M, Wübbeler JH, Grote J, Steinbüchel A (2011) Novel reaction of succinyl-CoA synthetase: activation of 3-sulfinopropionate to 3-sulfinopropionyl-CoA in *Advenella mimigardefordensis* strain DPN7T during degradation of 3,3′-dithiodipropionic acid. J Bacteriol 193:3078–3089

Simon EJ, Shemin D (1953) The preparation of *S*-succinyl-coenzyme A. J Am Chem Soc 75:2520

Slater S, Houmiel KL, Tran M, Mitsky TA, Taylor NB, Padgette SR, Gruys KJ (1998) Multiple β-ketothiolases mediate poly (β-hydroxyalkanoate) copolymer synthesis in *Ralstonia eutropha*. J Bacteriol 180:1979–1987

Thompson JD, Gibson TJ, Plewniak F, Jeanmougin F, Higgins DG (1997) The ClustalX windows interface: flexible strategies for multiple sequence alignment aided by quality analysis tools. Nucleic Acids Res 24:4876–4882

Tseng HC, Martin CH, Nielsen DR, Prather KL (2009) Metabolic engineering of *Escherichia coli* for enhanced production of (*R*)- and (*S*)-3-hydroxybutyrate. Appl Environ Microbiol 10:3137–3145

Vandamme P, Coenye T (2004) Taxonomy of the genus Cupriavidus: a tale of lost and found. Int J Syst Evol Microbiol 54:2285–2289

Volodina E, Schürmann M, Lindenkamp N, Steinbüchel A (2014) Characterization of propionate CoA-transferase from *Ralstonia eutropha* H16. Appl Microbiol Biotechnol 98:3579–3589

Weber K, Osborn M (1969) The reliability of molecular weight determinations by dodecyl sulfate-polyacrylamide gel electrophoresis. J Biol Chem 244:4406–4412

Yang SY, Li J, He XY, Cosloy SD, Schulz H (1988) Evidence that the fadB gene of the fadAB operon of Escherichia coli encodes 3-hydroxyacyl-Coenzyme A (CoA) epimerase, Δ3-cis, Δ2-trans-enoyl-CoA isomerase, enoyl-CoA hydratase in addition to 3-hydroxyacyl-CoA dehydrogenase. J Bacteriol 170:2543–2548

Characterisation of a recombinant β-xylosidase (xylA) from *Aspergillus oryzae* expressed in *Pichia pastoris*

Narin Kirikyali, Jonathan Wood and Ian F Connerton*

Abstract

β-xylosidases catalyse the hydrolysis of short chain xylooligosaccharides from their non-reducing ends into xylose. In this study we report the heterologous expression of *Aspergillus oryzae* β-xylosidase (XylA) in *Pichia pastoris* under the control of the glyceraldehyde-3-phosphate dehydrogenase promoter. The recombinant enzyme was optimally active at 55°C and pH 4.5 with K_m and V_{max} values of 1.0 mM and 250 μmol min^{-1} mg^{-1} respectively against 4-nitrophenyl β-xylopyranoside. Xylose was a competitive inhibitor with a K_i of 2.72 mM, whereas fructose was an uncompetitive inhibitor reducing substrate binding affinity (K_m) and conversion efficiency (V_{max}). The enzyme was characterised to be an exo-cutting enzyme releasing xylose from the non-reducing ends of β-1,4 linked xylooligosaccharides (X_2, X_3 and X_4). Catalytic conversion of X_2, X_3 and X_4 decreased (V_{max} and k_{cat}) with increasing chain length.

Keywords: *Aspergillus oryzae*; Xylose; β-xylosidase; Enzyme kinetics; Protein expression

Introduction

Xylanolytic and cellulolytic enzymes encoded by filamentous fungi have been employed in several industrial applications for improving digestibility in animal feed, production of sweeteners, pharmaceuticals, additive chemicals for biofuel production and for the replacement of hazardous chemicals in textile and paper manufacture (Michelin et al. 2012). As a consequence it has been a necessity for enzymes involved in cellulose and hemicellulose hydrolysis to be individually identified and characterised in order to utilise them in the process of converting waste agricultural materials into valuable products with greater efficiency.

Hemicellulose is comprised of a linear main chain β-1,4 linked D-xylose backbone with short lateral side chains of different lower molecular weight sugar residues (Dyk and Pletschke 2012). Enzymatic hydrolysis of hemicellulose commences with the removal of side chains that block the sites where xylanases cleave the xylan backbone. Endo-1, 4-β-xylanase enzymes cleave the glycosidic bonds in a selective manner depending on the chain length, degree of branching of substrate molecules and the presence of alternative carbohydrate moieties (Polizeli et al. 2005). Cleavage of the xylan backbone yields xylooligosaccharides and the final trimming is carried out by β-xylosidase, whereby short chain oligosaccharides and xylobiose are hydrolysed from the non-reducing termini to release xylose monomers (Polizeli et al. 2005; Teng et al. 2011).

Among xylanolytic enzymes, endo-xylanases and β-xylosidases have attracted attention as they commence and complete the breakdown of hemicellulose fraction respectively (Kulkarni et al. 1999). For comprehensive hydrolysis β-xylosidases play an important role in the removal of xylooligosaccharides from the catalytic environment, which assists by the elimination of the end-product inhibitors of endo-xylanases.

The gene encoding for XylA was previously identified by Kitamoto et al. (1999) and was reported to be responsible for the rapid browning of soy sauce. In addition Kitamoto et al. (1999) were interested in the antisense inhibition of XylA expression in order to hinder the translation in *Aspergillus oryzae* KBN616 to produce a mutant strain that could be used in Japanese food industry. However, XylA is a potentially efficient candidate for the facilitation of hydrolysis of hemicellulose applications in industrial processes. The work presented here reports the expression

* Correspondence: ian.connerton@nottingham.ac.uk
Division of Food Sciences, School of Biosciences, University of Nottingham, Sutton Bonington Campus, Loughborough LE12 5RD, UK

of a β-xylosidase from *Aspergillus oryzae* in *Pichia pastoris* and the kinetic characterisation of the recombinant enzyme.

Materials and methods

Construction of expression vector

The β-xylosidase encoding gene (*xylA*) was kindly provided by Noriyuki Kitamoto (Aichi Industrial Technology Institute, Japan). The gene sequence appears in the DDBJ, EMBL, and GenBank nucleotide sequence databases under accession number AB013851. The gene was sub-cloned into pCR®2.1 and subsequently into the *Pichia pastoris* expression vector pPpHis4_GAP_BglII (TU Graz) under control of the glyceraldehyde-3-phosphate dehydrogenase (GAP) promoter.

Transformation and expression of recombinant β-xylosidase in *Pichia pastoris*

Plasmid DNA (10 μg) containing the xylA gene sequence was linearized with *Bgl*II for integration at the GAP locus and was transformed into *P. pastoris* GS115 using Invitrogen *Pichia* Easycomp kit (as per manufacturers' guidelines). Positive transformants, displaying the His^+ phenotype on Regeneration Dextrose medium (RD) agar plates (1 M sorbitol, 2% dextrose, 1.34% yeast nitrogen base, 4×10^{-5}% biotin, 0.005% amino acids (L-glutamic acid, L-methionine, L-lysine, L-leucine, L-isoleucine) and 1.5% agar), before transfer onto RD plates containing 50 μg ml^{-1} bromo-4-chloro-3-indolyl β-D-xylopyranoside (X-xyl) (Sigma-Aldrich, UK) and incubated at 30°C for 2 days. Functional expression of the β-xylosidase under the control of GAP promoter was tested by cleavage of xylopyranoside from the synthetic indicator X-xyl.

A single colony displaying the highest level of blue precipitate was sub-cultured from the RD plate onto a YPD plate (1% yeast extract, 2% peptone, 2% glucose, 2% agar) and incubated at 30°C for 3 days. A seed culture was raised using a single colony inoculated into 50 ml YPD liquid medium, incubated at 28°C overnight. One hundred millilitres of YPD broth was then inoculated to 1 OD_{600nm} and incubated for 72 h at 28°C. One ml of culture was removed every 24 h to test expression levels using the synthetic 4-nitrophenyl β-xylopyranoside (PNPX) as described in section 2.4. Following incubation the cells were harvested by centrifugation at 4000 × g for 10 min.

Purification of recombinant β-xylosidase

Following centrifugation the culture supernatant was concentrated using a Sartorius Sartocon Slice then diafiltered with 10 volumes of Tris-salt buffer (10 mM Tris, 50 mM NaCl, pH 7.5). The concentrate was then stabilised using 30% (w/v) sucrose based on protein concentration and frozen for long term storage at −20°C. Prior to enzymatic assays the 30% (w/v) sucrose was removed from the recombinant enzyme concentrates using a Vivaspin concentrator (GE Healthcare, UK) with a 10 kDa molecular weight cut off membrane filter and the filtrate was washed with Tris-salt buffer (10 mM Tris, 50 mM NaCl, pH 7.5).

Enzyme assays using synthetic substrates

Assays for β-xylosidase activity were performed by measuring the pNP released from p-nitrophenyl glycoside synthetic substrates 4-nitrophenyl-β-D-xylopyranoside (PNPX), 4-nitrophenyl-β-D-glucopyranoside (PNPG) and 4-nitrophenyl-α-L-arabinofuranoside (PNPAf) in a final volume of 4 ml for 20 min in 50 mM sodium phosphate buffer pH 6.0 at 50°C. Reactions were terminated by the addition of 1 M Na_2CO_3 and the amount of released pNP was measured at OD_{410nm}. One unit (U) of β-xylosidase activity is defined as the amount of enzyme required to release 1 μmol of pNP per min under assay conditions. Kinetic parameters (K_m and V_{max}) were determined by the measurement of activity against pNPX using different substrate concentrations (0.5 - 12 mM) using the standard assay procedure. Enzyme assays were performed in triplicate and are presented as mean values with standard error.

Enzyme assays using xylooligosaccharides

Activities against xylobiose, xylotriose and xylotetraose were determined at varying substrate concentrations (0.25 – 4 mg ml^{-1}) in a final volume of 1 ml for 10 min in 50 mM sodium phosphate buffer pH 6.0 at 50°C. All assays were carried out in triplicate and were terminated by the addition of 1 M Na_2CO_3. Reaction products were separated according to molecular size by HPLC (Dionex ICS-3000 SP) with CarboPacTM PA20 column (3 × 150 mm) and a gradient of 10 – 50 mM sodium hydroxide was applied for 20 min at a flow rate of 1 ml min^{-1}. The products were quantified on the basis of standard peak areas from various concentrations of control xylose, xylobiose, xylotriose and xylotetraose solutions. Enzyme assays were performed in triplicate and are presented as mean values with standard deviations.

Inhibition studies

To investigate the effect of end product xylose on catalytic activity, reactions were carried out in the presence of various xylose concentrations from 1 mM to 80 mM using synthetic substrate concentrations of either 1 mM or 4 mM pNPX. To further confirm the type of inhibition, kinetic constants were determined from experiments carried out using fixed inhibitor concentrations of 5 mM xylose at varying substrate concentrations from 0.25 mM to 8 mM under standard assay conditions.

The effects of monosaccharide sugars (20 mM glucose, mannose, galactose, arabinose, fructose and xylose), metal ions and chemicals (10 – 20 mM LiCl, KCl, $ZnCl_2$, SDS, EDTA and DTT) on enzyme activity were tested using 50 mM sodium phosphate buffer pH 6, 1 mM pNPX and 2 µg of enzyme at 50°C for 10 min in a final volume of 4 ml and measured optical density of released p-nitrophenyl at OD_{410n}m.

Determination of protein concentration

Protein concentrations were determined by the standard assay procedure using Pierce Coomassie® Plus Protein Assay Reagent. Sample diluents were used as the blank and the absorbance measured at OD_{595nm}. All assays were performed in triplicate and the OD_{595nm} readings of unknown sample were compared against BSA protein standard series which covered the range of concentrations between 50 and 1500 µg ml^{-1}.

Determination of molecular mass by SDS-PAGE

SDS-PAGE was performed using 8% polyacrylamide gels according to the method described by Laemmli (1970). Protein bands were stained with colloidal Coomassie Blue. Bands from SDS-PAGE were excised and were subjected to trypsin digestion prior to mass spectrometry analysis.

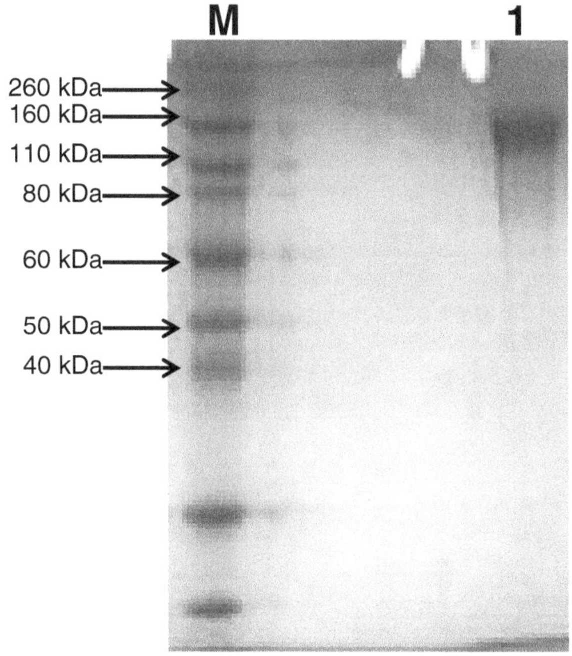

Figure 1 SDS-PAGE of recombinant β-xylosidase. Lane M, pre-stained Novex-Invitrogen protein marker. Lane 1, recombinant β-xylosidase.

Mass spectrometry

Analyses of samples were carried out by LC-ESI-tandem MS on a Q-TOFII mass spectrometer fitted with a nanoflow ESI (electrospray ionization) source (Waters Ltd, UK). Peptides were separated on a PepMap C18 reverse phase, 75 µmi.d., 15-cm column (LC Packings) and delivered online to the MS via a CapLC HPLC system. Sequence interpretation for individual peptides was performed using the PepSeq MASCOT tool of the MassLynx™ 4.0 software package (Waters).

Results

Characterisation of recombinant β-xylosidase

The β-xylosidase gene is contained within an open reading frame of 2397 nucleotides with no introns, which encodes a protein of 798 amino acids. A putative signal peptide was identified by SignalP software, thus the mature protein was predicted to be 778 amino acids with a molecular mass of 84.7 kDa. The recombinant enzyme was recovered from *Pichia pastoris* culture supernatant at approximately 100 mg L^{-1}. NetNGly 1.0 predicted 12 potential N-glycosylation sites for β-xylosidase, and consistent with this, the recombinant enzyme was heterogeneous with a molecular mass estimated between 153 and 165 kDa on SDS-PAGE (Figure 1). The recombinant protein was excised from SDS-PAGE and the masses and protein sequences of tryptic peptides were determined using mass spectrometry to confirm the protein product was exo-1,4 β-xylosidase originating from *Aspergillus oryzae.*

Determination of optimal conditions

The optimum temperature was determined by incubation of the recombinant enzyme in the presence of 50 mM sodium phosphate buffer pH 6.0 and 2 mM 4-nitrophenyl β-xylopyranoside (PNPX) for 15 min at varying temperatures between 20°C and 90°C. The enzyme displayed optimum activity at 55°C and retained more than 90% catalytic activity between the ranges of 50 – 60°C (Figure 2A).

The optimum pH of enzymatic activity was assayed in phosphate buffer system of varying pH values from 2–9 in the presence of 2 mM PNPX. The enzyme displayed activity within a narrow pH range, with an optimum of pH 4.5 and at least 65% activity from pH 3 – 6; less than 5% activity was observed at pH 7 – 9 (Figure 2B).

Substrate specificity and kinetic analysis

The substrate specificity of the recombinant enzyme was determined using various 4-nitrophenyl glycoside synthetic substrates and xylooligosaccharides. Recombinant XylA hydrolysed 4-nitrophenyl β-xylopyranoside efficiently but had trace hydrolytic activities against 4-nitrophenyl-β-D-glucopyranoside or 4-nitrophenyl-α-L-arabinofuranoside. Specific activities were determined as 150, 2 and 0.9 U mg^{-1} for PNPX, PNPAf, PNPG respectively.

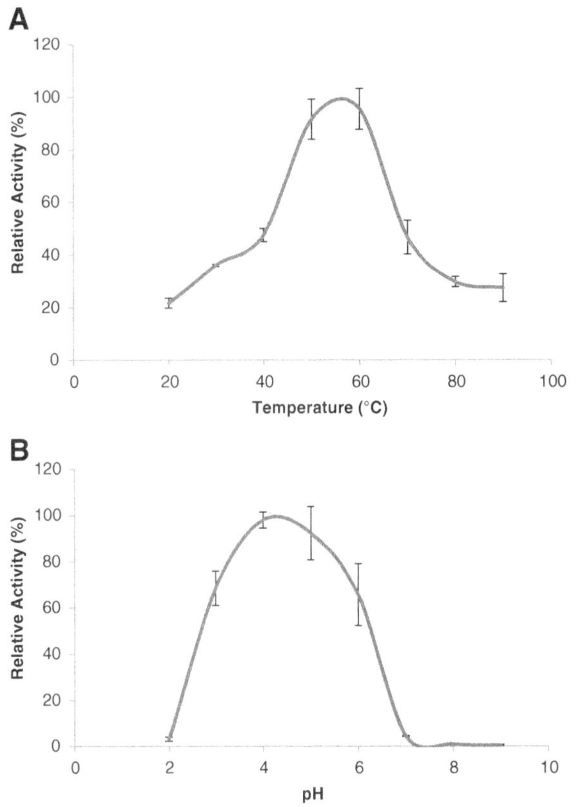

Figure 2 The effect of temperature and pH on recombinant β-xylosidase. (A) The optimal temperature was measured in 50 mM sodium phosphate buffer (pH 6.0) in the presence of 1 mM PNPX at various temperatures for 15 minutes. **(B)** The optimum pH was determined by incubating at 50°C for 15 minutes in the presence of 2 mM PNPX in sodium phosphate buffer varying pH.

Kinetic constants for synthetic and natural substrates were determined using Hanes plots (Table 1). The enzyme exhibited K_m and V_{max} values of 1.0 mM and 250 μmol min^{-1} mg^{-1} respectively for the hydrolysis of 4-nitrophenyl β-xylopyranoside. Addition of the reaction product xylose resulted in enzyme inhibition. Kinetic constants were determined using Hanes plots from experiments carried out using a fixed inhibitor concentration of 20 mM xylose at varying PNPX concentrations (0.5 - 4 mM). The K_m was altered whilst the V_{max} remained unchanged, suggesting that xylose is a competitive inhibitor. A K_i of 2.7 mM for xylose was determined from Dixon plots. The addition of 20 mM fructose also had an inhibitory effect. Hanes plots indicated that the catalytic conversion of substrate into product was altered in terms of a reduction in the V_{max} and substrate affinity (K_m), exhibiting the characteristics of uncompetitive inhibition.

The degradation of various xylooligosaccharides (X_2, X_3 and X_4) by recombinant XylA was analysed by HPLC. Xylose was released from all substrates and the rate of xylose released decreased with increasing chain length of the xylooligosaccharide. Table 1 shows reductions in the catalytic conversion parameters V_{max} and k_{cat} with respect to the increasing chain length of the xylooligosaccharides. The relative affinity of XylA towards the natural substrate xylotetraose (X_4) was significantly greater than xylobiose (X_2) or xylotriose (X_3) with respect to the K_m values.

Effect of carbohydrates on catalytic activity

The catalytic activity of recombinant β-xylosidase was measured using 1 mM pNPX as substrate in the presence of 20 mM sugar solutions to test whether the enzyme was inhibited or enhanced (Figure 3A). In the presence of 20 mM xylose and fructose the catalytic activity was reduced to 18% and 26% respectively, with none of the others (arabinose, mannose, galactose, glucose and sucrose) showing any change in catalytic activity.

Effect of metal ions and chemical compounds on enzyme activity

The effects of various metal ions and reagents on β-xylosidase activity were assayed at 10 mM and 20 mM concentrations (Figure 3B). Mosrt notably the addition of Zn^{2+} (10 mM) enhanced enzyme activity by 80%. The detergent SDS at 20 mM reduced the catalytic activity by 40%.

Discussion

A β-xylosidase encoding gene (*xylA*) from *Aspergillus oryzae* KBN616 was expressed in a soluble, active form under control of the constitutive GAP promoter in

Table 1 Kinetic analysis of synthetic and natural substrates determined in 50 mM sodium phosphate buffer (pH 6.0) at 50°C

Substrate	K_m (mM)	V_{max} (μmol min^{-1} mg^{-1})	k_{cat} (sec^{-1})	k_{cat}/Km (mM^{-1} s^{-1})
pNPX	1 ± 0.3	250 ± 0.001	352.7	352.7
pNPX + 20 mM Xylose	2.9 ± 0.5	250.5 ± 23	353.4	121.9
pNPX + 20 mM Fructose	0.1 ± 0.06	14.5 ± 3	20.6	206
Xylobiose (X_2)	2.6 ± 0.3	25.5 ± 0.1	36.0	13.8
Xylotriose (X_3)	3.07 ± 0.3	21.3 ± 0.3	30.1	9.7
Xylotetraose (X_4)	0.62 ± 0.4	14.5 ± 0.003	20.5	33.1

± standard deviation of the means of triplicate data.

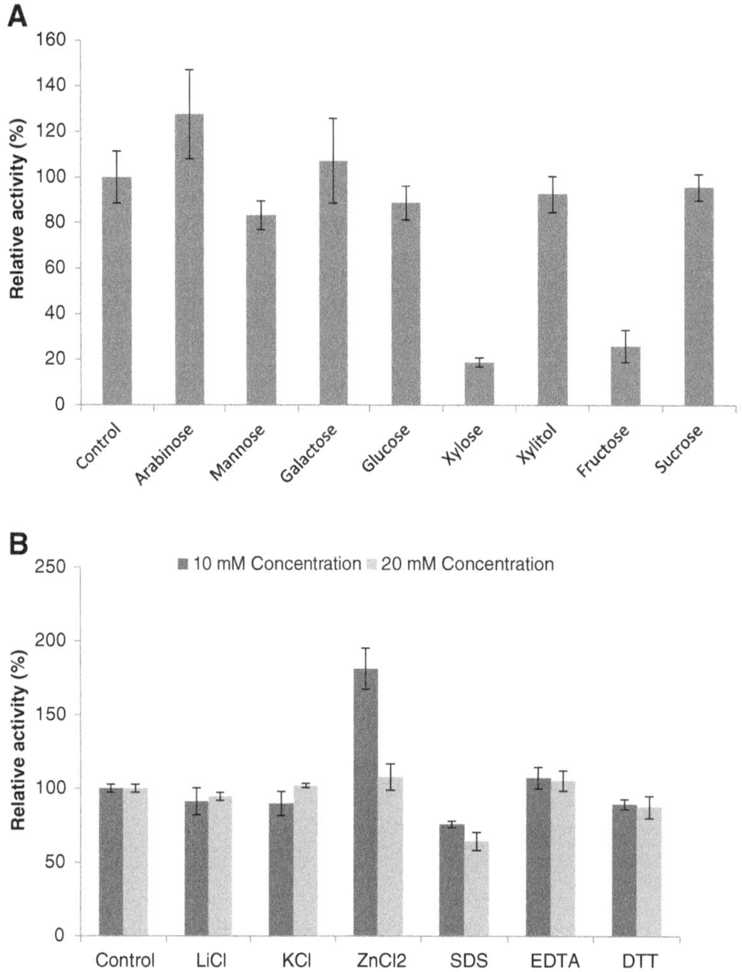

Figure 3 Enzyme activity profiles in the presence of (A) 20 mM carbohydrates and (B) metal ions and chemical compounds within reaction mixtures. All reactions were performed in 50 mM sodium phosphate buffer (pH 6.0) and 2 mM PNPX at 50°C. Control reaction performed in the absence of carbohydrates and chemicals under identical conditions.

Pichia pastoris. The predicted presence of a native signal sequence (SignalP) was confirmed through secretion of the mature protein by the expression host. XylA was predicted to have a molecular mass of 86.4 kDa and 12 potential N-linked glycosylation sites. Kitamoto *et al.* (1999) previously identified purified the native enzyme from *A. oryzae* culture supernatant, in which the enzyme produced a single protein band with an apparent molecular mass of 110 kDa on SDS-PAGE. However the molecular mass range of XylA determined by SDS-PAGE from *P. pastoris* was 153 to 165 kDa indicating differences in post-translational modification consistent with the predicted glycosylation sites. Heterologous proteins expressed in *P. pastoris* are subject to glycosylation and several plant cell wall degrading enzymes expressed in *P. pastoris* are reported to be hyper-glycosylated, including β-xylosidase from *Paecilomyces thermophila* (Juturu and Wu 2013), cellobiose dehydrogenase from *Neurospora crassa* (Zhang et al. 2011) and endo-xylanase from *Actinomadura* sp. S14 (Sriyapai et al. 2011).

The biochemical properties of the recombinant β-xylosidase closely match the native enzyme with respect to the observed optimal pH range (pH 4.5 – 5) and temperature (55°C) for enzyme activity. These are comparable to other fungal xylosidases (Saha 2003; La Grange et al. 2001; Wakiyama et al. 2008; Zanoelo et al. 2004; Rasmussen et al. 2006), which exhibit optimal activities between pH 3–5 at 60°C.

The recombinant enzyme was most active against p-nitrophenyl-β-D-xylopyranoside (PNPX), with minimal activities towards 4-nitrophenyl-β-D-glucopyranoside (PNPG) and 4-nitrophenyl-α-L-arabinofuranoside (PNPAf). Activity towards a broad range of synthetic substrates by other fungal β-xylosidases has been reported, although maximum activity is generally towards PNPX (Margolles-Clark et al. 1997; Ohta et al. 2010; Wakiyama et al. 2008;

Katapodis et al. 2006). The exception to this is *Aspergillus awamori* X-100 β-xylosidase, which is reported to exhibit a greater k_{cat} against PNPAf (Eneyskaya et al. 2007). The recombinant enzyme exhibited kinetic constants for the hydrolysis of PNPX of 1.0 mM and 353 μmol min^{-1} mg^{-1} for K_m and V_{max} respectively. A range of kinetic constants have been reported for the hydrolysis of PNPX by fungal β-xylosidases but these values are similar to those reported for purified β-xylosidases from *Sporotrichum thermophile* (Katapodis et al. 2006) and *Fusarium proliferatum* (Saha 2003). The hydrolysis of various xylooligosaccharides (X_2, X_3 and X_4) was monitored by HPLC. In the presence of individual xylooligosaccharides, xylose was detected as an initial product of catalysis indicating that the recombinant β-xylosidase is an exo-cutting enzyme. The k_{cat} values for the xylooligosaccharides decrease with increasing chain length in the order of X_2 (36.0 sec^{-1}) > X_3 (30.1 sec^{-1}) > X_4 (20.5 sec^{-1}). However, the reduction in the K_m value for xylotetraose (X_4) results in a greater catalytic efficiency towards this substrate. The observed changes in the kinetic constants (K_m and k_{cat}) with respect to xylooligosaccharide chain length show similar patterns to those reported for the β-xylosidases originating from *Talaromyces emersonii, Trichoderma reesei* and *Aspergillus nidulans* (Rasmussen et al. 2006; Dilokpimol et al. 2011). In contrast *Neurospora crassa* β-xylosidase is reported to show a reduction in catalytic efficiency towards xylotetraose compared to shorter chain xylooligosaccharides (Kirikyali and Connerton, 2014). This is largely due to a relative reduction in the affinity of the *N. crassa* enzyme towards the xylotetraose substrate (>K_m value).

Similar to the data reported for the β-xylosidases from *Penicillium sclerotiorum* (Knob and Carmona 2009), *Talaromyces thermophilus* (Guerfali et al. 2009) and *Paecilomyces thermophila* (Yan et al. 2008) the presence of 20 mM concentrations of the metal ions Li^+, K^+ or Zn^{2+} had no effect on enzyme activity. The β-xylosidase of *Talaromyces thermophilus* has been reported to retain 44% activity in the presence of 10 mM of the detergent SDS (Guerfali et al. 2009), and similarly XylA retained catalytic activities of 75% and 60% respectively at 10 and 20 mM SDS.

Xylose has been determined to be a competitive inhibitor of recombinant *A. oryzae* β-xylosidase. In the presence of 20 mM xylose with varying substrate concentrations the K_m was altered with no corresponding effect on V_{max}. This is consistent with competitive inhibition in which the inhibitor interferes with the catalytic properties of enzyme by affecting substrate binding affinity by conferring a K_i of 2.7 mM. In this respect the recombinant enzyme displays similar characteristic to the β-xylosidases from *A. niger* (K_i 2.9 mM) (Gomez et al. 2001) and *T. Reesei* (K_i 2.4 mM) (Rasmussen et al. 2006). However, xylose tolerant β-xylosidases have been reported to exist with K_i values up to 200 mM (Yan et al. 2008; Zanoelo et al. 2004). Fructose has the novel characteristics of an uncompetitive inhibitor in which the inhibitor interacts with the enzyme-substrate complex to prevent product formation. In this case the binding of xylooligosaccharide to the active site of β-xylosidase creates a binding site for fructose. High substrate concentrations of substrate will increase the occupancy of the active site and the binding sites for fructose, and therefore the effective inhibition. The functional and physiological consequences of this finding should be considered when the enzyme has to function in the presence of mixed substrates.

Competing interests

JW has become an employee of Biocatalysts Ltd since completion of the project. The authors declare no other competing interests.

Authors' contributions

NK carried out the biochemical studies and drafted the manuscript. JW carried out the protein expression studies and helped to draft the manuscript. IFC conceived of the study, and participated in its design and coordination and helped to draft the manuscript. All authors read and approved the final manuscript.

Acknowledgements

This work has been supported by project funds from BBSRC and Biocatalysts Ltd. The authors would like to thank Noriyuki Kitamoto (Aichi Industrial Technology Institute, Japan) for kindly providing the *XylA* gene. We also would like to thank our technicians, Lorraine Gillet, Nicola Cummings for their advice and David Coles for his assistance with HPLC and Dr Susan Liddell for her assistance with Mass Spectrometry.

References

Dilokpimol A, Nakai H, Gotfredsen CH, Appeldoorn M, Baumann MJ, Nakai N, Schols HA, Hachem MA, Svensson B (2011) Enzymatic synthesis of β-xylosyl-oligosaccharides by transylosylation using two β-xylosidases of glycoside hydrolase family 3 from *Aspergillus nidulans* FGSC A4. Carbohydr Res 346:421–429

Dyk JSV, Pletschke BI (2012) A review of lignocellulose bioconversion using enzymatic hydrolysis and synergistic cooperation between enzymes – Factors affecting enzymes, conversion and synergy. Biotechnol Adv 30:1458–1480

Eneyskaya EV, Ivanen DR, Bobrov KS, Isaeva-Ivanova LS, Shabalin KA, Savel'ev AN, Golubev AM, Kulminskaya AA (2007) Biochemical and kinetic analysis of the GH 3 family β-xylosidase from *Aspergillus awamori* X-100. Arch Biochem Biophys 457:225–234

Gomez M, Isorna P, Rojo M, Estrada P (2001) Chemical mechanism of beta-xylosidase from *Trichoderma reesei* QM 9414: pH dependence of kinetic parameters. Biochimie 83:961–967

Guerfali M, Maalej I, Gargouri A, Belghith H (2009) Catalytic properties of the immobilized *Talaromyces thermophilus* β-xylosidase and its use for xylose and xylooligosaccharides production. J Mol Cat B 57:242–249

Juturu V, Wu JC (2013) Heterologous expression of β-xylosidase gene from *Paecilomyces thermophila* in *Pichia pastoris*. World J Microbiol Biotechnol 29:249–255

Katapodis P, Nerinckx W, Claeyssens M, Christakopoulos P (2006) Purification and characterization of a thermostable intracellular beta-xylosidase from the thermophilic fungus *Sporotrichum thermophile*. Pro Biochem 41:2402–2409

Kirikyali N, Connerton IF (2014) Heterologous expression and kinetic characterisation of *Neurospora crassa* β-xylosidase in *Pichia pastoris*. Enzyme Microb Technol 57:63–68

Kitamoto N, Yoshino S, Ohmiya K, Tsukagoshi N (1999) Sequence analysis, overexpression and antisense inhibition of a beta-xylosidase gene, *xylA*, from *Aspergillus oryzae* KBN616. Appl Environ Microbiol 65:20–24

Knob A, Carmona EC (2009) Cell-associated acid β-xylosidase production by *Penicillium sclerotiorum*. N Biotechnol 26:60–67
Kulkarni N, Shendye A, Rao M (1999) Molecular and biotechnological aspects of xylanases. FEMS Microbiol Rev 23:411–456
La Grange DC, Pretorius IS, Claeyssens M, Zyl WHV (2001) Degradation of xylan to D-xylose by recombinant *Saccharomyces cerevisiae* coexpressing the *Aspergillus niger* β-xylosidase (*xlnD*) and the *Trichoderma reesei* xylanase II (*xyn2*) genes. Appl Environ Microbiol 67:5512–5519
Laemmli UK (1970) Cleavage of structural proteins during the assembly of the head of bacteriophage T4. Nature 227:680–685
Margolles-Clark E, Ihnen M, Penttila M (1997) Expression patterns of ten hemicellulase genes of the filamentous fungus *Trichoderma reesei* on various carbon sources. J Biotechnol 57:167–179
Michelin M, Polizeli MLTM, Ruzene DS, Silva DP, Vicente AA, Jorge JA, Terenzi HF, Teixeria JA (2012) Xylanase and beta-xylosidase production by *Aspergillus ochraceus*: New perspectives for the application of wheat straw autohydrolysis liquor. Appl Biochem Biotechnol 166:336–347
Ohta K, Fujimoto H, Fujii S, Wakiyama M (2010) Cell-associated beta-xylosidase from *Aureobasidium pullulans* ATCC 20524: Purification, properties and characterization of the encoding gene. J Biosci Bioeng 110:152–157
Polizeli MLTM, Rizzatti ACS, Monti R, Terenzi HF, Jorge JA, Amorim DS (2005) Xylanases from fungi: properties and industrial applications. Appl Microbiol Biotechnol 67:577–591
Rasmussen LE, Sorensen HR, Vind J, Vikso-Nielsen A (2006) Mode of action and properties of the β-xylosidase from *Talaromyces emersonii* and *Trichoderma reesei*. Biotechnol Bioeng 94:869–876
Saha BC (2003) Purification and properties of an extracellular beta-xylosidase from a newly isolated *Fusarium proliferatum*. Bioresour Technol 90:33–38
Sriyapai T, Somyoonsap P, Matsui K, Kawai F, Chansiri K (2011) Cloning of a thermostable xylanase from *Actinomadura* sp. S14 and its expression in *Escherichia coli* and *Pichia pastoris*. J Biosci Bioeng 111:528–536
Teng C, Jia H, Yan Q, Zhou P, Jiang Z (2011) High-level expression of extracellular secretion of a β-xylosidase gene from *Paecilomyces thermophila* in *Escherichia coli*. Bioresour Technol 102:1822–1830
Wakiyama M, Yoshihara K, Hayashi S, Ohta K (2008) Purification and properties of an extracellular β-xylosidase from *Aspergillus japonicus* and sequence analysis of the encoding gene. J Biosci Bioeng 106:398–404
Yan QJ, Wang L, Jiang ZQ, Yang SQ, Zhu HF, Li LT (2008) A xylose tolerant β-xylosidase from *Paecilomyces thermophila*: Characterisation and its co-action with the endogenous xylanase. Bioresour Technol 99:5402–5410
Zanoelo FF, Polizeli MLTM, Terenzi HF, Jorge JA (2004) Purification and biochemical properties of a thermostable xylose-tolerant β-xylosidase from *Scytalidium thermophilum*. J Ind Microbiol Biotechnol 31:170–176
Zhang R, Fan Z, Kasuga T (2011) Expression of cellobiose dehydrogenase from *Neurospora crassa* in *Pichia pastoris* and its purification and characterisation. Protein Expr Purif 75:63–69

8

Elucidation of the recognition mechanisms for hemicellulose and pectin in *Clostridium cellulovorans* using intracellular quantitative proteome analysis

Shunsuke Aburaya[1], Kohei Esaka[1], Hironobu Morisaka[1,2], Kouichi Kuroda[1] and Mitsuyoshi Ueda[1,2*]

Abstract

Clostridium cellulovorans is an anaerobic, cellulolytic bacterium, capable of effectively degrading and metabolizing various types of substrates, including cellulose, hemicellulose (xylan and galactomannan), and pectin. Among Clostridia, this ability to degrade and metabolize a wide range of hemicellulose and pectin substrates is a unique feature; however, the mechanisms are currently unknown. To clarify the mechanisms of hemicelluloses and pectin recognition and metabolism, we carried out a quantitative proteome analysis of *C. cellulovorans* cultured with these substrates. *C. cellulovorans* was cultured in the medium of glucose (control), xylan, galactomannan (Locus bean gum, LBG), or pectin for 36 h. Xylan and galactomannan were used to search for the common recognition mechanisms of hemicellulose, and pectin was used to search for unique recognition systems in *C. cellulovorans*. Using an isobaric tag method and liquid chromatograph/mass spectrometer equipped with a long monolithic silica capillary column, we identified 734 intracellular proteins from all substrates. We performed KEGG analyses and cluster analyses of the resulting proteins. In the KEGG analyses, we found common degradation mechanisms for hemicellulose and pectin. In the cluster analysis corresponding to the genome analysis, we detected substrate-specific clusters that include genes involved in substrate recognition, substrate degradation, and metabolism. Combining the results of the KEGG analyses and cluster analyses, we propose the mechanisms involved in the recognition and metabolism of hemicellulose and pectin in *C. cellulovorans*.

Keywords: *Clostridium cellulovorans*, Proteome analysis, Monolithic column, Substrate recognition, Hemicellulose, Pectin, Metabolism

Introduction

There has been enormous recent interest in biorefinery using non-edible biomass as a source for renewable fuels (Lynd et al. 1999). To achieve this aim, it is necessary to convert various biological substrates into the desired chemicals and fuels (Saha 2003). Terrestrial biomass consists of cellulose, hemicellulose, and pectin (Cosgrove 2005). Cellulose is a polymer of β-glucose that constitutes the main backbone of the plant cell wall, and composes of about 50% of plant cell walls. Hemicellulose forms a primary network with cellulose microfibrils in various conformations, and the hemicellulose binds with pectin. Polysaccharides belonging to hemicellulose can be extracted from plant cell walls with alkali solution. Common hemicellulose polysaccharides include xyloglucan, xylan, and mannan. Pectin is a family of polysaccharides that include homogalacturonan, rhamnogalacturonan-I, and rhamnogalacturonan-II. Polysaccharides belonging to pectin can be extracted with hot water or chelating agents. Homogalacturonan is a major component of pectin. It is believed that pectin crosslinking involves calcium and boron ions (Cosgrove 2005).

Of these three components of the plant cell walls, the degradation of cellulose has been best studied. For example, *Clostridium thermocellum* can metabolize cellulose

*Correspondence: miueda@kais.kyoto-u.ac.jp
[1] Division of Applied Life Sciences, Graduate School of Agriculture, Kyoto University, Sakyo-ku, Kyoto, Japan
Full list of author information is available at the end of the article

and is used for the degradation of cellulose to glucose (Prawitwong et al. 2013). *C. thermocellum* belongs to the Family 4 (Cluster III) of Clostridia (Rainey and Stackebrandt 1993; Collins et al. 1994), and genetically engineered strains of *C. cellulolyticum* can metabolize cellulose and xylan among hemicellulose, and are used for producing isobutanol from cellulose (Higashide et al. 2011). However, available strains cannot utilize all forms of hemicellulose and pectin. For maximally efficient biorefinery, utilization of hemicellulose and pectin is also essential.

Clostridium cellulovorans is a gram-positive, mesophilic, and cellulosome-producing anaerobe of the Family 1 (Cluster I) of Clostridia that can degrade and directly metabolize various kinds of carbohydrates such as cellulose, hemicellulose (xylan and mannan), and pectin (Petitdemange et al. 1984; Tamaru et al. 2010). Furthermore, *C. cellulovorans* alters its production of secreted enzymes, depending on which substrates are available (Morisaka et al. 2012; Matsui et al. 2013; Esaka et al. 2015). Thus, *C. cellulovorans* can distinguish among varied substrates and adapt to survive in different environments.

In Clostridia, the systems by which cells recognize different substrates have been widely studied. For example, *C. thermocellum* was reported to recognize the availability of cellulose through the coordinated action of sigma and anti-sigma factors with cellulose binding domains (Nataf et al. 2010). *C. cellulolyticum* was reported to recognize cellulose availability with a carbon catabolite repression mechanism and xylan availability with a two-component system (TCS) (Celik et al. 2013; Xu et al. 2013). These findings imply that substrate recognition systems can show great variety between species. However, the substrate recognition systems of *C. cellulovorans*, which can degrade and directly metabolize a greater variety of substrates than other Clostridia studied, are so far unknown. This suggests that *C. cellulovorans* may be an attractive model organism for the study of substrate recognition systems.

In the present study, we carried out an "intracellular" proteome analysis using an LC–MS/MS system equipped with a long monolithic silica capillary column (500 cm) (Aoki et al. 2013; Morisaka et al. 2012) in order to elucidate substrate recognition systems of *C. cellulovorans*. As substrates, we used xylan, galactomannan (locust bean gum; LBG), and pectin. Studies on the recognition and metabolism of galactomannan and pectin have not been performed in cellulosome-producing Clostridia because the commonly used model organisms, *C. cellulolyticum* and *C. thermocellum*, are unable to metabolize these substrates. Based on our findings, we propose metabolic systems for each substrate and molecular models of substrate recognition system in *C. cellulovorans*. We further suggest that hemicellulose and pectin were degraded outside of the cell with constitutively produced enzymes, such as cellulosomal proteins. We could clarify that the TCS response regulator AraC was used in the substrate recognition system common to both xylan and galactomannan. Interestingly, for pectin, *C. cellulovorans* regulated the hydrolase/isomerase pathway and the lyase/5-dehydro-4-deoxy-gluconate pathway via the Crp transcriptional regulator and the IclR family transcriptional regulator, respectively.

Materials and methods

Cell culture and media

C. cellulovorans 743B (ATCC35296) was grown anaerobically as previously described (Sleat et al. 1984), differing only in the carbon source, which was replaced by 0.3% (w/v) glucose (Nacalai Tesque, Kyoto, Japan), 0.3% (w/v) xylan from beechwood (Sigma, MO, USA), 0.3% (w/v) pectin from apple (Sigma), or 0.3% (w/v) LBG from *Caretonia silliqua* seeds (Sigma).

Estimating the growth of the anaerobic bacteria

The growth curves of *C. cellulovorans* on each medium were determined by bacterial protein estimation, as previously described (Raman et al. 2009) with small modifications. Cells from 1 mL of cell culture were collected by centrifugation (13,000×*g*, 4°C, 10 min). The cell pellets were washed with 1 mL of phosphate-buffered saline (pH 7.4; PBS) and incubated with 800 μL of sodium deoxycholate (2%) for 20 min at 37°C. Two hundreds microliters of trichloroacetic acid (24%) was added to the suspension, which was then centrifuged (13,000×*g*, 4°C, 10 min). One hundred microliters of resolubilization solution (5% SDS, 2 N NaOH) was added to the suspension and vigorously mixed. The protein concentration was measured using a protein assay bicinchoninate kit (Nacalai Tesque), with bovine serum albumin used as a standard.

Sample preparation for quantitative proteome analysis

C. cellulovorans was grown in 50 mL cultures to late-logarithmic phase (36 h). Cells were concentrated by centrifugation (6,000×*g*, 4°C, 10 min), and the supernatant was discarded. Cell pellets were collected by centrifugation (13,000×*g*, 4°C, 10 min) and washed with 500 μL PBS, and centrifuged again. Cells were resuspended with 500 μL of lysis buffer [2% (w/v) 3-(3-cholamidopropyl) dimethylammonio-1-propanesulfonate, 10 mM dithiothreitol, 1% (v/v) protease inhibitor cocktail for bacterial cell lysis (Sigma), 7 M urea, and 2 M thiourea in 50 mM Tris–HCl (Nacalai Tesque)]. The cells were disrupted with a Bioruptor UCD-250T (Cosmo Bio, Tokyo, Japan) at 250 W, 15 s on-and-off cycles for 10 min, on ice. The solution was centrifuged (13,000×*g*, 4°C, 20 min), and the supernatant was collected and subjected to ultrafiltration with an Amicon Ultra-0.5 Centrifugal Filter Unit (10 kDa, Millipore, MA, USA) and buffer-exchanged with 200 mM triethyl ammonium bicarbonate (TEAB; Sigma). Proteins were reduced by adding tris (2-carboxyethyl) phosphine to 10 mM from a 200 mM stock, and the reaction was allowed to proceed for 60 min at 55°C. Following the incubation, 5 μL of iodoacetamide (375 mM) was added, and the reaction continued for 30 min at room temperature in the dark. Proteins were precipitated by adding 1 mL of ice-cold acetone and incubating the solution overnight at −20°C. The precipitated proteins were resuspended with 100 μL of 200 mM TEAB, 2 μg of sequencing grade modified trypsin (Promega, WI, USA) was added, and incubated overnight at 37°C. The four samples for proteome analysis (glucose, xylan, pectin, and galactomannan) were labeled using a tandem mass tag (TMT) 6-plex labeling kit (Thermo Fisher Scientific, MA, USA) with reporters at $m/z = 126$, 127, 129, and 130, respectively, in 41 μL acetonitrile. After 60 min at room temperature, 8 μL of 5% (w/v) hydroxylamine was added to each tube and mixed for 15 min. As an internal standard for quantification, a mixture of tryptic fragments from all substrates was combined with TMT-131 (reporter at $m/z = 131$). Aliquots were pooled and evaporated under vacuum, then dissolved in 60 μL of formic acid (0.1%) and subjected to LC–MS/MS analysis.

LC–MS/MS analysis

Proteome analysis was performed using an LC (Ultimate 3000; Thermo Fisher Scientific)–MS/MS (LTQ Orbitrap Velos Mass Spectrometer Thermo Fisher Scientific) system equipped with a long monolithic silica capillary column. Tryptic digests were separated by reverse-phase chromatography using a monolithic silica capillary column (500 cm long, 0.1 mm ID) (Morisaka et al. 2012), at a flow rate of 500 nL/min. A gradient was achieved by changing the ratio of the two eluents: eluent A, 0.1% (v/v) formic acid; eluent B, 80% acetonitrile containing 0.1% (v/v) formic acid. The gradient started with 5% B, increased to 45% B for 600 min, further increased to 95% B to wash the column for 140 min, returned to the initial condition, and was held for re-equilibration of the column. The separated analytes were detected using a mass spectrometer with a full scan range of 350–1,500 m/z (resolution 60,000), followed by 10 data-dependent higher-energy c-trap dissociation (HCD) MS/MS scans acquired for TMT reporter ions by using 40% normalized collision energy in HCD with 0.1 ms activation time in quantitative proteome analysis and with a full scan range of 350–1,500 m/z (resolution 60,000), followed by 10 data-dependent collision-induced dissociation (CID) MS/MS scans in a qualitative proteome analysis. An electrospray ionization (ESI) voltage was set at 2.3 kV. Triplicate analyses were performed for each sample in three independent experiments, and the collected data were reviewed for protein identification and quantification in a quantitative proteome analysis. Single analyses were performed for each sample of the three independent experiments in a qualitative proteome analysis. The collected data were reviewed for protein identification and quantification. Blank runs were inserted between runs of different samples.

Data analysis

Data analysis was performed using Proteome Discoverer 1.4 (Thermo Fisher Scientific). Protein identification was performed using the Mascot algorithm against the *C. cellulovorans* protein database (4,254 sequences) from NCBI (National Center for Biotechnology Information, http://www.ncbi.nlm.nih.gov/), with a precursor mass tolerance of 20 ppm and a fragment ion mass tolerance of 50 mmu in quantitative proteome analysis, and mass tolerance of 2.0 Da and a fragment ion mass tolerance of 0.8 Da in qualitative proteome analysis. Carbamidomethylation of cysteine was set as a fixed modification for quantitative proteome analysis, and carbamidomethylation of cysteine and a TMT 6-plex at the N-terminus and lysine were set as fixed modifications for qualitative proteome analysis. Protein quantification was performed using the reporter ions quantifier with the TMT 6-plex method. The data were then filtered with a cut-off criteria of q value ≤ 0.05, corresponding to a 5% false discovery rate (FDR) on a spectral level. Proteins with no missing values in three replicates were accepted as identified proteins for the protein quantification analysis. In the qualitative proteome analysis, proteins with scores >10 were accepted as identified proteins. Global median normalization was performed to normalize the quantity of each tryptic digest injected into the mass spectrometer. The heat map was generated using Cluster 3.0 (de Hoon et al. 2004), which can carry out hierarchical cluster analysis (HCA). Euclidean distance was used to measure the similarities of the protein profile patterns within the clustering analysis. To visualize the clustering results from Cluster 3.0, Java TreeView software was used (Saldanha 2004). The pathway map was constructed from Kyoto Encyclopedia of Genes and Genomes (KEGG; http://www.genome.jp/kegg/) (Kanehisa and Goto 2000), and identified proteins were adapted to this map.

Results

To identify substrate recognition systems of *C. cellulovorans*, we carried out a quantitative proteome analysis of cells grown anaerobically on media with the carbon sources-glucose, xylan, galactomannan (LBG), or pectin. Using a quantitative proteome approach based on isobaric tagging, we obtained protein profiles from cells grown on each carbon source. The workflow of the "intracellular" proteome analysis is illustrated in Figure 1.

Growth confirmation

To determine the growth of *C. cellulovorans* on each of the substrates, we conducted bacterial protein estimation (Figure 2). As has previously been shown, *C. cellulovorans* can grow on xylan, pectin, or galactomannan as the sole carbon source (Sleat et al. 1984), although other cellulosome-producing Clostridia, such as *C. thermocellum* and *C. cellulolyticum*, cannot grow on pectin and galactomannan (Petitdemange et al. 1984; Prawitwong et al. 2013). Growth on glucose, xylan, and pectin was slower than that on galactomannan, but cells were collected from all cultures at similar growth phases. From the growth analysis, we selected a culture time of 36 h, as cells were in the late-logarithmic phase, which was appropriate for proteome analysis in terms of growth phase and protein concentration.

Qualitative and quantitative proteome analysis

LC–MS/MS equipped with a long monolithic silica capillary using a tandem mass tag (TMT) 6-plex isobaric tag column was employed for intracellular quantitative proteome analysis. *C. cellulovorans* has 4,254 protein-encoding genes in its genome (Tamaru et al. 2010). For protein identification, we constructed a protein database built from the genome of *C. cellulovorans*. In total, we could identify 734 proteins from all samples within our cutoff criteria (Additional file 1: Table S1). To correct for variations in the amount of TMT-labeled peptides infused into the mass spectrometer, we normalized all quantitative data to the median value of that analysis. For all data analysis, we used these normalized relative quantification values. We checked for reproducibility of quantitative protein profiles among three biological replicates for each substrate by HCA (Figure 3). Each array was clustered individually based on each biological replicate. Our results indicate that each sample was reproducibly quantified,

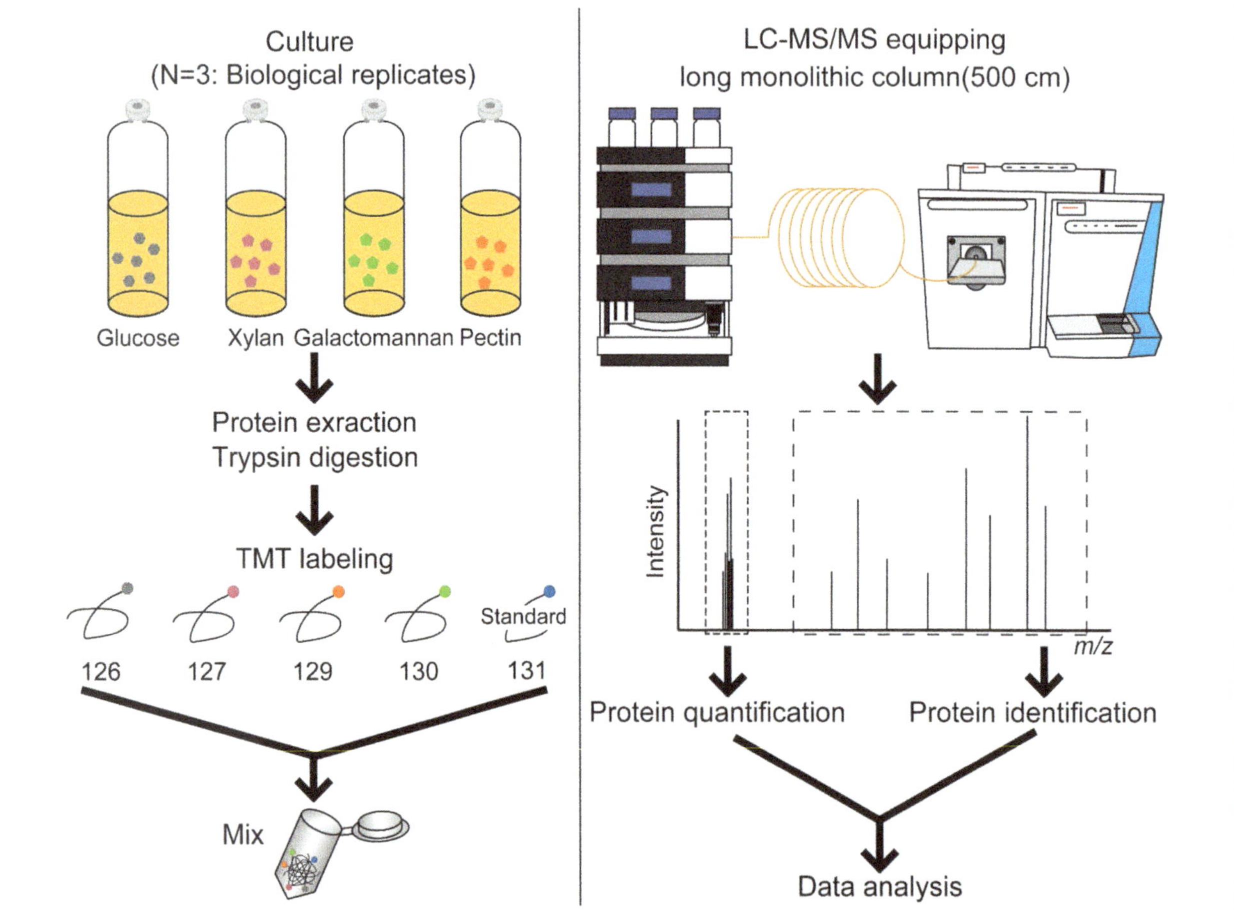

Figure 1 Experimental procedure of *C. cellulovorans* intracellular quantitative proteome analysis. Proteins prepared from cell lysates of *C. cellulovorans* grown in the presence of glucose, xylan, galactomannan (LBG), or pectin were individually reductive-alkylated and digested with trypsin. Tryptic fragments were labeled with tandem mass tags (TMTs). The labeled peptides were mixed and injected into the LC–MS/MS system with a long monolithic silica capillary column for mass measurement and collected data were used for protein quantification.

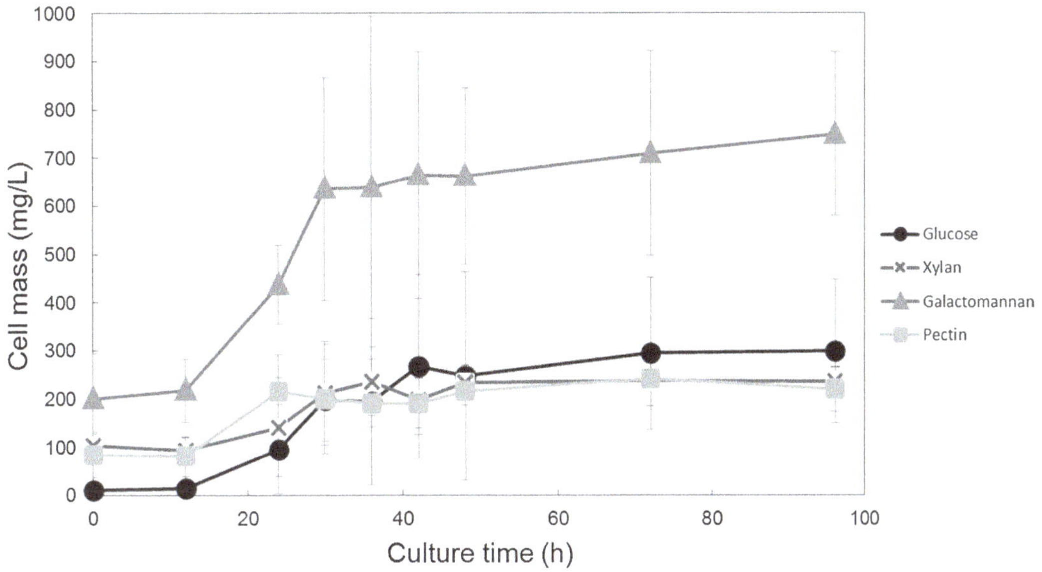

Figure 2 Confirmation of growth of *C. cellulovorans* cultured with four different substrates. Growth of *C. cellulovorans* was measured by estimation of protein in cell lysates to determine an appropriate culture time (Raman et al. 2009). At 36 h, *C. cellulovorans* appears to be in the late-logarithmic phase in all four substrates (glucose, xylan, galactomannan, pectin). These conditions were used for the proteome analysis. *Error bars* indicate SD ($n = 3$).

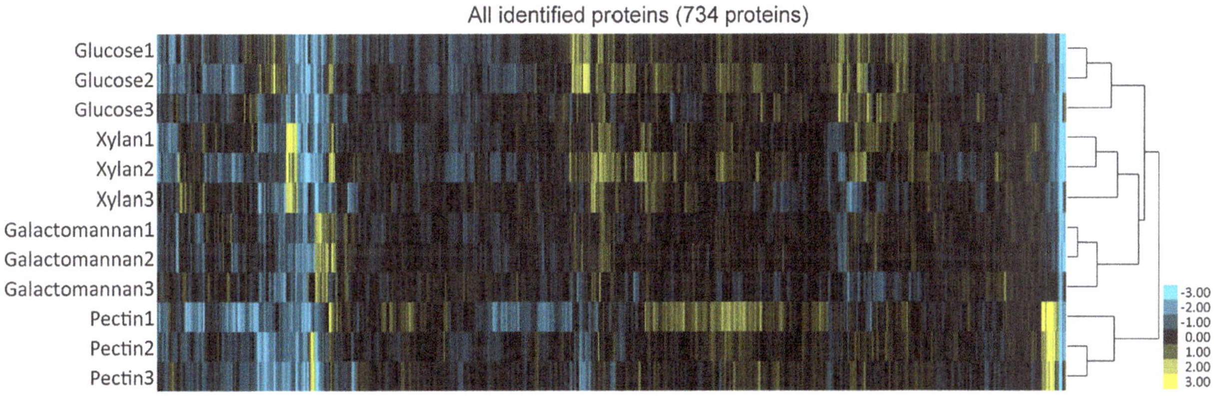

Figure 3 Hierarchical clustering analysis represented proteome profiles of *C. cellulovorans* with four different substrates. To standardize the data, the quantitative data were normalized using the median for each condition. The quantitative proteome data from three biological replicates of each substrate were used for the hierarchical clustering analysis. Each array was clustered in response to substrates, and each biological replicate was grouped together. This strategy ensures that the proteome analysis was biologically reproducible. *Color bar* indicates changes in protein abundance. Increased and decreased protein levels are shown in *yellow* and *blue*, respectively.

and analytical and biological replication of results was ensured. Furthermore, we separately carried out a qualitative proteome analysis to compare the results of quantitative proteome analysis (Additional file 2: Table S2).

Substrate-specific proteins

To discover substrate-specific proteins (those that showed a significant change between different growth substrates), we carried out an empirical Bayes moderated *t* test (Smyth 2004). *P* values were adjusted with the Benjamini–Hochberg method to avoid the problem of multiple testing. The thresholds that we adopted were FDR-adjusted *p* value of <0.05 and fold-change of protein ratio >2.0 compared to glucose. Proteins for which the levels significantly changed were defined by comparison between glucose and each polysaccharide (xylan, galactomannan, and pectin) at these thresholds. All substrate-specific proteins detected are shown in Table 1. Using KEGG analysis

Table 1 Substrate-specific proteins

Specificity	Locus	Description in NCBI	Description in KEGG	Vs. Glucose	
				Log$_2$ (fold change)	FDR-adjusted *p* value
Xylan	Clocel_0589	Alpha-L-fucosidase	Alpha-L-fucosidase 2	2.67	2.07E−04
	Clocel_0590	Xylose isomerase	Xylose isomerase	3.35	1.43E−04
	Clocel_0591	Transaldolase	Transaldolase	3.52	1.46E−06
	Clocel_0592	Xylulokinase	Xylulokinase	2.85	7.29E−05
	Clocel_1085	Dinitrogenase iron–molybdenum cofactor biosynthesis protein	–	2.32	4.24E−03
	Clocel_1151	Methyl-accepting chemotaxis sensory transducer	Methyl-accepting chemotaxis protein	1.96	1.70E−02
	Clocel_1430	Glycoside hydrolase family protein	Alpha-D-xyloside xylohydrolase	1.90	9.48E−03
	Clocel_2573	Hypothetical protein Clocel_2573	Chemotaxis protein CheX	1.60	3.95E−03
	Clocel_2592	Two component transcriptional regulator, AraC family	–	3.72	1.49E−04
	Clocel_2595	Xylan 1,4-beta-xylosidase	Xylan 1,4-beta-xylosidase	4.70	4.53E−06
	Clocel_2596	Sugar ABC transporter periplasmic protein	Ribose transport system substrate-binding protein	4.62	1.06E−03
	Clocel_2597	Inner-membrane translocator	Ribose transport system permease protein	4.01	7.36E−05
	Clocel_2598	ABC transporter	Ribose transport system ATP-binding protein	4.34	1.89E−05
	Clocel_2881	PTS system lactose/cellobiose-specific transporter subunit IIB	PTS system, cellobiose-specific IIB component	1.49	1.70E−02
	Clocel_2940	Putative phosphate transport regulator	Hypothetical protein	1.44	2.45E−02
	Clocel_3175	phoH family protein	phoH-like ATPase	1.64	3.84E−02
	Clocel_3761	ATP:guanido phosphotransferase	–	1.31	1.37E−02
	Clocel_3762	UvrB/UvrC protein	–	1.42	3.45E−02
	Clocel_4277	Aldo/keto reductase	–	1.08	4.66E−02
Galactomannan	Clocel_0034	Glycoside hydrolase family protein	Alpha-D-xyloside xylohydrolase	1.73	1.04E−02
	Clocel_0391	Glycosyltransferase	–	1.83	1.96E−02
	Clocel_0684	Thiazole biosynthesis protein ThiH	2-Iminoacetate synthase	1.57	4.69E−03
	Clocel_2225	Cobalamin (vitamin B12) biosynthesis CbiX protein	–	1.39	4.52E−02
	Clocel_2259	pfkB domain-containing protein	2-Dehydro-3-deoxygluconokinase	1.31	1.04E−02
	Clocel_2697	Sialate *O*-acetylesterase	Sialate *O*-acetylesterase	1.69	1.69E−02
	Clocel_2800	Alpha-galactosidase	Alpha-galactosidase	2.99	5.49E−05
	Clocel_2962	Inosine-5′-monophosphate dehydrogenase	IMP dehydrogenase	1.31	1.96E−02
	Clocel_3175	phoH family protein	phoH-like ATPase	1.98	1.14E−02
	Clocel_3194	Mannose-6-phosphate isomerase	Mannose-6-phosphate isomerase	3.77	6.86E−05
	Clocel_3196	Glycosidase-like protein	Beta-1,4-mannooligosaccharide/beta-1,4-mannosyl-*N*-acetylglucosamine phosphorylase	2.65	5.29E−03
	Clocel_3198	*N*-acylglucosamine 2-epimerase	Mannobiose 2-epimerase	3.19	2.10E−03
	Clocel_3200	Binding-protein-dependent transport system inner membrane protein	Multiple sugar transport system permease protein	4.76	8.61E−04
	Clocel_3201	Extracellular solute-binding protein	Raffinose/stachyose/melibiose transport system substrate-binding protein	4.23	5.49E−05
	Clocel_3205	Glycoside hydrolase family 2	Beta-mannosidase	2.55	1.51E−03
	Clocel_3657	Xylan 1,4-beta-xylosidase	Xylan 1,4-beta-xylosidase/alpha-*N*-arabinofuranosidase	1.79	4.26E−02
	Clocel_3857	ABC transporter	Multiple sugar transport system ATP-binding protein	2.15	1.03E−03

Table 1 continued

Specificity	Locus	Description in NCBI	Description in KEGG	Vs. Glucose	
				Log$_2$ (fold change)	FDR-adjusted *p* value
	Clocel_4053	LPXTG-motif cell wall anchor domain-containing protein	Sialate *O*-acetylesterase	2.28	3.79E−02
	Clocel_4087	Aldose 1-epimerase	Aldose 1-epimerase	1.45	2.68E−02
	Clocel_4088	Galactose-1-phosphate uridylyltransferase	UDP-glucose–hexose-1-phosphate uridylyltransferase	3.38	1.83E−04
	Clocel_4089	UDP-glucose 4-epimerase	UDP-glucose 4-epimerase	2.09	1.82E−03
	Clocel_4277	Aldo/keto reductase	–	1.08	4.26E−02
Pectin	Clocel_0048	Transcriptional regulator, AbrB family	Transcriptional pleiotropic regulator of transition state genes	1.91	2.50E−03
	Clocel_0322	tatD family hydrolase	tatD DNase family protein	1.74	8.66E−03
	Clocel_0513	Extracellular solute-binding protein	Raffinose/stachyose/melibiose transport system substrate-binding protein	2.12	2.50E−03
	Clocel_0519	Glycogen/starch/alpha-glucan phosphorylase	Starch phosphorylase	1.88	1.37E−02
	Clocel_1243	Extracellular solute-binding protein	Multiple sugar transport system substrate-binding protein	2.53	9.20E−05
	Clocel_1892	Acetate kinase	Acetate kinase	1.31	6.43E−03
	Clocel_2210	Nicotinate-nucleotide–dimethylbenzimidazole phosphoribosyltransferase	Nicotinate-nucleotide–dimethyl-benzimidazole phosphoribosyltransferase	1.08	1.47E−02
	Clocel_2214	ATP:corrinoid adenosyltransferase BtuR/CobO/CobP	Coblalamin adenosyltransferase	1.37	4.25E−02
	Clocel_2222	Precorrin-3B C(17)-methyltransferase	Precorrin-3B C17-methyltransferase	1.70	6.63E−03
	Clocel_2225	Cobalamin (vitamin B12) biosynthesis CbiX protein	–	1.95	4.32E−03
	Clocel_2227	Precorrin-4 C(11)-methyltransferase	Precorrin-4/cobalt-precorrin-4 C11-methyltransferase	2.11	7.92E−03
	Clocel_2250	Altronate dehydratase	Altronate hydrolase	4.07	1.02E−05
	Clocel_2251	Mannitol dehydrogenase domain-containing protein	Tagaturonate reductase	4.50	4.39E−06
	Clocel_2253	Crp family transcriptional regulator	–	2.37	2.50E−03
	Clocel_2254	Glycoside hydrolase family protein	–	4.36	4.39E−06
	Clocel_2255	Major facilitator superfamily protein	Oligogalacturonide transporter	4.33	4.02E−05
	Clocel_2256	Glycosyl hydrolase family protein	Unsaturated rhamnogalacturonyl hydrolase	4.19	1.49E−05
	Clocel_2259	pfkB domain-containing protein	2-Dehydro-3-deoxygluconokinase	3.75	1.49E−06
	Clocel_2262	Short-chain dehydrogenase/reductase SDR	Gluconate 5-dehydrogenase	3.21	2.11E−02
	Clocel_2263	4-Deoxy-L-threo-5-hexosulose-uronate ketol-isomerase	4-Deoxy-L-threo-5-hexosulose-uronate ketol-isomerase	3.15	4.92E−03
	Clocel_2403	Glucosamine/fructose-6-phosphate aminotransferase	Glucosamine–fructose-6-phosphate aminotransferase (isomerizing)	1.45	4.57E−03
	Clocel_2737	Small GTP-binding protein	–	1.10	2.98E−02
	Clocel_3380	LPXTG-motif cell wall anchor domain-containing protein	–	4.03	1.27E−04
	Clocel_3909	Quorum-sensing autoinducer 2 (AI-2), LuxS	*S*-ribosylhomocysteine lyase	1.19	2.34E−02
	Clocel_4088	Galactose-1-phosphate uridylyltransferase	UDP-glucose–hexose-1-phosphate uridylyltransferase	1.62	1.73E−02
	Clocel_4277	Aldo/keto reductase	–	1.31	8.08E−03

To standardize data, obtained data were normalized with the median for all identified quantitative data. *P* values were adjusted with the Benjamini–Hochberg method to avoid the problem of multiple testing. For thresholds, we adopted an FDR-adjusted *p* value of <0.05 and fold change of protein ratio >2.0, compared to glucose.

and cluster analysis based on genome analysis, we focused mainly on metabolism-related and substrate recognition-related proteins for further analysis.

Profiles of metabolism-related proteins

First, we focused on profiles of metabolism-related proteins, such as enzymes involved in substrate degradation and metabolism, and other characteristic metabolic pathway. *C. cellulovorans* is known to change production of carbohydrate-related enzymes secreted into media from exoproteome analyses (Esaka et al. 2015; Matsui et al. 2013) and alternation of production of different metabolic pathways depending on which substrates are available in culture is also predicted.

Degradation and metabolism of each substrate

We constructed a substrate degradation pathway from KEGG pathway maps, and presented the fold change of each protein (Figure 4). For xylan degradation- and metabolism-related proteins (Figure 4a), the levels of three proteins (Clocel_0590, 0592, and 2595) were significantly elevated in the presence of xylan (Table 1), but production of Clocel_2900 (Endo-1, 4-beta xylanase) was not specifically elevated (Additional file 1: Table S1).

For galactomannan degradation- and metabolism-related proteins (Figure 4b), the levels of 9 proteins (Clocel_2259, 2800, 3194, 3196, 3198, 3205, 4087, 4088, and 4089) were significantly elevated in the presence of galactomannan (Table 1). By contrast, the levels of mannanase (Clocel_1134 and 4119) were not significantly elevated (Additional file 1: Table S1).

For pectin degradation- and metabolism-related proteins (Figure 4c), levels of 8 proteins (Clocel_2250, 2251, 2254, 2256, 2259, 2262, 2263, and 3380) were significantly elevated in the presence of pectin (Table 1). For the degradation and metabolism of pectin, both the "hydrolase/isomerase pathway" and the "lyase/5-dehydro-4-deoxy-gluconate pathway" were found (Richard and Hilditch 2009). Clocel_2254 is a member of the glycoside hydrolases (GH) 28 family (Cantarel et al. 2009), which is known to have endogalacturonase and exogalacturonase activities. Additionally, Clocel_3380 is a member of the polysaccharide lyase (PL) 9 family. BLAST analysis indicates that the protein has exogalacturonate lyase activity.

Genome and cluster analyses

To identify candidates related to substrate recognition systems, we performed genomic analysis based on *C. cellulolyticum* TCS-related gene clusters (Xu et al. 2013). For the thresholds of TCS-regulated cluster identification, we applied two criteria. First, a cluster must contain more than two components of a TCS pathway: an integral membrane histidine kinase (sensor histidine kinase), a transcriptional regulator (response regulator), and an extracellular solute-binding protein (sugar binding protein). Second, degradation-, metabolism-, or transport-related gene loci must be located within 2 of the TCS genes identified above. Using this threshold, we identified 14 candidate clusters related to TCS.

Combining the information of our genome analysis and the substrate-specific proteins, we identified xylan- and galactomannan-specific clusters that include TCS-related genes. We identified clusters corresponding to two substrates; the xylan-specific cluster included Clocel_2592, 2595, 2596, 2597, and 2598 (Table 1), and Clocel_2593 and Clocel_2594 (Additional file 2: Table S2). The galactomannan-specific cluster included Clocel_3194, 3196, 3198, 3200, 3201, and 3205 (Table 1), and Clocel_3195, 3197, 3199, 3202, 3203, 3204 (Figure 5; Additional file 2: Table S2). Each cluster includes three components of a TCS: a transcriptional regulator AraC, an integral membrane histidine kinase, and an extracellular solute-binding protein. We suggest that these genes are common components of a hemicellulose recognition system.

From pectin, we found increased levels of proteins Clocel_2250, 2251, 2253, 2254, 2255, 2256, 2259, 2262, and 2263 (Table 1) and Clocel_2249, 2257, 2258, and 2260 (Additional file 2: Table S2). We suggest that this is a large cluster for the degradation and metabolism of pectin. For the degradation and metabolism of pectin, hydrolase/isomerase and lyase/5-dehydro-4-deoxy-gluconate pathways are known. The pectin-specific cluster which we identified contains genes related to both of these pathways, as well as for transporting pectin.

We also identified a xylose-metabolic cluster containing proteins Clocel_0589, 0590, 0591, and 0592 (Table 1) that had increased levels specifically in the presence of xylan. A galactose-metabolic cluster containing proteins Clocel_4087, 4088, and 4089 (Table 1) and Clocel_4090 (Additional file 2: Table S2) had elevated levels specifically in the presence of galactomannan. The xylose-metabolic cluster contained two xylose metabolism-related proteins, Clocel_0590 and 0592 (Table 1; Figure 4a), transaldorase (Clocel_0591; Table 1), which is related to the pentose phosphate pathway, and alpha-fucosidase (Clocel_0589; Table 1). In the galactose-metabolic cluster, galactose metabolism-related genes are present (Figure 4b) and increased levels of proteins Clocel_4087, 4088, and 4089 (Table 1) and Clocel_4090 (Additional file 2: Table S2) were found.

Discussion

In total, we identified 734 proteins using our "intracellular" quantitative proteome analysis, and we were able to identify substrate-specific proteins using statistical analysis. Our method provided reproducibility, as evaluated

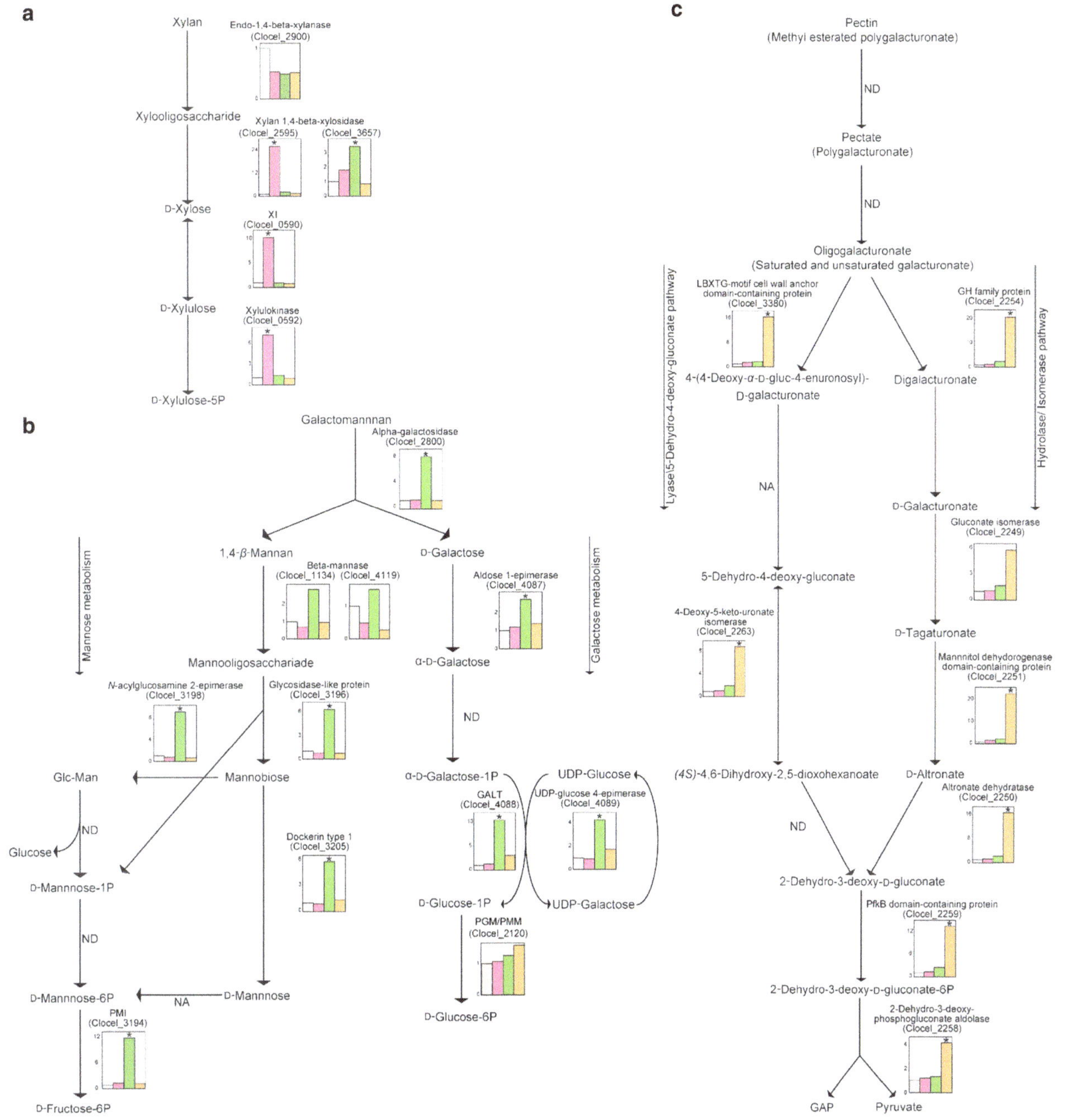

Figure 4 Degradation and metabolism pathways for each substrate, constructed from KEGG analysis. Each substrate degradation and metabolism pathway is shown: **a** xylan, **b** galactomannan (LBG), **c** pectin. For each protein, the fold change compared to glucose is shown. *Asterisk* denotes that protein levels are significantly elevated; as threshold, fold change >2.0, and FDR-adjusted *p* value <0.05 are adopted. *ND* protein not detected in this analysis. *NA* genes commonly assigned to the pathway, but not annotated. Fold change in glucose-grown cells is shown with a *gray bar* (=1); fold change in xylan-grown cells is shown with a *pink bar*; fold change of galactomannan-grown cells is shown with a *green bar*; fold change in pectin-grown cells is shown with an *orange bar*. *XI* xylose isomerase, *PMI* phosphomannose isomerase, *GALT* galactose-1-phosphate uridyltransferase, *PGM/PMM* phosphoglucomutase/phosphomannmutase alpha/beta/alpha domain I, *1P* 1-phosphate, *6P* 6-phosphate.

by HCA (Figure 3). For substrate-specific proteins, we carried out genome analysis and cluster analysis following genome analysis on the basis of the results from our proteome analysis.

We first focused on metabolic pathways of each substrate in *C. cellulovorans* by KEGG analysis. For xylan metabolism, it is interesting to note that intracellular levels of Clocel_2595 (xylosidase) increased, but levels

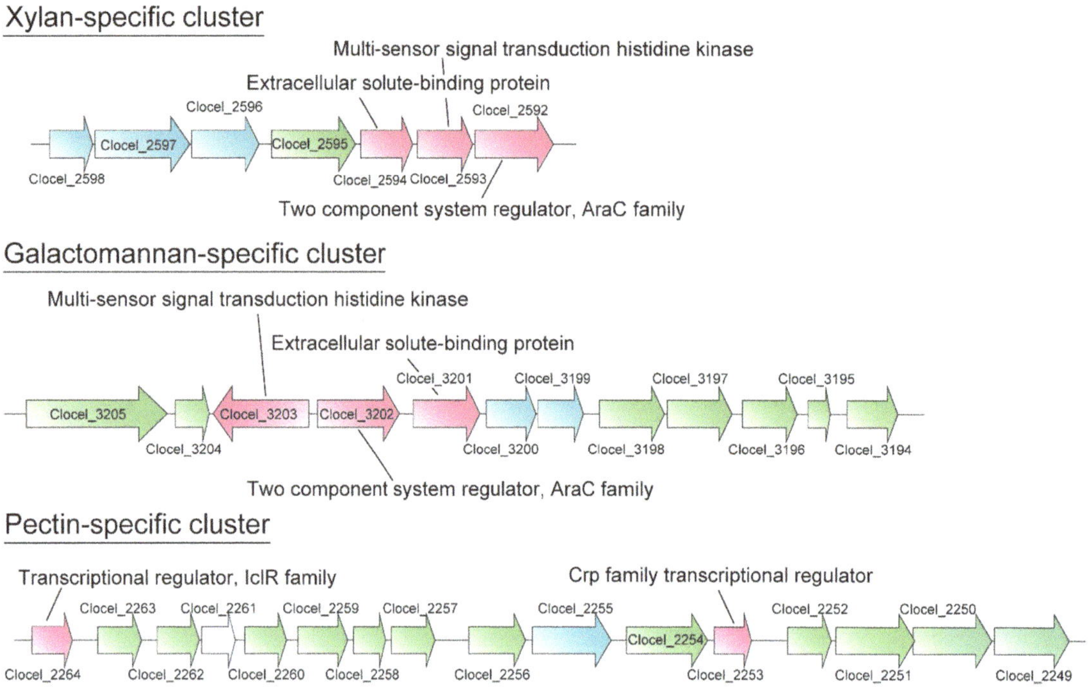

Figure 5 Candidates for gene clusters related to substrate recognition. We carried out genome analysis and identified candidates for gene clusters involved in TCS. Combining these results and substrate specific proteins, we found substrate-specific gene clusters related to recognition of each substrate. *Green arrows* indicate metabolism-related proteins, *pink arrows* indicate signal transduction-related proteins, and *blue arrows* indicate transport-related proteins. *White arrows* indicate pseudogenes.

of Clocel_2900 (xylanase) were not enhanced. Although Clocel_2595 (xylosidase) and some xylanases were categorized as xylan-specific in "exoproteome" analysis (Matsui et al. 2013), *C. cellulovorans* cannot grow in medium containing xylose as the sole carbon source (Sleat et al. 1984). Combining our findings with the inability of *C. cellulovorans* to grow in xylose medium, we predict that *C. cellulovorans* metabolizes xylooligosaccharides intracellularly. We hypothesize the following pathway for xylan metabolism: *C. cellulovorans* degrades xylan to xylooligosaccharides extracellularly and transports them into the cytoplasm. Next, xylooligosaccharides are degraded to xylose in the cell, and xylose is metabolized and used by the pentose phosphate pathway. Indeed, we observed increased levels of Clocel_0591 (transaldolase), which is related to the pentose phosphate pathway, in xylan medium.

For galactomannnan metabolism, we focused on Clocel_3196 (glycosidase-like protein), which is a member of the GH130 family and is annotated as beta-1,4-mannooligosaccharide/beta-1,4-mannosyl-*N*-acetylglucosamine phosphorylase (Kawahara et al. 2012). Similarly annotated genes are part of mannan metabolic pathways in other bacteria (Senoura et al. 2011), and we observed increased levels of proteins in this pathway (Clocel_3198, 3205) in the presence of galactomannan. However, Clocel_3197, which is known to play an important role in the mannan metabolic pathway (Esaka et al. 2015), was only identified in our qualitative proteome analysis (Additional file 2: Table S2). This protein has been also detected with "exoproteome" analysis (Esaka et al. 2015). It is assumed that this protein acts both intracellularly and extracellularly through an unknown mechanism. In addition, we identified increased levels of alpha-galactosidase (Clocel_2800). In a previous study, this protein was identified in an "exoproteome" analysis, and it is predicted to degrade galactose side chains both intracellularly and extracellularly. Specifically, the extracellular collaboration between mannase and alpha galactosidase rapidly degrades galactomannan. Next, galactose side chains that were not degraded extracellularly are degraded by "intracellular" alpha galactosidase, and the resulting mannooligosaccharides and galactose are further metabolized.

We predicted a pectin metabolism pathway in *C. cellulovorans* by combining the results of the present study and the previously reported "exoproteome" analysis (Matsui et al. 2013). Extracellularly, pectin is degraded by pectin lyase (Clocel_0873, 1172, and 3380) to oligogalacturonate with both saturated and unsaturated ends. In parallel, pectin is also de-esterified to pectate (Clocel_0211 and 3114), resulting in simpler extracellular degradation pathways. The resulting metabolites are transported into the cell and metabolized with the hydrolase/isomerase and lyase/5-dehydro-4-deoxy-gluconate pathways. In *Erwinia chrysathemi*, which is a well-studied pathogenic bacterium, pectin is degraded using similar systems; i.e. pectin is degraded to oligogalacturonides outside of the cell. Next, oligogalacturonides are transported into cells and metabolized in the hydrolase/isomerase and lyase/5-dehydro-4-deoxy-gluconate pathways (Hugouvieux-Cotte-Pattat et al. 1996). While known for *E. chrysathemi*, this system has not been reported in other cellulosome-producing Clostridia. *C. cellulovorans* commonly degrades all sugar-substrates (polysaccharides) to oligosaccharides, not to monosaccharide, outside of the cell. Degradation of those oligosaccharides to monosaccharides and further metabolism take place intracellularly. Similar degradation systems have not been reported in Clostridia although *C. cellulolyticum* can metabolize xylose as a sole carbon source. Furthermore, in the presence of pectin, we found increased levels of an oligogalacturonides (esterified-oligogalacturonates) transporter (Clocel_2255). In the current analysis, we did not identify pectinesterase "intracellularly", although it had previously been identified in an "exoproteome" analysis (Matsui et al. 2013). Together, these results suggest that oligogalacturonates are transported into *C. cellulovorans*, and degraded to monogalacturonates and further metabolized intracellularly. In addition to the degradation and metabolism of each substrate, levels of four proteins related to the synthesis of vitamin B12 were significantly elevated in cells cultured on pectin (Clocel_2214, 2222, 2225, and 2227). We are unable to find any pectin-specific proteins reported to have a requirement for vitamin B12 (Sañudo-Wilhelmy et al. 2014). For the degradation and metabolism of pectin, *C. cellulovorans* uses pectin lyase (Clocel_2259) (EC 4.2.1.7), which showed elevated levels in our analysis. We speculate that this enzyme may require vitamin B12.

Within the pectin clusters, a Crp family transcriptional regulator (Clocel_2253) was identified upstream of the hydrolase/isomerase pathway, and levels of this regulator were significantly increased. Crp family transcriptional regulators are known to regulate gluconate-related genes in *Escherichia coli* (Peekhaus and Conway 1998), and we propose that this protein similarly regulates the hydrolase/isomerase pathway in *C. cellulovorans*. Upstream of *Clocel_2263*, an IclR family transcriptional regulator encoded genes (*Clocel_2264*) was annotated (Figure 5). In our quantitative proteome analysis, we did not identify this protein, but in the qualitative proteome analysis, this protein was identified and displayed higher peptide spectrum matches (PSMs) in pectin than in xylan and galactomannan (Additional file 2: Table S2). The KdgR protein regulates the degradation of pectin in *Erwinia chrysanthemi*, and is in the IclR family transcriptional regulator group (Molina-Henares et al. 2006). BLAST analysis indicates that Clocel_2264 has similarity with the KdgR protein (e = 5E-48). Based on these findings, we proposed that IclR family transcriptional regulators control the expression of the lyase/5-dehydro-4-deoxy-gluconate pathway.

Based on the metabolism-related protein profiles and cluster analysis performed in the present study, we propose molecular models for substrate recognition of *C. cellulovorans* in hemicellulose (xylan and galactomannan) and pectin (Figure 6). For hemicellulose, polysaccharides (xylan and galactomannan) are first degraded to oligosaccharides (xylooligosaccharides and mannooligosaccharides) in the extracellular environment, and these oligosaccharides are recognized by extracellular solute binding proteins (Clocel_2594 in Additional file 2: Table S2; Clocel_3201 in Table 1). Oligosaccharide-binding proteins then transduce signals to integral membrane histidine kinases (Clocel_2593 and 3203 in Additional file 2: Table S2). Next, intracellular transcriptional regulators (Clocel_2592 in Table 1; Clocel_3202 in Additional file 2: Table S2) are phosphorylated by the activation of integral membrane histidine kinases. Finally, genes related to the degradation and metabolism of hemicellulose are upregulated, allowing *C. cellulovorans* to metabolize substrates more efficiently (Figure 6a).

For pectin, two regulators (Clocel_2253 in Table 1; Clocel_2264 in Additional file 2: Table S2) work in substrate recognition. We predict that oligosaccharides derived from pectin or other pectin metabolites bind to these regulators allowing oligosaccharides to be transported into the cell (Figure 6b). This is a similar system to the pectin-recognition system in *E. chrysathemi* (Molina-Henares et al. 2006), but *E. chrysathemi* does not have two regulators operating hydrolase/isomerase and lyase/5-dehydro-4-deoxy-gluconate pathways.

The xylan recognition system of *C. cellulovorans* is very similar to that of *C. cellulolyticum*, but based on previous research (Xu et al. 2013), the composition of clusters is clearly different. In the *C. cellulolyticum* cluster, metabolism-related enzymes were not included. We therefore predict that these genes may not be regulated by TCS (Xu et al. 2013). However, the *C. cellulovorans* cluster includes metabolism- and transport-related genes. Thus, we speculate that the recognition system for xylan in *C. cellulovorans* regulates both the production of

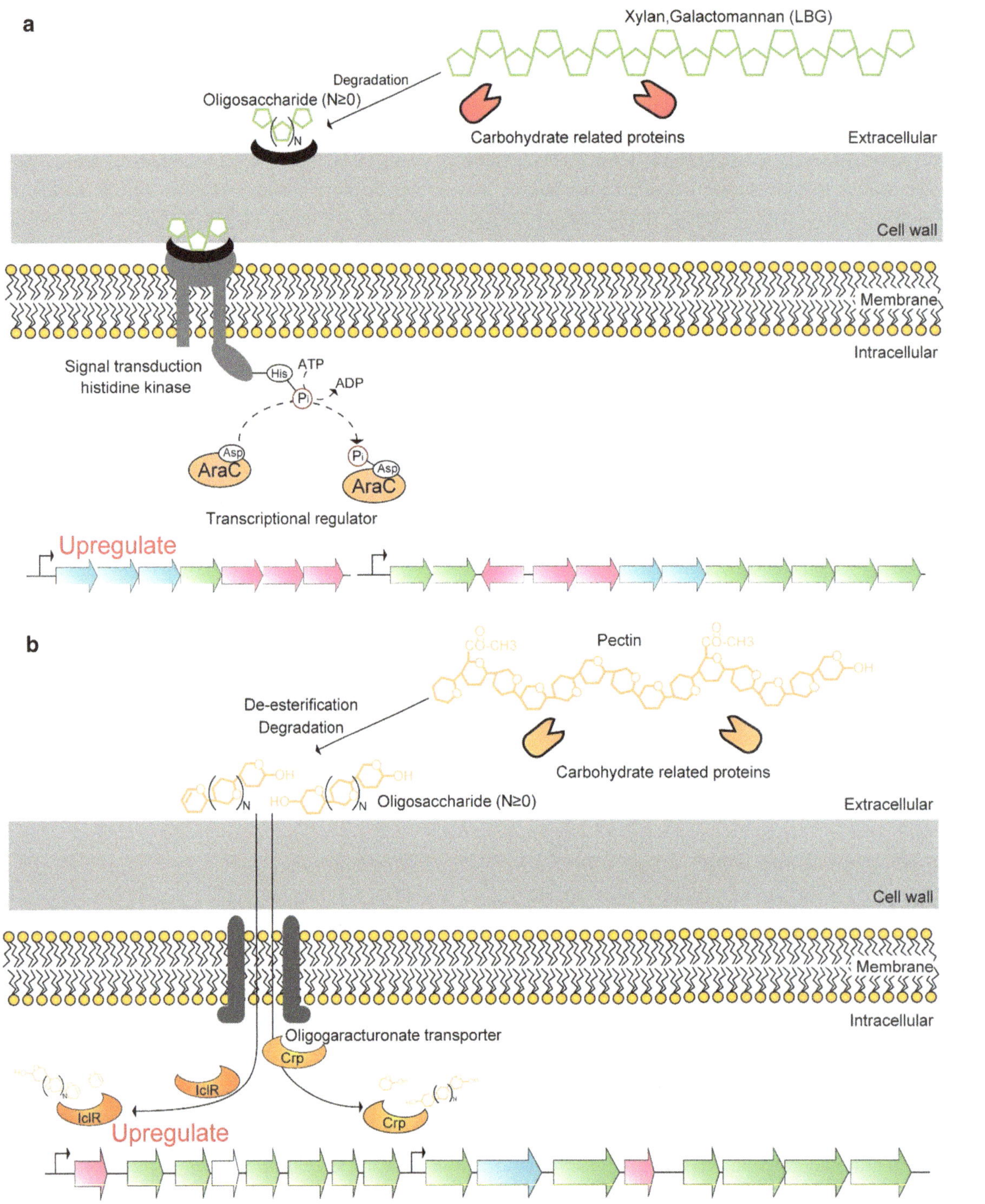

Figure 6 Proposed substrate recognition systems of hemicellulose and pectin in *C. cellulovorans*. For hemicellulose, polysaccharides are degraded to derived oligosaccharides outside of the cell, and extracellular solute-binding proteins bind these substrates. Solute-binding proteins induce signal-to-signal transduction integral membrane histidine kinases. Next the activated kinases phosphorylate transcriptional regulator AraC, and target genes (shown in Figure 5) are upregulated (**a**). For pectin, polysaccharides are de-esterified by pectinesterase and degraded by pectin lyase. Next, derived oligosaccharides are transported into cell, and these or other metabolites bind to a transcriptional regulator. Then, genes belonging to the target cluster (shown in Figure 5) are upregulated (**b**).

degradation- and transport-related enzymes and metabolism-related enzymes.

Substrate recognition systems in *C. cellulovorans* are one of the survival strategies. In other bacteria, monosaccharides such as glucose and xylose are mainly transported, but this bacterium recognizes oligosaccharides of hemicellulose and pectin, and can optimize degradation-, metabolism-, and transport-related proteins' profiles according to the substrate faster than other bacteria. This feature maybe suitable for competing with other anaerobic bacteria.

In conclusion, we performed "intracellular" quantitative proteome analysis of *C. cellulovorans* with glucose, xylan, galactomannan, and pectin and identified 734 proteins with LC–MS/MS equipped long monolithic silica capillary columns. We focused on substrate-recognition-, degradation-, and metabolism-related proteins, and carried out cluster analysis based on genome analysis together with the intracellular proteome analysis and the previous "exoproteome" analysis (Matsui et al. 2013; Esaka et al. 2015). Based on our results, we propose a mechanism for the recognition of hemicellulose and pectin in *C. cellulovorans*. For hemicellulose recognition, *C. cellulovorans* uses TCS and regulates all substrate-related genes with a single regulator. In the pectin recognition system, *C. cellulovorans* regulates hydrolase/isomerase and lyase/5-dehydro-4-deoxy-gluconate pathways with two regulators, Clocel_2253 and Clocel_2264, respectively. Judging from the presence of an oligogalacturonate transporter in the pectin regulatory clusters, the regulation of pectin metabolism-related genes appears to be controlled by the oligogalacturonates derived from pectin, or other metabolic products. This system has not been described for other cellulosome-producing Clostridia. Due to the multiplicity of regulator proteins involving substrate recognition, we suggest that *C. cellulovorans* has more distinct promoter regions that respond to each substrate. This information should be helpful for the elucidation of the environmental sensing strategy of *C. cellulovorans*.

Additional files

Additional file 1: Table S1. The 734 proteins identified in quantitative proteome analysis. Proteome analytes were injected to LC–MS/MS system. Obtained data were used for protein identification with Proteome Discoverer software 1.4. Three independent biological experiments were carried out, and proteins with no missing values were accepted as identified protein. As a result, 734 proteins were successfully identified.

Additional file 2: Table S2. All proteins identified in the qualitative proteome analysis. Supplemental qualitative proteome analysis was carried out to strengthen the quantitative proteome analysis. Three independent biological experiments were carried out, and proteins with MASCOT scores >10 were accepted as identified protein. Score: MASCOT score; PSMs: Peptides spectrum matches.

Abbreviations

ATCC: American type culture collection; CID: collision induced dissociation; EC: enzyme commission number; ESI: electrospray ionization; FDR: false discovery rate; GH: glycoside hydrolases; HCA: hierarchical cluster analysis; HCD: higher-energy c-trap dissociation; HPLC: high performance liquid chromatography; ID: internal diameter; KEGG: Kyoto encyclopedia of genes and genomes; LBG: locust bean gum; LC–MS: liquid chromatography–mass spectrometry; NCBI: National Center for Biotechnology Information; PBS: phosphate buffered saline; PL: polysaccharide lyase; SDS: sodium dodecyl sulfate; TCS: two component system; TEAB: triethyl ammonium bicarbonate; TMT: tandem mass tag

Authors' contributions

SA and KE generated the strains used. SA, KE and HM performed most of the mass measurement. SA, KE and HM performed most of the statistical analyses. SA, KE, HM, KK and MU designed the study and drafted the manuscript. All authors read and approved the final manuscript.

Author details

[1]Division of Applied Life Sciences, Graduate School of Agriculture, Kyoto University, Sakyo-ku, Kyoto, Japan. [2]Kyoto Integrated Science and Technology Bio-Analysis Center, Shimogyo-ku, Kyoto, Japan.

Acknowledgements

This research was supported by JST, CREST. This work was also partially supported by the Program for Promotion of Basic and Applied Researches for Innovations in Bio-oriented Industry (BRAIN).

Compliance with ethical guidelines

Competing interest

The authors declare that they have no competing interest.

References

Aoki W, Ueda T, Tatsukami Y, Kitahara N, Morisaka H, Kuroda K et al (2013) Time-course proteomic profile of *Candida albicans* during adaptation to a fetal serum. Pathog Dis 67(1):67–75

Cantarel BL, Coutinho PM, Rancurel C, Bernard T, Lombard V, Henrissat B (2009) The carbohydrate-active enzymes database (CAZy): an expert resource for glycogenomics. Nucleic Acids Res 37(suppl 1):D233–D238

Celik H, Blouzard J-C, Voigt B, Becher D, Trotter V, Fierobe H-P et al (2013) A two-component system (XydS/R) controls the expression of genes encoding CBM6-containing proteins in response to straw in *Clostridium cellulolyticum*. PLoS One 8(2):e56063

Collins M, Lawson P, Willems A, Cordoba J, Fernandez-Garayzabal J, Garcia P et al (1994) The phylogeny of the genus Clostridium: proposal of five new genera and eleven new species combinations. Int J Syst Bacteriol 44(4):812–826

Cosgrove DJ (2005) Growth of the plant cell wall. Nat Rev Mol Cell Biol 6(11):850–861

de Hoon MJ, Imoto S, Nolan J, Miyano S (2004) Open source clustering software. Bioinformatics 20(9):1453–1454

Esaka K, Aburaya S, Morisaka H, Kuroda K, Ueda M (2015) Exoproteome analysis of *Clostridium cellulovorans* in natural soft-biomass degradation. AMB Express 5(1):1–8

Higashide W, Li Y, Yang Y, Liao JC (2011) Metabolic engineering of *Clostridium cellulolyticum* for production of isobutanol from cellulose. Appl Environ Microbiol 77(8):2727–2733

Hugouvieux-Cotte-Pattat N, Condemine G, Nasser W, Reverchon S (1996) Regulation of pectinolysis in *Erwinia chrysanthemi*. Annu Rev Microbiol 50:213–257

Kanehisa M, Goto S (2000) KEGG: kyoto encyclopedia of genes and genomes. Nucleic Acids Res 28(1):27–30

Kawahara R, Saburi W, Odaka R, Taguchi H, Ito S, Mori H et al (2012) Metabolic mechanism of mannan in a ruminal bacterium, *Ruminococcus albus*, involving two mannoside phosphorylases and cellobiose

2-epimerase: discovery of a new carbohydrate phosphorylase, β-1,4-mannooligosaccharide phosphorylase. J Biol Chem 287(50):42389–42399
Lynd LR, Wyman CE, Gerngross TU (1999) Biocommodity engineering. Biotechnol Prog 15(5):777–793
Matsui K, Bae J, Esaka K, Morisaka H, Kuroda K, Ueda M (2013) Exoproteome profiles of *Clostridium cellulovorans* grown on various carbon sources. Appl Environ Microbiol 79(21):6576–6584
Molina-Henares AJ, Krell T, Guazzaroni ME, Segura A, Ramos JL (2006) Members of the IclR family of bacterial transcriptional regulators function as activators and/or repressors. FEMS Microbiol Rev 30(2):157–186
Morisaka H, Matsui K, Tatsukami Y, Kuroda K, Miyake H, Tamaru Y et al (2012) Profile of native cellulosomal proteins of *Clostridium cellulovorans* adapted to various carbon sources. AMB Express 2(1):1–5
Nataf Y, Bahari L, Kahel-Raifer H, Borovok I, Lamed R, Bayer EA et al (2010) *Clostridium thermocellum* cellulosomal genes are regulated by extracytoplasmic polysaccharides via alternative sigma factors. Proc Natl Acad Sci USA 107(43):18646–18651
Peekhaus N, Conway T (1998) Positive and negative transcriptional regulation of the *Escherichia coli* gluconate regulon gene gntT by GntR and the cyclic AMP (cAMP)-cAMP receptor protein complex. J Bacteriol 180(7):1777–1785
Petitdemange E, Caillet F, Giallo J, Gaudin C (1984) *Clostridium cellulolyticum* sp. nov., a cellulolytic, mesophilic: species from decayed grass. Int J Syst Bacteriol 34(2):155–159
Prawitwong P, Waeonukul R, Tachaapaikoon C, Pason P, Ratanakhanokchai K, Deng L et al (2013) Direct glucose production from lignocellulose using *Clostridium thermocellum* cultures supplemented with a thermostable β-glucosidase. Biotechnol Biofuels 6(1):184
Rainey FA, Stackebrandt E (1993) 16S rDNA analysis reveals phylogenetic diversity among the polysaccharolytic clostridia. FEMS Microbiol Lett 113(2):125–128
Raman B, Pan C, Hurst GB, Rodriguez M Jr, McKeown CK, Lankford PK et al (2009) Impact of pretreated switchgrass and biomass carbohydrates on *Clostridium thermocellum* ATCC 27405 cellulosome composition: a quantitative proteomic analysis. PLoS One 4(4):e5271
Richard P, Hilditch S (2009) D-Galacturonic acid catabolism in microorganisms and its biotechnological relevance. Appl Microbiol Biotechnol 82(4):597–604
Saha BC (2003) Hemicellulose bioconversion. J Ind Microbiol Biotechnol 30(5):279–291
Saldanha AJ (2004) Java Treeview—extensible visualization of microarray data. Bioinformatics 20(17):3246–3248
Sañudo-Wilhelmy SA, Gómez-Consarnau L, Suffridge C, Webb EA (2014) The role of B vitamins in marine biogeochemistry. Ann Rev Mar Sci 6:339–367
Senoura T, Ito S, Taguchi H, Higa M, Hamada S, Matsui H et al (2011) New microbial mannan catabolic pathway that involves a novel mannosylglucose phosphorylase. Biochem Biophys Res Commun 408(4):701–706
Sleat R, Mah RA, Robinson R (1984) Isolation and characterization of an anaerobic, cellulolytic bacterium, *Clostridium cellulovorans* sp. nov. Appl Environ Microbiol 48(1):88–93
Smyth GK (2004) Linear models and empirical bayes methods for assessing differential expression in microarray experiments. Stat Appl Genet Mol Biol 3(1)
Tamaru Y, Miyake H, Kuroda K, Nakanishi A, Kawade Y, Yamamoto K et al (2010) Genome sequence of the cellulosome-producing mesophilic organism *Clostridium cellulovorans* 743B. J Bacteriol 192(3):901–902
Xu C, Huang R, Teng L, Wang D, Hemme CL, Borovok I et al (2013) Structure and regulation of the cellulose degradome in *Clostridium cellulolyticum*. Biotechnol Biofuels 6(1):73

Biodegradation of variable-chain-length *n*-alkanes in *Rhodococcus opacus* R7 and the involvement of an alkane hydroxylase system in the metabolism

Jessica Zampolli[1,2], Elena Collina[2], Marina Lasagni[2] and Patrizia Di Gennaro[1*]

Abstract

Rhodococcus opacus R7 is a Gram-positive bacterium isolated from a polycyclic aromatic hydrocarbon contaminated soil for its versatile metabolism; indeed the strain is able to grow on naphthalene, *o*-xylene, and several long- and medium-chain *n*-alkanes. In this work we determined the degradation of *n*-alkanes in *Rhodococcus opacus* R7 in presence of *n*-dodecane (C12), *n*-hexadecane (C16), *n*-eicosane (C20), *n*-tetracosane (C24) and the metabolic pathway in presence of C12. The consumption rate of C12 was 88%, of C16 was 69%, of C20 was 51% and of C24 it was 78%. The decrement of the degradation rate seems to be correlated to the length of the aliphatic chain of these hydrocarbons. On the basis of the metabolic intermediates determined by the R7 growth on C12, our data indicated that *R. opacus* R7 metabolizes medium-chain *n*-alkanes by the primary alcohol formation. This represents a difference in comparison with other *Rhodococcus* strains, in which a mixture of the two alcohols was observed. By GC-MSD analysis we also identified the monocarboxylic acid, confirming the terminal oxidation.
Moreover, the *alkB* gene cluster from *R. opacus* R7 was isolated and its involvement in the *n*-alkane degradation system was investigated by the cloning of this genomic region into a shuttle-vector *E. coli-Rhodococcus* to evaluate the alkane hydroxylase activity. Our results showed an increased biodegradation of C12 in the recombinant strain *R. erythropolis* AP (pTipQT1-*alk*R7) in comparison with the wild type strain *R. erythropolis* AP. These data supported the involvement of the *alkB* gene cluster in the *n*-alkane degradation in the R7 strain.

Keywords: *Rhodococcus*; *n*-alkanes degradation; Alkane hydroxylase; AlkB; Enzymatic expression

Introduction

The problems associated with contaminated sites are assuming rising prominence in many countries (Vidali, 2001), due to the increase of anthropogenic pollution in the environment. Anthropic activities produce a wide amount of pollutants, such as aliphatic and aromatic polycyclic hydrocarbons; soil and water microorganisms can selectively degrade these xenobiotic compounds as the only carbon and energy source. The knowledge of the metabolic pathways of these microorganisms allows the reclamation of polluted sites. There are a lot of studies on the metabolism of these compounds in Gram-negative bacteria while Gram-positive bacteria have not been investigated to the same extent. Members of the *Rhodococcus* genus, found in many environmental niches, have a marked ability to metabolize a wide variety of xenobiotic compounds, and it is well recognized that species of this genus are key participants in the recycling of complex organic compounds (Finnerty, 1992; Larkin et al., 2005; Martínková et al., 2009). The genetic rearrangement and plasticity of the *Rhodococcus* genome have appeared to play a significant role in its adaptation to a wide variety of environmental contaminants (Leahy and Colwell, 1990; Van der Meer et al., 1992; Kulakov et al., 1998; Larkin et al., 2006; McLeod et al., 2006).

Alkane degraders are bacteria that have a very versatile metabolism, so that they can use as carbon source many other compounds in addition to alkanes (Smits et al., 2002; Margesin et al., 2003; Harayama et al., 2004). The ability of these bacteria of degrading *n*-alkanes is well

* Correspondence: patrizia.digennaro@unimib.it
[1]Department of Biotechnology and Biosciences, University of Milano-Bicocca, Piazza della Scienza 2, 20126 Milano, Italy
Full list of author information is available at the end of the article

established, but relatively little is known about the molecular characteristics of their alkane-degradative systems.

Alkanes are saturated, linear molecules whose chain length can vary from 1 (in methane) to more than 50 carbon atoms. They are the major components of petroleum fuels which can be commonly found in contaminated environments (So and Young, 1999), indeed they constitute about 20-50% of crude oil, depending on the source of the oil. In addition, alkanes (predominantly long-chain compounds) are produced throughout the biosphere by living organisms (plants, algae and bacteria) as a waste product, a structural element, a defense mechanism, or as a chemoattractant (van Beilen et al., 2003). The aerobic degradation of these molecules starts by oxidation of one of the terminal methyl groups to generate the corresponding primary alcohol by alkane hydroxylases (AHs). The oxidation can occur on various positions: terminal or subterminal, to final conversion to a fatty acid (van Beilen et al., 2003; Ji et al., 2013). So far some alkane-degradative systems of a small number of Gram-negative bacteria have been well characterized, such as those of *Alkanivorax, Pseudomonas* and *Acinetobacter* (Alonso and Roujeinikova, 2012; Ratajczak et al., 1998; van Beilen and Funhoff, 2007; Wang and Shao, 2013). The *alk* system found in *Pseudomonas putida* GPo1, which degrades *n*-alkanes from *n*-pentane to *n*-dodecane, remains the most extensively characterized alkane hydroxylase system. It is a three-component alkane hydroxylase complex consisting of a particulate nonheme integral membrane alkane monooxygenase (AlkB) and two soluble proteins, rubredoxin (AlkG) and rubredoxin reductase (AlkT) (Alonso and Roujeinikova, 2012).

Much less is known about the alkane-degradative systems of Gram-positive bacteria. Homologs of *alkB* gene were amplified from psychrotrophic *Rhodococcus* sp. Q15 (Whyte et al., 2002), *Gordonia* sp. strain SoCg (Lo Piccolo et al., 2011) and *Rhodococcus opacus* B4 (Sameshima et al., 2008).

The *Gordonia* sp. strain SoCg unique *alkB* gene was analyzed by functional heterologous expression in *E. coli* and in *S. coelicolor,* which were shown to oxidize *n*-hexadecane to the correspondent primary alcohol, 1-hexadecanol, but no expression occurred in presence of long-chain *n*-triacontane. Similarly, the *Rhodococcus opacus* B-4 *alkB1* and *alkB2* genes were expressed heterologously in two *E. coli* recombinants which were able to convert *n*-alkanes (*n*-pentane to *n*-hexadecane) to their corresponding alcohols in anhydrous organic solvents. The start codons of two genes were changed from GTG to ATG to express them in *E. coli. Rhodococcus* sp. strain Q15 was shown to be able to mineralize *n*-dodecane and *n*-hexadecane and bioconvert them into corresponding primary and secondary alcohols at 5°C. The utilization of potential metabolic intermediates indicated that Q15 oxidizes alkanes by both the terminal oxidation pathway and the subterminal oxidation pathway (Whyte et al., 1998).

Rhodococcus opacus R7 is a Gram-positive bacterium isolated from a polycyclic aromatic hydrocarbon contaminated soil for its ability to grow on naphthalene. It is characterized by a versatile metabolism, indeed the strain is able to grow on naphthalene, *o*-xylene, and several long- and medium-chain *n*-alkanes. In previous studies, the genes involved in the degradation of naphthalene and *o*-xylene were identified, and the related metabolic pathways were characterized (Di Gennaro et al., 2010). In this work we determined the degradation of *n*-alkanes in *Rhodococcus opacus* R7, performing biodegradation kinetics in presence of *n*-dodecane (C12), *n*-hexadecane (C16), *n*-eicosane (C20), *n*-tetracosane (C24), and the metabolic pathway in presence of C12. Moreover, the *alkB* gene cluster from *R. opacus* R7 was isolated and its involvement in *n*-alkane degradation system was investigated by cloning and expression of this genomic region.

Materials and methods

Bacterial strains, growth conditions and general procedures

Bacterial strains and plasmids used in this study are listed in Table 1. *Rhodococcus opacus* R7 (CIP 107348), isolated for its ability to grow on naphtalene and *o*-xlyene (Di Gennaro et al., 2001), was grown on M9 mineral medium (Maniatis et al., 1982), supplemented with naphthalene or *o*-xylene or different *n*-alkanes in an atmosphere saturated with these compounds, as the only carbon and energy source. Growth of R7 was performed in 100 ml-flasks with 20 ml of M9 mineral medium in presence of *n*-hexane (C6), *n*-octane (C8), *n*-decane (C10), *n*-dodecane (C12), *n*-hexadecane (C16), *n*-eicosane (C20), *n*-tetracosane (C24), *n*-hexatriacontane (C36) at the concentration of 1 g/l.

n-Eicosane (C20), *n*-tetracosane (C24), *n*-hexatriacontane (C36) were added in flasks as hexane solution evaporated over night or as finely ground powder.

Rhodococcus erythropolis AP, isolated in our laboratory (CIP 110799) for its ability to grow on diesel fluel (Maffei, 2004), was maintained on M9 mineral medium in a saturated atmosphere of diesel fuel at 30°C (Table 1). *Escherichia coli* DH5α was grown on Luria-Bertani medium at 37°C. To select *E. coli* and *Rhodococcus* trasformants antibiotics (ampicillin at 100 μg/ml for *E. coli* and tetracycline at 25 μg/ml for *Rhodococcus*) were added in the cultural medium.

Genomic DNA of *R. opacus* R7 was extracted from R7 cells as reported by Di Gennaro et al., 2010. DNA manipulation, enzymatic digests, ligation and PCR reactions were performed using standard molecular techniques (Sambrook and Russell, 1989). For recombinant plasmids extraction NucleoSpin Plasmid Kit by Machery and Nagel was used according to the manufacturer's instruction.

Table 1 Bacterial strains and plasmids used in this study

Strain or plasmid	Description	Reference or source
Strains		
Rhodococcus opacus R7	Long-medium-chain *n*-alkane degrader, *alkB*$^+$, *nar*$^+$, *gen*$^+$, *o-xyl*$^+$	Di Gennaro et al. 2001; Di Gennaro et al. 2010
Rhodococcus erythropolis AP	Diesel Fuel degrader	Maffei 2004
Escherichia coli DH5α	*dlacZ*ΔM15, *recA*1, *endA*1, *gyrA*96, *thi*-1, *hsd*R17(rK$^-$, mK$^+$), *supE*44, *relA*1, *deoR*, Δ(*lacZYA-argF*)U169	Promega
Escherichia coli DH5α (pDrive-*alk*R7)	*E. coli* DH5α containing the cloning vector pDrive, *alkB* fragment, Amp^r	This study
Escherichia coli DH5α (pTipQT1-*alk*R7)	*E. coli* DH5α containing the recombinant expression vector pTipQT1, *alkB* fragment, Thio^r, Amp^r	This study
Rhodococcus erythropolis AP (pTipQT1-*alk*R7)	*R. erythropolis* AP containing the recombinant expression vector pTipQT1, *alkB* fragment, Thio^r, Tc^r	This study
plasmids		
pDrive	*E. coli* cloning vector, Amp^r	Qiagen
pDrive-*alk*R7	pDrive containing *alkB* fragment from *R. opacus* R7	This study
pTipQT1	Shuttle-vector Amp^r *E. coli* – Tc^r *Rhodococcus* spp.	Nakashima and Tamura 2004b
pTipQT1-*alk*R7	pTipQT1 containing *alkB* fragment from *R. opacus* R7	This study

Growth curves and *n*-alkanes biodegradation

Growth experiments on R7 strain were performed in presence of *n*-alkanes (1 g/l) ranging from C6 to C36 in 20 ml of M9 with OD_{600} 0.1.

To determine *R. opacus* R7 *n*-alkanes degradation, the strain was inoculated in 20 ml of M9 with OD_{600} 0.1 and supplemented with 1 g/l of *n*-alkanes (C12, C16, C20, C24). To determine abiotic loss, uninoculated flasks were also prepared. The flasks were incubated at 30°C for a maximum of 72 h. Every 24 h a flask was sacrified to evaluate *n*-alkanes degradation by determination of the residual *n*-alkane by gas chromatography–mass spectrometry (GC-MSD) after extraction with *n*-hexane. Bacterial growth was determined by OD_{600} measurements. When the strain was inoculated on *n*-alkanes with more than 16 carbon atoms, the bacteria formed clumps which included *n*-alkanes (van Beilen et al., 2002). For optical density measurements, clumps were resuspended by strong agitation.

Chromatographic analyses

Residual *n*-alkanes were extracted from the inoculated and uninoculated flasks using *n*-hexane. It occurred in 20 min by strong manual agitation (in test tubes), after washing of the flasks to take away any residual *n*-alkanes on their wall. The suspension was centrifuged at 4000 rpm for 10 min and after 10 min of settling, 2 ml were drawn from organic phase. It was conserved at −20°C in vial with teflon-coated screw caps. The organic phase extracted was diluted 1:100 for GC injection. Residual *n*-alkanes were analysed by GC-MSD using a Technologies 6890 N Network GC System, interfaced with 5973 Network Mass Selective Detector (MSD) (Agilent Technologies). A ZB-5MS capillary column was used (5% diphenyl-95% dimethylpolysiloxane 60 m x 0.25 mm, 0.25 μm; Alltech).

Analyses were carried out in split-less injection mode using helium as carrier gas at 99.99%. The injector port was set at 250°C. The oven temperature was programmed from 60°C for 3 min, then 15°C min^{-1} to 280°C, holding this temperature for 10 min. Electron impact ionization spectra were obtained at 70 eV, with recording of mass spectra from *m*/*z* 42 to 550 amu, which allows 3.5 scans s^{-1}. All analyses were carried out with three replicates, and the mean values obtained are reported.

Analysis of the metabolic intermediates from the *n*-alkane pathway

The metabolic intermediates resulting from incubation of *R. opacus* R7 on C12 were analyzed by GC-MSD after extraction with ethyl acetate. It occurred in 20 min by strong manual agitation (in test tubes) in presence of 1.5 ml HCl 1 M. The suspension was centrifuged at 4000 rpm for 10 min. After centrifugation, 1.5 ml of samples were taken from organic phase and transferred into 1.5 ml vials. For derivatization the solution was stripped under a gentle stream of nitrogen, 10 μL of derivatizing agent TMSI and Pyridine (Supelco) and hexane up to 1 ml were added. The solutions were heated at 70°C for 30 min. The derivatized samples were analyzed by GC-MSD and the extracted samples were conserved at − 20°C in vial with teflon-coated screw caps. The registered mass spectra were compared with those of the spectra library (NIST) of the instrument, and the identification of the metabolites was confirmed by injection and analysis of the corresponding standard compound.

Kinetic modeling

The residual *n*-alkane concentrations as a function of time were modeled according to the Monod-type kinetic models as reported by Simkins and Alexander (1984). In particular, considering the substrate scarce water solubility and the biomass initial concentration (growth *viz.* exposition runs), the logistic and first-order kinetic models were applied. For logistic model, the maximum specific growth rate, μ_{max} (h^{-1}), was estimated from the microbial growth curve, plotting the natural logarithm of OD_{600} vs time and determining the slope of the linear part of the graph (Zhukov et al., 2007). The inverse yield parameter, q (ppm OD_{600}^{-1}) (Simkins and Alexander, 1984), was estimated as the ratio of the degraded *n*-alkane to OD_{600} increase. A non-linear regression was used for the fitting of the logistic model to the experimental data. The goodness of fit was evaluated on the basis of the determination coefficient, the standard error of the estimated parameter and the p-value.

For first-order kinetic model, the natural logarithm of substrate concentration *vs* time was plotted and the slope of the least-squares-regression line was determined.

Identification and cloning of the *alkB* gene cluster

The isolation of the complete coding region of the *alkB* gene cluster from *R. opacus* R7 was obtained from purified genomic DNA of the strain by PCR-amplification.

To identify the genomic region, two primers (F1, 5′-AAGGCCATGGGGCGTTAGAGCACCGCAGCTAAT-3′ and R1, 5′-ACAGCATATGACCTAGCGGGCGGCCGCGACCCG-3′) were designed on the basis of the more conserved sequences between the *alkB* gene cluster of different bacteria belonging to the *Rhodococcus* genus deposited in Genbank DataBase. It was carried out using the following program: 95°C for 3 min; 95°C for 30 sec, 68°C for 45 sec, 72°C for 5 min, for 35 cycles; and 72°C for 3 min.

DNA fragments were purified from agarose gel by the NucleoSpin Extraction II Kits by Machery and Nagel. The eluted fragment was sequenced by automated sequencing (Eurofins MWG). The nucleotide sequence of the *alkB* gene cluster of *R. opacus* R7 was deposited in the GenBank DataBase (KJ573524).

The 3.0 Kb fragment, containing the *alkB* gene cluster, was then amplified by PCR using the primers *Nco*I-*alk*R7 5′-AAGGCCATGGACGTGACGACGTCGGATATC-3′ and *Nde*I-*alk*R7 5′-ACAGCATATGACCTAGCGGGCGGCCGCGAC-3′ in order to generate the *Nco*I-*Nde*I ends. The 3.0 Kb fragment was cloned as PCR product into the pDrive vector (Qiagen). The ligation mixture was used to transform *E. coli* DH5α by electroporation and the recombinant clones were selected on LB agar supplemented with ampicillin 100 μg/ml and IPTG 1 mM and 5-bromo-4-chloro-3-indolyl-β-D galactopyranoside (X-Gal) 40 μg/ml. White colonies were isolated and plasmid (pDrive-*alk*R7) of the recombinant clones was extracted and verified with digestion *Nco*I/*Nde*I.

Construction of the recombinant strain *R. erythropolis* AP (pTipQT1-*alk*R7) and activity of the *alkB* system

The *alkB* insert was ligated as *Nco*I-*Nde*I fragment into a shuttle-vector *E. coli-Rhodococcus*, pTipQT1 (Nakashima and Tamura 2004a). The ligation mixture was used to transform *E. coli* DH5α by electroporation with standard procedures (Sambrook and Russell, 1989) and the recombinant clones were selected on LB agar supplemented with ampicillin (100 μg/ml) at 37°C. Ampicillin-resistant clones were selected and the recombinant plasmid (pTipQT1-*alk*R7) was isolated. The same recombinant plasmid was used as shuttle-vector to transform others *Rhodococcus* spp. strains by electroporation. We chose as host *Rhodococcus erythropolis* AP selected in our laboratory (Maffei, 2004) because the expression of the *Ptip* system efficiency was more performant in *erythropolis* species (Nakashima and Tamura, 2004b). Plasmids were introduced into strains of *Rhodococcus* spp. by electroporation using a Gene Pulser II (Biorad, Italia) set at 2.50 kV, 600 Ω, 25 μF (Treadway et al., 1999) in presence of about 1 μg DNA. Immediately after electroporation, 2.5 ml recovery broth (LB medium with 1.8% sucrose) were added and cells were incubated at 30°C for 4 h. Cells were plated on LB supplemented with tetracycline 25 μg/ml and grown at 30°C for 3–4 days. The stability of the recombinant plasmid in the *R. erythropolis* AP was verified after extraction and restriction analysis of the plasmid from all *R. erythropolis* AP (pTipQT1-*alk*R7) clones. All the clones were able to maintain the recombinant plasmid.

Moreover, before biodegradation experiments, a representative amount of colonies of the recombinant strain underwent the same procedure and we verified the maintaining of the plasmid in the following generations.

Recombinant strain *R. erythropolis* AP (pTipQT1-*alk*R7) was used for biodegradation experiments in presence of *n*-alkane C12 to evaluate the activity of the *alkB* system.

Chemicals

All chemicals used were of analytical grade. All organic solvents used were high-performance liquid chromatographic (HPLC) grade supplied from Fluka. Naphtalene, *o*-xylene and *n*–alkanes, 1-dodecanol, 2-dodecanol, dodecanoic acid were supplied from Sigma.

Results

Growth of *R. opacus* R7 on medium- and long-chain *n*-alkanes

Rhodococcus opacus R7 was isolated from a soil contaminated by polycyclic aromatic hydrocarbons, and the strain was able to grow on naphthalene, *o*-xylene (Di

Gennaro et al., 2001; Di Gennaro et al., 2010) as well as on various medium- and long-chain *n*-alkanes.

The metabolism on *n*-alkanes was investigated in presence of medium- and long-chain *n*-alkanes. Growth of R7 in M9 mineral broth in presence of *n*-hexane (C6), *n*-octane (C8), *n*-decane (C10), *n*-dodecane (C12), *n*-hexadecane (C16), *n*-eicosane (C20), *n*-tetracosane (C24), *n*-hexatriacontane (C36) (1 g/l) supplied as the only carbon and energy source was tested in 96 h. An increasing biomass accumulation was observed by spectrophotometric analysis at OD_{600} every 24 h. *R. opacus* R7 grew with *n*-alkanes ranging in length from C10 to C36, but no growth was observed in the range of C6-C8 *n*-alkanes (Table 2).

Although R7 growth on *n*-alkanes ranging in length from C20 to C36 was difficult to determine, due to the low solubility of solid *n*-alkanes in water, a certain growth was observed around the solid *n*-alkane agglomerates formed and on the wall of the flasks. Before the growth measurement, a strong agitation of the flasks was necessary to observe a homogeneous cells suspension (van Beilen et al., 2002).

The capacity of *R. opacus* R7 to degrade *n*-alkanes ranging in length from C12 to C24 was investigated as a function of time.

Biodegradation kinetics of medium- and long-chain *n*-alkanes C12-C24

To investigate the capacity of *R. opacus* R7 to degrade *n*-alkanes, kinetic runs were performed in M9 mineral medium supplemented with C12, C16, C20, C24 as the sole carbon and energy source; growth was followed at different times up to 72 h (Figure 1, a, b, c, d). The increase in biomass was quite the same for the four substrates, indeed OD_{600} reached about 0.5-0.7. GC-MSD analyses of *n*-alkane residues were performed after extraction from broths; the results showed that substrate consumption occurred parallel to biomass increase.

The consumption rate of C12 was 88%, of C16 was 69%, of C20 was 51% and of C24 it was 78%. The consumption

Table 2 Degradation of *n*-alkanes by *Rhodococcus opacus* R7

n-Alkane	Carbon atoms	*Rhodococcus opacus* R7
Hexane	6	-
Octane	8	-
Decane	10	+
Dodecane	12	+
Hexadecane	16	+
Eicosane	20	+
Tetracosane	24	+
Hexatriacontane	36	+

-, no growth; +, growth.

of C12 was higher than longer *n*-alkanes: the decrement of degradation rate of these hydrocarbons seems to be correlated to the length of the aliphatic chain.

Abiotic controls were performed in presence of each hydrocarbon. The amount of abiotic loss was about 20% for C12, about 24% for C16, about 18% for C20, and about 26% for C24, respectively.

n-Alkane residual concentrations as a function of time followed a logistic curve, typical of a biodegradation process where a scarcely soluble organic compound is degraded by a growing microbial culture. From the experimental data treatment, values of the maximum specific growth rate, μ_{max}, and of the semisaturation constant, K_S, were estimated (Table 3). As expected, K_S increased from C12 to C24, indicating a decrease in the biodisponibility of the *n*-alkane for the biomass. On the contrary, μ_{max} was almost constant for C12, C16 and C20 and of the order of 0.015 h^{-1}, while it approached 0.032 h^{-1} for C24. This last value could indicate that more enzymatic systems were involved in the degradation of C24 or that a different up-take mechanism could be hypothesized.

Metabolic intermediates of *n*-alkane C12 degradation

Metabolic intermediates of *n*-alkanes in *R. opacus* R7 were determined from cultural broths during the growth on *n*-dodecane as reference substrate, chosen on the basis of the kinetic parameters determined for biodegradation runs. R7 growth on C12 was followed during the exposition to the hydrocarbon as a function of time. At time 0 h, 3 h, 5 h, 8 h and 24 h, culture samples were acidified, extracted, and analyzed by GC-MSD as reported in Material and Methods. Cultural broth analysis showed the presence of 1-dodecanol and the corresponding carboxylic acid (Figure 2) which were identified by comparison with the relative standards. We identified the 1-dodecanol only and not the mixture of 1-dodecanol and 2-dodecanol, as reported in literature for many cases of *Rhodococcus* strains (Whyte et al., 1998; Whyte et al., 2002). To confirm these data, both 1-dodecanol and 2-dodecanol were supplied separately as carbon and energy source and growth was observed only in presence of 1-dodecanol and no growth was observed in presence of 2-dodecanol. Moreover, we observed growth when the dodecanoic acid was supplied as the only carbon and energy source.

Identification and sequencing of the *R. opacus* R7 *alkB* gene cluster involved in *n*-alkanes degradation

The *alkB* gene cluster from *R. opacus* R7 genomic DNA was identified (Figure 3) and sequenced. This region was isolated by PCR with primers designed on the basis of the alignment of conserved sequences from the *alkB* gene region from different *n*-alkane degrading *Rhodococcus*

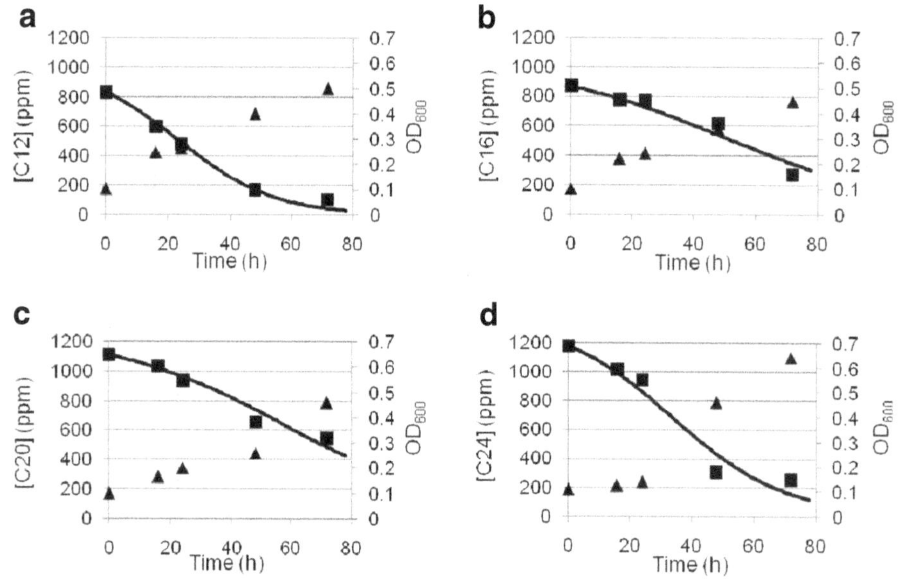

Figure 1 Kinetic analyses of *n*-dodecane (C12) (a), *n*-hexadecane (C16) (b), *n*-eicosane (C20) (c), *n*-tetracosane (C24) (d) degradation in *R. opacus* R7. Cells of *R. opacus* R7 were exposed to the *n*-alkane after growth on naphthalene and a flask each day was sacrified for the extraction and the GC-MSD determination of the residual hydrocarbon.

bacteria. The genomic 3.0 Kb fragment was isolated and sequenced. From the sequence analysis, the covering region contained four consecutive ORFs homologues to the *alkB* gene cluster components: *alkB* encoding for an alkane monooxygenase, *rubA* encoding for a rubredoxin, *rubB* encoding for a second rubredoxin, and *rubred* encoding for a rubredoxin reductase.

These sequences isolated from R7 strain were compared with the ones of other bacteria belonging to the *Rhodococcus* genus. In comparison with *Rhodococcus opacus* B4 (Sameshima et al., 2008), benzene-tolerant capable of growing on *n*-alkanes, *alkB* gene showed an identity of 94% (the corresponding protein AlkB 91%), *rubA, rubB*, and *rubred* an identity of 90%, 95% and 86% (RubA 91%, RubB 97%, RubRed 85%), respectively. In comparison with *Rhodococcus* sp. BCP1 (Cappelletti et al., 2011), capable of growing on volatile and medium-chain *n*-alkanes, *alkB* gene presented an identity of 83% (AlkB 80%), *rubA, rubB* and *rubred* an identity of 86%, 84% and 69% (RubA 79%, RubB 84%, RubRed 60%), respectively. Compared to *Rhodococcus jostii* RAH1 (McLeod et al., 2006), able to degrade polychlorinated biphenyls (PCBs), *alkB* gene presented an identity of 94% (AlkB 91%), *rubA, rubB* and *rubred* an identity of 99%, 94% and 91% (RubA 98%, RubB 97%, RubRed 91%), respectively.

Table 3 Biodegradation kinetic parameters for *n*-alkanes metabolism in *R. opacus* R7

Substrate consumption			Cellular growth		
Substrate	μ_{max}/K_S (1/h ppm)	R^2	μ_{max} (h^{-1})	R^2	K_S (ppm)
C12	$(6.3 \pm 0.3)\ 10^{-5}$	0.988	0.013 ± 0.002	0.972	210
C16	$(3.1 \pm 0.2)\ 10^{-5}$	0.938	0.0129 ± 0.0007	0.997	420
C20	$(2.7 \pm 0.2)\ 10^{-5}$	0.952	0.017 ± 0.004	0.956	650
C24	$(3.9 \pm 0.4)\ 10^{-5}$	0.952	0.032 ± 0.009	0.905	810

Construction of the recombinant strain *R. erythropolis* AP (pTipQT1-*alk*R7) and activity of the *alkB* system

The *alkB* gene cluster was isolated from R7 genomic DNA as PCR product and cloned into the pDrive vector giving the plasmid pDrive-*alk*R7. The region was isolated as *NcoI*/*NdeI* fragment and cloned into the shuttle-vector *E. coli-Rhodococcus* pTipQT1. The recombinant plasmid pTipQT1-*alk*R7 was isolated from *E. coli* DH5a and transferred into *Rhodococcus erythropolis* AP, because the *Ptip/regulator* system of the shuttle-vector is more efficient in *Rhodococcus erythropolis* species (Nakashima and Tamura 2004b). In order to verify the expression of the *alkB* gene cluster under the control of the *Ptip/regulator* system, experiments with resting cells of *Rhodococcus erytropolis* AP(pTipQT1-*alk*R7) exposed to C12 were performed. The expression of the *alk* region was determined comparing the biodegradation kinetics of the recombinant strain *R. erythropolis* AP (pTipQT1-*alk*R7) with the wild type strain *R. erythropolis* AP. Results are reported in Figure 4. The percentage of biodegradation in 6 h was near 80% in the recombinant strain and 37% in the wild type strain. From kinetics analysis we can observe that the initial

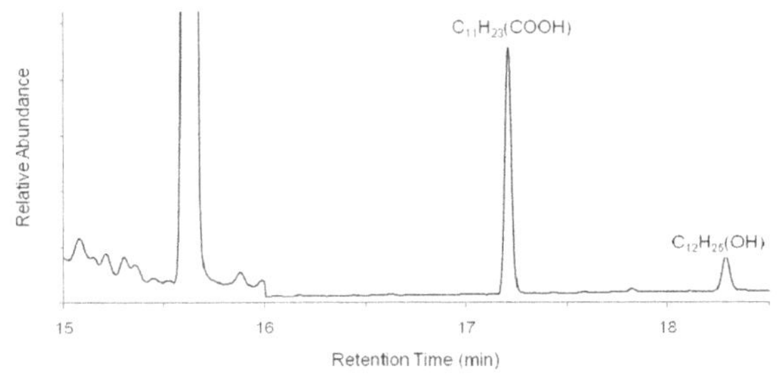

Figure 2 Cultural broth analysis by GC-MSD. Cells of *R. opacus* R7 were exposed to *n*-dodecane and at different times, the cultural broth were analysed in GC-MSD. The intermediate metabolites identified are reported in the graph.

degradation rate was higher in the recombinant strain with respect to the wild type strain indicating a difference in the activity levels of the *alkB* system. This difference was confirmed by a statistical test for the comparison of the slopes of the regression lines at 95% significance level.

Discussion

In this paper *Rhodococcus opacus* R7, a strain with a versatile metabolism, was characterized for the ability to degrade variable-chain-length *n*-alkanes.

Many alkane-degrading bacteria have been isolated and the enzyme systems that oxidize *n*-alkanes up to C16 have been characterized (Rojo, 2009; van Beilen and Funhoff, 2007, Wentzel et al., 2007). Long-chain *n*-alkanes are more persistent in the environment than the shorter but few data are available in particular concerning the metabolism of these compounds in Gram-positive bacteria (Whyte et al., 2002; Smits et al., 2002; Lo Piccolo et al., 2011).

R. opacus R7 showed the ability to grow on medium and long chain *n*-alkanes ranging from C10 to C36. Biodegradation kinetics suggested that R7 strain degraded the *n*-alkanes with a lower number of carbon atoms to a greater extent (88% C12, 69% C16, 51% C20, 78% C24). Moreover, μ_{max} was almost constant for C12, C16 and C20 and about 0.015 h^{-1}, while it approached 0.032 h^{-1} for C24. To our knowledge, no literature data are available concerning specific growth rates on single *n*-alkanes, but only Zhukov et al. reported data in presence of a mixture of *n*-alkanes (Zhukov et al., 2007). The decrement of degradation percentage of these hydrocarbons seems to be inversely correlated to the length of the aliphatic chain as already reported (van Beilen et al., 2003; Lo Piccolo et al., 2011). An exception was C24; in fact, when growing on this hydrocarbon, R7 strain showed a higher increase in OD_{600}, although the degradation percentage was similar to that observed for C12, and a higher μ_{max}. These data suggested that other enzymatic systems could be involved in C24 degradation. Another hypothesis could be a different up-take mechanism, supported by the experimental observation of a massive adhesion of R7 cells to C24 powder (van Beilen et al., 2003; Rapp et al., 2003; Lo Piccolo et al., 2011).

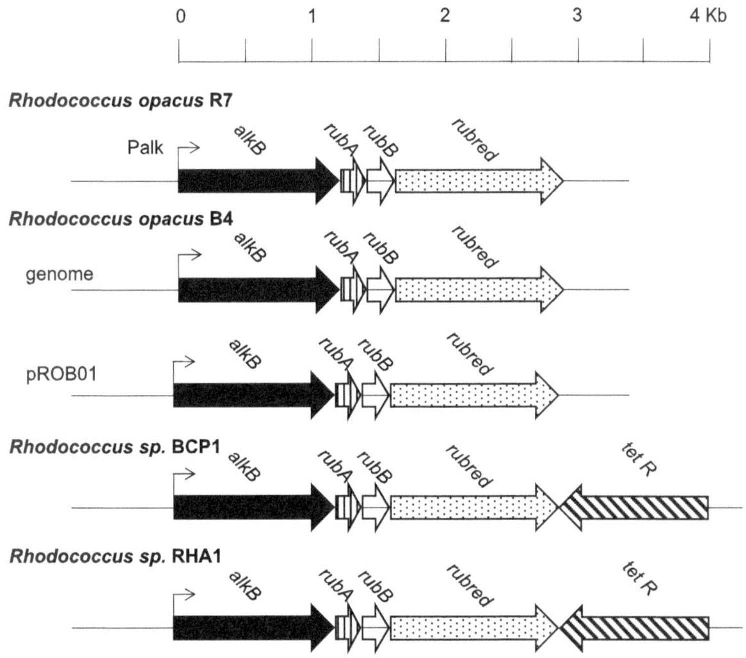

Figure 3 **Gene organization of the *alkB* gene cluster of *R. opacus* R7 and comparison with equivalent clusters from other alkane-degrading *Rhodococcus* bacteria.**

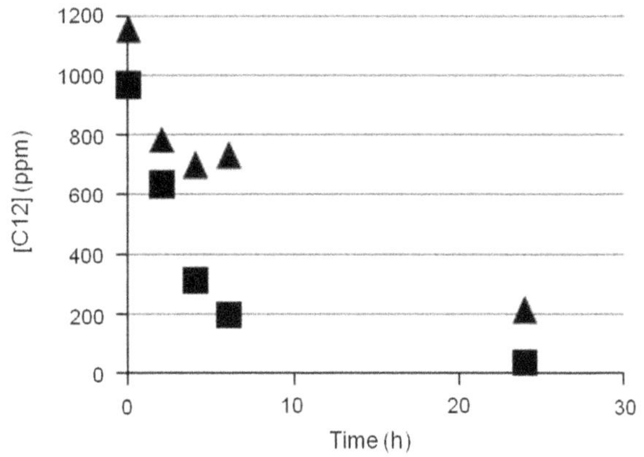

Figure 4 Kinetic analyses of *n*-dodecane degradation in *R. erytropolis* AP (pTipQT1-*alk*R7) recombinant strain (square in the graph). Cells of the recombinant strain were grown on rich medium and then, after induction with thiostrepton, were collected and exposed to *n*-dodecane. For comparison, kinetics of *n*-dodecane degradation in *R. erytropolis* AP (triangle in the graph) without the cloned fragment expressing the *alk*B gene, is also showed.

Many strains from different microbial genera are able to grow on *n*-alkanes. Alkanes are usually activated by a terminal oxidation to the corresponding primary alcohol, which is further oxidized by the alcohol and the aldehyde dehydrogenases to the corresponding acid. The resulting fatty acid enters into the beta-oxidation cycle. In some cases *n*-alkanes are metabolized via terminal as well as subterminal oxidation to the corresponding secondary alcohol. The secondary alcohol is converted to the corresponding ketone which is oxidized by a monooxygenase to an ester that is successively hydrolyzed by an esterase to an alcohol and a fatty acid (van Beilen et al., 2003; Ji et al., 2013). On the basis of the metabolic intermediates determined by the R7 growth on C12, our data indicated that *R. opacus* R7 metabolizes medium-chain *n*-alkanes by the primary alcohol formation because no trace of the secondary alcohol was found and because the strain is not able to grow on this compound. This result highlights a difference in comparison with many *Rhodococcus* strains in which a mixture of the two alcohols was observed (Whyte et al., 1998 and references therein). In the GC-MSD analysis we also identified the monocarboxylic acid, confirming therefore the activation by a terminal oxidation of the *n*-alkane as reported for *Rhodococcus* sp. MS11 by Rapp et al., 2003.

Moreover, the *alkB* gene cluster from *R. opacus* R7 was isolated and its involvement in the *n*-alkane degradation was studied. In this paper we identified the *alkB* gene cluster including an *alkB* gene, encoding for an alkane hydroxylase, two rubredoxins A and B, and a gene encoding for a rubredoxin reductase that are all required in the catalytic process as electron transfer proteins. Phylogenetic analysis of R7 *alkB* gene cluster showed a significant similarity with the homologous system of *R. opacus* B4 (Sameshima et al., 2008) and a lower identity with *Rhodococcus* sp. BCP1 (Cappelletti et al., 2011).

Although four types of *n*-alkane aerobic degradation pathways have been identified to date, the number of alkane hydroxylases that have been isolated, characterized and analyzed by structural biology techniques remains limited. The most widely characterized alkane degradation system is the AlkB of *Pseudomonas putida* GPo1 (Smits et al., 2002; van Beilen et al., 2003). Researchers have also isolated and cloned novel genes encoding AlkB from bacteria belonging to *Rhodococcus* genus, such as *Rhodococcus opacus* B4 (Sameshima et al., 2008) or *Rhodococcus* sp. Q15 (Whyte et al., 2002), but more information in these bacteria needs to be addressed. Literature data report the involvement of the alkane monoxygenase in the *n*-alkane biodegradation demonstrated by the changing of the GTG coding sequence for the expression in *E. coli* of this enzyme (Sameshima et al., 2008). Other authors were unable to show function heterologous expression of *alkB* genes in *Pseudomonas* or *E. coli* expression system, principally because the functional expression requires proper synthesis, correct folding, and proper assembly, and these are not always ensured for rhodococcal proteins (Whyte et al., 2002). On these bases, the *alkB* region identified from *R. opacus* R7 was cloned for the first time into a shuttle-vector *E. coli-Rhodococcus* and the alkane hydroxylase activity was evaluated in *R. erythropolis* AP. This strain was chosen not only because it belongs to the *Rhodococcus* genus but also because it belongs to the *erythropolis* species. In fact, the *Ptip/regulator* system of the shuttle-vector showed the best performance when expressed in *erythropolis* species as reported by Nakashima and Tamura, 2004b. In this way, we were able to overcome difficulties concerning gene expression in heterologous bacteria. Our results showed an increased biodegradation of C12 in the recombinant strain *R. erythropolis* AP (pTipQT1-*alk*R7) in comparison with the wild type strain *R. erythropolis* AP. These data supported the involvement of the *alkB* gene cluster in the *n*-alkane degradation in R7 strain. Considering the biodegradation kinetics for the C12-C24 substrate range, we can hypothesize that for the medium-chain-length *n*-alkanes an *alkB* like system plays the main role, while for *n*-alkanes longer than C20, other alkane hydroxylases could be involved.

Competing interests
The authors declare that they have no competing interests.

Authors' contributions
JZ carried out the molecular biology studies and chemical analysis, participated in the sequence alignment and drafted the manuscript. EC participated in the design of the study and performed kinetic analysis. ML

performed statistical data treatments and interpretation of the data. PDG participated in the design of the study and the coordination and helped to draft the manuscript. All authors read and approved the final manuscript.

Acknowledgment
We thank Prof. T. Tamura from Okkaido University of Japan to for supplying the shuttle-vector pTipQT1.

Author details
[1]Department of Biotechnology and Biosciences, University of Milano-Bicocca, Piazza della Scienza 2, 20126 Milano, Italy. [2]Department of Earth and Environmental Sciences, University of Milano-Bicocca, Milano, Italy.

References

Alonso H, Roujeinikova A (2012) Characterization and two-dimensional crystallization of membrane component alkB of medium-chain alkane hydroxylase system from *Pseudomonas putida* GPo1. Appl Environ Microbiol 78:7946–7953

Cappelletti M, Fedi S, Frascari D, Ohtake H, Turner RJ, Zannoni D (2011) Analyses of both the alkB gene transcriptional start site and alkB promoter-inducing properties of *Rhodococcus* sp. strain BCP1 Grown on n-alkanes. Appl Environ Microbiol 77:1619–1627

Di Gennaro P, Rescalli E, Galli E, Sello G, Bestetti G (2001) Characterization of *Rhodococcus opacus* R7, a strain able to degrade naphthalene and *o*-xylene isolated from a polycyclic aromatic hydrocarbon-contaminated soil. Res Microbiol 152:641–651

Di Gennaro P, Terreni P, Masi G, Botti S, De Ferra F, Bestetti G (2010) Identification and characterization of genes involved in naphthalene degradation in *Rhodococcus opacus* R7. Appl Microbiol Biotechnol 87:297–308

Finnerty WR (1992) The biology and genetics of the genus *Rhodococcus*. Annu Rev Microbiol 46:193–218

Harayama S, Kasai Y, Hara A (2004) Microbial communities in oil-contaminated seawater. Curr Opin Biotechnol 15:205–241

Ji Y, Mao G, Wang Y, Bartlam M (2013) Structural insights into diversity and n-alkane biodegradation mechanisms of alkane hydroxylases. Front Microbiol 4:1–13

Kulakov LA, Delacroix VA, Larkin MJ, Ksenzenko VN, Kulakova AN (1998) Cloning of new *Rhodococcus* extradiol dioxygenase genes and study of their distribution in different *Rhodococcus* strains. Microbiology 144:955–963

Larkin MJ, Kulakov LA, Allen CC (2005) Biodegradation and *Rhodococcus*-masters of catabolic versatility. Curr Opin Biotechnol 16:282–290

Larkin MJ, Kulakov LA, Allen CC (2006) Biodegradation by members of the genus *Rhodococcus*: biochemistry, physiology, and genetic adaptation. Adv Appl Microbiol 59:1–29

Leahy J, Colwell R (1990) Microbial biodegradation of hydrocarbons in environment. Microbiol Rev 54:305–315

Lo Piccolo L, De Pasquale C, Fodale R, Puglia AM, Quatrini P (2011) Involvement of an alkane hydroxylase system of *Gordonia* sp. strain SoCg in degradation of solid *n*-alkanes. Appl Environ Microbiol 77:1204–1213

Maffei F (2004) Individuazione dei fattori che stimolano la biodegradazione di idrocarburi presenti in suoli contaminati. University of Milano-Bicocca, Thesis

Maniatis T, Fritsch EF, Sambrook J (1982) Molecular cloning: a laboratory manual. Cold Spring Harbor Laboratory, Cold Spring Harbor, New York

Margesin R, Labbé D, Schinner F, Greer CW, Whyte LG (2003) Characterization of hydrocarbon-degrading microbial populations in contaminated and pristine alpine soils. Appl Environ Microbiol 69:3085–3092

Martínková L, Uhnáková B, Pátek M, Nešvera J, Křen V (2009) Biodegradation potential of the genus *Rhodococcus*. Environ Intern 35:162–177

McLeod MP, Warren RL, Hsiao WWL, Araki N, Myhre M, Fernandes C, Miyazawa D, Wong W, Lillquist AL, Wang D, Dosanjh M, Hara H, Petrescu A, Morin RD, Yang G, Stott JM, Schein JE, Shin H, Smailus D, Siddiqui AS, Marra MA, Jones SJM, Holt R, Brinkman FSL, Miyauchi K, Fukuda M, Davies JE, Mohn WW, Eltis LD (2006) The complete genome of *Rhodococcus* sp. RHA1 provides insights into a catabolic powerhouse. Proc Natl Acad Sci 103:15582–15587

Nakashima N, Tamura T (2004a) A novel system for expressing recombinant proteins over a wide temperature range from 4 to 35°C. Biotechnol Bioeng 86:136–148

Nakashima N, Tamura T (2004b) Isolation and characterization of a rolling-circle-type plasmid from *Rhodococcus erythropolis* and application of the plasmid to multiple-recombinant-protein expression. Appl Environ Microbiol 70:5557–5568

Rapp P, Lotte H, Gabriel-Jurgens E (2003) Degradation of alkanes and highly chlorinated benzenes, and production of biosurfactants, by a psychrophilic *Rhodococcus* sp. and genetic characterization of its chlorobenzene dioxygenase. Microbiology 149:2879–2890

Ratajczak A, Geiβdörfer W, Hillen W (1998) Expression of alkane hydroxylase from *Acinetobacter* sp. strain ADP1 is induced by a broad range of *n*-alkanes and requires the transcriptional activator AlkR. J Bacteriol 180:5822–5827

Rojo F (2009) Degradation of alkanes by bacteria. Environ Microbiol 11:2477–2490

Sambrook J, Russell DW (1989) Molecular cloning. A laboratory manual. Cold Spring Harbor Laboratory, Cold Spring Harbor, New York

Sameshima Y, Honda K, Kato J, Omasa T, Ohtake H (2008) Expression of *Rhodococcus opacus alkB* genes in anhydrous organic solvents. J Biosci Bioeng 106:199–203

Simkins S, Alexander M (1984) Model for mineralization kinetics with the variable of substrate concentration and population density. Appl Environ Microbiol 47:1299–1306

Smits THM, Balada SB, Witholt B, van Beilen JB (2002) Functional analysis of alkane hydroxylases from Gram-negative and Gram-positive bacteria. J Bacteriol 184:1733–1742

So CM, Young LY (1999) Isolation and characterization of a sulfate-reducing bacterium that anaerobically degrades alkanes. Appl Environ Microbiol 65:2969–2976

Treadway SL, Yanagimachi KS, Lankenau E, Lessard PA, Stephanopoulos G, Sinskey AJ (1999) Isolation and characterization of indene bioconversion genes from *Rhodococcus* strain I24. Appl Microbiol Biotechnol 51:786–793

van Beilen JB, Funhoff EG (2007) Alkane hydroxylases involved in microbial alkane degradation. Appl Microbiol Biotechnol 74:13–21

van Beilen JB, Smits THM, Whyte LG, Schorcht S, Röthlisberger M, Plaggemeier T, Engesser KH, Witholt B (2002) Alkane hydroxylase homologues in Gram-positive strains. Environ Microbiol 4:676–682

van Beilen JB, Li Z, Duetz WA, Smits THM, Witholt B (2003) Diversity of alkane hydroxylase system in the environment. Oil Gas Sc Technol 58:427–440

Van der Meer JR, de Vos WM, Harayama S, Zehnder AJB (1992) Molecular mechanisms of genetic adaptation to xenobiotic compounds. Microbiol Rev 56:677–694

Vidali M (2001) Bioremediation. An overview. Pure Appl Chem 73:1163–1172

Wang W, Shao Z (2013) Enzymes and genes involved in aerobic alkane degradation. Front Microbiol 4:1–7

Wentzel A, Ellingsen TE, Kotlar H-K, Zotchev SB, Throne-Holst M (2007) Bacterial metabolism of long-chain *n*-alkanes. Appl Microbiol Biotechnol 76:1209–1221

Whyte LG, Hawari J, Zhou E, Bourbonnière L, Inniss WE, Greer CW (1998) Biodegradation of variable-chain-length alkanes at low temperatures by a psychrotrophic *Rhodococcus* sp. Appl Environ Microbiol 64:2578–2584

Whyte LG, Smits THM, Labbé D, Witholt B, Greer CW, van Beilen JB (2002) Gene cloning and characterization of multiple alkane hydroxylase systems in *Rhodococcus* strains Q15 and NRRL B-16531. Appl Environ Microbiol 68:5933–5942

Zhukov DV, Murygina VP, Kalyuzhnyi SV (2007) Kinetics of degradation of aliphatic hydrocarbons by bacteria *Rhodococcus ruber* and *Rhodococcus erythropolis*. Appl Biochem Microbiol 43:587–592

10

The use of nano-sized acicular material, sliding friction, and antisense DNA oligonucleotides to silence bacterial genes

Yuya Mitsudome[1], Mamiko Takahama[1], Jun Hirose[2] and Naoto Yoshida[1*]

Abstract

Viable bacterial cells impaled with a single particle of a nano-sized acicular material formed when a mixture containing the cells and the material was exposed to a sliding friction field between polystyrene and agar gel; hereafter, we refer to these impaled cells as penetrons. We have used nano-sized acicular material to establish a novel method for bacterial transformation. Here, we generated penetrons that carried antisense DNA adsorbed on nano-sized acicular material (α-sepiolite) by providing sliding friction onto the surface of agar gel; we then investigated whether penetron formation was applicable to gene silencing techniques. Antisense DNA was artificially synthesized as 15 or 90mer DNA oligonucleotides based on the sequences around the translation start codon of target mRNAs. Mixtures of bacterial cells with antisense DNA adsorbed on α-sepiolite were stimulated by sliding friction on the surface of agar gel for 60 s. Upon formation of *Escherichia coli* penetrons, β-lactamase and β-galactosidase expression was evaluated by counting the numbers of colonies formed on LB agar containing ampicillin and by measuring β-galactosidase activity respectively. The numbers of ampicillin resistant colonies and the β-galactosidase activity derived from penetrons bearing antisense DNA (90mer) was repressed to 15% and 25%, respectively, of that of control penetrons which lacked antisense DNA. Biphenyl metabolite, ring cleavage yellow compound produced by *Pseudomonas pseudoalcaligenes* penetron treated with antisense oligonucleotide DNA targeted to *bphD* increased higher than that lacking antisense DNA. This result indicated that expression of bphD in *P. pseudoalcaligenes* penetrons was repressed by antisense DNA that targeted *bphD* mRNA. Sporulation rates of *Bacillus subtilis* penetrons treated with antisense DNA (15mer) targeted to *spo0A* decreased to 24.4% relative to penetrons lacking antisense DNA. This novel method of gene silencing has substantial promise for elucidation of gene function in bacterial species that have been refractory to experimental introduction of exogenous DNA.

Keywords: Antisense oligonucleotide DNA; BphD; Gene silencing; β-Galactosidase; β-Lactamase; Sepiolite; Sliding friction; Spo0A

Introduction

RNA with substantial structure, such as double-stranded RNA, is a poor template for protein synthesis (Lee et al. 1993; Fire et al. 1998; Hamilton and Baulcombe 1999; Elbashir et al. 2001). Knowledge of this has been used effectively to block translation in cell-free extracts. In vitro, hybrid-arrested translation has been used to prevent the synthesis of specific proteins and has been used to detect recombinant DNA molecules that contain sequences complementary to a given mRNA. Stephenson and Zamecnik (1978) first demonstrated that DNA oligonucleotides complementary to the reiterated 3′- and 5′- terminal nucleotides of Rous sarcoma virus 35S RNA inhibit the translation of RNA in a cell-free system; they also prevent virus production by chicken fibroblast grown in tissue cultures (Stephenson and Zamecnik, 1978; Zamecnik and Stephenson 1978). This method is a valuable tool for sequence-specific inhibition of gene expression, and also useful for functional genomics, target validation and treatment of diseases (Bennett and Cowsert 1999; Dean 2001; Lavery and King 2003). Presumably, mechanisms involving steric blockage

* Correspondence: a04109u@cc.miyazaki-u.ac.jp
[1]Department of Biochemistry and Applied Biosciences, University of Miyazaki, 1-1 Gakuen, Kibanadai-Nishi, Miyazaki 889-2192, Japan
Full list of author information is available at the end of the article

of the ribosomes are responsible for inhibition of transcription (Kurreck 2003). Izant and Weintraub (1984; 1985) examined the effects of complementary antisense RNA on the expression of the herpes simplex virus I thymidine kinase (TK) gene in mouse cells. Microinjection of recombinant DNA plasmid that directs transcription of the TK gene with reverse polarity causes the mouse cells themselves to produce the antisense RNA. TK activity is diminished 4- to 5-fold in cells that produce such antisense RNA; this finding indicated that antisense nucleotides may be used to inhibit expression of specific genes. Molecular biologists use antisense DNA or RNA fragments as powerful tools to block expression of specific target genes, and thereby assess gene function.

Upon bacterial gene silencing method, Jayaraman et al. (1981) synthesized a series of deoxyribooligonucleoside methylphosphonates, which are nonionic oligonucleotide analogs. The base sequences of these compounds are complementary to the Shine-Dalgarno sequence found at the 3′ end of bacterial 16S rRNA. These oligonucleoside methylphosphonates inhibit both protein synthesis and colony formation by mutant *E. coli*; importantly, these permeable mutants contain negligible quantities of lipopolysaccharide. Experiments were done on the growth of *E. coli* in broth in the presence of the oligonucleotide methylphosphonate; culture growth was inhibited up to 50%, relative to controls, by the oligonucleotide. Nakashima *et al.* (2006) used *E. coli* to produce a gene silencing effect on bacterial species; they found that 100 bp of paired-termini antisense RNA enable gene silencing to be dramatically effective than short length antisense RNA.

For an oligonucleotide to display an antisense effect, it must be stable until it forms a double helix with the target mRNA. Phosphodiester oligo DNA (D-oligo DNAs), which are frequently used as antisense oligonucleotides, may be digested by nuclease. To stabilize antisense oligonucleotides, different types of oligonucleotide analogs (*e.g.,* phosphorothioates oligonucleotide DNA, methylphosphoate oligonucleotide DNA) have been developed (Akhtar et al. 1991; Zhao et al. 1993; Verma and Eckstein 1998). However, analogs of D-oligo DNA are disadvantageous for gene silencing because they are very costly to synthesize. Additionally, other data indicate that antisense DNA exhibits efficacy in vivo when administered at the relatively high concentration of 5 mM (Whitesell et al. 1991). The high doses are necessary in vivo because of the susceptibility of normal DNA oligomers to rapid degradation by serum nucleases. More likely candidates for in vivo applications are modified oligonucleotides that are resistant to nucleases because the nucleotide structure itself is altered (e.g., alkylphosphonates, phosphorothioates, 2′-*O*-alkylnucleotides) or because protecting groups are present at the 3′ and 5′ ends of DNA oligomers (Wickstrom 1992). Which such modifications most effectively balance cost, in vivo and intracellular mobility, stability, and most importantly, binding affinity and specificity, remain to be determined.

An antisense oligonucleotide is preferably between 15 and 30 mer in length and specifically hybridizes with an mRNA around the translation start codon (Uhlmann and Peyman 1990; Good et al. 2001). Recently, Nakashima *et al.* successfully repressed the β-galactosidase gene by engineering expression of a 100 mer antisense oligonucleotide that was complementary to the mRNA near the ribosome binding site and the translation start codon (Nakashima and Tamura, 2009).

As shown in Figure 1, a bacterial cell impaled by a nano-sized acicular material (penetron) was formed when bacterial cells and nano-sized acicular materials were exposed to a friction field in hydrogel. This phenomenon is called the Yoshida effect (Yoshida and Sato 2009). Penetrons readily take up exogenous DNA; consequently, penetrons are applicable to a genetic transformation technique called tribos transformation (Yoshida and Sato 2009). Transformation efficiency of *E. coli* penetrons is 10^7 colony forming units (cfu) per 1 μg of plasmid DNA (Yoshida et al. 2007). Nano-sized acicular materials, e.g., chrysotile and sepiolite, have high affinity for nucleic acid. Notably, DNA adsorbed on nano-sized acicular materials is protected from enzyme-mediated nucleolytic degradation and remains stable (Somiya et al. 2012).

We reasoned that penetrons that had been mixed with artificially synthesized antisense DNA could be used for bacterial gene silencing techniques. In this study, to investigate whether bacterial penetron was utilized to gene silencing technique, we introduced artificially synthesized antisense oligonucleotide DNA into bacterial cells by using sliding friction and sepiolite. We used *Escherichia coli* JM109 harboring pUC18 DNA and *Pseudomonas pseudoalcaligenes* KF707 to represent Gram-negative bacteria. We synthesized 15 and 90 mer antisense DNA oligonucleotides; each was targeted to the translation start codon of β-galactosidase α-fragment gene or β-lactamase gene coded in pUC18 DNA. Silencing of the β-galactosidase α-fragment was evaluated by measurement of enzyme activity; silencing of β-lactamase was evaluated by numerical comparisons of ampicillin resistant colonies. *Pseudomonas pseudoalcaligenes* KF707 was isolated by Furukawa and Miyazaki (1986) as a biphenyl degrading bacterium (1986). The genes associated with biphenyl metabolism and their functions are well studied (Furukawa and Miyazaki 1986; Furukawa and Arimura, 1987; Taira et al. 1992). We synthesized 15 and 90 mer antisense DNA oligonucleotides; each was targeted to the 2-hydroxy-6-oxo-6-phenylhexa-2,4-dienoate hydrolase gene (*bphD*). We then evaluated expression of *bphD*. *Bacillus subtilis*

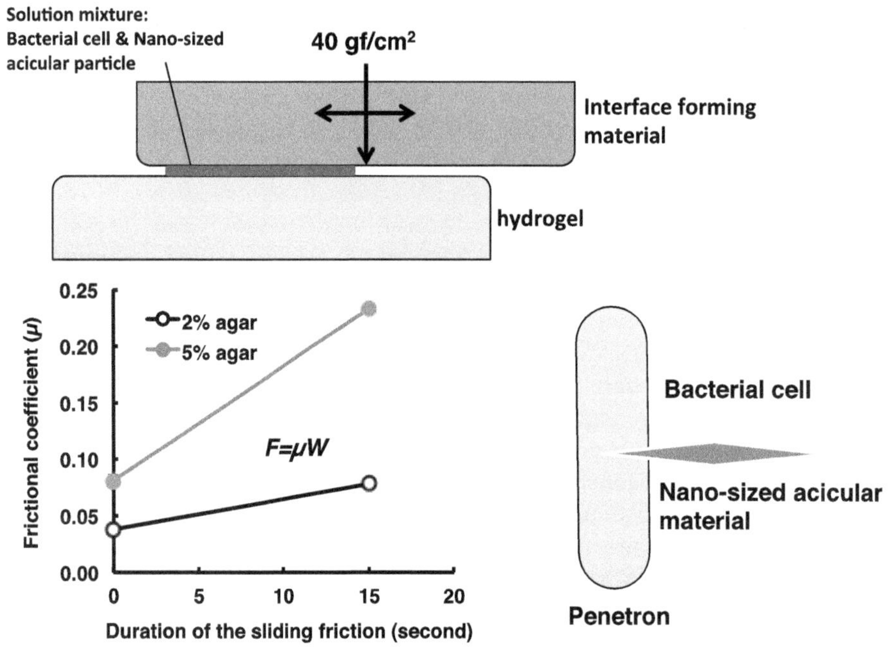

Figure 1 Schematic representation of the Yoshida effect. The Yoshida effect is defined as the formation of complexes called penetrons, which are bacterial cells, each impaled by a single nano-sized acicular material in a friction field formed at a hydrogel interface. The hydrogel, interface forming material, nano-sized acicular material, bacterial cells, sliding friction, and an energy source to provide the friction force are each essential to the formation of penetrons. The hydrogel (e. g., agar, gellan gum, κ-karagenan) involved shear stress at more than 2.1 N. The interface forming material could comprise polymer material such as polystyrene, polyethylene, or acrylonitrile butanediene rubber. The optical vertical reaction force against hydrogel was around 40 gf/cm^2. Multi-walled carbon nanotube, maghemite, or α-sepiolite are each nano-sized acicular materials that can generate a Yoshida effect. The sliding friction force can be represented with the following formula: $F = \mu W$, where μ and W denote the frictional coefficient and the vertical reaction force, respectively. When using 2 and 5% agar hydrogel, frictional coefficient value increased 0.038 to 0.078 and 0.081 to 0.233 respectively by given sliding stimulus for 15 seconds. The rapid increase in frictional resistance was essential for the Yoshida effect.

str. 169 (Kunst et al. 1997) represents Gram-positive spore-forming bacterium. We also synthesized antisense DNA oligonucleotides (15 and 90 mer) targeting the translation start codon of sporulation initiation protein gene (*Spo0A*). Gene silencing of *spo0A* was evaluated by spore formation rate and accumulation of dipicolinic acid in spores.

Materials and methods

Bacterial strains and growth condition

Escherichia coli JM109 (el4$^-$, *rec*A1, *end*A1, *gyr*A96, *thi*-1, *hsd*R17, *sup*E44, *rel*A1, Δ(*lac-pro*AB), [F', *tra*D36 *pro*AB, *lac*I^q ZΔM15]) harboring pUC18 DNA was grown aerobically in Luria-Bertani (LB) broth containing 50 μg/ml ampicillin at 30°C for 18 hrs (Sambrook et al. 1989; Shiloach et al. 1996).

Pseudomonas pseudoalcaligenes KF707 which has been deposited in the NITE Patent Microorganisms Depositary under the accession number FERM P-8297 was grown in a defined medium (pH 7.0) containing (in grams per liter): K_2HPO_4, 4.3; KH_2PO_4, 3.4; $(NH_4)_2SO_4$, 2.0; $MgCl_2$, 0.16; $MnCl_2 \cdot 4H_2O$, 0.001; $FeSO_4 \cdot 7H_2O$, 0.006; $CaCl_2 \cdot 2H_2O$, 0.026; and $Na_2MoO_4 \cdot 2H_2O$, 0.002. For 1.5% agar plating medium (Wako Chemical Co., Japan), biphenyl was introduced as a vapor by placing crystals on the lid of each petri dish. Each dish was sealed with polyethylene tape. Strain KF707 cells were inoculated onto agar plating medium and incubated at 30°C for 4 days (Furukawa and Miyazaki 1986).

Bacillus subtilis str.168 (Kunst et al. 1997) was cultured on LB agar medium that contained 1% glucose at 37°C for 18 hrs.

Preparation of sepiolite solution

Sepiolite was used as the nano-sized acicular material in each experiment to form penetrons. Sepiolite particles (Wako Chemical Co., Japan) suspended in distilled water at final concentrations of 50 μg/ml or 1 mg/ml for Gram-negative and Gram-positive bacteria respectively were incubated in suspension at 60°C for 24 hrs, and the sepiolite solution was then used.

Design of antisense DNA oligonucleotides

To evaluate the effects on gene silencing of DNA oligonucleotides in *E. coli*, two mRNAs—β-lactamase (*bla*) and β-galactocidase (*lacZα*)—were targeted. To assess the effects in *Pseudomonas* and *Bacillus*, the mRNAs encoding hydrolase (*bphD*), which is associated with biphenyl metabolism (Triscari-Barberi et al. 2012) and sporulation initiation protein (*spo0A*) were targeted, respectively. The sequence of each 15 and 90 bp DNA oligonucleotide was designed based on the sequence around the translation start codon in the respective targeted mRNA. The sequence of each antisense DNA oligonucleotide DNA and the respective abbreviations are listed in Table 1.

Introduction of antisense oligonucleotide DNA into *E. coli* penetrons and evaluation of gene silencing effect on β-lactamase

A 50 μl aliquot of sepiolite solution containing 0.2 μM of antisense oligonucleotide DNA (B-LAC15, B-LAC90), 200 mM NaCl, and an equivalent volume of *E. coli* culture broth were streaked onto the surface of 2% agar hydrogel containing LB contents and 100 μg/ml of ampicillin. Sliding friction between a polystyrene streak bar (SARSTED, Germany) and agar hydrogel was applied to these solution mixtures for 60 sec. The agar hydrogel was incubated at 37°C for 18 hrs, and then ampicillin resistant colonies were counted. The number of ampicillin resistant colony formed on the agar hydrogel treated by same method without antisense oligonucleotide DNA was used as the 100% control.

Introduction of antisense DNA oligonucleotides into *E. coli* penetrons and evaluation of gene silencing effect on β-galactosidase α-fragment

A 100 μl mixture containing 50 μl of sepiolite solution with 0.2 μM of antisense DNA oligonucleotide (LacZ15, LacZ90), 200 mM NaCl and 50 μl of *E. coli* culture broth was streaked onto the surface of a 2% agar hydrogel containing 1 mM β-D-thiogalactopyranoside (IPTG) and 200 mM NaCl. Each of these mixtures and control mixtures that were identical except that they lacked any antisense oligonucleotides were individually exposed to the sliding friction between a polystyrene streak bar (SARSTED) and agar hydrogel for 60 sec. Each agar hydrogel coated with cells was then incubated at 37°C for 3 hrs; *E. coli* penetrons on agar hydrogel were then resuspended in 2 ml of NaCl solution. A standard colorimetric method (Miller, 1972; Griffith and Wolf 2002) was then used to measure β-galactosidase activity in control and oligonucleotide-treated *E. coli* penetrons. Values of β-galactosidase activity in the control *E. coli* penetrons were designated as the 100% standard.

SDS-PAGE and analysis of protein bands

A 100 μl mixture containing 50 μl of sepiolite solution with 0.2 μM of antisense DNA oligonucleotide (LacZ15, LacZ90), 200 mM NaCl and 50 μl of *E. coli* culture broth was streaked onto the surface of a 2% agar hydrogel containing 1 mM IPTG and 200 mM NaCl. Again, each mixture was exposed to the sliding friction field between a streak bar and agar hydrogel for 60 sec. The cell-coated agar hydrogels were each incubated at 37°C for 3 hrs; control and oligonucleotide-treated *E. coli* penetrons were collected from the surface of the agar hydrogels with 2 ml of 200 mM NaCl and a cell scraper. Proteins were extracted from collected *E. coli* penetrons by sonication and analyzed by SDS-PAGE according to the Leammli protocol (Laemmli, 1970). Each proteins gel was stained and then destained; electrophoresis image analysis software (NIH ImageJ, Version 1.41) (http://rsb.info.nih.gov/ij/) was then used to measure the concentration of protein in each band that corresponded to β-galactosidase.

Introduction of antisense oligonucleotide DNA into *P. pseudoalcaligenes* penetrons and evaluation of gene silencing effect on *bphD* expression

Fresh *P. pseudoalcaligenes* cells (10 mg/ml) were suspended in 50 μl of sepiolite solution containing 200 mM NaCl and 0.2 μM of antisense DNA oligonucleotide (bphD15 or bphD90) or no oligonucleotides. Each 50 μl mixture was spread onto the surface of a 2% agar hydrogel containing 200 mM NaCl. Each mixture was exposed to sliding friction between a streak bar and agar hydrogel for 60 sec to form *P. pseudoalcaligenes* penetrons. Control and oligonucleotide-treated penetrons were collected from the surface of each agar hydrogel with 2 ml of 200 mM NaCl and a cell scraper. Collected *P. pseudoalcaligenes* penetrons were washed with 200 mM phosphate buffer (pH 7.0); that was used to adjust the turbidity of each sample of *P. pseudoalcaligenes* penetrons to an absorbance of 0.2 at 600 nm. Each 1.0 ml suspension of *P. pseudoalcaligenes* penetrons was added to a respective 20 ml volume of 0.1 μM biphenyl solution; each such mixture was then incubated at 30°C. After a 40 min and 80 min incubation, the absorbance value at 403 nm of each centrifuged reaction mixture was determined with a spectrophotometer (UV-1800, Shimazu, Japan) (Triscari-Barberi et al. 2012).

Introduction of antisense oligonucleotide DNA into *B. subtilis* penetrons and evaluation of gene silencing effect on *spo0A* expression

Fresh *B. subtilis* vegetative cells (10 mg/ml) were suspended in 50 μl of sepiolite solution containing 200 mM NaCl and 0.2 μM of antisense Spo0A15 or Spo0A90 DNA oligonucleotide or no oligonucleotides. The 50 μl mixture was spread onto the surface of a 5% agar hydrogel containing 10 g/L NaCl, 0.016 g/L $MnCl_2$, 0.12 g/L $MgSO_4$, 0.016 mg/L $FeSO_4$, 0.11 mg/L $CaCl_2$, and 8.3 μl/L HCl. The mixture was exposed to the sliding friction between the

Table 1 Sequence of each DNA oligonucleotide used in this study

Strain	Target gene	Name of antisense oligonucleotide	Accession no.	Position	Sequence of antisense oligonucleotide DNA (5′-3′)	Length of antisense oligonucleotide DNA (bp)
Escherichia coli JM109 (pUC18)	β-lactamase (*bla*)	B-LAC15	L08752	876-891	ACT***CAT***ACTCTTCCT	15
		B-LAC90		851-940	AGGCAAAATGCCGCAAAAAAGGGAATAAGGGCGACACGGAAAT GTTGAATACT***CAT***ACTCTTCCTTTTTCAATATTATTGAAGCATTTAT	90
	β-galactosidase (*lacZα*)	LacZ15		221-207	GGT***CAT***AGCTGTTTC	15
		LacZ90		185-274	CTGCAGGTCGACTCTAGAGGATCCCCGGGTACCGAGCTCGAATT CGTAATCATGGT***CAT***AGCTGTTTCCTGTGTGAAATTGTTATCCGCT	90
Pseudomonas pseudoalcaligenes KF707	Hydrolase (*bphD*)	BphD15	X66123	178-164	GTGCGGT***CAT***TTTTC	15
		BphD90		231-142	GGTAGAACTTTCGGTGAGTGCGGT***CAT***TTTTCATCCTTTAAGTGAGT GGAGTGGAAACTGGTCAGGCGCAGCGCTTCATGCCGCGCCGGG	90
Bacillus subtilis str.168	Sporulation initiation protein (*spo0A*)	Spo0A15	AL009129	2518821-2518835	CTC***CAC***GTTTCTTCC	15
		Spo0A90		2518768-2518857	CTTAACAGGCTTACCAGCTCTCGATTATCATCAGCAACACAAACT TTAATTTTCTC***CAC***GTTTCTTCCTCCCCAAATGTAGTTAACAGGA	90

Bold italic letter indicates sequence complementary to the translation codon (AUG or GUG) in a target mRNA.

streak bar and agar hydrogel for 60 sec to form penetrons. The *B. subtilis* penetrons formed on the surface of the agar hydrogel were incubated at 30°C for 72 hrs, and were then collected from the surface of the agar hydrogel with 2 ml of 200 mM NaCl and a cell scraper. The number of vegetative cells and spores was counted under a confocal microscope (ZEISS, Germany) and spore formation rate (spores/spores + vegetative cells) was calculated.

To alternatively evaluate the effect of antisense oligonucleotide DNA on spore formation rate, we measured the amount of dipicolinic acid (DPA) production by *B. subtilis* penetrons (Janssen et al. 1958; Rotman and Fields 1968). The *B. subtilis* penetrons formed on the surface of the agar hydrogel were incubated at 30°C for 72 hrs, and were collected from the surface of the agar hydrogel with 2 ml of 200 mM NaCl and a cell scraper. The *B. subtilis* penetrons were washed by centrifugation, suspended in 1 ml of distilled water, boiled for 20 min, cooled in ice for 15 min and then exposed to DPA extract solution. A spectrophotometer (UV-1800, Shimazu, Japan) was then used to measure the absorbance at 440 nm of a mixture of 0.4 ml of centrifugated DPA extract solution, 0.2 ml of distilled water, and 0.2 ml of coloring reaction solution containing 1.0 g/L L-Cysteine, 6.8 g/L $FeSO_4 \cdot 7H_2O$, 3.2 g/L $(NH_4)_2SO_4$, and 50 mM sodium acetate buffer (pH4.6). Reagent grade DPA (Wako Chemical Co., Japan) was used as the standard.

Results

Evaluation of gene silencing effect on β-lactamase

The number of ampicillin resistant colonies derived from penetrons exposed to the antisense DNA oligonucleotides, B-LAC15 or B-LAC90, decreased to 42% or 15% of the controls respectively (Figure 2). These results indicated that the antisense oligonucleotides bound to the complementary region of the mRNA encoding β-lactamase, which catalyzes the hydrolysis of the β-lactam ring in the ampicillin molecule, and that the oligonucleotides inhibited translation of the bound mRNAs. The oligonucleotides LacZ15 and LacZ90, which were synthesized as antisense oligonucleotides targeted to the mRNA encoding the β-galactosidase α-fragment, were not associated with the mRNA encoding β-lactamase; nevertheless the number of ampicillin resistant colonies derived from penetrons exposed to LacZ15 or LacZ90 decreased to 72% of the controls. The mRNA encoding β-lactamase assumes to involve semi-complementary sequence with oligonucleotide LacZ15 and LacZ90.

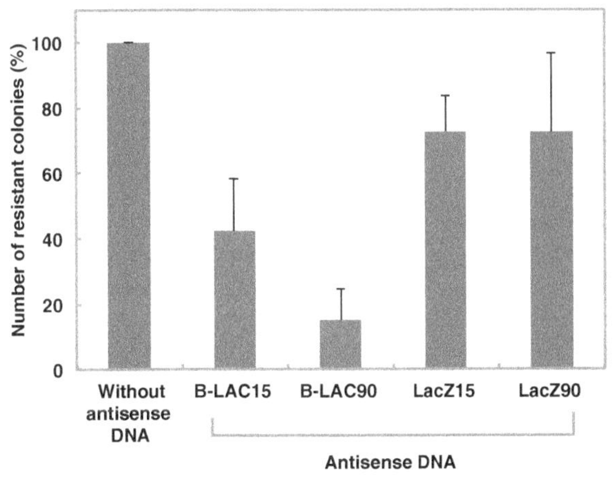

Figure 2 Gene silencing effects of antisense DNA oligonucleotides (B-LAC15, 90) on β-lactamase expression in *Escherichia coli* JM109 (pUC18). Values indicate the means ± standard deviation of three independent experiments.

Evaluation of gene silencing effect on β-galactosidase α-fragment

β-galactosidase activity in *E. coli* penetrons that had been exposed to antisense DNA oligonucleotides (LacZ15 or LacZ90) was reduced to 38.4 and 26.0% of that in control *E. coli* penetrons respectively (Figure 3A). These antisense DNAs would have inhibited translation of targeted mRNA by binding complementary sequences in the 5'-region of the targeted mRNA and resulted in depression of β-galactosidase activity. The β-galactosidase activity in *E. coli* penetrons that were exposed to B-LAC15 or B-LAC90 was reduced to 84% of that in control *E. coli* penetrons. This result was attributed to the possibility that the mRNA encoding β-galactosidase carried complementary sequences similar to the B-LAC15 and B-LAC90 sequences.

Coomassie-stained protein bands that corresponded to the β-galactosidase expressed in *E. coli* penetrons into which LacZ90 or LacZ15 had been injected exhibited 2330 and 1971 as peak intensities respectively (Figure 3B). In the mean time, the protein bands corresponding to the β-galactosidase expressed in control *E. coli* penetrons exhibited a higher intensity at 15893. This result indicated that β-galactosidase expression in *E. coli* penetrons in which LacZ90 or LacZ15 had been injected was depressed to 14.7 and 12.4% of that in control *E. coli* penetrons respectively.

Evaluation of gene silencing effect on hydrolase

The biphenyl solution added to *P. pseudoalcaligenes* penetrons in which antisense DNA oligonucleotide DNA had been injected and that added to control *P. pseudoalcaligenes* penetrons were incubated at 30°C (reaction solution). Absorbance value at 403 nm of reaction solution after a 40 min incubation was defined as 100%. Absorbance value at 403 nm of the control reaction solution after an 80 min incubation was 122%. Absorbance values at 403 nm of the BphD15 and BphD90 reaction solutions after an 80 min incubation were 195 and 160% respectively (Figure 4). The results showed that the gene

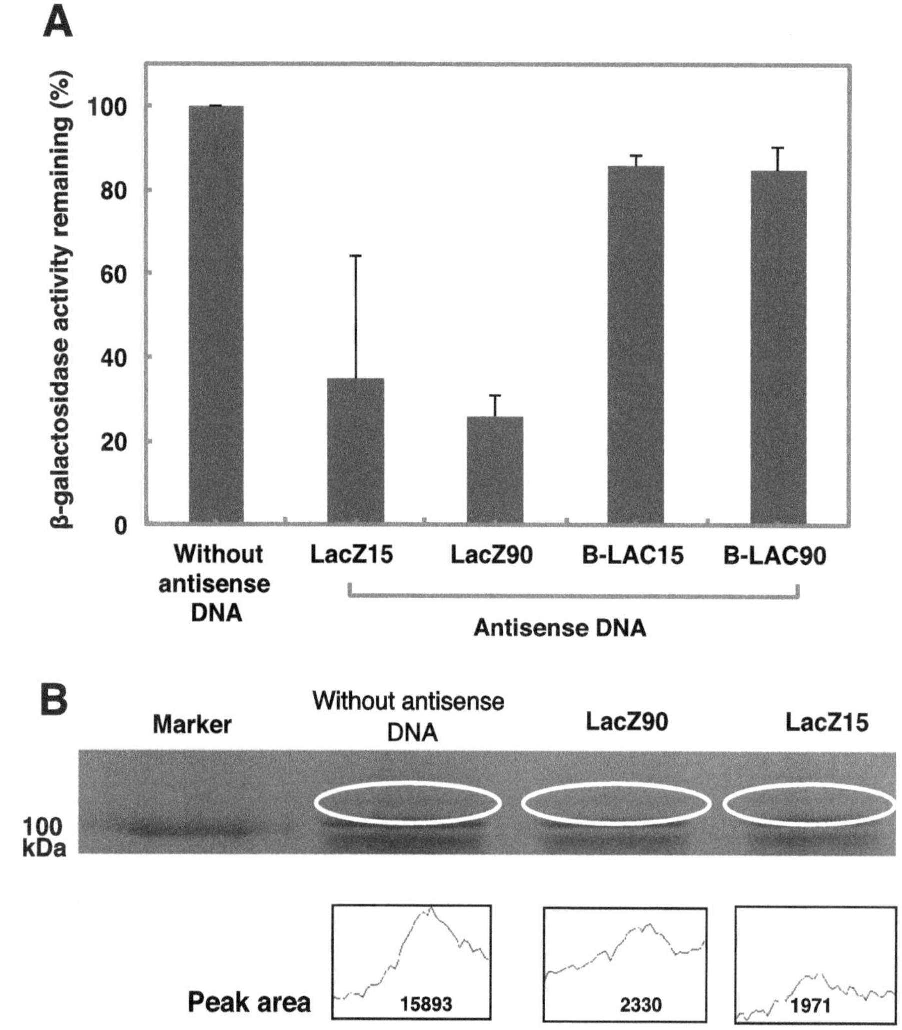

Figure 3 Gene silencing effects of antisense DNA oligonucleotides (LacZ15, 90) on β -galactosidase activity in *Escherichia coli* JM109 (pUC18). (A) Values indicate the means ± standard deviation of three independent experiments. **(B)** Comparison of β -galactosidase band intensity on SDS-PAGE gels from *E. coli* bearing or lacking antisense DNA (LacZ15, 90). The peak area indicates the intensity of Coomassie-stained protein bands that correspond to β-galactosidase.

silencing cause by the 15 mer oligonucleotide was stronger than that cause by the 90 mer oligonucleotide.

Evaluation of gene silencing effects on sporulation formation rates

SpoOA is a signal protein involved in initiation of *Bacillus* sp. sporulation. SpoOA is activated by phosphorylation and serves as a translation regulation factor for one set of sporulation mRNAs (Errington, 1993). Phosphorylation-mediated activation of spoOA is essential for switching of vegetative cells to spore cells, thus gene silencing of *spoOA* should cause repression of differentiation to spore cells. Spore formation rates of *B. subtilis* penetrons injected with antisense oligonucleotides (SpoOA15 or SpoOA90) was repressed to 24.4% and 35.8% of that of control *B. subtilis* penetrons respectively (Figure 5A). Alternatively, the amount of dipicolinic acid (Janssen et al. 1958; Rotman and Fields 1968) production of *B. subtilis* penetrons in which antisense oligonucleotide SpoOA15 and SpoOA90 was injected was repressed to 42.2 and 65.3% of control *B. subtilis* penetrons (Figure 5B). Presumably, these antisense DNAs bound to complementary sequences within the targeted *spoOA* mRNA and inhibited the translation of spoOA, thereby resulting in depression of spore formation in *B. subtilis* penetrons.

Discussion

The most common methods used to explore gene function involve gene deletion (gene knockout) or inactivation of target mRNA (gene knockdown). Antisense methods are among the most commonly used for such research

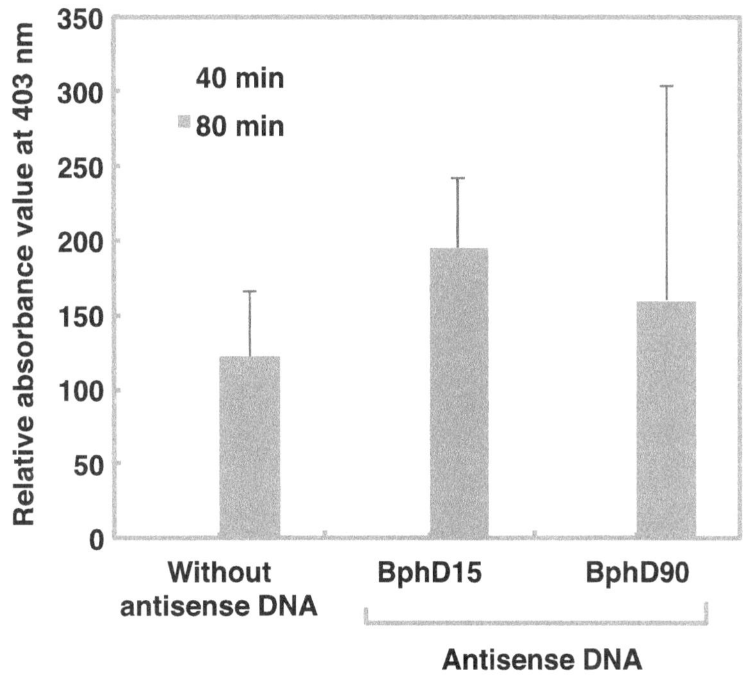

Figure 4 Gene silencing effects of BphD15 and 90 antisense oligonucleotide DNA on hydrolase activity in *Pseudomonas pseudoalcaligenes* KF707. Values indicate the means ± standard deviation of three independent experiments.

purposes. Antisense molecules, as experimental materials, involve ribozymes and antisense RNA generated from expression vectors. Here, we focused on a method involving artificially synthesized DNA oligonucleotides. Antisense RNAs expressed from vectors have lead to many achievements in biochemical research. However, construction of such expression vectors is technically time consuming and low amount of antisense RNA can be expressed is still problematic. The advantages of our proposed gene silencing technique, which involves antisense DNA oligonucleotides, are that specially prepared equipment and special techniques are not required and that many genes can be targeted. In proposed gene silencing method using nano-sized acicular material and antisense DNA, penetrons that lack antisense DNA or penetrons with antisense DNA complementary to off-target genes must be used as controls for any experiments.

β–lactamase and β–galactosidase α-fragment genes are encoded in pUC18 DNA, and expression of the genes was induced by ampicillin, which bears a β–lactam ring, and IPTG, respectively. Comparisons among effects on gene silencing of β–lactamase and β–galactosidase α-fragment and resulting from oligonucleotides of different lengths demonstrated that the 90 mers were more effective than 15 mers. This result may depend on secondary structure of targeted mRNA or stability of injected antisense oligonucleotide DNA.

The *P. pseudoalcaligenes* strain used in our experiments was isolated from soil near a biphenyl-producing factory in Kitakyushu, Japan (Furukawa and Miyazaki 1986). *P. pseudoalcaligenes* KF707 can degrade 4, 4′-dichlorobiphenyl quickly, but it degrades 2, 5, 2′, 5′-tetrachlorobiphenyl poorly (Mondello et al. 1997). A catabolic pathway responsible for biphenyl degradation by *P. pseudoalcaligenes* KF707 has been proposed by Furukawa and Miyazaki (1986). This strain expresses 2, 3-dioxygenase, which introduces molecular oxygen at the 2, 3-position of a nonchlorinated or less-chlorinated ring in polychlorinated biphenyl to produce a dihydrodiol (Kimura et al. 1997). The dihydrodiol is then dehydrogenated to a 2, 3-dihydroxybiphenyl by a dihydrodiol dehydrogenase. The 2, 3-dihydroxybiphenyl is then cleaved at the 1,2-position by a dihydroxybiphenyl dioxygenase to produce 2-hydroxy-6-oxo-6-phenylhexa-2,4-dienoate (HPDA). The *meta* cleavage compound, HPDA, is then hydrolyzed to the chlorobenzoic acid and 2-oxopent-4-enoate by a hydrolase which is the product of *bphD*. Here we measured the activity of this hydrolase by monitoring the conversion of the HPDA to benzoic acid by measuring the decrease in absorbance at 403 nm. When biphenyl was degraded by *P. pseudoalcaligenes* penetrons bearing the *bphD* antisense DNA target, the HPDA-associated absorbance values at 403 nm were significantly higher than those of the control *P. pseudoalcaligenes* penetrons. These data indicated that antisense DNA inhibited translation of *bphD* mRNA and depressed expression of hydrolase (bphD), and that this inhibition resulted in the accumulation of HPDA.

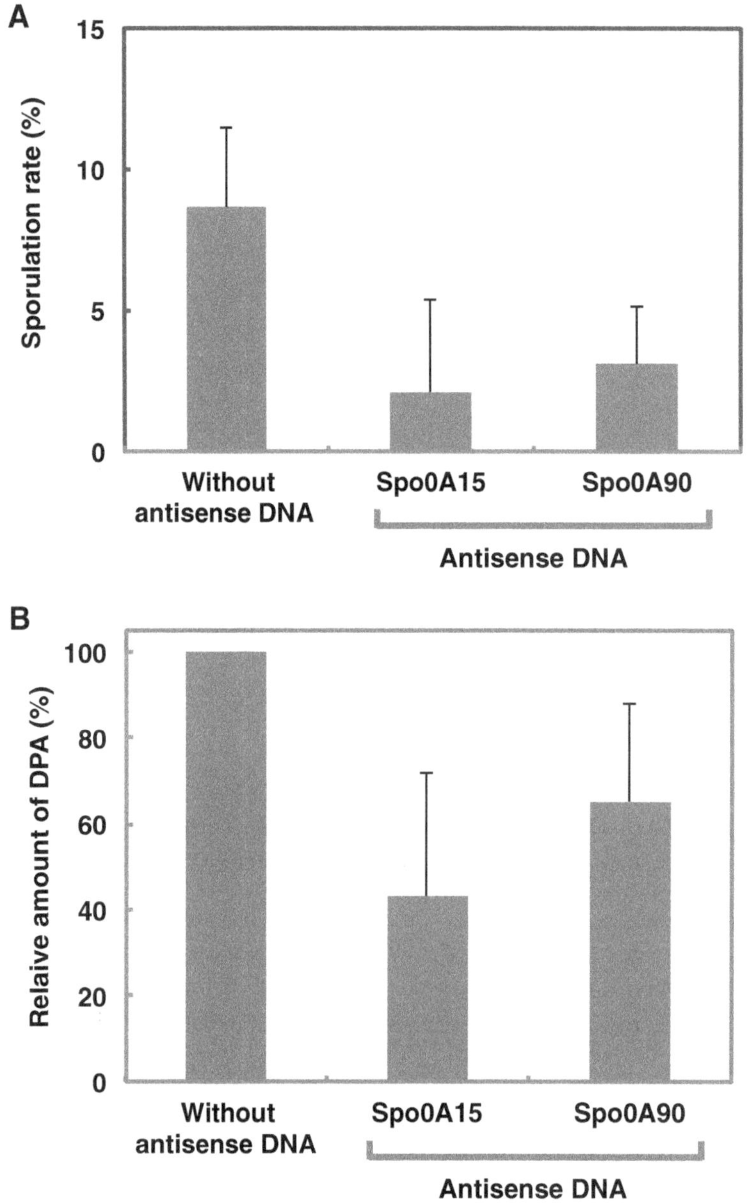

Figure 5 Gene silencing effects of Spo0A15 and 90 antisense oligonucleotides DNA on (A) spore formation and (B) dipicolinic acid accumulation in *Bacillus subtilis* str. 168. Values indicate the means ± standard deviation of three independent experiments.

Currently, the nine genes involving *spo0A* are known as the genes associated with initiation of sporulation in *B. subtilis* (Tan and Ramamurthi 2014). If any one of the sporulation proteins is nonfunctional because of genetic mutation, sporulation does not initiate even under starvation conditions. Activation of the transcription factor encoded by spo0A is essential for initiation of sporulation in *B. subtilis* (LeDeaux et al. 1995). External signals that result from environmental stress induce phosphorylation of Spo0A, and phosphorylated Spo0A activates the σ^F, σ^E subunit of RNA polymerase (Molle et al. 2003). As expected, the sporulation rate among *B. subtilis* penetrons bearing Spo0A15 or 90 oligonucleotides was lower than that among controls. We presumed that the Spo0A15 and 90 antisense DNA inhibited translation of *spo0A* mRNA and repressed initiation of sporulation. Dipicolinic acid (DPA) was accumulated intracellularly during sporulation, and we could measure DPA accumulation to evaluate sporulation rates. Notably, spo0A15 or spo0A 90 antisense DNA decreased DPA accumulation and repressed endospore formation. The effects of the 15 mer on expression of hydrolase and spo0A were greater than those of the 90 mer. These results were consistent with results reported by many other researchers (Krötz et al. 2003; Falzarano et al. 2014).

It is noteworthy that the effects on gene silencing caused by the antisense DNA and nano-sized acicular materials were transient, thus this method affected only resting cells. In these experiments, we used antisense DNA oligonucleotides that were chemically stable and reasonable to synthesize. Many published results demonstrate that the length of antisense oligonucleotides that are effective for gene silencing ranges from 15 to 25 residues (Uhlmann and Peyman 1990; Mizuta et al. 1999). The antisense DNAs with sequences complementary to sequences around the target AUG translation initiation codon should provide favorable results for gene silencing. A mechanism by which antisense DNA oligonucleotides are released from α-sepiolite was proposed previously (Yoshida and Ide 2008). Our novel gene silencing method described in this paper was more effective with Gram-negative bacteria than Gram-positive bacteria. We suspect that Gram-positive penetrons were harder to generate than Gram-negative penetrons because the cell wall of Gram-positive bacteria is thicker than that of Gram-negative bacteria. Until now, researchers have used *E. coli* mutants with increased permeability properties to introduce antisense DNA oligonucleotides when evaluating the gene silencing effects of antisense DNA (Jayaraman et al. 1981). The advantage of penetron technology for gene silencing is that the technique is simple and applicable to different bacterial species, and many such species have been unable to directly uptake oligonucleotide DNA until now. Once the experimental protocol is established for many bacterial species, the proposed gene silencing technique will be a powerful tool for the elucidation of physiological functions of biological molecules. In any case, our findings demonstrated the substantial potential of antisense DNA as a new class of gene silencing molecules for not only bacteriology but also molecular biology, physiology, developmental biology involving clinical research.

Competing interest
The authors declare that they have no competing interest.

Author's contribution
YM carried out the measurement of β-lactamase and β-galactosidase activity in *E. coli* and sporulation rate in *B. subtilis*. MT carried out the measurement of hydrolase activity in *P. pseudoalcaligenes*. JH participated in the design of the study on gene silencing in *P. pseudoalcaligenes*. NY participated in the design, coordination of the study and contributed in the drafting of the manuscript. All authors read and approved the final manuscript.

Acknowledgments
This work was financially supported by the Research Funding from the Faculty of Agriculture, University of Miyazaki.

Author details
[1]Department of Biochemistry and Applied Biosciences, University of Miyazaki, 1-1 Gakuen, Kibanadai-Nishi, Miyazaki 889-2192, Japan. [2]Department of Applied Chemistry, University of Miyazaki, 1-1 Gakuen, Kibanadai-Nishi, Miyazaki 889-2192, Japan.

References
Akhtar S, Kole R, Juliano RL (1991) Stability of antisense DNA oligodeoxynucleotide analogs in cellular extracts and sera. Life Sci 49:1793–1801
Bennett CF, Cowsert LM (1999) Application of antisense oligonucleotides for gene functionalization and target validation. Curr Opin Mol Ther 1:359–371
Dean NM (2001) Functional genomics and target validation approaches using antisense oligonucleotide technology. Curr Opin Biotechnol 12:622–625
Elbashir SM, Harborth J, Lendeckel W, Yalcin A, Weber K, Tuschl T (2001) Duplexes of 21-nucleotide RNAs mediate RNA interference in cultured mammalian cells. Nature 411:494–498
Errington J (1993) *Bacillus subtilis* sporulation: regulation of gene expression and control of morphogenesis. Microbiol Rev 57:1–33
Falzarano MS, Passarelli C, Ferlini A (2014) Nanoparticle delivery of antisense oligonucleotides and their application in the exon skipping strategy for duchenne muscular dystrophy. Nucleic Acid Ther 24:87–100
Fire A, Xu S, Montgomery MK, Kostas SA, Driver SE, Mello CC (1998) Potent and specific genetic interference by double-stranded RNA in *Caenorhabditis elegans*. Nature 391:806–811
Furukawa K, Arimura N (1987) Purification and properties of 2,3-dihydroxybiphenyl dioxygenase from polychlorinated biphenyl-degrading *Pseudomonas pseudoalcaligenes* and *Pseudomonas aeruginosa* carrying the cloned bphC gene. J Bacteriol 169:924–927
Furukawa K, Miyazaki T (1986) Cloning of a gene cluster encoding biphenyl and chlorobiphenyl degradation in *Pseudomonas pseudoalcaligenes*. J Bacteriol 166:392–398
Good L, Awasthi SK, Dryselius R, Larsson O, Nielsen PE (2001) Bactericidal antisense effects of peptide-PNA conjugates. Nat Biotechnol 19:360–364
Griffith KL, Wolf RE Jr (2002) Measuring beta-galactosidase activity in bacteria: cell growth, permeabilization, and enzyme assays in 96-well arrays. Biochem Biophys Res Commun 290:397–402
Hamilton AJ, Baulcombe DC (1999) A species of small antisense RNA in posttranscriptional gene silencing in plants. Science 286:950–952
Izant JG, Weintraub H (1984) Inhibition of thymidine kinase gene expression by anti-sense RNA: a molecular approach to genetic analysis. Cell 36:1007–1015
Izant JG, Weintraub H (1985) Constitutive and conditional suppression of exogenous and endogenous genes by anti-sense RNA. Science 229:345–352
Janssen FW, Lund AJ, Anderson LE (1958) Colorimetric assay for dipicolinic acid in bacterial spores. Science 127:26–27
Jayaraman K, McParland K, Miller P, Ts'o PO (1981) Selective inhibition of *Escherichia coli* protein synthesis and growth by nonionic oligonucleotides complementary to the 3′ end of 16S rRNA. Proc Natl Acad Sci U S A 78:1537–1541
Kimura N, Nishi A, Goto M, Furukawa K (1997) Functional analyses of a variety of chimeric dioxygenases constructed from two biphenyl dioxygenases that are similar structurally but different functionally. J Bacteriol 179:3936–3943
Krötz F, de Wit C, Sohn HY, Zahler S, Gloe T, Pohl U, Plank C (2003) Magnetofection-a highly efficient tool for antisense oligonucleotide delivery in vitro and in vivo. Mol Ther 7:700–710
Kunst F, Ogasawara N, Moszer I, Albertini AM, Alloni G, Azevedo V, Bertero MG, Bessieres P, Bolotin A, Borchert S, Borriss R, Boursier L, Brans A, Braun M, Brignell SC, Bron S, Brouillet S, Bruschi CV, Caldwell B, Capuano V, Carter NM, Choi SK, Codani JJ, Connerton IF, Cummings NJ, Daniel RA, Denizot F, Devine KM, Dusterhoft A, Ehrlich SD et al (1997) The complete genome sequence of the gram-positive bacterium *Bacillus subtilis*. Nature 390:249–256
Kurreck J (2003) Antisense technologies. Improvement through novel chemical modifications. Eur J Biochem 270:1628–1644
Laemmli UK (1970) Cleavage of structural proteins during the assembly of the head of bacteriophage T4. Nature 227:680
Lavery KS, King TH (2003) Antisense and RNAi: powerful tools in drug target discovery and validation. Curr Opin Drug Discov Devel 6:561–569
LeDeaux JR, Yu N, Grossman AD (1995) Different roles for KinA, KinB, and KinC in the initiation of sporulation in *Bacillus subtilis*. J Bacteriol 177:861–863
Lee RC, Feinbaum RL, Ambros V (1993) The *C. elegans* heterochronic gene lin-4 encodes small RNAs with antisense complementarity to lin-14. Cell 75:843–854
Miller JH (1972) Experiments in Molecular Genetics. Cold Spring Harbor Laboratory Press, N.Y., pp 352–355

Mizuta T, Fujiwara M, Hatta T, Abe T, Miyano-Kurosaki N, Shigeta S, Yokota T, Takaku H (1999) Antisense oligonucleotides directed against the viral RNA polymerase gene enhance survival of mice infected with influenza A. Nature Biotechnol 17:583–587

Molle V, Fujita M, Jensen ST, Eichenberger P, González-Pastor JE, Liu JS, Losick R (2003) The Spo0A regulon of *Bacillus subtilis*. Mol Microbiol 50:1683–1701

Mondello FJ, Turcich MP, Lobos JH, Erickson BD (1997) Identification and modification of biphenyl dioxygenase sequences that determine the specificity of polychlorinated biphenyl degradation. Appl Environ Microbiol 63:3096–3103

Nakashima N, Tamura T (2009) Conditional gene silencing of multiple genes with antisense RNAs and generation of a mutator strain of *Escherichia coli*. Nucleic Acids Res 37:e103

Nakashima N, Tamura T, Good L (2006) Paired termini stabilize antisense RNAs and enhance conditional gene silencing in *Escherichia coli*. Nucleic Acids Res 34:e138

Rotman Y, Fields ML (1968) modified reagent for dipicolinic acid analysis Anal. Biochem 22:168

Sambrook J, Fritsch EF, Maniatis T (1989) Molecular cloning: a laboratory manual, 2nd edn. Cold Spring Harbor Laboratory, Cold Spring Harbor, N. Y

Shiloach J, Kaufman J, Guillard AS, Fass R (1996) Effect of glucose supply strategy on acetate accumulation, growth, and recombinant protein production by *Escherichia coli* BL21 (lambdaDE3) and *Escherichia coli* JM109. Biotechnol Bioeng 49:421–428

Somiya Y, Higo N, Yoshida N (2012) Combination of rolling vibration and serpentinite induces the formation of penetration-intermediates. Geomicrobiol J 29:820–829

Stephenson ML, Zamecnik PC (1978) Inhibition of Rous sarcoma viral RNA translation by a specific oligodeoxyribonucleotide. Proc Natl Acad Sci U S A 75:285–288

Taira K, Hirose J, Hayashida S, Furukawa K (1992) Analysis of *bph* operon from the polychlorinated biphenyl-degrading strain of *Pseudomonas pseudoalcaligenes* KF707. J Biol Chem 267:4844–4853

Tan IS, Ramamurthi KS (2014) Spore formation in *Bacillus subtilis*. Environ Microbiol 6:212–225

Triscari-Barberi T, Simone D, Calabrese FM, Attimonelli M, Hahn KR, Amoako KK, Turner RJ, Fedi S, Zannoni D (2012) Genome sequence of the polychlorinated-biphenyl degrader *Pseudomonas pseudoalcaligenes* KF707. J Bacteriol 194:4426–4442

Uhlmann E, Peyman A (1990) Antisense oligonucleotides: a new therapeutic principle. Chem Rev 90:543–584

Verma S, Eckstein F (1998) Modified oligonucleotides: synthesis and strategy for users. Annu Rev Biochem 67:99–134

Whitesell L, Rosolen A, Neckers LM (1991) In vivo modulation of N-myc expression by continuous perfusion with an antisense oligonucleotide. Antisense Res Dev 1:343–350

Wickstrom E (1992) Strategies for administering targeted therapeutic oligodeoxynucleotides. Trends Biotechnol 10:281–287

Yoshida N, Ide K (2008) Plasmid DNA is released from nanosized acicular material surface by low molecular weight oligonucleotides: exogenous plasmid acquisition mechanism for penetration intermediates based on the Yoshida effect. Appl Microbiol Biotechnol 80:813–821

Yoshida N, Sato M (2009) Plasmid uptake by bacteria: a comparison of the methods and efficiencies. Appl Microbiol Biotechnol 83:791–798

Yoshida N, Nakajima-Kambe T, Matsuki K, Shigeno T (2007) Novel plasmid transformation method mediated by chrysotile, sliding friction, and elastic body exposure. Anal Chem Insights 2:9–15

Zamecnik PC, Stephenson ML (1978) Inhibition of Rous sarcoma virus replication and cell transformation by a specific oligodeoxynucleotide. Proc Natl Acad Sci U S A 75:280–284

Zhao Q, Matson S, Herrera CJ, Fisher E, Yu H, Krieg AM (1993) Comparison of cellular binding and uptake of antisense phosphodiester, phosphorothioate, and mixed phosphorothioate and methylphosphonate oligonucleotides. Antisense Res Dev 3:53–66

11

Reduced catabolic protein expression in *Clostridium butyricum* DSM 10702 correlate with reduced 1,3-propanediol synthesis at high glycerol loading

Mine Gungormusler-Yilmaz[1,2], Dmitry Shamshurin[3], Marine Grigoryan[3], Marcel Taillefer[4], Victor Spicer[5], Oleg V Krokhin[5], Richard Sparling[4] and David B Levin[1*]

Abstract

Higher initial glycerol loadings (620 mM) have a negative effect on growth and 1,3-propanediol (1,3-PDO) synthesis in *Clostridium butyricum* DSM 10702 relative to lower initial glycerol concentrations (170 mM). To help understand metabolic shifts associated with elevated glycerol, protein expression levels were quantified by LC/MS/MS analyses. Thirty one (31) proteins involved in conversion of glycerol to 1,3-PDO and other by-products were analyzed by multiple reaction monitoring (MRM). The analyses revealed that high glycerol concentrations reduced cell growth. The expression levels of most proteins in glycerol catabolism pathways were down-regulated, consistent with the slower growth rates observed. However, at high initial glycerol concentrations, some of the proteins involved in the butyrate synthesis pathways such as a putative ethanol dehydrogenase (CBY_3753) and a 3-hydroxybutyryl-CoA dehydrogenase (CBY_3045) were up-regulated in both exponential and stationary growth phases. Expression levels of proteins (CBY_0500, CBY_0501 and CBY_0502) involved in the reductive pathway of glycerol to 1,3-PDO were consistent with glycerol consumption and product concentrations observed during fermentation at both glycerol concentrations, and the molar yields of 1,3-PDO were similar in both cultures. This is the first report that correlates expression levels of glycerol catabolism enzymes with synthesis of 1,3-PDO in *C. butyricum*. The results revealed that significant differences in the expression of a small subset of proteins were observed between exponential and stationary growth phases at both low and high glycerol concentrations.

Keywords: *Clostridium butyricum*; 1,3-propanediol synthesis; Glycerol catabolism; Proteomics; Multiple reaction monitoring

Introduction

1,3-propanediol (1,3-PDO) and terephthalic acid (TPA) are used to produce polytrimethylene terephthalate (PTT), a high performance polymer that is used in apparel, upholstery, carpet, specialty resins and other applications, where properties such as softness, comfort-stretch and recovery, dye ability, and easy-care are desired. Purified 1,3-PDO may also be used in adhesives, resins, laminates, powder coatings, films, solvents (ink-jet dyes), medicines, cosmetics, engine coolants, detergents, insect repellents and fragrances (DuPont 2014). Demand for PTT was estimated to be approximately 400,000 to 800,000 tones in 2013. Therefore, low-cost production of 1,3-PDO is important for the competitiveness in this market (Nakamura and Whited 2003).

1,3-PDO may be produced by chemical or biotechnological techniques. Multinational companies, like Degussa (DuPont™) and Shell Chemicals, use chemical processes for 1,3-PDO production (DuPont 2014; Shell Chemicals 2014). These processes, however, are high-cost and environmentally hazardous. In contrast, microbial production of 1,3-PDO, for example from glycerol, is less costly and more environmentally sustainable. Although glycerol generated as a by-product of biodiesel production processes has become a "waste" disposal problem for the biodiesel industry, crude biodiesel-derived glycerol can be used as a substrate for microbial production of 1,3-PDO (Clomburg and Gonzalez 2013). Moon et al. (2010) studied the effects of the impurities in biodiesel-derived glycerol and showed that in spite of having inhibitory effects on cell growth, the

* Correspondence: david.levin@umanitoba.ca
[1]Department of Biosystems Engineering, Faculty of Engineering, University of Manitoba, Manitoba, Winnipeg, MB R3T 5 V6, Canada
Full list of author information is available at the end of the article

impurities only slightly affected volumetric yields, suggesting the effective use of biodiesel-derived glycerol for microbial production of 1,3-PDO. However, regardless of its grade, glycerol concentrations over a certain threshold concentration (approximately 600 mM) negatively impact cell growth rates (Zeng et al. 1994), reduce glycerol consumption, and reduce 1,3-PDO yields (Gungormusler et al. 2011a). Conversely, greater consumption of glycerol results in greater 1,3-PDO yields. Therefore, it is important to understand the metabolic impacts of glycerol concentrations that negatively effect cell growth and 1,3-PDO synthesis.

Microbial conversion of glycerol to 1,3-PDO and a variety of other fermentation end-products, such as ethanol, acetate, and butyrate, has been well known since the early 1930s (Werkman and Gillen 1932; Mickelson and Werkman 1940). Facultative anaerobes such as *Klebsiella pneumoniae, Klebsiella oxytoca, Citrobacter freundii* (Pflugmacher and Gottschalk 1994; Homann et al. 1990), *Citrobacter werkmanii* (Maervoet et al. 2012), *Escherichia blattae* (Ripoll et al., 2012), *Enterobacter aerogenes* and *Enterobacter (Pantoea) agglomerans* (Ito et al. 2005; Barbirato et al. 1995), and strict anaerobes such as *Clostridium beijerinckii* (Gungormusler et al. 2011b), *Clostridium butyricum, Clostridium diolis* (Kaur, et al. 2012), and *Clostridium pasteurianum* (Biebl 1991; Biebl et al. 1992; Gonzalez-Pajuelo et al. 2005) are the most studied wild-type microorganisms that produce 1,3-PDO. In addition, some species of *Lactobacillus,* such as *Lactobacillus diolivorans* (Pflugl et al. 2012), and an uncharacterized marine species were also reported as possible 1,3-PDO producers when crude glycerol was used as the main substrate (Abad and Turon 2012). Considering the volumetric productivities and substrate tolerances of the above mentioned bacteria, *C. butyricum* is amongst the best producers (Wilkens et al. 2012).

In this study, two important factors were taken into consideration for bacteria selection: 1) the tolerance of *C. butyricum* to both high concentrations of 1,3-PDO (up to 93 g/L; Wilkens et al. 2012) and glycerol (up to 129 g/L, Himmi et al. 1999); and 2) the non-pathogenic physiology of *C. butyricum,* which does not require special safety precautions to grow, in contrast to *K. pneumoniae* (Saxena et al. 2009). 1,3-PDO is synthesized predominantly during exponential growth by *C. butyricum* when it is cultured on glycerol as a sole carbon source. During stationary phase, under carbon excess, however, butyrate is synthesized more dominantly by *C. butyricum* (Wilkens et al. 2012). An understanding of the underlying mechanisms that regulate the formation of fermentation products throughout the phases of microbial growth is essential for development of strategies to maximize 1,3-PDO synthesis.

Proteomics together with end-product analysis can help elucidate possible rate-limiting steps in the metabolic pathways of glycerol degradation to 1,3-PDO. Preliminary studies suggested a growth inhibition after a certain threshold concentration of initial glycerol (620 mM). Hence, the aim of this study was to investigate the expression of key enzymes associated with synthesis of 1,3-PDO and other fermentation end-products (acetate, butyrate, lactate, ethanol, H_2, and CO_2) by *C. butyricum* during exponential and stationary phases, under low and high initial glycerol concentrations.

Protein levels were determined using label-free liquid chromatography (LC) based quantitative proteomics using one-dimensional (1D) and two-dimensional (2D) LC/MS/MS, and Multiple Reaction Monitoring (MRM; Walsh et al. 2009). Our findings suggest a direct correlation between expression levels of key enzymes involved in the glycerol degradation pathways to 1,3-PDO and other end-products. Moreover, our data show that the expression levels of enzymes involved in both the oxidative and reductive pathways of glycerol catabolism are lower at higher glycerol concentrations, with a concomitant reduction in both the glycerol utilisation rate and the cell growth rate.

Materials and methods

Organism

Clostridium butyricum DSM 10702, which is equivalent to *Clostridium butyricum* 5521 from NCBI, was obtained from the German Type Culture Collection (DSMZ), and was used in all experiments described in this work. The genome sequence used is available through GenBank with the accession no. NZ_ABDT00000000.1.

Media and growth

The fermentation medium used for the production of 1,3-PDO was as described in Gungormusler et al. (2011b). The substrate concentration varied between 4.6 and 85.6 g/L glycerol (50 to 930 mM glycerol) for the trials, whereas for proteomic analysis the media contained 15.6 and 57 g/L (170 mM and 620 mM, respectively) of pure glycerol (initial concentration). Three independent biological replicate cultures were used for each experiment. The triplicate cultures were grown in anaerobic serum bottles (110 mL; Bellco Glass Inc., Vineland, NJ) containing 50 mL of fermentation medium (pH 7.0). All chemicals were obtained from Sigma Chemical Company (St. Loius, MO). Crude glycerol (78-86% w/v glycerol) was obtained from REG Danville, LLC (Danville, IL), and contained major impurities: methanol (0.02% w/v), ash (7% w/v), total fatty acid content (1.0%, w/v) and water (6-13% w/v). All gases were purchased from Welder's Supplies LTD (Winnipeg, MB, Canada).

C. butyricum was activated from lyophilized cultures in fermentation media containing 1 g/L of glucose and incubated anaerobically at 37°C for 24 h. The activated pure strain was used as inoculum (10% v/v) throughout the experiments in 27 mL Balch tubes (with 10 mL working volume) or 110 mL serum bottles (50 mL working volume) with agitation at 100 rpm. The media used for anaerobic experiments were gassed (1 min) with N_2 gas and degassed (4 min) for 4 cycles before autoclaving (Islam et al. 2006). Glycerol was used as the sole carbon source for 1,3-PDO synthesis throughout the study. However, revival and growth of *C. butyricum* from glycerol stocks was carried out in rich complex medium containing glucose, as described above.

Cell growth, pH, and end-product analysis

Cell growth was monitored spectrophotometrically (Biochrom Ultrospec 500 Pro, UK) at 600 nm. End-product analyses were performed as previously described (Gungormusler et al. 2011b) using high performance liquid chromatography (Waters Breeze™ HPLC System, USA) with a Phenomenex Rezex RHM Monosaccharide (H^+) 300 x 7.8 mm ion exchange column, using a Waters 2414 Refractive Index Detector, a Waters 1515 Isocratic HPLC Pump, and a Waters 2707 Autosampler. Gas measurements and calculations were carried out as described in Rydzak et al. (2012). Data are presented as the means of three independent biological replicates.

Preparation of peptides for proteomic analysis

Cells grown in medium containing low (170 mM) and high (620 mM) initial glycerol concentrations were harvested for the proteomic analyses described below from exponential and stationary phase cultures. Two biological replicates were used for each 1D and MRM proteomic run, whereas a 2D run was carried out with one sample from the exponential phase of the low initial glycerol concentration cultures. Nine (9) samples in total were analyzed for protein express levels with the various methods, as described below.

All chemicals were sourced from Sigma Chemicals (St-Louis, MO), unless noted otherwise. HPLC-grade acetonitrile and de-ionized water were used for the preparation of eluents. The in-house designed standard peptides (P1-P6) (Krokhin and Spicer 2009) were custom synthesized by BioSynthesis Inc. (Lewisville, TX) and purified individually using reverse phase (RP) HPLC. Sequencing-grade modified trypsin (Promega, Madison, WI) was used for digestion. A modified "filter-aided sample preparation" (FASP) method (Wisniewski et al. 2009) was used to extract proteins from *C. butyricum* cells cultured with the two different initial glycerol concentrations, at both exponential and stationary time points. The FASP procedure is based on complete lysis of all cell components using sodium dodecylsulfate (SDS), followed by the subsequent substitution with urea.

Protein amounts to be subjected to digestion were monitored using micro-BCA assay (Thermo Scientific, Pierce, Rockford, IL). Digests were acidified with TFA and purified by RP-HPLC. Protein profiles in the samples were monitored via a SDS-polyacrylamide gel electrophoresis (PAGE) separation. Monitoring the amounts of protein during the digestion procedure is critically important, as a substrate/enzyme ratio of ~50:1 is required for efficient tryptic digestion. Tryptic digests were conducted in the same filter used throughout the experiment containing 1–2 mg/mL protein lysate and 50–100 μg of trypsin. The falcon tubes were replaced with clean 50 mL falcon tubes and incubated overnight in dark at room temperature. Following tryptic digestion, peptide concentrations were estimated using Nanodrop 2000 (Thermo Scientific, Pierce, Rockford, IL). The peptide samples were acidified with TFA to a final concentration of 0.5% and desalted using a 1x100 mm column packed with 5 μm Luna C18 (Phenomenex, Torrance, CA). Purified peptide mixtures were lyophilized and stored at −80°C prior to LC-MS analysis.

Both 1D and 2D LC/MS/MS analyses were employed via identical nano-flow RP-HPLC instrumental settings in the final stage of separation prior to MS acquisition. A splitless nano-flow 2D-Ultra system (Eksigent/ABSciex, Dublin, CA) with 10 μL sample injection via a 300 μm × 5 mm PepMap100 (Thermo Scientific) trap-column and a 100 μm × 200 mm analytical column packed with 5 μm Luna C18 (Phenomenex, Torrance, CA) was used.

Protein identification and quantitation using LC/MS/MS

Both the 1D and 2D LC/MS/MS analysis were performed using an ABSciex TripleTOF 5600 TOF-MS system (Applied Biosystems, Foster City, CA) equipped with a Nano-sprayIII ionization source. Each acquisition cycle included 250 ms survey MS acqusition (m/z 400–1500) and up to twenty 100 ms MS/MS measurements on the most intense parent ions (300 counts/s threshold, +2 to +4 charge state, m/z 100–1500 mass range for MS/MS). The 1D runs for stationary and exponential phases were performed as biological replicates of two hours runs each.

The resulting raw WIFF files were processed using a standard conversion script bundled with Analyst QS 1.6 into Mascot Generic File format (MGF). MGF files contain information on charge, m/z, retention time of fragmented peptide as well as a list of m/z values and intensities for the CID fragment ions. Mass measurements for the parent and daughter ions are used for peptide identification, while intensities of the daughter ions were employed for quantitation.

All peptide identifications were done with our in-house GPU-based engine (McQueen et al. 2012). Searches used the 50 most intense fragment peaks with 20 ppm and + −0.1 Da mass tolerances on the parent and fragment masses, respectively. The NCBI annotation of *C. butyricum* was used for peptide identification, with hypothetical tryptic peptides with up to one missed cleavage permitted. The fixed modification of cysteine residues +57.021 Da (cysteine protection with iodoacetamide) was applied to all searches.

The TripleTOF 5600 provides very consistent MS/MS acquisition with excellent linearity between intensity of the parent peptide and its resulting daughter ions. We used protein Total Ion Current (TIC) as a relative measure of protein expression within an experimental run, which is simply the sum of MS/MS fragment ion intensities for every identified member peptide. This value is expressed in a $\log_2$ scale for simple differential analysis and plotting.

1D and 2D LC/MS/MS Analyses

Table 1 shows an overview of protein and peptide identifications for the eight runs. Protein identification confidence expectation values were computed using a Bayes theorem application of its member peptide expectation values, following the design elucidated by Beavis & Fenyo in the X! tandem search engine (http://thegpm.org/docs/peptide_-protein_expect.pdf), and is effectively the $\log_{10}$ measure of the probability that the protein assignment has a better random answer. Thus an expectation value < -10 has a one-in-10-billion probability of being incorrect.

A generalized analysis system was developed applying a unified data format to quantitation results from a wide range of omics data, across a collection of experiments for a target organism. Values associated with genes/proteins are mapped into Higher Order Variables (HOV) based on broad biological classifications including KEGG modules, COG letters, Enzyme Class numbers and MetaCyc pathways, allowing us to rapidly detect and evaluate system-wide changes in expression levels. These HOV mappings are driven by contents of the "extract gene information" file generated in the IMG/ER annotation workflow (Markowitz et al. 2012).

The 1D collection of runs was initially filtered to contain only proteins with at least two peptides; this filtering yielded ~ 800 proteins for each growth condition. The expression measurement used for each protein was TIC, measured as the sum of the signal intensities for all peptides derived from a specific protein, and expressed on a $\log_2$ scale. Differential analysis was constructed by subtracting log2 protein intensity values between points in the growth curve at both glycerol concentrations, and between glycerol concentrations at both points in the growth curve. These were done in replicates, yielding a total of eight differential expression values.

Each set of detected proteins each forms a three-lobed histogram, with the central population of data points containing proteins observed in both states, while the outer histograms contain proteins observed only in one state, but not the other. It is important to emphasize that these are manifestations of the instrument's information-directed acquisition (IDA) detection limits, and do not represent the absolute presence or absence of a protein in a sample.

For this analysis, we considered only the common populations, containing approximately 500 proteins for each difference set. For each difference set the values were normalized to a mean of zero and standard deviation of one (Z-scores), which exhibit roughly Gaussian distributions: proteins with differences $|Z| > 1.65$ represent the outermost 10% of the population, and $|Z| > 1.96$ the outermost 5% of the population. Z-scores for replicates were simply averaged to simplify the analysis, but show reasonable correlations.

Parallels for these eight 1D-IDA proteomic runs were conducted as targeted acquisition via MRM. Since there was precise control of the amount of protein injected for each MRM run, (relatively small) difference means were shifted to 0 only, rather than normalize them to Z-scores. This approach gives fold-change measurements of protein expression levels across the states, which were comparable across the two proteomic methods used.

Sample preparation and analysis using Multiple Reaction Monitoring (MRM)

Tryptic digests of *C. butyricum* for MRM experiments (QTRAP 5500) were separated using a linear 0.87% acetonitrile per minute gradient (0.4-37.4% acetonitrile in 42.5 min) followed by washing and equilibration steps. Both 1D and 2D analyses were done using the GPU system, with the same search settings.

Stock solutions containing 25 fM/uL of each of six custom peptides and ~1 ug/uL of digests were used for sample preparation by 10-times dilution prior sample injection. 10 uL (IDA) and 20 uL (MRM experiment) sample

Table 1 An overview of protein and peptide identifications for the 2D and the eight 1D runs

RUN	MS/MS Spectra	Total Peptides	Non-Redundant Peptides	Proteins Log(e) < −3	Proteins Log(e) < −10
2D-EXP	301661	133611	17975	2115	1918
EXP1	34182	17317	5448	883	722
EXP2	35122	17716	5533	902	756
ST1	34476	16490	5457	882	728
ST2	34526	16088	5298	881	716
G60EXP1	28806	13304	5651	980	776
G60EXP2	27663	12461	5513	988	788
G60ST1	27827	13238	5696	954	768
G60ST2	26874	12781	5584	962	764

loop volumes provided injection of ~1 ug of digest (25 fM of standard peptides) and ~2 ug of digest (50 fM of standard peptides) for IDA and MRM runs, respectively.

MRM experiments were performed using QTRAP 5500 (Applied Biosystems, Foster City, CA) hybrid triple quadrupole/ion trap mass spectrometer equipped with a nano-Spray III ion source (2900 V spray voltage and 150°C interface temperature). 60-min MRM methods were used with fixed 10 msec dwell time for each transition and "unit" resolution for both Q1 and Q3 quadrupoles. Collision energy for each transition was calculated according to standard QTRAP 5500 settings.

Protein identification and quantitation using MRM

The MRM method included 280 transitions from 33 proteins (2–3 peptides per protein, 3–4 transitions per peptide; Table 2; Additional file 1). Retention times of identified peptides were assigned using Skyline. Skyline is a freely available application for building MRM and analyzing the resulting mass spectrometry data (MacLean et al. 2010).

The target enzymes were selected using in-house pathway tools and the NCBI *C. butyricum* 5521 genome sequence. The end-product synthesis patterns were then correlated with the proteomic analyses (Figure 1b, c and d). The enzymes selected for analysis (and their corresponding locus tags) included: glycerol dehydratase (CBY_0501 and 0502), 1,3-PDO dehydrogenase (oxidoreductase) (CBY_0500), glycerol dehydrogenase (CBY_3235 and 0508), dihydroxyacetone kinase (dhaK) (CBY_0505, 0506, 3691 and 3690), pyruvate-ferredoxin oxidoreductase (CBY_0233, CBY_1889 and 3642), FeFe hydrogenase (CBY_2300), acetyl CoA acetyltransferase (CBY_1290), 3-hydroxybutyryl-CoA dehydrogenase (CBY_3045), 3-hydroxybutyryl-coa dehydratase (CBY_3041), acyl-CoA dehydrogenase (CBY_3258 and 3042), phosphate butyryltransferase (CBY_2919), butyrate kinase (CBY_2920), butanol dehydrogenase (CBY_3747 and 3751), phosphate acetyltransferase (CBY_0205), acetate kinase (CBY_0206), lactate dehydrogenase (CBY_0742, 2341 and 2757), and ethanol dehydrogenase (CBY_3753) (Figure 2 and Table 2).

Results

Cell growth and end-product analysis

Initial trials to determine the effects of the grade of glycerol on the yields of 1,3-PDO from *C. butyricum* were carried out with 170 mM of initial glycerol (crude versus pure) concentration (data not shown). No significant difference between crude and pure glycerol substrates for the production of 1,3-PDO were observed. Accordingly, further experiments were carried out using pure glycerol. Subsequently, additional experiments were carried out with increasing the initial glycerol concentrations (ranging between 50 mM and 930 mM) in order to understand the threshold values of substrate inhibition (see Additional file 2).

Growth curves of *C. butyricum* cultured with different initial glycerol concentrations revealed a prolonged lag-phase as the initial glycerol concentration increased (see Additional file 2). Cultures with the lowest initial glycerol concentration (50 mM) had a shorter lag phase and entered stationary phase earlier than the cultures with higher initial glycerol concentrations. Cultures with 620 mM initial glycerol concentration displayed longer lag-phases and exponential growth phases, and negligible growth was observed by *C. butyricum* at the two highest initial glycerol concentration cultures (820 and 930 Mm).

1,3-PDO synthesis was observed in cultures with an initial glycerol concentration of up to 820 mM, but the concentrations of 1,3-PDO produced at the highest glycerol concentrations was very low (5 mM; see Additional file 2). Given the negligible growth and synthesis of 1,3-PDO at 820 mM initial glycerol, this concentration was considered to be inhibitory to *C. butyricum*. The highest initial glycerol concentration that was not inhibitory for *C. butyricum* was found to be 620 mM. Acetate, butyrate, lactate, ethanol, H_2 and CO_2 were produced in varying amounts during fermentation by most, but not all, cultures with non-inhibitory initial concentrations of glycerol. Butyrate, however, was the only end-product that was observed in all cultures with non-inhibitory initial concentrations of glycerol.

Cell growth curves, pH profiles, and end-product synthesis profiles for *C. butyricum* cultured with two different initial glycerol concentrations are shown in Figure 1. *C. butyricum* uses both reductive and oxidative pathways during the degradation of glycerol (See Discussion section for details). As seen in Figure 1b and 1d, 1,3-PDO was always the primary reduced product followed by butyrate. As both seen in Table 3 and Figure 1b, cultures containing an initial concentration of 620 mM glycerol were always in carbon excess (even after the cells had reached stationary phase, 30 hrs post-inoculation), whereas cultures containing an initial concentration of 170 mM glycerol became carbon limited by 20 hrs post-inoculation, causing growth arrest. However, the molar yields of 1,3-PDO from the two culture conditions were very similar: 0.55 mol 1,3-PDO/mol glycerol consumed for the carbon excess cultures (620 mM initial glycerol) versus 0.51 mol 1,3-PDO/mol glycerol consumed in the carbon limiting cultures (170 mM initial glycerol). Molar yields of butyrate were also similar under both culture conditions, reaching up to 0.22 mol butyrate/mol glycerol consumed in both media.

As stated in Table 3, the specific growth rate, was lower in the high glycerol concentration cultures, suggesting growth inhibition. Although high glycerol concentrations reduced growth of *C. butyricum*, they did not significantly affect the substrate-specific yield of 1,3-PDO (mol 1,3-PDO per mole glycerol consumed). A previous study with *C. butyricum* also reported decrease in the growth rate as the total osmolarity of the medium

Table 2 The average Z-scores for 1D comparative LC/MS proteomic analyses (four columns after description) and MRM results (last four columns) for the key proteins

Locus Tag	Description	ST170 mM-EXP170 mM	ST620 mM-EXP620 mM	EXP620 mM-EXP170 mM	ST620 mM-ST170 mM	ST170 mM-EXP170 mM	ST620 mM-EXP620 mM	EXP620 mM-EXP170 mM	ST620 mM-ST170 mM
CBY_0205	phosphate acetyltransferase	−0.04	**1.24**	−0.85	0.25	NQ	0.68	NQ	NQ
CBY_0206	acetate kinase	0.00	0.19	−1.00	−0.99	0.25	0.60	−0.73	−0.40
CBY_0233	pyruvate:ferredoxin oxidoreductase	−0.84	0.09	0.30	**1.32**	NQ	NQ	NQ	NQ
CBY_0500	1.3-propanediol dehydrogenase	−0.26	−0.47	0.06	−0.12	−0.20	−0.32	−0.61	−0.75
CBY_0501	glycerol dehydratase (activator)	−0.63	−0.52	**−1.02**	**−1.05**	−0.64	−0.59	**−1.56**	**−1.53**
CBY_0502	glycerol dehydratase[a]	−0.34	−0.87	0.24	−0.21	−0.71	−0.74	−0.63	−0.67
CBY_0505	dihydroxyacetone kinase. (phosphotransfer subunit)	−0.31	**−1.55**	**−1.14**	**−2.52**	0.18	−0.06	**−1.32**	**−1.57**
CBY_0506	dihydroxyacetone kinase (L subunit)	0.47	0.28	0.08	−0.15	0.52	−0.27	−0.15	−0.95
CBY_0508	Glycerol dehydrogenase[b]	0.07	−0.19	0.04	−0.24	−0.39	−0.17	**−1.97**	**−1.77**
CBY_0742	L-lactate dehydrogenase	0.19	−0.79	**1.08**	0.29	0.22	−0.59	0.26	−0.56
CBY_1290	acetyl-CoA acetyltransferase	0.58	0.06	**1.19**	0.83	0.23	0.27	0.68	0.71
CBY_1889	pyruvate:ferredoxin oxidoreductase	−0.35	0.00	0.32	0.75	−0.04	0.50	**−1.12**	−0.60
CBY_2300	FeFe hydrogenase-1	−0.27	**−1.46**	**1.57**	0.69	−0.27	−0.49	−0.43	−0.67
CBY_2341	L-lactate dehydrogenase	0.66	0.10	0.51	−0.02	**1.31**	0.95	0.13	−0.25
CBY_2757	L-lactate dehydrogenase	−1.08	−1.43	0.28	0.08	NQ	0.07	NQ	NQ
CBY_2919	phosphate butyryltransferase	0.60	0.16	**1.15**	0.86	0.36	0.59	0.00	0.22
CBY_2920	butyrate kinase	0.62	0.54	0.89	0.91	0.26	0.32	−0.38	−0.33
CBY_3041	3-hydroxybutyryl-CoA dehydratase	0.50	0.29	**1.17**	**1.11**	NQ	NQ	NQ	NQ
CBY_3042	acyl-coa dehydrogenase (short-chain specific)	0.20	−0.25	0.90	0.60	−0.06	−0.06	0.96	0.94
CBY_3045	3-hydroxybutyryl-coa dehydrogenase	0.24	0.20	**1.16**	**1.28**	0.31	0.06	−0.04	−0.30
CBY_3235	Glycerol dehydrogenase[c]	−0.02	−0.52	**−1.47**	**−2.23**	−0.34	−0.36	**−2.66**	**−2.70**
CBY_3258	acyl-coa dehydrogenase (short-chain specific)	0.13	0.41	−0.39	−0.20	0.16	0.64	**−1.51**	**−1.05**
CBY_3642	pyruvate:ferredoxin oxidoreductase	−0.05	0.29	0.02	0.35	−0.01	0.49	**−1.04**	−0.56
CBY_3690	dihydroxyacetone kinase.(L subunit)	0.57	0.75	**−1.75**	**−1.94**	0.13	0.40	**−2.44**	**−2.19**
CBY_3691	dihydroxyacetone kinase (DhaK subunit)	0.42	0.58	−0.79	−0.81	NQ	NQ	NQ	NQ
CBY_3747	NADPH-dependent butanol dehydrogenase	0.35	−1.12	−0.12	**−1.61**	−0.13	−0.22	**−1.01**	**−1.11**
CBY_3751	NADPH-dependent butanol dehydrogenase	0.23	0.08	0.36	0.26	0.53	0.18	−0.57	−0.94
CBY_3753	Ethanol dehydrogenase (aldehyde-alcohol dehydrogenase 2)	**−2.58**	**−1.08**	**2.61**	**4.78**	−0.27	−0.12	−0.15	−0.01
CBY_9999	glycerol uptake facilitator protein	NQ	NQ	NQ	NQ	NQ	NQ	NQ	NQ

NQ: Not quantifiable. [a]Wrongly annotated as pyruvate formate-lyase.[b]Wrongly annotated as 3-dehydroquinate synthase.[c]Wrongly annotated as 3-dehydroquinate synthase.
The values indicate the differences between the samples taken from different growth phases (EXP; exponential and ST; stationary) and/or different initial glycerol concentrations (170 mM and 620 mM) of *C. butyricum* cells. Numbers in bold indicate differences in protein measurements that are greater that 2 fold.

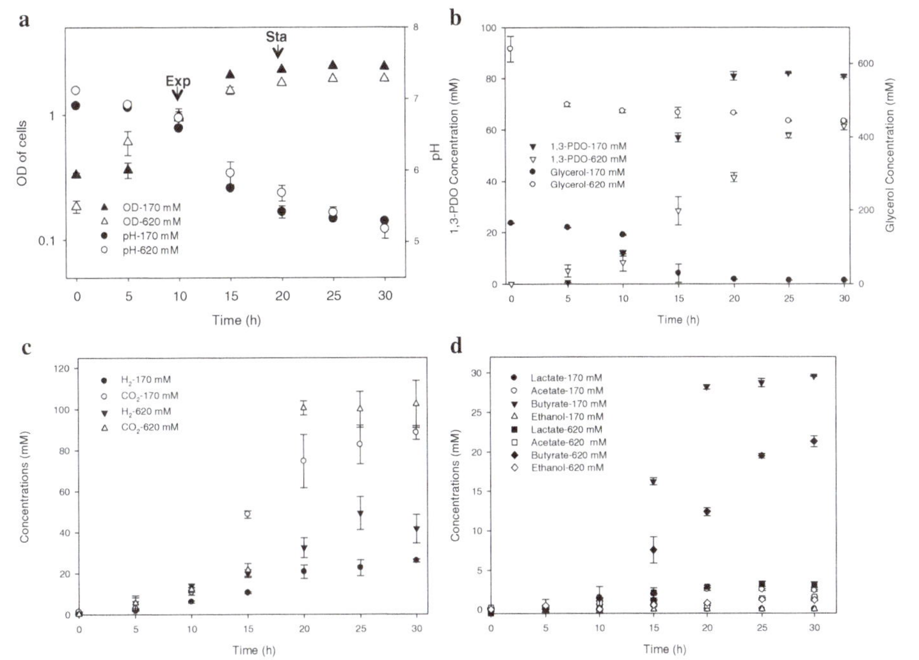

Figure 1 Growth curves, pH profiles, glycerol utilization, and end-product synthesis profiles of *C. butyricum* cultures with low (170 mM) and high (620 mM) initial glycerol concentrations. a) OD of cells in log scale versus pH values, **b)** residual glycerol concentrations versus 1,3-PDO concentrations, **c)** H_2 and CO_2 concentrations, **d)** lactate, acetate, butyrate and ethanol concentrations.

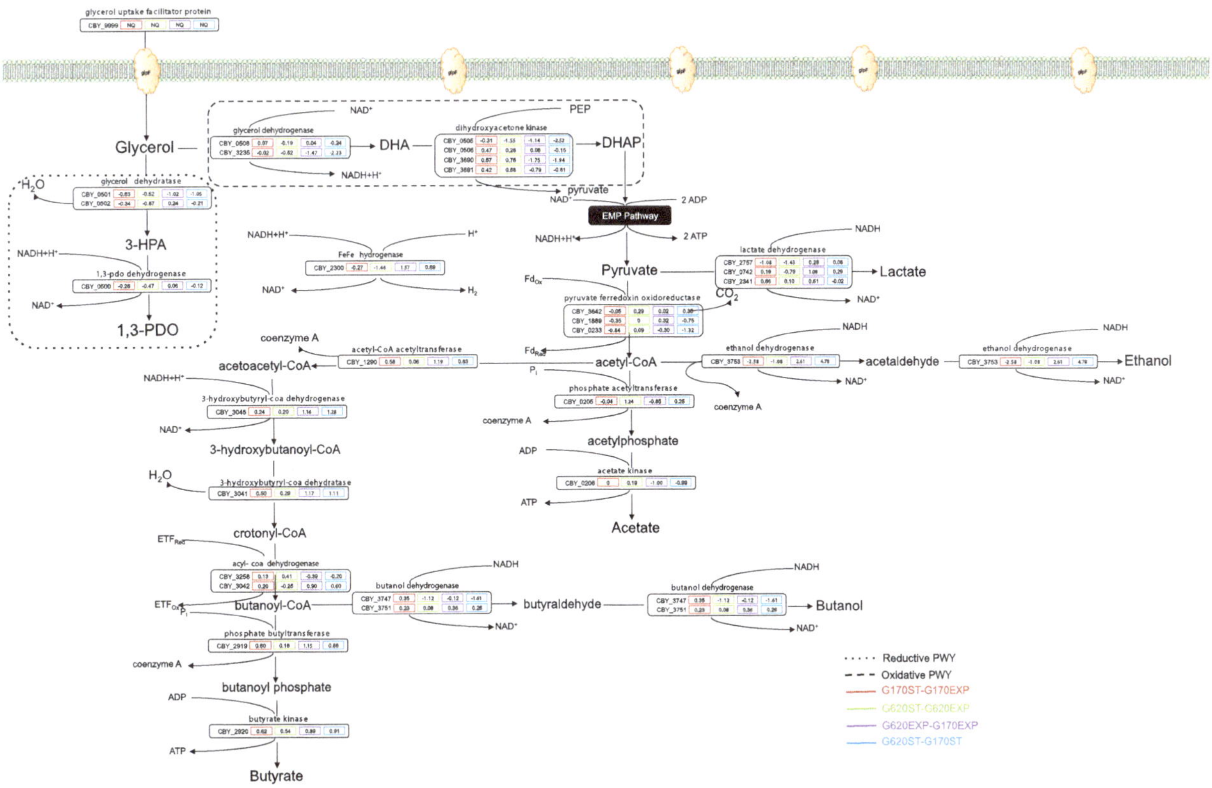

Figure 2 Differential protein measurements for the proteins in the suggested glycerol catabolism pathways in *C. butyricum*, in average Z-scores for exponential vs. stationary growth phases in media containing low (170 mM) versus high (620 mM) initial glycerol concentrations. Proteins with significant differences (Z-score > 1 or < −1) in expression between states were shown in bold.

Table 3 *C. butyricum* growth parameters, with aqueous and gaseous end-product concentrations for low (170 mM) vs high (620 mM) initial glycerol concentrations after 20 hours of fermentation

Parameter/Glycerol Concentration	170 mM	620 mM
C Balance	1.05	1.08
Specific Growth Rate (h^{-1}) [a]	0.175	0.113
Substrate utilization (%)	92.1	28.3
Y PDO/s (mol /mol)	0.51 ± 0.01	0.55 ± 0
PDO (g/L h)	0.31 ± 0.01	0.16 ± 0.01
PDO (g/L)	6.15 ± 0.13	3.13 ± 0.11
Y Buty/s (mol /mol)	0.18 ± 0	0.22 ± 0.01
Buty (g/L)	2.46 ± 0.05	1.28 ± 0.37
Y Acet/s (mol /mol)	0.02 ± 0	0
Acet (g/L)	0.23 ± 0.02	0.04 ± 0
Y H_2/s (mol /mol)	0.13 ± 0.02	0.19 ± 0.01
H_2 (mmol/L)	20.84 ± 3.29	32.47 ± 4.76
Y CO_2/s (mol /mol)	0.47 ± 0.08	0.60 ± 0.01
CO_2 (mmol/L)	74.76 ± 12.84	100.58 ± 3.42

[a]Growth rate was calculated from the increase in number of cells with time during the exponential phase. Carbon recovery was calculated from the ratio of (carbon out)/(carbon in) where the formula $C_4H_7O_2N$ (Zeng et al., 1993) was used for *C. butyricum* biomass; Y, Yield; PDO, 1,3-PDO; Buty, butyrate; Acet, acetate. Three independent biological replicates were used for each experiment.

increased (Biebl 1991), which may also be observed in 620 mM glycerol.

Protein identification in the core metabolic pathways

One 2D LC/MS/MS run (20 x 1 h runs) (Dwivedi et al., 2008) was conducted on an exponential phase sample from cells grown in low glycerol (170 mM) (see Additional file 1), primarily to survey the microorganism's proteome for future cross-organism and RNAseq studies. Detailed discussion of these results are outside the scope of this paper, but it does help validate the 1D-quant approach: while this instrument run set gave almost 9 times as many MS/MS spectra as the 1D runs, the number of highly confident proteins (log(e) < −10) identified increased only by a factor of 2.6 (~1900 proteins). This supports the notion that this genome is sufficiently simple (<4000 genes) to apply 1D quantitation analysis for accessing biologically relevant processes, especially catabolic processes.

Changes in the proteome during the exponential and stationary phases of *C. butyricum* cells, cultured in low and high initial glycerol concentrations, were further investigated using 1D label-free LC/MS/MS proteomics. A total of 3071 genes were annotated in the *C. butyricum* genome (Table 1) with total TIC scores spanning from ~ 12 to 27 units (in $\log_2$ scale; see Additional file 1) for 170 mM glycerol exponential phase, 170 mM glycerol stationary phase, 620 mM glycerol exponential phase, and 620 mM glycerol stationary phase, respectively.

The first objective of this work was to elucidate the key enzymes involved in the conversion of glycerol to 1,3-PDO and other fermentation end-products by 1D LC/MS/MS analysis on the basis of observed TIC scores. The second objective was to compare the expression levels of the targeted proteins using MRM, which is one of the most sensitive methods for quantification of known analytes (Kitteringham et al. 2009). All the enzymes of interest were expressed at detectable levels under all the conditions studied (see Additional file 1).

The intra-sample replicates and the cross sample replicates for both 1D and MRM analysis showed a linearity that demonstrates the similarity of the data, indicating that the data for 1D and MRM are highly correlated. The proteins selected for MRM were mainly found around the centroid (see Additional file 2) and below the centroid (see Additional file 2) when cross comparisons between glycerol concentrations and/or between growth phases were used, respectively. These results demonstrated that the selected proteins did not change in the same way in response to glycerol concentrations and growth phases. Specifically (see Additional file 2), the data generated from the intra-treatment replicate sample pairs for the two different initial glycerol concentrations were highly similar for both stationary and exponential phase samples. On the basis of these results and similar correlation trends, average values of the differences between states were used (Table 2). In addition, comparison of data derived from MRM analyses of exponential versus stationary phases, resulted in data that clustered more towards to the middle of the overall data set, indicating little change in the overall protein complement of core catabolic proteins examined, as the cells transitioned from exponential to stationary phase. Whereas comparison of data derived from MRM analyses of samples from low versus high initial glycerol concentrations resulted in data that were more spread-out across the overall data set (see Additional file 2).

Significant differences in the expression levels of the key proteins involved in glycerol catabolism, 1,3-PDO synthesis (CBY_0500, 0501 and 0502) were observed in cultures with different initial glycerol concentrations (Table 2). Specifically, a decrease was detected in CBY_0501 in higher glycerol concentration. Likewise, proteins corresponding to CBY_0505, 3235, 3690 and 3691 associated with the energy yielding oxidative arm of glycerol metabolism were also reduced in significant amounts in the presence of high glycerol. The TIC scores of the genes CBY_0506 and 3691 related to the synthesis of putative dihydroxyacetone kinase only slightly decreased in amounts during the stationary phase of growth in high glycerol relative to the low glycerol concentration in the stationary phase of growth (see Additional file 1).

In general, enzymes involved in butyrate synthesis were detected in higher amounts for both growth phases in high glycerol cultures, except for acyl-CoA dehydrogenase (CBY_3258), which showed lower expression levels in both exponential and stationary growth phases in high initial glycerol concentration cultures. Phosphate acetyltransferase and acetate kinase (CBY_0205 and 0206) also showed lower amounts in high glycerol except for the stationary phase in CBY_0205 (Table 2).

Transportation of glycerol through the cell membrane can be carried out by (GlpF), which forms a highly selective trans-membrane channel that conducts glycerol and small, uncharged organic molecules (Stroud et al. 2003), thus maintaining the membrane potential. A putative glycerol uptake facilitator, CBY_9999, was detected in the proteome of *C. butyricum* (see Additional file 1) according to the 1D-IDA analysis. Although these findings cannot be supported by MRM results, it is important to point out that the expression levels (as measured by the TIC scores) of this protein were above detectable levels. However, when the initial glycerol concentration of the culture was altered, only slight changes in the expression of CBY_9999 were observed.

According to the proteome analyses, expression levels of most *C. butyricum* enzymes did not change significantly under the two growth conditions tested (Table 2). However, a key subset of proteins involved in the synthesis of 1,3-PDO from glycerol did display significant differences, including enzymes involved in glycerol oxidation, glycerol reduction to 1,3-PDO production, and butyrate synthesis were up or down regulated in the high glycerol concentration cultures as discussed above.

Discussion

Glycerol uptake

Glycerol is an osmolyte, and changes observed in cells grown in media with different initial glycerol concentrations may be explained by possible changes in osmotic pressure in the cell (Kubiak et al., 2012). Several studies have reported on the effect of various glycerol concentrations (Zhang et al. 2007; Sattayasamitsathit et al. 2011; Metsoviti et al. 2012; Kang et al. 2013). High initial glycerol concentrations have been observed to result in lower glycerol consumption rates, with prolonged fermentation times. In addition, depending on the species and operational conditions, certain threshold concentrations of glycerol (e.g. >60 g/L for *C. butyricum*, as reported in this study) inhibit bacterial growth.

A study by Sattayasamitsathit et al. (2011) showed that as the initial glycerol concentration increased above 60 g/L (up to 100 g/L), *Klebsiella pneumoniae* SU6 cells started to consume less glycerol, indicating a possible growth inhibition as a result of the higher concentration of certain impurities in the crude glycerol. *Citrobacter freundii* produced lower concentrations of 1,3-PDO (45.9 g/L) as the initial glycerol concentration increased to 100 g/L, suggesting growth inhibition (Metsoviti et al. 2012). A decrease in 1,3-PDO production (26%) with a concomitant increase in ethanol production (15%) was reported by *Lactobacillus panis* when the initial glycerol concentration was doubled from 150 mmol/L to 300 mmol/L (Kang et al. 2013). The authors suggested that glycerol fermentation was in competition with ethanol production and thus served as an alternate route for NAD regeneration. Our findings with *C. butyricum* are consistent with these results, suggesting an alteration in the regeneration routes of NADH as the initial glycerol concentration increases. In contrast to *L. panis,* which synthesized more ethanol as the initial glycerol concentration increased, we observed an increase in lactate synthesis by *C. butyricum.*

Glycerol can be transported through the membrane either by passive diffusion or via the glycerol facilitator protein (a glycerol permease), GlpF (da Silva et al. 2009; Sun et al. 2008). Therefore, the transportation through the membrane is done by diffusion and ATP utilization is not required. The results from the proteomic analysis (1D-IDA only) showed that GlpF (CBY_9999; Figure 2) was detected in cells grown in both low and high initial glycerol concentrations (except for one biological replicate in the exponential phase of the high glycerol concentration cultures). However, the TIC scores for this protein, were near the lower detection limit when detected, indicating that the facilitator protein may not play a major role at either of the glycerol concentrations tested.

Reductive pathways of glycerol degradation

The yields and productivities that can be achieved from the degradation of a specific substrate are adjusted such that ATP gain and thermodynamic efficiency are optimal for the respective growth conditions. Although the thermodynamic efficiency of ATP synthesis through acetate production pathways was reported to be higher in *C. pasteurianum* (85%) than the efficiency of acetate production pathways (63%), butyrate was always produced suggesting an incompatibility of high efficiency with the entropy requirements of the cells (Thauer et al. 1977). In addition, reducing equivalents are formed simultaneously in the form of reduced NADH during the formation of biomass (via the oxidation of glycerol), since glycerol has a higher degree of reduction per carbon than the cell mass ($CH_{1.9}O_{0.5}N_{0.2}$). In this case, the ability of the microorganism to maintain an overall redox balance would also be relevant to its ability to produce a product more reduced than glycerol by the consumption of the reducing equivalents (Nakamura and Whited 2003). When compared with sugars, such as glucose, this degradation to pyruvate generates twice the number of reducing equivalents. Regeneration of oxidized nicotinamide adenine dinucleotide (NAD^+) requires the formation of a more reduced end-product, such as 1,3-PDO, to serve as an electron sink (Nakamura and Whited 2003; Clomburg and Gonzalez 2013).

C. butyricum is a strict anaerobe, and maintains a balanced redox by utilizing two pathways (Figure 2). In the reductive pathway, glycerol is first dehydrated to the toxic intermediate 3-hydroxypropionaldehyde (3-HPA) and then 3-HPA is reduced to 1,3-PDO by the regeneration of oxidized NAD^+ using 1,3-PDO dehydrogenase. A study to increase the NADH availability in *K. pneumoniae* was carried-out by Ma et al. (2012) with the introduction of an *fdh* from *Candida boidinii* into the microorganism. The results were in agreement with our statement and demonstrated that the increase of NADH availability could efficiently improve glycerol metabolism, suggesting that increased 1,3-PDO yields are not only controlled by expression levels of key enzymes, but also by the intracellular concentration of NADH.

On the basis of the results (Figure 2), the total flux of carbon through both the glycerol reductive and oxidative branches are reduced in the presence of 620 mM glycerol. This is consistent with the proteomic results which show greater than 2-fold reduction in expression for key enzymes from both branches: putative glycerol dehydratase (CBY_0501), glycerol dehydrogenase (CBY_3235), and DhaK (CBY_0505 and 3690), which are key enzyme components in the reductive and oxidative pathways affected by high glycerol concentrations. Interestingly, many of these are located in the same genome neighbourhood (except for CBY_3691, 3690 and 3235), possibly forming a transcriptional operon. This putative operon would manage both reductive and oxidative branches, leading to balanced pathway utilization.

Abbad-Andaloussi et al. (1996) reported that glycerol dehydratase prevents intracellular accumulation of 3-HPA, therefore it is important to have a high glycerol dehydratase activity in order to increase the 1,3-PDO yields. According to the authors, based on excess NADH measurements, glycerol dehydratase limits the activity of 1,3-PDO dehydrogenase and it is the rate-limiting step in this fermentation. In this paper, glycerol dehydratase levels (CBY_0502) were also always higher than 1,3-PDO dehydrogenase levels (CBY_0500) based on their TIC scores (see Additional file 1). Gonzalez et al. (2013) used 2D gel electrophoresis coupled to MS analysis to compare the differentially expressed proteins in *Clostridium* sp. native strain (IBUN 158B) in two phases of growth (exponential and stationary) during 1,3-PDO synthesis. Thirty-two proteins were identified in the Gonzalez et al. (2013) study, including 1,3-PDO dehydrogenase, which was detected in both exponential and stationary phases. Our results are consistent with these results.

A recent study by Zhu et al. (2014) reported the effects of altered oxido-reduction potential (ORP) in protein expression by combined proteomic profiling and flux balance analysis in *K. oxytoca* cultures. The authors stated that the ratio of metabolic fluxes through reductive and oxidative pathways of glycerol degradation were sensitive to changes in extracellular ORP, and reductive pathways were channelled by low ORP. In addition, 2D gel electrophoresis showed that low ORP correlated with increased expression levels of an NADPH-dependent hypothetical oxidoreductase (HOR), but did not alter expression of 1,3-propanediol oxidoreductase. In our study, although the TIC values of 1,3-propanediol oxidoreductase (as also known as 1,3-propanediol dehydrogenase) were significantly high, suggesting the importance of this enzyme in the production of 1,3-PDO, the expression levels of the enzyme only slightly changed in both conditions studied.

Oxidative pathways of glycerol degradation

In the oxidative pathway, glycerol is first oxidized to dihydroxyacetone (DHA) (also known as glycerone) using NAD^+ as a cofactor by glycerol dehydrogenase. Then DHA is phosphorylated to dihydroxyacetonephosphate (DHAP) using a phosphoenolpyruvate dependent DHA kinase. In some clostridia glycerol can also be utilized via a glycerol kinase and glycerol-P-dehydrogenase pathway (Gonzalez-Pajuelo et al. 2006). Interestingly, a putative glycerol-P-dehydrogenase was detected in relatively high abundance in the proteome of *C. butyricum*, but it is not believed to utilize this pathway for glycerol oxidation as the TIC scores of the putative glycerol kinase was below detectable levels (see Additional file 1). Wang et al. (2003) compared the differences in growth phases in fed-batch fermentation by *K. pneumoniae* using 2D gel electrophoresis. The authors stated that this strain also uses a PEP-dependent DhaK as opposed to the ATP-dependent enzyme previously reported for *K. pneumoniae* and *C. freundii*.

Following DHAP production, it enters the Embden-Meyerhof-Parnas (EMP) pathway for glycolysis, resulting in a net gain of 2 moles of ATP and 1 mole of NADH per mole of DHAP. Following the EMP pathway, pyruvate can be reduced to lactate or oxidized to acetyl-CoA plus CO_2. Electrons from this oxidation can be used for hydrogen production allowing the cells to conserve further energy (ATP) through the conversion of acetyl-CoA to acetate through acetyl-CoA phosphotransferase and acetate kinase. In the presence of hydrogen, there are excess reducing equivalents in the pool that can further be converted into other reduced compounds in order to maintain the redox balance.

Initially, the conversion of pyruvate to acetyl-coA by the utilization of pyruvate-ferrodoxin oxidoreductase results in the formation of CO_2 and a reduced ferredoxin (Figure 2). This reduced ferredoxin is then used as an electron transporter for the formation of butanoyl-CoA by acyl-CoA dehydrogenase, which is then further converted to butyrate following two more steps. As a result, a ratio of 1 CO_2 to 1/2 butyrate (mol:mol) is maintained following the degradation of pyruvate. Butyrate production leads to 0.5 mole of

ATP produced whereas acetate production leads to 1 mole of ATP produced for every pyruvate consumed. Considering that acetate production does not consume NADH as opposed to butyrate, this pathway favours 1,3-PDO production. However, under the experimental conditions, in the presence of hydrogen in the atmosphere, butyrate was produced at significantly higher levels than acetate (Figure 1d). Our data are consistent with the 2D gel electrophoresis-MS proteomic study of Gonzalez et al. (2013), which showed that the enzymes in the butyrate synthesis pathways (CBY_3041, 3045, 3258, 3042, 2919, and 2920) were present in both exponential and stationary growth phases.

All the by-products of glycerol catabolism (lactate, ethanol, butanol, H_2, and particularly butyrate) compete with 1,3-PDO for reducing equivalents (NADH) during their synthesis. While lactate production from pyruvate uses 1 mole of NADH, ethanol production from acetyl-CoA through acetaldehyde results in the consumption of 2 moles of NADH. In addition, ethanol synthesis results in the formation of CO_2 and a reduced ferredoxin. As seen in Figure 2, pyruvate catabolism to acetyl-CoA generates a reduced ferredoxin that can be re-oxidized in the formation of H_2 or possibly for butyrate synthesis (ETF_{Red} to ETF_{Ox}). Although a putative hydrogenase (CBY_2676) was annotated in the genome of *C. butyricum*, in our study, it was below detectable levels in the proteome. Thus, butyrate might be the strongest competitor for NADH.

Theoretically, while acetic acid formation leads to the formation of 1 mole of excess NADH per mole of acetic acid formed, butyrate and lactate synthesis do not result in excess NADH per mole of the end-product formed, since NADH is utilized by the dehydrogenases (Kubiak et al. 2012). The excess NADH might be utilized if butanol or ethanol pathways are active for the conversion of acetyl-CoA into these solvents. The ratio of acetate to butyrate can be related to the metabolic status of the cell, and the increase in the acetate:butyrate ratio (acetate concentration/butyrate concentration) may act as an indicator of a faster microbial growth. Our results were consistent with this statement. As seen in Table 3, growth of *C. butyricum* was faster in low initial glycerol concentration cultures. Accordingly, the acetate:butyrate ratio was higher in low initial glycerol concentration cultures (1.17) than high initial glycerol concentration cultures (0.25).

The effects of various end-products synthesized during the oxidative degradation of glycerol on *C. butyricum* growth has been analyzed in some detail (Colin et al. 2000; Barbirato et al. 1996, Biebl 1991). (Colin et al. 2000) showed that *C. butyricum* cells had better tolerance to higher concentrations of internally produced 1,3-PDO (83.7 g/L) than externally added 1,3-PDO (65 g/L). Accordingly, the authors related the inhibitory effects of the diols (1,3-PDO, 1,2-PDO, 2,3-BD) to the general phenomenon of alcohol inhibition in microbial fermentations. Another study by Biebl (1991) reported the concentrations of acetate (27 g/L), butyrate (19 g/L), and 1,3-PDO (60 g/L) that were inhibitory to *C. butyricum* and showed that 1,3-PDO was the main inhibitor. However, the acetate:butyrate ratio was dependent on the environmental conditions, such as pH and glycerol concentration, and could reduce the production of 1,3-PDO. Chatzifragkou et al. (2012) compared the effects of high 1,3-PDO concentrations (biologically produced versus externally added, up to 70 g/L) in batch and continuous cultures. The results showed that *C. butyricum* was less resistant to the inhibitory effects of diols in batch cultures than in continuous cultures.

Although neither butanol nor ethanol were detected (Figure 1d), the *C. butyricum* genome contains the putative genes that encode the enzymes required for synthesis of these end-products during glycerol fermentation (Saint-Amans et al. 2001). In addition, the proteomic analyses revealed that the enzymes involved in ethanol synthesis were expressed and were detected at levels above detectable limits (TIC scores around 22.6, see Additional file 1). However, end-product analysis revealed the fact that expression of a protein does not necessarily correlate with the formation of the expected end-product, most likely because of post-translational regulation such as allostery (Monod et al. 1963). Similarity, despite the lack of butanol production, the enzymes responsible for butanol synthesis were also expressed in the proteome. As seen in Table 2, one putative alcohol dehydrogenase (CBY_3747) was down-regulated, while another putative alcohol dehydrogenase (CBY_3751) was up-regulated, in the higher initial glycerol concentration cultures (Figure 1c). The putative ethanol synthesizing alcohol dehydrogenase, AdhE (encoded by CBY_3753), was expressed at higher levels during both exponential and stationary phase in high initial glycerol concentration cultures than in low glycerol concentration cultures. Initial trials to determine the threshold for ethanol consumption by *C. butyricum* showed that the microorganism can consume ethanol up to a concentration of 25 mM (data not shown) and this could possibly explain the lack of net ethanol production.

A second possible explanation this observation could be the competition for NADH. A mutant strain of *C. butyricum* (W5) with a blocked butyrate synthesis pathway revealed that during glucose fermentation, increased hydrogen production leads to decreased ethanol formation, even though the opposed was expected. The authors suggested that a bifurcating hydrogenase encoded by *C. butyricum* W5 may compete for NADH (Cai et al. 2011).

In addition to a lack of ethanol, there was also no evidence for lactate production in our low glycerol concentration cultures. However, up to 2.9 mM of lactate was produced when initial glycerol concentration was increased (Figure 1d). This could be explained by the higher expression levels of two of the genes encoding LDHs (CBY_0742 and CBY_2341) during exponential phase in the high glycerol concentration cultures, based on their TIC scores. On the other hand, for all possible genes encoding LDHs (CBY_2757, CBY_0742, and CBY_2341), the expression levels did not change significantly during stationary phase in the high glycerol concentration cultures.

In *C. butyricum*, under excess substrate, $NADP^+$:ferredoxin oxidoreductase (CBY_3749) can be utilised to transfer the electrons from ferredoxin to NAD^+ instead of being released as H_2, contributing to additional 1,3-PDO formation (Leja et al. 2011). Our proteomic analysis, however, showed the presence of an annotated single sub-unit, ferredoxin-dependent, iron-only hydrogenase (CBY_2300) in relatively high abundance (see Additional file 1), which suggests that the observed H_2 production in *C. butyricum* can be carried out by the oxidation of reduced ferredoxin derived from the oxidation of pyruvate to acetyl-CoA and CO_2 through pyruvate ferredoxin oxidoreductase (CBY_0233, 1889 and 3642). A hydrogenase can also act as a safety valve to maintain the redox balance by using the excess reducing equivalents due to the fact that this is an exergonic reaction and no energy is conserved (Madigan et al. 2009).

In the proteome of *C. butyricum*, when the initial glycerol concentration of glycerol was increased, most of the pathways for the synthesis of soluble end-products (1,3-PDO, lactate, butanol, butyrate, acetate and ethanol), except for butyrate, lactate and ethanol (CBY_3045, 0742 and 3753), were down-regulated in both growth phases (Table 2). Furthermore, the key proteins that control flux of glycerol through the reductive and oxidative branches were down-regulated which is consistent with the lower 1,3-PDO concentrations in high glycerol concentration cultures (Figure 1b). In addition, based on the TIC scores, the putative ethanol dehydrogenase was expressed in considerable amounts in the proteome (CBY_3753, see Additional file 1), even though there were low concentrations of ethanol in the cultures with high initial glycerol concentration and no detectable production of ethanol was observed in cultures with low initial glycerol concentration.

The low concentration of ethanol in the high glycerol cultures could be attributed to the lower rate of glycerol consumption and lower growth rate in the presence of high glycerol concentration (Table 3). The expression levels of enzymes in a recombinant *E. coli* engineered to synthesize 1,3-PDO were analyzed by 2D gel electrophoresis. The results showed that as the initial glycerol concentration of the cultures was reduced, the major enzymes, expression of all increased and synthesis of 1,3-PDO was enhanced (Jin and Lee 2008). Our observations were consistent with these findings.

This paper correlates expression levels of key enzymes involved in glycerol catabolism and 1,3-PDO synthesis with end-product synthesis rates. This is the first study reporting the changes in the proteomic profiles of *C. butyricum* cells under different culture conditions using comparative proteomic analysis. The results of this study showed that higher glycerol concentrations can reduce the expression levels of enzymes involved in both pathways (oxidative and reductive), consistent with lower growth rate and glycerol utilisation rate. The results indicate that there is an optimum initial loading of glycerol for fastest growth rate and 1,3-PDO production. However, there is stability in the proportion of 1,3-PDO produced relative to other end-products generated under the growth conditions and growth phases tested. Nevertheless, the proteomic results described can serve as baseline for optimization of fermentation or selection of targets for genetic engineering for end-product optimization which could lead to more efficient production processes with low cost downstream processes and higher final yields of the main product.

Additional files

Additional file 1: Data from 2D, 1D LC/MS/MS, and MRM proteomic analyses. Expression levels for each protein are indicated as the Total Ion Current (TIC), measured as the sum of the signal intensities for all peptides associated with each specific protein, and expressed on a log2 scale.

Additional file 2: Growth curves and end-product synthesis profiles of *C. butyricum* cultures with increasing (50, 170, 300, 620, 820, 930 mM) initial glycerol concentrations. a) Growth curves, b) 1,3-PDO, butyrate, acetate, and lactate concentrations, c) H_2 and CO_2 concentrations. Correlation between the Z-scores of all the proteomics data between intra- and cross-replicate samples for both 1D-IDA (closed circles) and MRM (open circles) analyses. Differences between a) 170 and 170 mM, b) 620 and 620 mM, c) 170 and 620 mM, d) 170 and 620 mM initial glycerol concentrations. ST, Stationary phase; EXP, Exponential phase; R1, Replicate one; R2, Replicate two. Correlation between the Z-scores of intra- and cross-replicate samples: a) 1D, b) MRM, c) 1D versus MRM analysis for key proteins in exponential versus stationary growth phase, in cultures with low (170 mM) versus high (620 mM) initial glycerol concentrations. ST, Stationary phase; EXP, Exponential phase; R1, Replicate one; R2, Replicate two.

Competing interests
The authors declare that they have no competing interests.

Acknowledgements
The authors would like to acknowledge Prof. Dr. Nuri Azbar for his valuable contribution and insightful discussions. The authors would like to thank Dr. Graham Alvare for his technical assistance with bioinformatics. This work was funded by The Scientific And Technological Research Council Of Turkey (TUBITAK, BIDEB 2214 and CAYDAG Project No: 109Y150), by Genome Canada, through the Applied Genomics Research in Bioproducts or Crops (ABC) program for the grant titled, "Microbial Genomics for Biofuels and CoProducts from Biorefining Processes", and by the Province of Manitoba, through the Manitoba Research Innovation Fund (MRIF).

Author details
[1]Department of Biosystems Engineering, Faculty of Engineering, University of Manitoba, Manitoba, Winnipeg, MB R3T 5 V6, Canada. [2]Department of Bioengineering, Ege University, Izmir, Turkey. [3]Manitoba Centre for Proteomics and Systems Biology, University of Manitoba, Winnipeg, Manitoba, Canada. [4]Department of Microbiology, University of Manitoba, Winnipeg, Manitoba, Canada. [5]Department of Physics and Astronomy, University of Manitoba, Winnipeg, Manitoba, Canada.

References

Abad S, Turon X (2012) Valorization of biodiesel derived glycerol as a carbon source to obtain added-value metabolites: Focus on polyunsaturated fatty acids. Biotechnol Adv 30:733–741, doi: 10.1016/j.biotechadv.2012.01.002

Abbad-Andaloussi S, Guedon E, Spiesser E, Petitdemange H (1996) Glycerol dehydratase activity: the limiting step for 1,3-propanediol production by *Clostridium butyricum* DSM 5431. Lett Appl Microb 22(31):311–314, doi: 10.1111/j.1472-765X.1996.tb01168.x

Barbirato F, Camarasa-Claret C, Grivet JP, Bories A (1995) Glycerol fermentation by a new 1,3-propanediol-producing microorganism:*Enterobacter agglomerans*. Appl Microbiol Biotechnol 43(5):786–793, doi: 10.1007/BF02431909

Barbirato F, Grivet JP, Soucaille P, Bories A (1996) 3-Hydroxypropionaldehyde: An inhibitory metabolite of the glycerol fermentation by enterobacterial species. Appl Environ Microbiol 62:1448–1451

Biebl H (1991) Glycerol fermentation of 1,3-propanediol by *Clostridium butyricum*. Measurement of product inhibition by use of a pH-auxostat. Appl Microbiol Biotechnol 35:701–705, doi: 10.1007/BF00169880

Biebl H, Marten S, Hippe H, Deckwer W (1992) Glycerol conversion to 1,3-propanediol by newly isolated clostridia. Appl Microbiol Biotechnol 36:592–597, doi: 10.1007/BF00183234

Cai G, Jin B, Saint C, Monis P (2011) Genetic manipulation of butyrate formation pathways in *Clostridium butyricum*. J Biotechnol 155(3):269–274, doi: 10.1016/j.jbiotec.2011.07.004

Chatzifragkou A, Aggelis G, Gardeli C, Galiotou-Panayotou M, Komaitis M, Papanikolaou S (2012) Adaptation dynamics of *Clostridium butyricum* in high 1,3-propanediol content media. Appl Microbiol Biotechnol 95:1541–1552, doi: 10.1007/s00253-012-4003-6

Clomburg JM, Gonzalez R (2013) Anaerobic fermentation of glycerol: a platform for renewable fuels and chemicals. Trends Biotechnol 31:20–28, doi: 10.1016/j.tibtech.2012.10.006

Colin T, Bories A, Moulin G (2000) Inhibition of *Clostridium butyricum* by 1,3-propanediol and diols during glycerol fermentation. Appl Microbiol Biotechnol 54:201–205

da Silva GP, Mack M, Contiero J (2009) Glycerol: a promising and abundant carbon source for industrial microbiology. Biotechnol Adv 27(1):30–39, doi: 10.1016/j.biotechadv.2008.07.006

DuPont (2014) DuPont™, The miracles of science™., http://www.dupont.com/products-and-services/fabrics-fibers-nonwovens/fibers/brands/dupont-sorona/articles/how-dupont-sorona-is-made.html. Accessed 17 January 2014

Dwivedi RC, Spicer V, Harder M, Antonovici M, Ens W, Standing KG, Wilkins JA, Krokhin OV (2008) Practical implementation of 2D HPLC scheme with accurate peptide retention prediction in both dimensions for high-throughput bottom-up proteomics. Anal Chem 80(18):7036–7042, doi: 10.1021/ac800984n

Gonzalez NC, Vallejo AF, Sanchez-Gomez M, Montoya D (2013) Protein identification in two phases of 1,3-propanediol production by proteomic analysis. J Proteomics 89:255–264, doi: 10.1016/j.jprot.2013.06.011

Gonzalez-Pajuelo M, Meynial-Salles I, Mendes F, Andrade JC, Vasconcelos I, Soucaille P (2005) Metabolic engineering of *Clostridium acetobutylicum* for the industrial production of 1,3-propanediol from glycerol. Metab Eng 7(5–6):329–336, doi: 10.1016/j.ymben.2005.06.001

Gonzalez-Pajuelo M, Meynial-Salles I, Mendes F, Soucaille P, Vasconcelos I (2006) Microbial conversion of glycerol to 1,3-propanediol: Physiological comparison of a natural producer, *Clostridium butyricum* VPI 3266, and an engineered strain, *Clostridium acetobutylicum* DG1(pSPD5). Appl Environ Microbiol 72(1):96–101, doi: 10.1128/AEM.72.1.96-101.2006

Gungormusler M, Gonen C, Azbar N (2011a) 1,3-Propanediol production potential by a locally isolated strain of *Klebsiella pneumoniae* in comparison to *Clostridium beijerinckii* NRRL B593 from waste glycerol. J Polym Environ 19:812–817, doi: 10.1007/s10924-011-0326-0

Gungormusler M, Gonen C, Azbar N (2011b) Continuous production of 1,3-propanediol using raw glycerol with immobilized *Clostridium beijerinckii* NRRL B-593 in comparison to suspended culture. Bioprocess Biosyst Eng 34:727–733, doi: 10.1007/s00449-011-0522-2

Himmi EH, Bories A, Barbirato F (1999) Nutrient requirements for glycerol conversion to 1,3 propanediol by *Clostridium butyricum*. Biores Technol 67(2):123–128, http://www.sciencedirect.com/science/article/pii/S0960852498001096

Homann T, Tag C, Biebl H, Deckwer W, Schink B (1990) Fermentation of glycerol to 1,3-propanediol by *Klebsiella* and *Citrobacter* strains. Appl Microbiol Biotechnol 33:121–126, doi: 10.1007/BF00176511

Islam R, Cicek N, Sparling R, Levin D (2006) Effect of substrate loading on hydrogen production during anaerobic fermentation by *Clostridium thermocellum* 27405. Appl Microbiol Biotechnol 72(3):576–583, doi: DOI:10.1007/s00253-006-0316-7

Ito T, Nakashimada Y, Senba K, Matsui T, Nishio N (2005) Hydrogen and ethanol production from glycerol-containing wastes discharged after biodiesel manufacturing process. J Biosci Bioeng 100:260–265, doi: 10.1263/jbb.100.260

Jin LH, Lee JH (2008) Change in proteomic profiles of genetically modified 1,3-propanediol-producing recombinant *E-coli*. J Microbiol Biotech 18(8):1439–1444

Kang TS, Korber DR, Tanaka T (2013) Glycerol and environmental factors: effects on 1,3-propanediol production and NAD + regeneration in *Lactobacillus panis* PM1. J Appl Microbiol 115(4):1003–1011, doi: 10.1111/jam.12291

Kaur G, Srivastava AK, Chand S (2012) Mathematical modelling approach for concentration and productivity enhancement of 1,3-propanediol using *Clostridium diolis*. Biochem Eng J 68:34–41, doi: 10.1016/j.bej.2012.07.004

Kitteringham NR, Jenkins RE, Lane CS, Elliot VL, Kevin Park B (2009) Multiple reaction monitoring for quantitative biomarker analysis in proteomics and metabolomics. J Chrom B 877:1229–1239, doi: 10.1016/j.jchromb.2008.11.013

Krokhin OV, Spicer V (2009) Peptide retention standards and hydrophobicity indexes in reversed-phase high-performance liquid chromatography of peptides. Anal Chem 91:9522–9530, doi: 10.1021/ac9016693

Kubiak P, Leja K, Myszka K, Celinska E, Spychana M, Szymanowska-Powanowska D, Czaczyk K, Grajek W (2012) Physiological predisposition of various *Clostridium* species to synthetize 1,3-propanediol from glycerol. Process Biochem 47:1308–1319, doi: 0.1016/j.procbio.2012.05.012

Leja K, Czaczyk K, Myszka K (2011) Biotechnological synthesis of 1,3-propanediol using *Clostridium* ssp. Afr J Biotechnol 10:11093–11101, doi: 10.5897/AJB11.873

Ma Z, Shentu X, Bian Y, Yu X (2012) Effects of NADH availability on the *Klebsiella pneumoniae* strain with 1,3-propanediol operon over-expression. J Basic Microbiol 53:348–354, doi: 10.1002/ jobm.201100580

MacLean BT DM, Shulman N, Chambers M, Finney GL, Frewen B, Kern R, Tabb DL, Liebler DC, MacCoss MJ (2010) Skyline: an open source document editor for creating and analyzing targeted proteomics experiments. Bioinformatics 26 (7):966–968, doi: 10.1093/bioinformatics /btq054

Madigan M, Martinko J, Dunlap PV, Clark DP (2009) Brock Biology of Microorganisms, 12th edn. Pearson Education, San Francisco, CA

Maervoet VET, Beauprez J, De Maeseneire SL, Soetaert WK, De Mey M (2012) *Citrobacter werkmanii*, a new candidate for the production of 1,3-propanediol: strain selection and carbon source optimization. Green Chem 14:2168–2178, doi: 10.1039/C2GC35369E

Markowitz VM, Chen IA, Palaniappan K, Chu K, Szeto E, Grechkin Y, Ratner A, Jacob B, Huang J, Williams P, Huntemann M, Anderson I, Mavromatis K, Ivanova NN, Kyrpides NC (2012) IMG: the integrated microbial genomes database and comparative analysis system. Nucleic Acids Res 40(D1):D115–D122, doi: 10.1093/nar/gkr1044

McQueen P, Spicer V, Rydzak T, Sparling R, Levin D, Wilkins JA, Krokhin O (2012) Information-dependent LC-MS/MS acquisition with exclusion lists potentially generated on-the-fly: case study using a whole cell digest of *Clostridium thermocellum*. Proteomics 12:1160–1169, doi: 10.1002/pmic.201100425

Metsoviti M, Zeng AP, Koutinas AA, Papanikolaou S (2012) Enhanced 1,3-propanediol production by a newly isolated *Citrobacter freundii* strain cultivated on biodiesel-derived waste glycerol through sterile and non-sterile bioprocesses. J Biotechnol 163(4):408–418, doi: 10.1016/j.jbiotec. 2012.11.018

Mickelson MN, Werkman CH (1940) The Dissimilation of Glycerol by Coli-Aerogenes Intermediates. J Bacteriol 39(6):709–715

Monod J, Changeux JP, Jacob F (1963) Allosteric proteins and cellular control systems. J Mol Biol 6:306–329, doi: 10.1016/S0022-2836(63)80091-1

Moon C, Ahn JH, Kim SW, Sang BI, Um Y (2010) Effect of biodiesel-derived raw glycerol on 1,3-propanediol production by different microorganisms. Appl Biochem Biotechnol 161:502–510, doi: 10.1007/s12010-009-8859-6

Nakamura CE, Whited GM (2003) Metabolic engineering for the microbial production of 1,3-propanediol. Curr Opin Biotechnol 14:454–459, doi: 10.1016/j.copbio.2003.08.005

Pflugl S, Marx H, Mattanovich D, Sauer M (2012) 1,3-Propanediol production from glycerol with *Lactobacillus diolivorans*. Bioresour Technol 119:133–140, doi: 10.1016/j.biortech.2012.05.121

Pflugmacher U, Gottschalk G (1994) Development of an immobilized cell reactor for the production of 1,3-propanediol by *Citrobacter freundii*. Appl Microbiol Biotechnol 786–793, doi: 10.1007/BF00221225

Ripoll V, Belkhodja S, Santos VE, García-Ochoa F (2012) Enhanced 1,3-propanediol production by glycerol fermentation pathway by *Escherichia blattae* using different pH strategies. New Biotechnol 29(Supplement):S52–S53

Rydzak T, McQueen PD, Krokhin O, Spicer V, Ezzati P, Dwivedi RC, Shamshurin D, Levin DB, Wilkins JA, Sparling R (2012) Proteomic analysis of Clostridium thermocellum core metabolism: relative protein expression profiles and growth phase-dependent changes inprotein expression. BMC Microbiol 12:214, doi: 10.1186/1471-2180-12-214

Saint-Amans S, Girbal L, Andrade J, Ahrens K, Soucaille P (2001) Regulation of carbon and electron flow in *Clostridium butyricum* VPI 3266 grown on glucose-glycerol mixtures. J Bacteriol 183:1748–1754, doi: 10.1128/JB.183.5.1748-1754.2001

Sattayasamitsathit S, Methacanon P, Prasertsan P (2011) Enhance 1,3-propanediol production from crude glycerol in batch and fed-batch fermentation with two-phase pH-controlled strategy. Electron J Biotechnol 14(6), Open Access. doi: 10.2225/vol14-issue6-fulltext-6

Saxena RK, Anand P, Saran S, Isar J (2009) Microbial production of 1,3-propanediol: Recent developments and emerging opportunities. Biotechnol Adv 27:895–913, doi: 10.1016/j.biotechadv.2009.07.003

Shell Chemicals (2014), http://www.shell.com/global/products-services/solutions-for-businesses/chemicals/about-shell-chemicals/our-history.html

Stroud RM, Miercke LJW, O'Connell J, Khademi S, Lee JK, Remis J, Harries W, Robles Y, Akhavan D (2003) Glycerol facilitator GlpF and the associated aquaporin family of channels. Curr Opin Struct Biol 13:424–431, doi: 10.1016/S0065-3233(03)63011-1

Sun YQ, Qi WT, Teng H, Xiu ZL, Zeng AP (2008) Mathematical modeling of glycerol fermentation by *Klebsiella pneumoniae*: Concerning enzyme-catalytic reductive pathway and transport of glycerol and 1,3-propanediol across cell membrane. Biochem Eng J 38(1):22–32, doi: 10.1016/j.bej.2007.06.002

Thauer RK, Jungermann K, Decker K (1977) Energy Conservation in Chemotrophic Anaerobic Bacteria. Bacteriol Rev 41:100–180

Walsh GM, Lin S, Evans DM, Khosrovi-Eghbal A, Beavis RC, Kast J (2009) Implementation of a data repository-driven approach for targeted proteomics experiments by multiple reaction monitoring. J Proteomicss 72:838, doi: 10.1016/j.jprot.2008.11.015

Wang W, Sun J, Hartlep M, Deckwer WD, Zeng AP (2003) Combined use of proteomic analysis and enzyme activity assays for metabolic pathway analysis of glycerol fermentation by *Klebsiella pneumoniae*. Biotechnol Bioeng 83:525–536, doi: 10.1002/bit.10701

Werkman CH, Gillen GF (1932) Bacteria producing trimethylene glycol. J Bacteriol 23:167–182

Wilkens E, Ringel AK, Hortig D, Willke T, Vorlop KD (2012) High-level production of 1,3-propanediol from crude glycerol by *Clostridium butyricum* AKR102a. Appl Microbiol Biotechnol 93:1057–1063, doi: 10.1007/s00253-011-3595-6

Wisniewski JR, Zougman A, Nagaraj N, Mann M (2009) Universal sample preparation method for proteome analysis. Nat Methods 6:359–U60, doi: 10.1038/nmeth.1322

Zeng AP, Biebl H, Schlieker H, Deckwer WD (1993) Pathway analysis of glycerol fermentation by *Klebsiella pneumoniae*: Regulation of reducing equivalent balance and product formation. Enzyme Microb Technol 15:770, doi: 10.1016/0141-0229(93)90008-P

Zeng AP, Ross A, Biebl H, Tag C, Gunzel B, Deckwer WD (1994) Multiple product inhibition and growth modeling of *Clostridium butyricum* and *Klebsiella pneumoniae* in glycerol fermentation. Biotechnol Bioeng 44(8):902–911, doi: 10.1002/bit.260440806

Zhang GL, Ma B, Xu XL, Li C, Wang L (2007) Fast conversion of glycerol to 1,3-propanediol by a new strain of *Klebsiella pneumoniae*. Biochem Eng J 37 (3):256–260, doi: 10.1016/j.bej.2007.05.003

Zhu Y, Li D, Bao GH, Wang SH, Mao SM, Song JN, Li Y, Zhang YP (2014) Metabolic changes in *Klebsiella oxytoca* in response to low oxidoreduction potential, as revealed by comparative proteomic profiling integrated with flux balance analysis. Appl Env Microbiol 80(9):2833–2841, doi: 10.1128/AEM.03327-13

12

A psychrotolerant strain of *Serratia marcescens* (MTCC 4822) produces laccase at wide temperature and pH range

Gaurav Singh Kaira, Kusum Dhakar and Anita Pandey*

Abstract

A psychrotolerant bacterial strain of *Serratia marcescens*, originally isolated from a glacial site in Indian Himalayan Region (IHR), has been investigated for laccase production under different culture conditions. The bacterial strain was found to grow between 4 to 45°C (opt. 25°C) and 3 to 14 pH (opt. 5 pH) on prescribed growth medium, coinciding with production of laccase in laccase producing medium. However, the production of laccase was more consistent toward alkaline pH. Laccase enzyme was partially purified using gel filtration chromatography. The molecular mass of laccase was determined ~53 kDa on native PAGE. The Km and Vmax values were determined to be 0.10 mM and 50.00 μM min^{-1}, respectively, with ABTS. Inoculum size (4.0% v/v at 1.5 O.D.) resulted in significantly higher production of laccase. Carbon and nitrogen sources also affected the laccase production significantly. All the carbon sources enhanced laccase production, xylose being the best enhancer ($P < 0.01$). Among nitrogen sources, organic sources were found to act as inhibitors ($P < 0.01$), and among the in-organic sources only sodium nitrate enhanced the laccase production. Low molecular weight organic solvents significantly ($P < 0.01$) enhanced laccase production up to 24 h of incubation with a decline in later incubation period. Production of laccase by the psychrotolerant bacterium in wide range of temperature and pH is likely to have inference in biotechnological processes.

Keywords: Laccase; Psychrotolerant; *Serratia marcescens*; Organic solvents; Indian Himalayan Region (IHR)

Introduction

Laccases (benzenediol: oxygen oxidoreductase, EC 1.10.3.2) are multicopper oxidases that catalyze oxidation of an extensive array of recalcitrant phenolic and nonphenolic compounds with simultaneous reduction of molecular oxygen to water (Rivera-Hoyos et al. 2013). Catalytic diversity of laccases are exploited for a number of industrial and environmental applications such as dye effluent decolorization, bio-bleaching, xenobiotics bioremediation, biosensors and food industry (Couto and Herrera 2006). Despite the naturally evolved recalcitrance of lignin, the enzymatic approach to depolymerization also stretched out to eukaryotic and prokaryotic domains of life form. In nature, laccases are found to be widely distributed among plants, fungi, insects and bacteria (Alexandre and Zhulin 2000; Madhavi and Lele 2009).

* Correspondence: anita@gbpihed.nic.in
Biotechnological Applications, G. B. Pant Institute of Himalayan Environment and Development, Kosi-Katarmal, Almora 263 643, Uttarakhand, India

Laccases have mainly been explored from fungal sources for their various applications, especially towards the lignin depolymerization. However, the potential of fungal laccases are limited by the metabolic pH and salt intolerance. Generally, the fungal laccases are known with their optimum activity between pH 2.0-4.0, that can be attributed to their growth adaptation towards acidic pH. Consequently, a need to increase this active pH range for laccase production through molecular engineering is being realized (Torres-Salas et al. 2013). Besides, relatively slow growth is another limitation towards the laccase production from fungal sources. In view of these limitations, laccase production from alternate sources, such as bacteria, is getting attention. Bacteria possess merits in terms of their ability to adapt to extreme physico-chemical conditions and feasibility for amenable recombinant expression (Joseph et al. 2014; Sharma et al. 2007; Sondhi et al. 2014). Industrial scale-up of bacterial laccases is still riddled due to its extracellular location and tolerance to extreme culture conditions. Being an extracellular enzyme, it has to tolerate the adverse environmental conditions pertaining to extremozyme category. To overcome all these limitations there is a need for identification of novel bacterial strains possessing the ability to produce extremozymes under extreme conditions. The tolerance towards the wide range of pH and temperature is an emblem of microbial strains making them potential candidates for various biotechnological processes.

Serratia marcescens, a gram negative bacterium from the family *Enterobacteriaceae,* has been recognized for its various applications in degradation (Manangeeswaran et al. 2007; Perestelo et al. 1994; Verma and Madamwar 2003). *S. marcescens* H30 has recently been studied for its potential in formation of one of the bulk chemicals (2,3-Butanediola) of biotechnological as well as industrial relevance (Zhang et al. 2014). However, studies based on its optimization for production of laccase under various physico-chemical and nutritional conditions are still negligible. Psychrotolerant strains of *S. marcescens* have been frequently isolated from high altitudes, including glacial sites, in the author's laboratory. The focus of the present study is on the production, purification and characterization of laccase from a psychrotolerant strain of *S. marcescens* (MTCC 4822) at a wide pH and temperature range.

Material and methods

Study site and the bacterium

The bacterium was isolated from the soil collected from a glacial site in the IHR (altitude: 3040 m above mean sea level; state Uttarakhand). Bacterial isolation was carried out following 10-fold serial dilutions and pour-plate method on Tryptone Yeast extract (TY) agar. The pure culture was maintained on TY agar slants at 4°C and in 10% glycerol at −20°C. All the experiments were conducted after raising fresh cultures. The bacterium was characterized on the basis of cultural (oxygen requirement), morphological (colony morphology and pigmentation), microscopic (Gram reaction and cell morphology), biochemical (utilization of carbon sources and enzyme activity), physiological (temperature, pH and salt tolerance) and molecular (16S rRNA gene sequence) methods. The bacterial isolate and its nucleotide sequence have been deposited in the Microbial Type Culture Collection and Gene Bank (IMTECH), Chandigarh, India and NCBI, Bathesda, Maryland, US, respectively.

Inoculum preparation and laccase activity

A distinct bacterial colony from 24 h old agar plate culture was aseptically inoculated in 250 ml Erlenmeyer flask containing 50 ml modified Kirk and Farrell (1987) broth medium (Dhakar and Pandey 2013) pH 4.50 ± 0.5. The mother culture was raised at 25°C for 24 h under static and aerobic conditions. 1.0% v/v (O.D.$_{600nm}$ 1.50 ± 0.25) inoculum from mother culture was inoculated in 20 ml of the respective medium and incubated in the static condition. Following incubation, broth culture was centrifuged at 8000 rpm at 4°C for 10 min, the supernatant was treated as crude enzyme. Laccase activity was determined by ABTS (2, 2'-azino-bis (3-ethylbenzothiazoline-6-sulphonic acid)). Reaction mixture contained 0.10 M citrate- phosphate buffer (pH 2.5), ABTS (2 mM) and crude enzyme. Following two minutes of incubation at room temperature, activity was measured at 420 nm (Han et al. 2005). Enzyme unit was defined as 1 μM of ABTS oxidized per min under above standard assay conditions.

The growth curve on Kirk and Farrell medium with the laccase production up to 48 h (every 6 h) of incubation was recorded at 25°C.

Partial purification and characterization of laccase

Bacterial culture, grown at 25°C for 48 h, was centrifuged at 8000 rpm at 4°C for 10 min. Supernatant was used for further process. Chilled acetone was added in 1:1 volume and kept at −20°C, overnight. Then, it was centrifuged at 12000 rpm for 20 min at 4°C. The pellet was re-dissolved in citrate- phosphate buffer (pH 2.6). The sample was further purified using gel filtration chromatography. Column (10 X 1 cm) was prepared using Sephadex G- 75 (Sigma). It was equilibrated with the citrate-phosphate buffer (pH 2.5). The fractions were collected every 3 min; the flow rate of the column was 1 mL min^{-1}. Enzyme activity (ABTS assay) and total protein concentration by Lowry's method was estimated at every step of the purification. The partially purified sample was used for the determination of Km and Vmax. Enzyme activity was determined with the different concentration of ABTS (0.05 mM to 1.00 mM). Lineweaver Burk plot was drawn between 1/(V μM min^{-1}) and 1/([S] mM) concentration. Km and Vmax values were calculated using the graph. Molecular weight of the enzyme was determined using native polyacrylamide gel consisting 12.50% separating and 4% of stacking gel. After electrophoresis, gel was incubated in 0.10% ABTS in citrate- phosphate buffer (pH 2.5) for 20 min at room temperature. Green color band due to the oxidation of ABTS appeared on the gel. Protein marker was also run with the sample to determine the approximate molecular mass of the enzyme.

Effect of the inoculum size, the physico-chemical and nutritional conditions, and the organic solvents on laccase production

The inoculum size

The laccase producing medium was inoculated with the different inoculum size ranging from 0.125 to 5.00% (0.125, 0.25, 0.50, 1.00, 2.00, 4.00, 5.00%), separately. The broth culture was incubated at 25°C in static conditions and observations were taken at 24 and 48 h of incubation.

The physico-chemical and nutritional conditions

In physico-chemical conditions, the laccase production was studied at different temperatures ranged from 5–45°C (at an interval of 10°C) and at the pH range from 3–13 (at an interval of 2 units). 1 N HCl and 1 N NaOH were used for maintaining the initial medium pH. For nutritional conditions, eight carbon (glucose, fructose, maltose, xylose, galactose, sucrose, starch and cellulose) and eight nitrogen (ammonium nitrate, ammonium chloride, ammonium sulfate, yeast, casein, peptone, urea, sodium nitrate) sources (0.20%) were supplemented in the medium, separately. Glucose and ammonium nitrate, being the ingredients in the original medium, were considered as control.

The organic solvents

Five low molecular weight organic solvents, namely methanol, ethanol, acetone, iso-propanol and iso-amyl alcohol were added to the medium in four concentrations (0.50, 1.00, 1.50 and 2.00%), separately, after 12 h of incubation.

All these experiments were performed at 25°C (excluding effect of temperature) for 48 h under static conditions.

Statistical analysis

All the experiments were done in triplicate. Results were analyzed using one way ANOVA and two way repeated measures with Post-hoc Tukey's test. The significance level was $P < 0.05$ and $P < 0.01$.

Results

The bacterium and the growth curve

The bacterium developed blood red colonies on TY agar plates and showed short and single Gram negative rods under microscope. In carbon utilization experiments, it could utilize 7 out of 12 tested sugars. It was positive for catalase and oxidase, and negative for lactose and starch hydrolysis. The bacterium was aerobic in nature and could grow between 4 to 45°C (opt. 25°C) and 3 to 14 pH (opt. 5–7 pH). The growth was observed restricted and without pigment beyond 35°C. It could tolerate salt up to 20%. In molecular characterization, it showed 98.8% similarity to *Serratia marcescens* N4-5. The bacterium has been allotted accession number (MTCC 4822) by IMTECH, Chandigarh, India and the nucleotide sequence accession number EF 462913 by NCBI. The phenotypic and genotypic characters of the bacterium are summarized in Table 1.

In the growth curve, drawn in laccase medium at an interval of 6 h up to 48 h of incubation, the maximum biomass was attained at 24 h that afterwards showed decline up to 48 h. While the laccase production was recorded maximum at 18 h of incubation that remained almost steady up to 30 h with slight increase during further incubation up to 48 h. An increase in the enzyme production was again recorded at the later stage of stationary phase (48 h) with the highest (208.83 UL^{-1}) enzyme titer (Figure 1).

Table 1 Phenotypic and genotypic characters of *Serratia marcescens*

Characters	Description
Colony morphology	Round, smooth, convex (2–5 mm dia.) colonies with red pigment
Microscopic features	single Gram -ve short rods with 0.5-0.8 μm dia and 0.9-2.0 μm length
Biochemical characters	Carbohydrate utilization: +ve for dextrose, fructose, mannose, maltose, mannitol, sucrose and trehalose and –ve for arabinose, galactose, rahmnose, raffinose and melibiose
	Enzyme activity: +ve for catalase and oxidase, –ve for lactose and starch hydrolysis
Physiological characters	Temperature tolerance 4–45°C (opt. 25°C), Produces pigment till 35°C, pH tolerance 3–14 (opt. 5–7), salt tolerance up to 20%
Molecular characters	98.8% similarity with *Serratia marcescens* N4-5
Accessions	Bacterial accession no. MTCC 4822
	Nucleotide sequence no. EF035134

Partial purification and characterization of laccase

Native PAGE revealed the approximate molecular mass of the bacterial laccase 53 kDa (Figure 2). The comparative graph drawn for protein and enzyme with fractions is presented in Figure 3. The enzyme activity was recorded in the fractions 5 and 6. The Km and Vmax value of the partially purified enzyme was 0.10 mM and 50.00 μM min^{-1}, respectively, with ABTS. The enzyme activity and protein content estimated at every step of the process is presented in Table 2 and the Lineweaver Burk plot is shown in Figure 4.

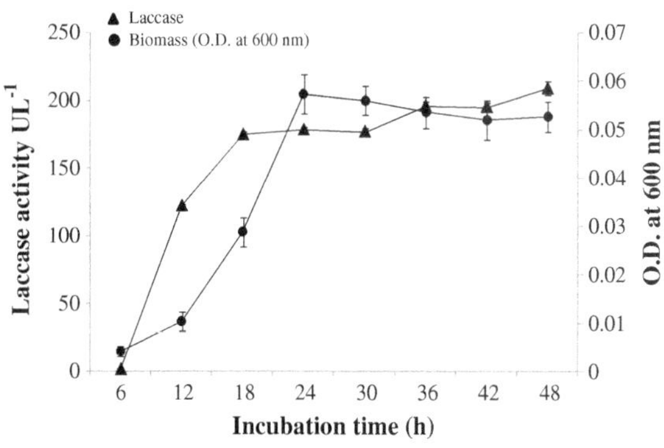

Figure 1 Laccase activity with corrosponding biomass of *Serratia marcescens* at 25°C.

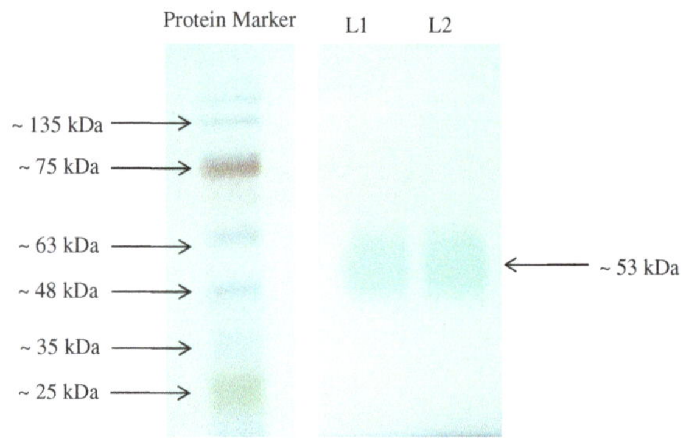

Figure 2 Native PAGE analysis: Standard protein marker; L1 and L2 contain laccase enzyme with a band of approximate ~ 53 kDa.

Effect of the inoculum size, the physico-chemical and nutritional conditions, and the organic solvents on laccase production

The inoculum size

The inoculum size of bacterial culture, ranged from 0.125 to 5.00% (v/v), showed significant differences on laccase production (Figure 5). Out of 7 inoculum concentrations used, 4.00% resulted in maximum laccase production, that was significantly higher ($P < 0.05$) from 2.00% level in two way repeated ANOVA. The effect of bacterial inoculum size also affected the biomass, although the differences were not statistically significant.

The physico-chemical and nutritional conditions

Among the physico-chemical factors, the incubation temperature and medium pH vitally affected the bacterial milieu conditions and its protein machinery. Between the temperature ranges from 5 to 45°C, the highest production of laccase was recorded at 25°C. Production of laccase at this temperature was significantly higher at $P < 0.01$ level for all the temperatures, while laccase production at 5 and 45°C were not significantly different ($P < 0.05$) from each other. From 25°C onward, a steep decline in laccase production was recorded deviating the temperature on either side (<25°C<). While it decreased by 50% at 35°C ($P < 0.01$), it became negligible at 45°C (Figure 6a). The pH of the medium, considered from 3.0-13.0, also influenced the laccase production significantly ($P < 0.05$). The enzyme production was recorded highest at pH 5.0 and lowest at pH 3.0 and appeared to be limited in the stationary growth phase. The laccase production at pH 5.0 was significantly higher ($P < 0.01$) at 24 h in comparison to all the treatments under consideration. The values obtained for pH 5.0 to pH 11.0 at 48 h of incubation were not statistically significant. The enzyme production tended to increase from neutral to pH 13.0, showing significantly higher values ($P < 0.05$). Variation in production of laccase was influenced by pH with respect to growth phases, the stationary and the exponential growth phase, are shown in Figure 6b. The respective values for biomass, in relation to the temperature and pH, are presented in Table 3.

The influence of carbon and nitrogen sources on production of laccase enzyme is summarized in Table 4. All the tested carbon sources showed positive influence on laccase production in basal medium, being statistically highest for xylose (228.4 ± 8.7 UL^{-1}) and lowest for glucose (178.3 ± 3.1 UL^{-1}) ($P < 0.01$). However, the enhancement recorded for laccase production in case of xylose was not significantly higher in comparison of the enhancing effect

Table 2 Partial purification of laccase from *Serratia marcescens*

Sample	Volume (ml)	Total protein (mg)	Total enzyme (Units)	Specific activity (Units/mg)	Purification
Crude	200.0	32.00	52.00	1.60	1.00
Acetone precipitated	5.00	1.10	5.50	5.00	3.10
Gel filtration	3.00	0.01	0.08	6.70	4.10

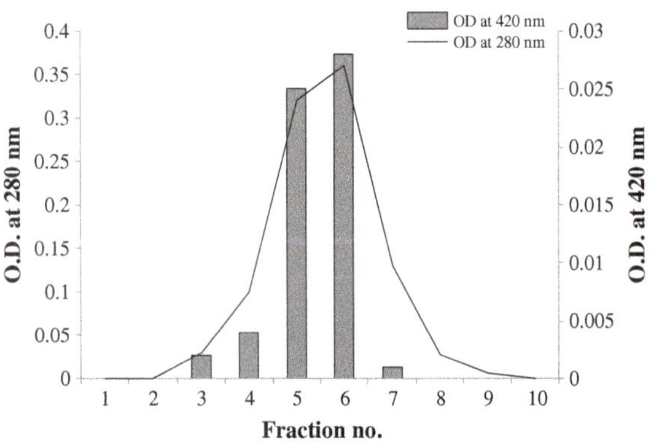

Figure 3 Elution profile of gel filtration chromatography.

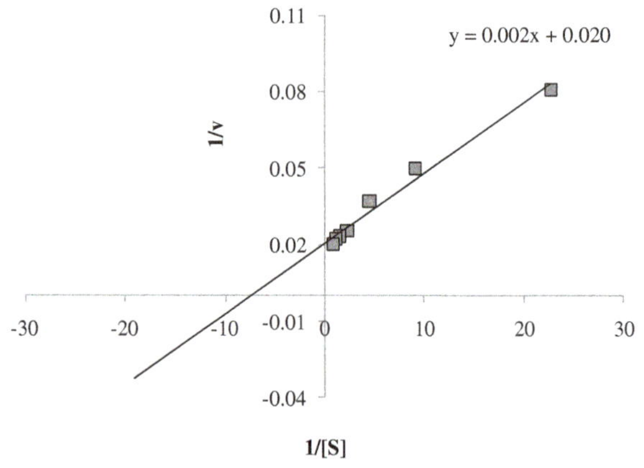

Figure 4 Line weaver-Burk plot for determination of Km and Vmax of partially purified enzyme.

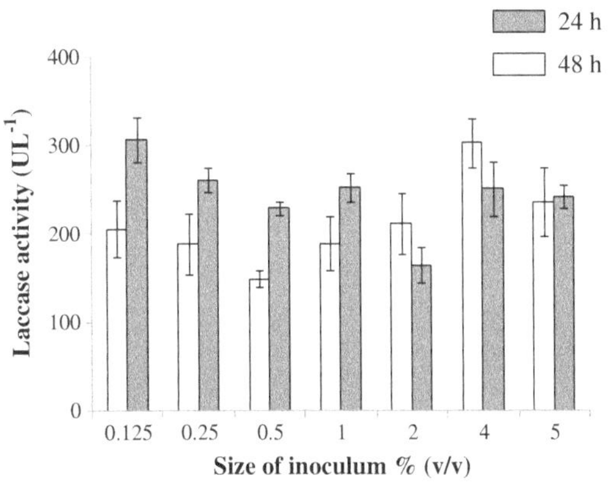

Figure 5 Effect of inoculum size on laccase production by *Serratia marcescens* at 25°C.

of fructose, maltose and starch (P < 0.01). In case of nitrogen supplementation, organic and inorganic nitrogen sources showed varying responses. Organic sources acted as inhibitors while the sodium nitrate enhanced laccase production in comparison to control (Table 4).

The organic solvents

Low molecular organic weight solvents behaved as significant enhancers of laccase production in comparison to control during the early phase (24 h) of incubation (P < 0.01). All the solvents enhanced approximately 2–3 folds production at 0.5 to 2.0% concentration. The enhancement declined at later stage (48 h) in case of all the solvents giving values comparable to control, and in case of methanol even being significantly inhibitory (P < 0.05).

Discussion

Microorganisms are ubiquitous in nature and their distribution is largely governed by environmental specificities. The microbial diversity of IHR is increasingly gaining attention mainly due to their ecological resilience based applications, such as production of secondary metabolites often higher at suboptimal conditions (Mishra et al. 2012; Rinu et al. 2012; Yarzábal 2014). Amongst bacteria, species of *Bacillus*, *Pseudomonas* and *Serratia* have been isolated and studied for their various biotechnological applications (Pandey et al. 1999, 2001, 2006; Malviya et al. 2012; Rinu and Pandey 2009; Selvakumar et al. 2008; Trivedi et al. 2007; Trivedi and Pandey 2008). *Serratia marcescens* MTCC 4822, characterized in the present study for its tolerance to low temperature, wide pH and high salt concentration, has been investigated for laccase production- an enzyme of ecological as well as biotechnological relevance. Absence of pigment production at higher temperature appeared to be in response to particular environmental stimuli that can be considered as a microbial indicator of climate variation. The change in pigment production can be linked to the ability of the bacterial cells to alter their metabolic strategy in response to the environmental conditions. Species of *Serratia* spp are known to give optimum growth from 20-37°C and at pH 7–9 (Giri et al. 2004). *S. marcescens* is characterized by production of a non-diffusible red pigment, named as prodigiosin. The optimum conditions for growth and pigment formation in *S. marcescens* biovar A2/A6 has been reported 30°C in a neutral to slightly alkaline medium (Hardjito et al. 2002).

The bacterial strain produced laccase during its entire growth phase as soon as its metabolism enabled it to outlive the lag phase. Upon entering the exponential phase (6–24 h), the bacterium produced extracellular laccase leading to a steep rise of the laccase activity in the nutrient medium. The growth as well as the enzyme production became steady indicating the onset of stationary phase (24–30 h) of the bacterium, however, it again increased at the later stage of stationary phase. The increase–decrease-increase phenomenon pertaining to the production of laccase can be attributed to the utilization of the nutrients in the medium. The increase in laccase production at the second stage, towards the end of stationary phase, is an indication of the self digestion metabolic activities of the bacterium that has been related to the oxidative properties of the enzyme (Nystrom 2004).

The molecular mass of the laccase from various fungi has been reported between 50 to 140 kDa (Rivera-Hoyos et al. 2013). In a recent case study based on a bacterium

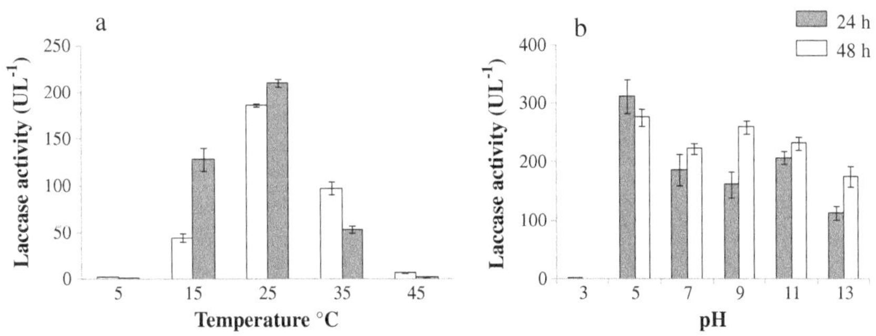

Figure 6 Effect of physico-chemical factors a) temperature and b) pH on laccase production by *Serratia marcescens*.

Table 3 Effect of temperature and pH on bacterial biomass at 24 and 48 h of incubation

Temperature (°C)	24 h	48 h	pH	24 h	48 h
5	0.014 ± 0.001	0.002 ± 0.001	3	0.004 ± 0.001	0.011 ± 0.002
15	0.024 ± 0.003	0.048 ± 0.003	5	0.069 ± 0.008	0.103 ± 0.012
25	0.058 ± 0.006	0.057 ± 0.001	7	0.057 ± 0.003	0.144 ± 0.008
35	0.082 ± 0.003	0.062 ± 0.007	9	0.061 ± 0.001	0.184 ± 0.013
45	0.053 ± 0.006	0.037 ± 0.004	11	0.081 ± 0.015	0.225 ± 0.024
			13	0.0257 ± 0.061	0.467 ± 0.023

(*Micrococcus* sp.) it was found to be 23 kDa (Joseph et al. 2014). The molecular mass of the laccase in the present study was relatively on the higher side (53 kDa). Km values for laccases, in general, have been known ranging between 5 to 500 μM and were recorded 100 μM in the present bacterium. The data on the characterization of laccase, such as molecular mass and Km, provide useful knowledge for further applied studies on protein engineering and phylogeny of the enzyme.

The inoculum size of the bacterium, measured as optical density, affected laccase production. The oxidative catalytic properties of laccase are known to help in self-digestion process of cytoplasmic and outer membranes. The enzyme profile of the bacterium under study was observed to be greatly affected by the final population size and accumulation of end products in the culture medium. Smaller inoculum size (0.125-0.500%) was required for higher enzyme activity in exponential as well as stationary phase. Inoculum size beyond 1% although triggered the enzyme production in stationary phase, it was not sustained at the later stage, i.e., stationary phase. The increase in inoculum size beyond the optimum level is likely to result in rapid depletion of indispensible nutrients. It subsequently may also contribute towards the accumulation of toxic metabolic substances in the growth medium. Similar observations on inoculum size of the bacterium with respect to their metabolic activities have been reported in earlier study (Darah et al. 2013).

Temperature and pH are known to be two of the most important factors affecting the enzyme production. Optimum production of laccase from strains of *Serratia marcescens* has been reported in mesophilic temperature range with neutral to alkaline pH (Chandra et al. 2011; Verma and Madamwar 2003). *Serratia marcescens*, in the present study, exhibited the peculiarity with respect to its ability to produce laccase in a wider temperature (5–45°C) and pH (5–13) range. The extracellular enzymes produced by the microorganisms growing under extreme conditions are known to possess the ability to maintain their activities under harsh conditions contrary to the intracellularly located enzymes which only perform at cytosolic neutral pH. While the ability of the bacterium to produce at wider temperature and pH range is an important feature of ecological relevance for its own survival under harsh conditions, it is also likely to be beneficial for industrial applications such as food, pharmaceutical and textiles. To the best of our knowledge this is the first report on extracellular laccase production by *S. marcescens* isolated from a glacial site under mountain ecosystem of IHR. Albeit, molecular studies on the enzyme under the entire pH scale needs further advance investigation in order to understand the biochemical mechanism of extracellular laccases.

A range of nutritional supplements, mainly the carbon and nitrogen sources, are considered for enhancing the enzyme production from microbial sources. All the carbon sources used in the present study were observed as the enhancers of laccase production probably due to their easy metabolism. In case of nitrogen sources, in-organic and organic sources exhibited variable effects. Two of the inorganic nitrogen sources, sodium nitrate and ammonium nitrate (control), influenced the bacterial biomass in inverse manner in relation to laccase production. The other

Table 4 Effect of carbon and nitrogen sources on laccase production at 25°C after 24 h of incubation

Carbon Source	O.D. at 600 nm	Laccase activity (UL^{-1})	Nitrogen source	O.D. at 600 nm	Laccase activity (UL^{-1})
Fructose	0.047 ± 0.005	207.9 ± 3.0	Ammonium Chloride	0.222 ± 0.020	6.0 ± 0.5
Maltose	0.042 ± 0.008	208.4 ± 1.8	Ammonium sulfate	0.135 ± 0.110	3.9 ± 0.4
Sucrose	0.043 ± 0.006	195.1 ± 0.4	Yeast extract	0.912 ± 0.044	17.4 ± 0.5
Xylose	0.048 ± 0.003	228.4 ± 8.7	Casein	0.192 ± 0.023	3.4 ± 0.4
Galactose	0.062 ± 0.008	193.3 ± 4.5	Peptone	0.791 ± 0.074	14.3 ± 0.4
Cellulose	0.037 ± 0.003	200.5 ± 0.5	Urea	0.081 ± 0.031	6.7 ± 0.2
Starch	0.069 ± 0.001	206.6 ± 1.0	Sodium nitrate	0.051 ± 0.004	193.4 ± 6.5

Control: O.D. at 600 nm = 0.040 ± 0.008; Laccase activity = 178.3 ± 3.1 UL^{-1}.

nitrogen sources promoted growth significantly inhibiting the production of laccase. This indicated the inability of the bacterium to metabolize the nitrogen in form of nitrate, limiting the nitrogen in the medium and indirectly stimulating the laccase expression. Yeast extract, among organic sources, is usually considered as a strong enhancer of laccase production, the inhibitory effect observed due to addition of yeast extract in the present study needs attention in view of the importance of supplements with respect to individual organisms and the specific growth conditions. In general, nitrogen sources are known to regulate the laccase production more effectively in comparison to carbon sources (Madhavi and Lele 2009).

Besides nutritional sources, other supplements are also considered for their effects on laccase production from bacterial sources. Several phenolic and non phenolic compounds have been recognized for the purpose (Muthukumarasamy and Murugan 2014), the use of low molecular weight organic solvents has been rare (Dhakar and Pandey 2013; Lomascolo et al. 2003). In the present study the five organic solvents, selected on the basis of their low molecular weight, found to be efficient enhancers of laccase production. Besides, these organic solvents also possess degradable properties with no harmful effects and are also cost effective.

Extremophiles are becoming the major attraction of the scientists due to their rare and extra ordinary physico-chemical properties. The adaptability of such microorganisms towards broad range of temperature and pH makes them an unbeatable tool for the various purposes in biotechnology. They are likely to be applicable to the processes of biotechnological relevance where the physico-chemical conditions fluctuate frequently and the microorganisms are required to adapt to the fluctuations. Besides the biotechnological relevance, these microorganisms contribute to the geochemical cycles and play crucial role in the dynamics of environmental processes. Combination of peculiar physico-chemical characters, such as tolerance to wide temperature and pH ranges and high salt concentration, along with the production of laccase under these fluctuations, *Serratia marcescens* (MTCC 4822) may serve as a model organism for specific pharmaceutical, biomedical and biotechnological applications.

Competing interests

The authors declare that they have no competing interests.

Authors' contributions

GSK and KD performed the kinetics, purification, characterization and optimization of the enzyme, and drafted the manuscript. AP isolated, characterized and identified the bacterium, designed the study, and finalized the manuscript. All authors read and approved the final manuscript.

Acknowledgements

The Director, GB Pant Institute of Himalayan Environment and Development, Almora, is gratefully acknowledged for extending the facilities. Ministry of Environment, Forests and Climate Change, Govt. of India is thanked for financial support.

References

Alexandre G, Zhulin IB (2000) Laccases are widespread in bacteria. Trends Biotechnol 18:41–42

Chandra R, Singh R, Yadav S (2011) Effect of bacterial inoculum ratio in mixed culture for decolourization and detoxification of pulp paper mill effluent. J Chem Technol Biotechnol 87:436–444

Couto SR, Herrera JLT (2006) Industrial and biotechnological applications of laccases: A review. Biotechnol Adv 24:500–513

Darah I, Nisha M, Lim SH (2013) Enhancement of polygalacturonase production from *Enterobacter aerogenes* NBO2 by submerged fermentation. Adv Stud Biol 5(5):173–189

Dhakar K, Pandey A (2013) Laccase Production from a temperature and pH tolerant fungal strain of *Trametes hirsuta* (MTCC, 11397). Enz Res doi.org/10.1155/2013/869062

Giri AV, Anandkumar N, Muthukumaran G, Pennathur G (2004) A novel medium for the enhanced cell growth and production of prodigiosin from *Serratia marcescens* isolated from soil. BMC Microbiol 4:11

Han MJ, Choi HT, Song HG (2005) Purification and characterization of laccase from the white rot fungus *Trametes versicolor*. J Microbiol 43(6):555–560

Hardjito L, Huq A, Colwell R (2002) The influence of environmental conditions on the production of pigment by *Serratia marcescens*. Biotechnol Bioprocess Eng 7:100–104

Joseph B, Kuddus M, Ramteke PW (2014) Extracellular alkaline thermostable laccase from *Micrococcus* species: Partial purification and characterization. Curr Biotechnol 3:145–151

Kirk TK, Farrell RL (1987) Enzymatic "combustion": the microbial degradation of lignin. Ann Rev Microbiol 41:465–505

Lomascolo A, Record E, Herpoel-Gimbert I, Delattre M, Robert JL, Georis J, Dauvrin T, Sigoillot J-C, Asther M (2003) Over-production of laccase by a monokaryotic strain of *Pycnoporus cinnabarinus* using ethanol as inducer. J Appl Microbiol 94:618–624

Madhavi V, Lele SS (2009) Laccase: Properties and Applications. BioResources 4(14):1694–1717

Malviya MK, Sharma A, Pandey A, Rinu K, Sati P, Palni LMS (2012) *Bacillus subtilis* NRRL B-30408: A potential inoculant for crops grown under rainfed conditions in the mountains. J Soil Sci Plant Nutr 12(4):811–824

Manangeeswaran M, Ramalingam VV, Kumar K, Mohan N (2007) Degradation of indulin, a kraft pine lignin, by *Serratia marcescens*. J Environ Sci Health B 42(3):321–327

Mishra PK, Bisht SC, Bisht JK, Bhatt JC (2012) Cold tolerant PGPRs as bioinoculants for stress management. In: Maheshwari DK (ed) Bacteria in Agrobiology: Stress Management. Springer, Berlin, Germany, pp 95–118

Muthukumarasamy NP, Murugan S (2014) Production, purification and application of bacterial laccase: A review. Biotechnol 13(5):196–205

Nystrom T (2004) Stationary-phase physiology. Annu Rev Microbiol 58:161–181

Pandey A, Durgapal A, Joshi M, Palni LMS (1999) Influence of *Pseudomonas corrugata* inoculation on root colonization and growth promotion of two important hill crops. Microbiol Res 154:259–266

Pandey A, Palni LMS, Hebbar KP (2001) Suppression of damping-off in maize seedlings by *Pseudomonas corrugata*. Microbiol Res 156:191–194

Pandey A, Trivedi P, Kumar B, Palni LMS (2006) Characterization of a phosphate solubilizing and antagonistic strain of *Pseudomonas putida* (B0) isolated from a sub-alpine location in the Indian Central Himalaya. Curr Microbiol 53(2):102–107

Perestelo F, Falcon MA, Camicero A, Rodriguez A, De la Fuente G (1994) Limited degradation of industrial, synthetic and natural lignins by *Serratia marcescens*. Biotechnol Lett 16:299–302

Rinu K, Pandey A (2009) *Bacillus subtilis* NRRl B-30408 inoculation enhances the symbiotic efficiency of *Lens esculenta* Moench at a Himalayan location. J Plant Nutr Soil Sci 172:134–139

Rinu K, Pandey A, Palni LMS (2012) Utilization of psychrotolerant phosphate solubilizing fungi under low temperature conditions of the mountain ecosystem. In: Satyanarayana T, Johri BN, Prakash A (eds) Microorganisms in Sustainable Agriculture and Biotechnology. Springer Science, Business Media, Dordrecht, pp 77–90

Rivera-Hoyos CM, Morales-Alvarez ED, Poutou Pinales RA, Pedroza- Rodriguez AM, Rodriguez- Vazquez R, Delgado- Boada JM (2013) Fungal laccases. Fun Biol Rev 27:67–82

Selvakumar G, Mohan M, Kundu S, Gupta AD, Joshi P, Nazim S, Gupta HS (2008) Cold tolerance and plant growth promotion potential of *Serratia marcescens* strain SRM (MTCC 8708) isolated from flowers of summer squash (*Cucurbita pepo*). Lett Appl Microbiol 46(2):171–175

Sharma P, Goel R, Capalash N (2007) Bacterial laccases. World J Microbiol Biotechnol 23(6):823–832

Sondhi S, Sharma P, Saini S, Puri N, Gupta N (2014) Purification and characterization of an extracellular, thermo-alkali-stable, metal tolerant laccase from *Bacillus tequilensis* SN4 Plos One 9(5): e96951

Torres-Salas P, Mate DM, Ghazi I, Plou FJ, Ballesteros AO, Alcalde M (2013) Widening the pH activity profile of a fungal laccase by directed evolution. Chem Bio Chem. doi:10.1002/cbic.201300102

Trivedi P, Pandey A (2008) Plant Growth Promotion Abilities and Formulation of *Bacillus megaterium* strain B 388 (MTCC6521) Isolated from a Temperate Himalayan Location. Ind J Microbiol 48(3):342–347

Trivedi P, Kumar B, Pandey A, Palni LMS (2007) Growth promotion of rice by phosphate solubilizing bioinoculants in a Himalayan location. In: Velazqez E, Rodriguez- Barrueco C (eds) Plant and Soil, Developments in Plant and Soil Sciences, Volume 102, First International Meeting on Microbial Phosphate Solubilization. Springer, Dordrecht, The Netherlands, pp 291–299

Verma P, Madamwar D (2003) Decolourization of synthetic dyes by a newly isolated strain of *Serratia marcescens*. World J Microbiol Biotechnol 19(6):615–618

Yarzábal LA (2014) Cold-Tolerant phosphate-solubilizing microorganisms and agriculture development in mountainous regions of the World. In: Khan MS et al. (eds) Phosphate Solubilizing Microorganisms, doi:10.1007/978-3-319-08216-5_5, © Springer International Publishing Switzerland

Zhang LY, Xu QM, Zhan SR, Li YY, Lin H, Sun SJ, Sha L, Hu KH, Guan X, Shen YL (2014) A new NAD(H)-dependent *meso*-2,3-butanediol dehydrogenase from an industrially potential strain *Serratia marcescens* H30. Applied Microbiol Biotechnol 98(3):1175–1184

13

NADH-dependent biosensor in *Saccharomyces cerevisiae*: principle and validation at the single cell level

Jan Dines Knudsen, Magnus Carlquist and Marie Gorwa-Grauslund*

Abstract

A reporter system was constructed to measure perturbations in the NADH/NAD^+ co-factor balance in yeast, by using the green fluorescent protein gene under the control of the *GPD2* promoter that is induced under conditions of excess of NADH. High fluorescence levels were obtained in a glycerol 3-phosphate dehydrogenase double deletion strain (*gpd1Δgpd2Δ*), which is deficient in the ability to regenerate NAD^+ via glycerol formation. The responsiveness of the reporter system to externally induced perturbations in NADH oxidation was also evaluated in the *gpd1Δgpd2Δ* strain background by addition of acetoin, as well as by introduction of a set of heterologous xylose reductases (XRs) having different selectivities for NADH. Addition of acetoin during cell proliferation under oxygen-limited conditions resulted in a more than 2-fold decrease in mean fluorescence intensity as compared to the control experiment. Strains carrying XRs with different selectivities for NADH could be distinguished at the single cell level, so that the XR with the highest selectivity for NADH displayed the lowest fluorescence. In conclusion, the designed system successfully allowed for monitoring perturbations in the cellular redox metabolism caused by environmental changes, or by heterologous gene expression. The reporter system displayed high resolution in distinguishing cytosolic NADH oxidation capacity and hence has potential to be used for high-throughput screening based on the fluorescence of single cells.

Keywords: *Saccharomyces cerevisiae*; Redox balance; NADH biosensor; Single cell analysis

Introduction

The relative concentrations of NADH/NAD^+ redox couple determine the thermodynamic feasibility of more than 100 reactions in cellular metabolism which places it at the core of the metabolic network of every organism including *Saccharomyces cerevisiae* (Förster et al. 2003). Efficient tools to measure the impact of redox enzymes or environmental perturbations on the cellular NADH levels are therefore needed (Murray et al. 2011). However quantification of redox couples in cells is complicated by several factors: the NADH and NAD^+ are compartmentalized in *S. cerevisiae* and the cofactors are to a large extent in a protein bound form (Veech et al. 1969; Bücher et al. 1972). Fluorescence imaging enables the distinction between free and bound redox couples, but not between NADH and NADPH (Zhang et al. 2002). In mammalian cells, coupled enzymatic reactions have been used to determine the ratio between the cofactors NADH/NAD^+ (Williamson et al. 1967; Theobald et al. 1997). A similar approach has been applied for *S. cerevisiae* where mannitol-1-phosphate 5-dehydrogenase (M1PDH) was introduced and the cytosolic free NADH/NAD^+ ratio was determined by the ratio between fructose-6-phosphate (F6P) and mannitol-1-phosphate (M1P) and the equilibrium constant for the reaction M1P + NAD ↔ F6P + NADH + H^+ (Canelas et al. 2008). The approaches are however often tedious and not applicable for high-throughput screening.

In *S. cerevisiae*, the cytosolic enzyme glycerol-3-phosphate dehydrogenase (GPD) 2 encoded by the *GPD2* gene plays a central role in redox metabolism. Under anaerobiosis, *GPD2* expression increases as the need for cytosolic reoxidation of NADH increases because Gpd2p acts as redox sink via the NADH-coupled reduction of dihydroxyacetone phosphate (DHAP) to glycerol 3-phosphate (G3P). G3P is further converted to

* Correspondence: marie-francoise.gorwa@tmb.lth.se
Division of Applied Microbiology, Department of Chemistry, Lund University, PO Box 124, SE 221 00 Lund, Sweden

glycerol (Eriksson et al. 1995; Ansell et al. 1997). The need for NADH reoxidation is thought to be linked, indirectly or indirectly, to the NADH/NAD⁺ ratio (Ansell et al. 1997). Glycerol production can also be initiated by the isoenzyme Gpdlp encoded by *GPD1*. However Gpdlp and Gpd2p are expressed under different growth conditions, as illustrated by their different cellular localization (Valadi et al. 2004). The physiological role of Gpdlp is also different from that Gpd2p as *GPD1* is induced under osmotic stress (Blomberg and Adler 1989). Although the genes can complement each other's functions, a double deletion renders the yeast cell osmosensitive as well as incapable of anaerobic growth, due to the inability to reoxidize cytosolic NADH arising from biomass formation (van Dijken and Scheffers 1986; Nissen et al. 1997).

We used this knowledge to design a GFP-based biosensor for cytosolic NADH/NAD⁺ levels by coupling the *GPD2* promoter (*GPD2*p) to the *yEGFP3* gene encoding yeast enhanced green fluorescent protein, an established yeast reporter of the GFP class 2 (Cormack et al. 1996; Cormack et al. 1997; Zacharias and Tsien 2005). The applicability of fluorescent proteins as reporters has previously been demonstrated through the measurement of various user-specified cell properties. For example, GFP has been used for protein localization (Valadi et al. 2004), intracellular pH measurement (Valkonen et al. 2013), for monitoring cell growth mode and cell membrane robustness (Carlquist et al. 2012). Dardalhon et al. (2012) also implemented a fluorescent glutathione redox reporter in *S. cerevisiae* using a yellow fluorescent protein sensor that reports on the nuclear and cytosolic glutathione redox state i.e the glutathione/glutathione disulfide (GSH/GSSG) state, which is NADPH dependent.

The functionality of the NADH/NAD⁺-ratio biosensor developed in this work was first evaluated by introducing it into *S. cerevisiae* strains with reduced NADH reoxidation capacity due to the deletion of Gpd1,2p encoding genes. The biosensor potential was then validated by evaluating the impact on cell fluorescence of acetoin addition and of the introduction of xylose reductases with different selectivities for NADH.

Table 1 Plasmids and yeast strains used in the present study

Strains and plasmids	Relevant features	Reference
Plasmids		
Yiplac211	*URA3*	(Gietz and Sugino 1988)
p424	*TRP1 Mumberg vector*	(Mumberg et al. 1995)
P425	*LEU2 Mumberg vector*	(Mumberg et al. 1995)
YEplacHXT	*URA3*	(Karhumaa et al. 2005)
pYGFP3	*URA3 ADH1p-yEGFP3-ADH1t*	(Cormack et al. 1997)
YIpJK01	*HIS3 GPD2p-yEGFP3-PGK1t*	This study
YEpJK01	*URA3 GPD2p-yEGFP3-PGK1t*	This study
YIpOB8	*URA3 TDH3p-XYL1-ADH1t* PGK1p-XYL2-PGK1t	(Bengtsson et al. 2009)
YIpDR7	*URA3 TDH3p-XYL1(N272D)-ADH1t PGK1p-XYL2-PGK1t*	(Runquist et al. 2010)
YIpCR01	*URA3 TDH3p-XYL1-ADH1t*	This study
YIpCR03	*URA3 TDH3p-XYL1 (N272D)-ADH1t*	This study
S. cerevisiae strains		
CEN.PK2-1C	*MATa ura3-52 trp1-289 leu2-3, 112 his3Δ* 1 MAL2-8C SUC2	EUROSCARF
CEN.PK113-7D	*MATa, MAL2-8ᶜ SUC2*	EUROSCARF
TMB4120	CEN.PK2-1C YEpJK01 *trp1::TRP1 leu2::LEU2 his3::HIS3*	This study
TMB4121	CEN.PK2-1C YEpJK01 *gpd1::TRP1 leu2::LEU2 his3::HIS3*	This study
TMB4122	CEN.PK2-1C YEpJK01 *trp1::TRP1 gpd2::LEU2 his3::HIS3*	This study
TMB4123	CEN.PK2-1C YEpJK01 *gpd1::TRP1 gpd2::LEU2 his3::HIS3*	This study
TMB4132	CEN.PK2-1C *his3*::YIpJK01 *trp1::TRP1 leu2::LEU2 ura3⁻*	This study
TMB4133	CEN.PK2-1C *his3*::YIpJK01 *gpd1::TRP1 gpd2:LEU2 ura3⁻*	This study
TMB4140	TMB4132 *ura3*::YIplac211	This study
TMB4141	TMB4132 *ura3*::YIpCR01	This study
TMB4143	TMB4132 *ura3*::YIpCR03	This study
TMB4144	TMB4133 *ura3*::YIplac211	This study
TMB4145	TMB4133 *ura3*::YIpCR01	This study
TMB4147	TMB4133 *ura3*::YIpCR03	This study

Methods

Strains

S. cerevisiae strains used in the study are summarized in Table 1. *Escherichia coli* NEB 5α (New England BioLabs, Ipswich, MA, USA) was used for subcloning. All strains were stored in 20% glycerol at -80°C. Yeast cells from freshly streaked YNB plates (6.7 g/L yeast nitrogen base without amino acids, 20 g/L glucose, 2% agar) were used for all cultivations.

Molecular biology methods

Plasmid DNA was prepared using the GeneJET Plasmid Miniprep Kit (Thermo Scientific, USA). Restriction and modification enzymes as well as T4 DNA ligase were obtained from the same manufacturer. The QIAquick gel extraction kit (QIagen, Hilden, Germany) was used for DNA extractions from agarose. All nucleotides were ordered at Eurofins (Germany). All genetic constructs were checked by sequencing (Eurofins, Germany). The primers used in the study are listed in Table 2.

Competent cells of *E. coli* NEB 5α were prepared and transformed by the method of Inoue et al. (1990) and yeast transformations were performed using the lithium acetate method (Gietz and Woods 2002). The cells were streaked out on selective medium afterwards. *E. coli* transformants were selected on Lysogene Broth (LB) plates (Sambrock and Russell 2001) with 50 μg/ml ampicillin (IBI Shelton Scientific, Shelton, USA). *S. cerevisiae* transformants were selected on YNB plates supplemented with the requested amino acids or base.

Construction of deletion and insertion cassettes

The *GPD1* deletion cassette was constructed by joining three PCR fragments through overlap extension PCR (Sandström et al. 2012). A nucleotide fragment homologous to the upstream region of *GPD1* was PCR amplified

Table 2 Primers used in the present study

Name	Description	Sequence
TRP1_gpd1Over_f	Forward primer for amplification of *TRP1* (p424 as template) with *GPD1* overhangs	TCCACAAACACAAATATTGATAATATAAAGAACGACATT ACTATATATATAATATAG
TRP1_gpd1Over_r	Reverse primer for amplification of *TRP1* (p424 as template) with *GPD1* overhangs	AGTATGATATGTTATCTTTCTCCAATAAATAGGCAAGTG CACAAACAATAC
TRP1_gpd1_veri_f	Amplification of the 5′ region of *GPD1* (CEN.PK113-7D as template)	TTCCATTCACATATCGTCTTTGG
	+ Verification of *GPD1* deletions	
5-OEGpd1_r	Amplification of the 5′ region of *GPD1* (CEN.PK113-7D as template)	TATATTATCAATATTTGTGTTTGTGG
3-OEGpd1_f	Amplification of the 3′ region of *GPD1* (CEN.PK113-7D as template)	TTTATTGGAGAAAGATAACATATCATAC
3-OEGpd1_r	Amplification of the 3′ region of *GPD1* (CEN.PK113-7D as template)	ATTTTCTTAGGACGCCGCAAAATATC
TRP1_gpd1_veri_r	Verification of *GPD1* deletions	GAGGAACTCTTGGTATTCTTGCC
LEU2_gpd2Over_f	Forward primer for amplification of *LEU2* (p425 as template) with *GPD2* overhangs	TTCCTTTTCCTTCGCTCCCCTTCCTTATCATCGACTACG TCGTAAGGCCG
LEU2_gpd2Over_r	Forward primer for amplification of *LEU2* (p425 as template) with *GPD2* overhangs	GATCAGAGGGGGAGGGGGGGGGAGAGTGTCGAGGAG AACTTCTAGTATATC
LEU2_gpd2OO_f	Forward primer for nested PCR for adding homology regions to the *GPD2* deletion casette	GTATTTTGGTAGATTCAATTCTCTTTCCCTTTCCTTTTCCT TCGCTCCC
LEU2_gpd2OO_r	Reverse primer for nested PCR for adding homology regions to the *GPD2* deletion casette	AAATTGGTTGGGGGAAAAAGAGGCAACAGGAAAGATCA GAGGGGGAGGG
LEU2_gpd2_veri_f	Verification of *GPD2* deletions	CGTGTATCTTCTAAGATTCAGTC
LEU2_gpd2_veri_r	Verification of *GPD2* deletions	CTAATGGCTCAACGTGATAAGG
GPD2p_XbaI_r	Cloning of *GPD2* from CEN.PK113-7D	CTTCTAGATTGATAAGGAAGGGGAG
GPD2p_SacI_f	Cloning of *GPD2* from CEN.PK113-7D	TCGAGCTCCGCAATGTTTCGTTGG
yEGFP_XbaI_f	Cloning of *yEGFP3* from pyEGFP3	AATCTAGAAGCATTAAAAAATGTCTAAAGGTGAAG
yEGFP_PstI_r	Cloning of *yEGFP3* from pyEGFP3	TTCCTGCAGAGTTATTTGTACAATTCATCCATAC
HIS3_flank_f	Cloning of *HIS*3 - Forward primer	CCAGGTATCGTTTGAACACGG
HIS3_flank_r	Cloning of *HIS*3 - Reverse primer	GCTCAGTTCAGCCATAATATG
LEU2_flank_f	Cloning of *LEU2* - Forward primer	GGATAATTATACTCTATTTCTCAAC
LEU2_r	Cloning of *LEU2* - Reverse primer	TTAAGCAAGGATTTTCTTAAC
TRP1_f	Cloning of *TRP1* - Forward primer	ATGTCTGTTATTAATTTCAC
TRP1_r	Cloning of *TRP1*- Reverse primer	CTATTTCTTAGCATTTTTG
YIplac211_f	Cloning of *GPD2*p-*yEGFP3*-*PGK1*t	TTTATCTTCGTTTCCTGC
YIplac211_r	Cloning of *GPD2*p-*yEGFP3*-*PGK1*t	AAAACTGTATTATAAGTAAATG
HIS3_O_f	Cloning of *HIS3*- expression cassette, for change of auxotrophic marker.	TTATAATACAGTTTTCCAGGTATCGTTTGAACACGG
HIS3_O_r	Cloning of *HIS3*-expression cassette, for change of auxotrophic marker.	GGAAACGAAGATAAATCGCTCAGTTCAGCCATAATATG

from genomic DNA of CEN.PK113-7D strain using the primer couple TRP1_*gpd1_veri_f*, 5-OEGpd1_r. The promoter, ORF and terminator of *TRP1* were PCR amplified from a Mumberg p424 vector using the primers TRP1_gpd1Over_f, TRP1_gpd1Over_r. The downstream region of *GPD1* was PCR amplified from genomic CEN.PK113-7D DNA using the primer couple 3-OEGpd1_f, 3-OEGpd1_r. The three fragments were all purified before being combined into one fragment using the nucleotides TRP1_gpd1Over_f, 3-OEGpd1_r.

The *GPD2* deletion cassette was constructed by nested PCR on a PCR fragment containing the promoter, ORF and terminator of *LEU2*. The template fragment was constructed using the primers LEU2_gpd2Over_f, LEU2_gpd2Over_r. The fragment was purified and then used as template using the primers LEU2_gpd2OO_f, LEU2_gpd2OO_r. The added regions contained around 50 bp regions homologous to the upstream and downstream region of *GPD2*, respectively.

Expression cassettes for curing tryptophan, leucine and histidine auxotrophies were constructed using CEN.PK113-7D as a template and corresponding primer couples (TRP1_f, TRP1_r, LEU2_flank_f, LEU2_r and HIS3_flank_f, HIS3_flank_r, respectively) to PCR amplify the promoter, gene and terminator of the respective genes.

Plasmid construction

Plasmid YEpJK01was constructed by ligation of three DNA fragments using T4 DNA ligase: 1) A 1.4 kb region directly upstream of the *GPD2* ORF that was PCR amplified from CEN.PK113-7D genomic DNA using the primers GPD2p_XbaI_r and GPD2p_SacI_f, digested with *Sac*I and *Xba*I and purified, 2). The *yEGFP3* fragment that had been PCR amplified from pYGFP3, using the primers yEGFP_XbaI_f and yEGFP_PstI_r, digested with *Xba*I and *Pst*I and finally purified, and 3) YEplacHXT digested with *Sac*I and *Pst*I. The ligation mix was used to transform *E. coli*.

Plasmid YIpJK01was constructed by excision of the *GPD2p-yEGFP3-PGK1t* cassette from YEpJK01 plasmid using the restriction enzymes *Sph*I and *Sac*I. The fragment was purified from agarose gel and ligated to YIplac211 plasmid that was previously digested with the same restriction enzymes. The auxotrophic marker of the resulting plasmid was changed from *URA*3 to *HIS3* by PCR amplifying the entire plasmid except for the *URA3* region. The amplification was done using the primers YIplac211_f, YIplac211_r. The fragment was then recombined with a PCR fragment containing a *HIS3* expression cassette using the *In-Fusion® HD Cloning Kit*. The *HIS3* cassette was obtained from CEN.PK113-7D genomic DNA using the nucleotides HIS3_O_f, HIS3_O_r. The primers were designed according to the *In-Fusion® HD Cloning Kit* guidelines. The entire YEpJK01 plasmid was transformed into *E. coli*.

For the construction of YIpCRO1 and YIpCRO3, the *XR*-expression cassettes were excised from YIpOB8 and YIpDR7, respectively, using *Hind*III. The vector YIplac211 was cleaved using the same enzyme. The relevant fragments were isolated and purified. Ligation of YIplac211 and *TDH3p-XYL1-ADH1t* yielded YIpCR01. Ligation of YIplac211 and *TDH3p-XYL1(N272D)-ADH1t* yielded YIpCR03.

Construction of yeast strains

TMB4120, TMB4121, TMB4122, TMB4223 (Table 1) were all constructed using CEN.PK2-1C as parental strain. The strains were made prototrophic by transforming sequentially the CEN.PK2-1C strain with (i) the required expression cassettes for *LEU2, TRP1* or *HIS3*, (ii) the *GPD1* and/or *GPD2* deletion cassette and (iii) YEpJK01plasmid.

TMB4132 and TMB4133 were constructed by transformation of CEN.PK2-1C with expression cassettes for *TRP1* and *LEU2* or the *GPD1 (TRP1)* and *GPD2 (LEU2)* deletions cassettes. Finally the strains were transformed with YIpJK01 that was linearized with *Nhe*I.

TMB4140, TMB4141, TMB4143, TMB4144, TMB4145 and TMB4147 were constructed by transforming TMB4132 or TMB4133 with YIplac211, YIpCR01 or YIpCR03 respectively. All vectors were linearized with *Apa*I prior to transformation.

Single gene deletions were verified by PCR amplification whereas double deletions were verified by PCR screening and enzymatic assay (described below). *GPD1* deletions were PCR verified using the nucleotide couple TRP1_gpd1_veri_f and TRP1_gpd1_veri_r. *GPD2* deletions were verified using the nucleotide couples LEU2_gpd2_veri_f and LEU2_gpd2_veri_r

Cell cultivation

For the initial Flourescence intensity (FI) measurements, single colony cells were taken from YNB plates not older than 2 weeks and inoculated in 250 ml shake flasks containing 25 ml YNB medium buffered to pH 5.5 with potassium hydrogen phthalate and complemented with 2% glucose. The cells were grown overnight at 180 rpm and 30°C.

For the FI measurement of XR strains, pre-cultivations and cultivations were done in YNB medium buffered to pH 5.5 with potassium hydrogen phthalate and complemented with 2% galactose and 2% xylose. The cells from the pre-cultivation step were washed with sterile water before inoculation. Oxygen-limited cultures were performed in 65 mL flasks with 45 mL medium and ergosterol and Tween-80 were added to the medium at a final concentration of 0.42 g/L and 0.01 g/L respectively (Andreasen and Stier 1953, 1954). Bottles were closed with rubber stoppers and cannulas were used for gas outlet and sampling, thus minimizing oxygen transfer. Cultures were stirred gently with magnetic stirrers and incubated at 30°C. Fermentation was performed in biological duplicates. Substrate consumption and product formation values were determined from samples withdrawn during 72 h of fermentation.

Analyses

Samples for OD_{600} were analyzed directly while samples for HPLC were kept at −20°C. Growth was monitored by measuring OD at 620 nm.

Xylose, galactose, glycerol, ethanol and xylitol were separated and quantified by high-pressure liquid chromatography (HPLC) (Waters, Milford, MA, USA). The compounds were separated with a HPX-87P (Bio-Rad, Hercules, CA, USA) ion exchange column). Separation was performed at 80°C, with H_2O as mobile phase at a flow rate of 0.6 mL/min. A refractive index detector (RID-6A; Shimadzu, Kyoto, Japan) was used for quantification.

A BD Accuri^{TM} C6 (Becton-Dickinson, NJ, USA) was used for the evaluation of the redox biosensor in the different deletions strains. Fresh cells were used. Samples for flow cytometry were centrifuged for 1 min at 3000g and 4°C, and resuspended in saline solution. The impact of the reporter gene copy number on fluorescence heterogeneity was evaluated on A BD FACSort (Becton-Dickinson, NJ, USA) flow cytometer. Fresh cells were harvested at the late exponential phase. Evaluation of the biosensor responsiveness to redox perturbations by heterologous expression of NADH-dependent oxidoreductases was performed with a BD FACSAria III (Becton-Dickinson, NJ, USA) flow cytometer. Samples were frozen in phosphate buffer, pH 7, 25% glycerol at -80°C before being thawed and analyzed. Events were recorded with a rate of approximately 1,000 events per second. Excitation wavelength for the lasers used was 488 nm. Fluorescence emission levels were measured using a band pass filter at 530/30 nm (FI). Processing and analysis of flow cytometry raw data was performed by using MatLab®R2010b (The MathWorks, Inc., Natick, MA, USA). The measurement files, exported as fcs files by the flow cytometer FACSAria III, were imported into MATLAB, using a "fcs data reader" routine (by L. Balkay, University of Debrecen, Hungary), available on MATLAB W File Exchange website.

Results

Construction and evaluation of the redox biosensor

In order to construct a biosensor that responds to cytosolic NADH/NAD^+ ratio, a 1.4 kb region upstream of the *GPD2* open reading frame that includes the whole *GPD2* promoter region (Eriksson et al. 1995), was cloned 18 bp upstream of the start codon of the *yEGFP3* gene (Cormack et al. 1997). The resulting *GPD2*p-*yEGFP3* expression cassette was introduced in the multicopy vector YEplacHXT generating plasmid YEpJK01. The YEpJK01 plasmid was then introduced into a control laboratory strain as well as in redox engineered strains carrying either single or double deletion of *GPD1* and *GPD2* genes, thereby generating strains TMB4120 (control), TMB4121 (*gpd1Δ*), TMB4122 (*gpd2Δ*) and TMB4123 (*gpd1Δ gpd2Δ*).

Strains were cultivated in batch mode under oxygen-limited conditions for 5 hours in defined mineral medium with glucose as sole carbon source, after which the FI was measured (Figure 1). At this time point, cells were in exponential growth phase and could be considered as being in pseudo-steady state with regards to regulation of

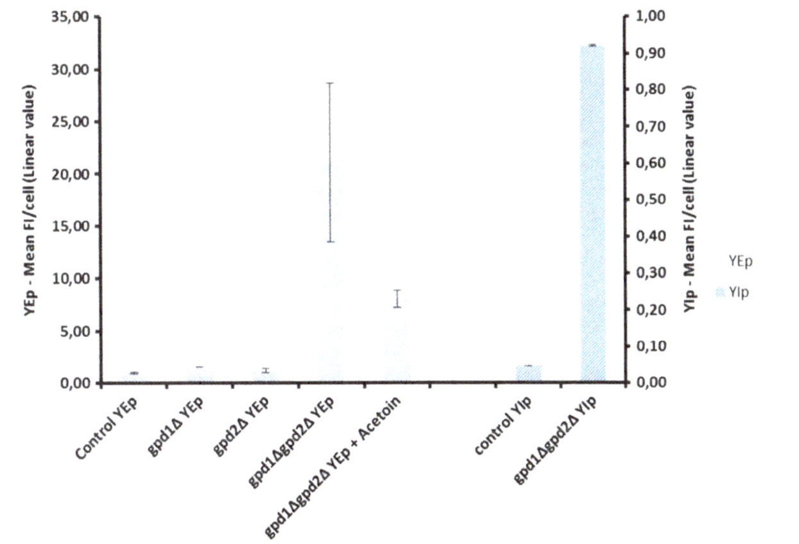

Figure 1 Mean fluorescence intensities (FI) for the different strains (wild type, *gpd1Δ, gpd2Δ and gpd1Δgpd2Δ*) carrying either the multicopy (YEp)- or the integrative (YIp)- based GFP reporter. Fluorescence intensities are given in linear values. All values are normalized to the control strain TMB4120 carrying YEp-based GFP reporter. Acetoin was added at a concentration of 0.5 g/L.

the glucose catabolic pathways. All redox engineered strain had higher FI than the control strain TMB4120. However the double deletion strain TMB4123 (*gpd1Δ gpd2Δ*) had the highest FI (21.5 fold higher than the control), whereas each single deletion had moderate increase in FI (1.5 and 1.2 fold for *gpd1Δ* and *gpd2Δ*, respectively) (Figure 1). To further evaluate the biosensor responsiveness to perturbations in $NADH/NAD^+$ ratio, the FI of TMB4123 (*gpd1Δ gpd2Δ*) was measured after addition of acetoin (3-hydroxy-2-butanone) during respiro-fermentative growth under oxygen-limited conditions. Acetoin is an alternative redox sink to the glycerol shunt/pathway through its NADH-dependent conversion to 2,3-butanediol (Bruinenberg et al. 1983; Björkqvist et al. 1997). The reaction is catalyzed by a NADH-dependent 2,3-butanediol dehydrogenase encoded by *BDH1* (Gonzalez et al. 2000). Indeed the addition of acetoin led to reduced FI as the relative fluorescence decreased from 21.5 to 8.0 (Figure 1). Altogether our results demonstrated that the constructed biosensor reported on the cells capacity to oxidize NADH, and more generally on the NAD(H) cofactor balance.

Impact of the reporter gene copy number on fluorescence intensity and heterogeneity

To investigate the impact of the copy number of the reporter gene *yEGFP3* and only relate the expression level to the transcription step, the reporter cassette was also transferred to an integrative vector (YIp) and introduced into the same control and *gpd1Δ gpd2Δ* strain backgrounds, generating strains TMB4140 (control) and TMB4144 (*gpd1Δ gpd2Δ*). Strains were cultivated in shake flasks with YNB media and 2% glucose and FI was recorded. Similar trend was observed as for the YEp based reporter constructs. But the FI levels were lower, with strain TMB4140 (control- YIp) only emitting 4% of the FI signal of TMB4120 (control- YEp). However, a clear difference in FI was again observed between the wild type and the double mutant, as TMB4144 (*gpd1Δ gpd2Δ*- YIp) displayed 21.5 fold higher FI than its corresponding control TMB 4140 (Figure 1). Also the heterogeneity of the fluorescence signal that can be measured by the standard deviation of the population was lower in the YIp system (Figure 2). Standard deviations of 44 was measured in TMB 4144 (*gpd1Δ gpd2Δ*- YIp) whereas a standard deviation of 172 was obtained in TMB4123 (*gpd1Δ gpd2Δ*- YEp). Thus, to reduce the influence of gene copy number and plasmid replication kinetics on FI output in the population, further experiments were performed using the YIp based sensor.

Evaluation of the biosensor responsiveness to redox perturbations by heterologous expression of NADH-dependent oxidoreductases

The potential of the *GPD2*p-*yEGFP3* redox biosensor to differentiate between reductases having different selectivities for NADH was also tested. For this purpose, two NAD(P)H-dependent xylose reductases (XRs) that reduces

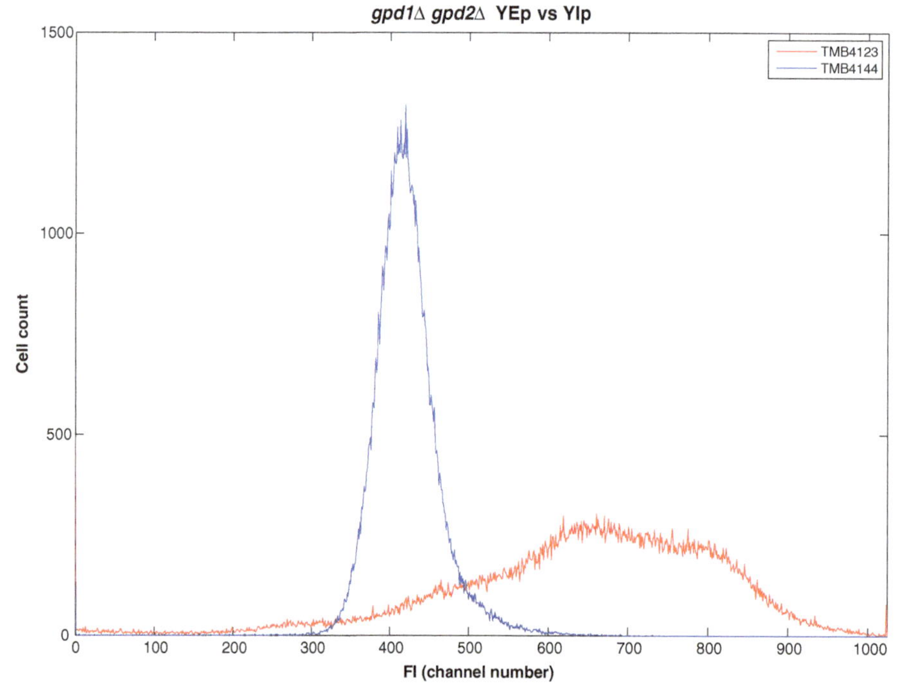

Figure 2 Histograms of the fluorescence intensity of strains TMB4123 (*gpd1Δ*gpd2Δ YEp-GFP, "red dash") and TMB4144 (*gpd1Δ*gpd2Δ YIp-GFP, "blue dash") that were grown aerobically on glucose.

xylose to xylitol using NADPH and NADH as cofactor were utilized: *Pichia stipitis* XR (wtXR) and the corresponding N272D mutant (mutXR) that display NADH/NADPH selectivity ratios of 0.04 and 0.80, respectively (Runquist et al. 2010). Both the non-engineered and the *gpd1Δ gpd2Δ* yeast backgrounds were chosen as hosts as they represent the two redox extremes i.e. the strain with the best ability to reoxidize cytosolic NADH and the one with the poorest. The YIp-based reporter and XR constructs were introduced in these strains generating strains TMB4140 (control, no XR), TMB4141 (control, wtXR), TMB4143 (control, mutXR), TMB4144 (*gpd1Δgpd2Δ*, no XR), TMB4145 (*gpd1Δgpd2Δ*, wtXR) and TMB4147 (*gpd1Δgpd2Δ*, mutXR).

Strains were cultivated under oxygen-limited conditions in sealed vials containing minimal medium with galactose as the carbon source and xylose as the substrate for the XR-catalysed reduction. Galactose was used because *S. cerevisiae* has a specific galactose uptake system, Gal2p (Cirillo 1968) that also transports xylose, and whose expression benefits xylose utilization (Johnston 1987; Hamacher et al. 2002). So, even though the yeast does not metabolize galactose as fast as glucose, the use of galactose allows the cell to have co-consumption of hexose and xylose (Bro et al. 2005; Garcia Sanchez et al. 2010). The growth profiles of the different strains are shown in Figures 3A and B. During the first 24 hours, *gpd1Δ gpd2Δ* strains reached an OD of around 3, which was only slightly lower than the corresponding control strains. During the next 48 hours, however, TMB4147 strain (*gpd1Δ gpd2Δ*, mutXR) continued growing and reached an OD of 5.2 whereas strain TMB4145 (*gpd1Δ gpd2Δ*, wtXR) only reached OD 3.7 while the OD of strain TMB4144 (*gpd1Δ gpd2Δ*, no XR) decreased to 2 (Figure 3B). Also TMB4147 strain carrying the XR with the highest selectivity for NADH (mutXR) almost

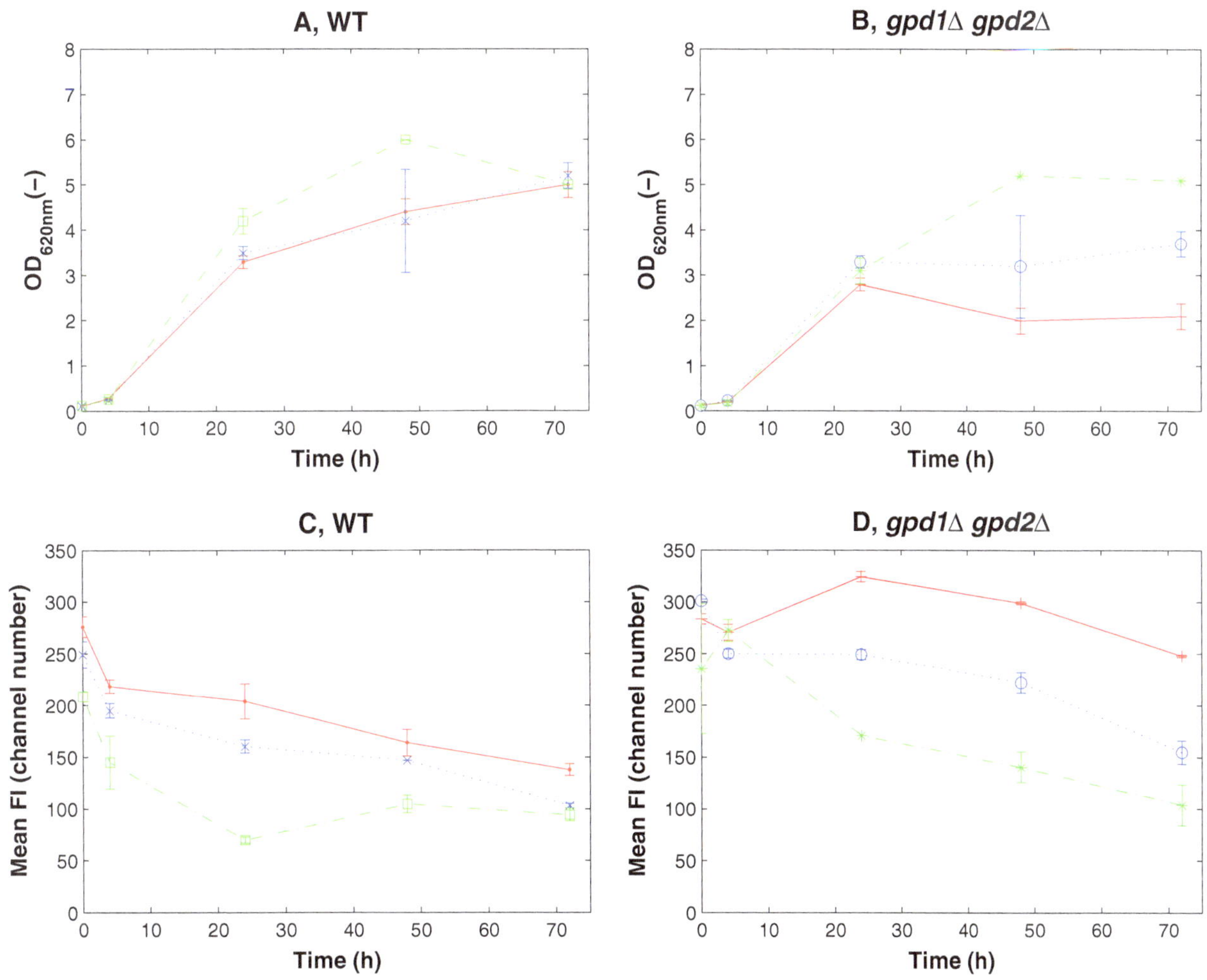

Figure 3 Optical density measured at 620 nm (A, B) and mean fluorescence intensity (C, D) over time for the control background (A, C) or the *gpd1Δ*gpd2Δ background (B, D) carrying no XR, wtXR or mutXR. Legend: TMB4140 (control, no XR "red dot"), TMB4141 (control, wtXR "blue x"), TMB4143 (control, mutXR(green square)), TMB4144 (*gpd1Δgpd2Δ*, no XR "red plus")), TMB4145 (*gpd1Δgpd2Δ*, wtXR "blue o")), TMB4147 (*gpd1Δgpd2Δ*, mutXR "green asterisk").

reached the same OD as the corresponding control strain TMB4143 (control, mutXR) (Figures 3A, B), which was not the case for the other strains carrying either no XR or wtXR. Improved NADH reoxidisation was expected to be coupled to the NADH-dependent xylose reduction to xylitol. Indeed high xylitol production, between 3 and 5 g/L, was measured during the first 48 hours of cultivations in all strains harboring a heterologous XR (Figure 4). Also the deletion strains produced xylitol faster, and in larger quantities than the control strains. Within the first 48 hours, there was no significant difference in xylitol production between strains carrying wtXR and mutXR, neither in the control strain background (Figure 4A) nor in the *gpd1Δ gpd2Δ* background (Figure 4B). However, the strains carrying mutXR reached slightly higher xylitol concentrations at the end of the cultivations. Xylitol was also produced in strains lacking heterologous XR, but to a significantly lower degree (<0.8 g/L after 70h). Xylitol may actually arise from the NADPH-dependent conversion of xylose by endogenous NADPH-dependent XR, such as the one encoded by *GRE3* gene (Träff et al. 2002). In the case of XR-carrying strains, xylitol may also originate from the NADPH-catalysed reaction of the introduced XR. Glycerol production was observed for the double deletion strains carrying the control or mutated XR (Figure 4C, D), which may be explained by the fact that *P. stipitis* XR can use DHAP as substrate, which leads to glycerol formation (Jeppsson et al. 2003).

Both the control and the *gpd1Δ gpd2Δ* strains showed the highest fluorescence at the beginning of the cultivation, followed by a gradual decrease until the end of fermentation. An inverse correlation was obtained between growth (Figure 3A, B) and the mean FIs (Figure 3C, D),

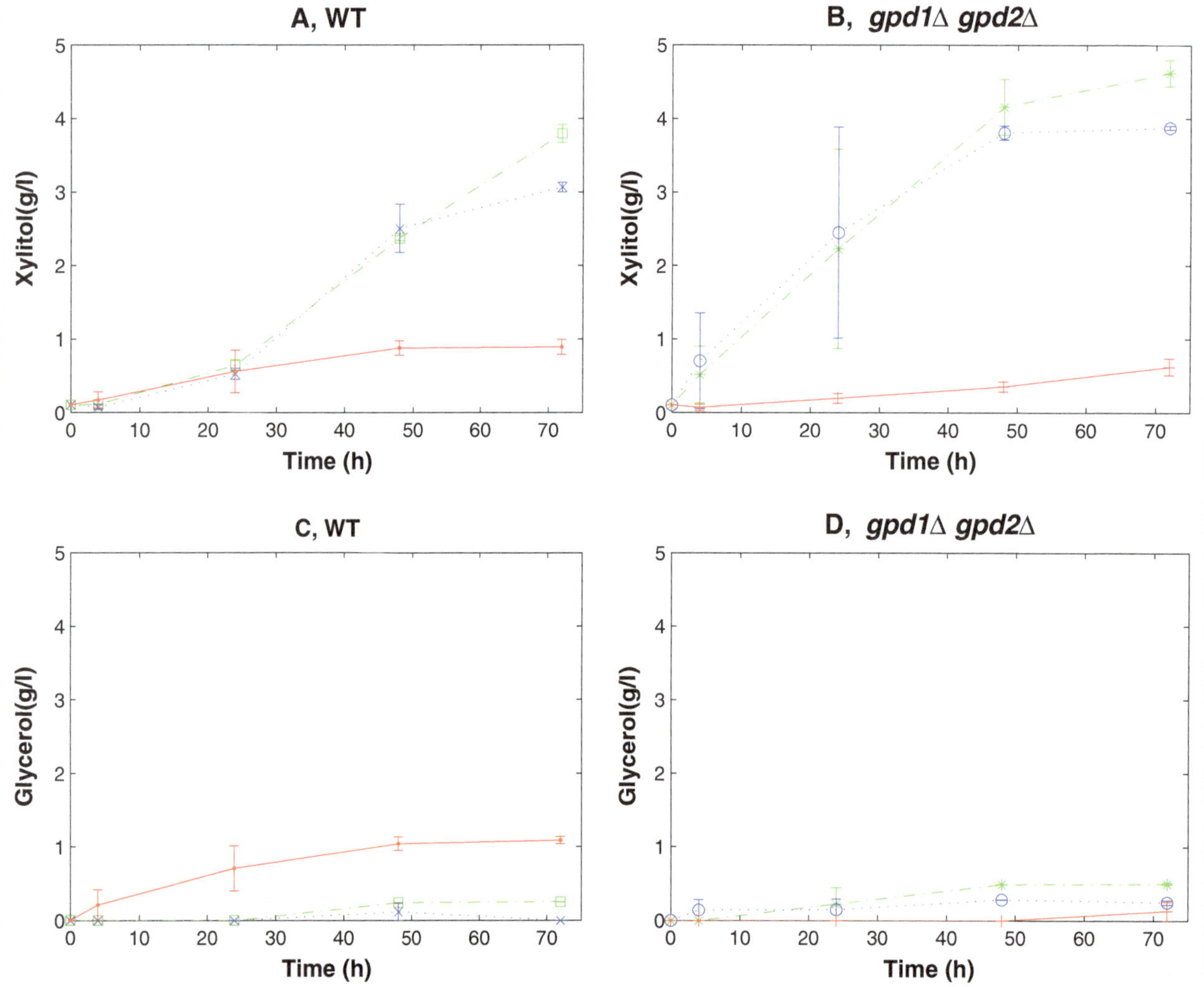

Figure 4 Extracellular xylitol (A, B) and glycerol (C, D) concentrations (in g/L) over time for the control background (A, C) or the *gpd1Δ*gpd2*Δ* background (B, D) carrying no XR, wtXR or mutXR. Legend: TMB4140 (control, no XR "red dot"), TMB4141 (control, wtXR "blue x"), TMB4143 (control, mutXR(green square)), TMB4144 (*gpd1Δgpd2Δ*, no XR "red plus"), TMB4145 (*gpd1Δgpd2Δ*, wtXR "blue o"), TMB4147 (*gpd1Δgpd2Δ*, mutXR "green asterisk").

with the best growing strains having the lowest fluorescence both in the control and the *gpd1Δ gpd2Δ* backgrounds. The difference in fluorescence was however less pronounced between the control strains than between the *gpd1Δ gpd2Δ* strains, which can be explained by the limited redox imbalance in the control strains.

The population histogram, i.e. the amount of single cells displaying a given FI, also clearly showed differences between the strains (Figures 5A, B). After 24 hours, the different populations displayed significantly different FIs, and were thus easily distinguished in an overlay plot. This was the case both in the control (Figure 5A) and in the *gpd1Δ gpd2Δ* (Figure 5B) strains, with mean FIs given in channel number units of 204 (control, no XR), 160 (control, wtXR), 70 (control, mutXR) or 325 (*gpd1Δ gpd2Δ*, no XR), 250 (*gpd1Δ gpd2Δ*, wtXR) and 171 (*gpd1Δ gpd2Δ*, mutXR), respectively. Some overlap between strains was observed in both strain backgrounds, but it was more pronounced for the wild type strains than for the deletions strains

Discussion

Microbial populations exist in dynamic equilibrium with constantly changing conditions, which requires fast and fluid regulatory networks for optimizing cell functions. Among these, an accurate metabolic response is fundamental for the cell to maintain redox homeostasis. In *S. cerevisiae*, it is well known that a skewed ratio of cytosolic NADH/NAD^+ can be compensated by glycerol production regulated by a fine-tuned expression of the *GPD2* gene (Ansell et al. 1997; Björkqvist et al. 1997). This served as basis for the construction of our *in vivo* redox state reporter that enables green fluorescence to be used as an indicator of the cells ability to reoxidize NADH, *i.e.* to the NADH/NAD^+ ratio of the cell. This

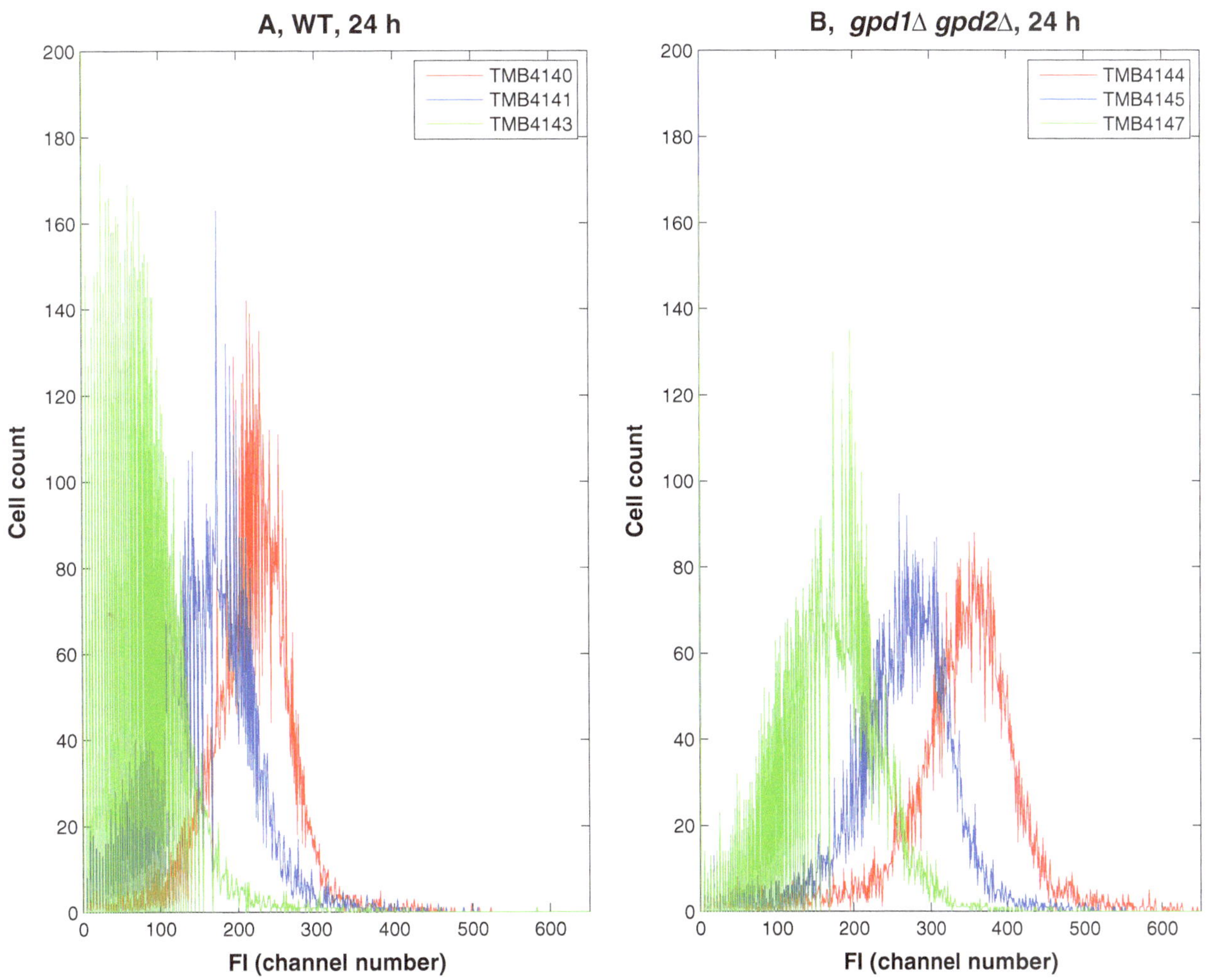

Figure 5 Fluorescence plots after 24 hours of growth. (A) Histograms of the FI of the control strains carrying no XR "red dash", wtXR "blue dash" or mutXR "green dash"). **(B)** Histogram of the FI of the *gpd1Δ gpd2Δ* strains carrying no XR (red dash), wtXR (blue dash) or mutXR (green dash). Legend TMB4140: control, no XR; TMB4141: control, wtXR; TMB4143: control, mutXR; TMB4144: *gpd1Δgpd2Δ*, no XR;TMB4145: *gpd1Δgpd2Δ*, wtXR; TMB4147:*gpd1Δgpd2Δ*, mutXR.

is, to our knowledge, the first biosensor to use the *GPD2* promoter as sensor for the cytosolic NADH/NAD$^+$ cofactor balance.

By placing a *GFP* gene downstream of the *GPD2* promoter, a direct relationship between green fluorescence intensity (FI) and glycerol formation or NADH accumulation was observed. In strains carrying double deletion of the *GPD* genes, the FI was significantly higher (21.5-fold) than for the control strain, indicating a strong activation of the *GPD2* promoter, and thereby verifying the lack of other enzymes to counteract the lacking NADH oxidation functionality. In parallel, FIs were in the same order of magnitude in single *GPD* deletion strain and in the control strain, also confirming the previously observed compensatory pattern between Gpdp isoenzymes in order to maintain NAD$^+$ regeneration via glycerol production (Ansell et al. 1997). The FI signal of the *gpd1Δ* strain was also slightly higher than the FI of the *gpd2Δ* strain and the control strain: 25% and 50%, respectively, as expression of the *GDP2*, gene whose promoter is used in the constructed system, should be higher in the *gpd1Δ* strain, as a consequence of *GPD2* compensating for the lacking *GPD1* functionality. Finally the high responsiveness of the constructed reporter supported the hypothesis that *GPD2* expression was directly connected to the cell need to oxidize NADH via glycerol formation. It cannot be excluded that FI may be correlated to more than the need for NADH oxidation via glycerol formation. Growth rate and the capacity to reoxidise NADH are closely related and thus FI will always be indirectly correlated to growth rate. As an example of this, the rescued growth rate of the *gpd1Δgpd2Δ* deletion strain by increased NADH oxidation led to significantly lower FI. However the oxygen concentration, which reflects the cells ability to replace the need for glycerol formation by aerobic respiration, was probably a more important parameter in that strain. Cultivating cells under oxygen-limited conditions could also have had implications for the FI measurements as fluorescent proteins require oxygen to mature. However, GFP expression in *S. cerevisiae* could previously be measured in cells that were cultivated under strict anaerobic conditions, most probably due to exposure to oxygen during the aerobic handling of samples. (Hebisch et al. 2013) previously demonstrated that the average maturation time in *E. coli* for GFP is around 6 min, which is enough time for GFP to mature during the preparation for the FI measurement in the present study.

There was a clear difference between placing the reporter cassette on a multicopy plasmid or on a plasmid that gets integrated in the chromosome. Copy number per cell varies between 50 and 200 when using YEp vectors as opposed to one when using YIp vectors (Romanos et al. 1992). So placing the GFP-based reporter construct on YEp plasmids led to a significantly higher FI than when using YIp-based construct. However, more defined fluorescence peaks were obtained when using the YIp system as opposed to the YEp system. This is in accordance with previous findings illustrating the stability issues of YEp-based systems compared to YIp-based systems (Ishii et al. 2009) and indicates lower degree of cell population heterogeneity in chromosomally integrated biosensors as compared to the YEp-based systems, which may later facilitate the distinction between any underlying subpopulations. Promoters can also have different levels of noise in the respective gene expression (Blake et al. 2003), which can lead to a higher heterogeneity of the signal. The low degree of heterogeneity in the YIp-based system further stresses the suitability of the *GPD2* promoter for analysis of populations at the single cell level.

A 2-fold lowering of the GFP levels was observed when acetoin was added to *GPD* double deletion strains, which can be attributed to the induced NADH-dependent reduction of acetoin (Gonzalez et al. 2000). Acetoin or acetaldehyde, that is reduced to ethanol with the concurrent oxidation of NADH, have previously been used to demonstrate the lack of alternative cytosolic reductive reactions in double deletion strain (Lidén et al. 1996; Ansell et al. 1997). In these studies, the involved oxidation reactions were efficient enough to recover sufficient redox balance for the cells to start growing. By monitoring FI instead, we wanted to investigate whether the constructed biosensor could monitor less drastic changes in the redox balance. To validate this hypothesis, we introduced XR enzymes with slightly different catalytic activity for NADH oxidation, namely two xylose reductases having NADH/NADPH selectivity ratios of 0.04 and 0.80 respectively (Runquist et al. 2010). As expected, the biggest difference in growth was observed in the double deletion strains between the control and the strain carrying the mutXR, since more carbon was made available for growth when xylitol formation from xylose was used as redox sink instead of glycerol formation from glucose. This is also part of the explanation for the lower glycerol formation in the TMB4141 and TMB4143 as opposed to the TMB4140, as the cells can produce xylitol as a redox sink in addition to glycerol. A higher xylitol production could have been expected for the mutXR than wtXR in the *gpd1Δgpd2Δ* strain, if only considering the NADH-dependent activity. However wtXR also has a significant activity with NADPH as a cofactor (Rizzi et al. 1988), which led to a non-negligible NADH-independent xylitol production. The maximum difference in OD was 2.5-fold after 48 hours of growth whereas the difference between the control strain and the strain carrying the wtXR was only 1.5 fold. So, in theory, the reductases could be distinguishable by repeating batch cultivations and using growth as selection. Assuming that after 48 hours one cell of TMB4144 (double deletion mutXR) grows to 2.5 times the cell density of the TMB4140 (double deletion no XR), it would take 27 batch cultivations of 48 hours each before one TMB4144 cell would outgrow 10^9 TMB4140 cells by a factor of 1:100 (according to the following calculation: $(1/2.5)^{27} \cdot 10^9 \approx 0.01$). Using fluorescence instead, the difference was clearly visible after a few hours. Increased xylitol formation and lower fluorescence were observed in strains carrying the wtXR and the mutXR. However, a clear difference in FIs was observed between the two strains whereas differences in xylitol levels were not that significant. The results, although qualitative, indicate that fluorescence was the more sensitive and faster method to discriminate between the two XR enzymes and their difference in NADH-catalyzed xylitol production and consequently their ability to affect the cytosolic NADH/NAD$^+$ ratio. It also opens the possibility for high-throughput screening of enzymes with increased selectivity for NADH. In our cases, similar XR levels were measured (data not shown). However, in future screening, an enzyme with lower specificity for NADH, but expressed at a higher level and giving a higher activity/per cell for NADH oxidation will lead to a lower FI. This limitation in the screening will, however, occur in any method based on the NADH pool and not only in the FI-based one. Another element that can be further considered when using a GFP-based reporter system for screening or analysis of the NAD(H) cofactor imbalance is the half-live of the chosen protein. The yEGFP3 protein used in the present study has been reported to have a half-life of about 7.5 hours (Mateus and Avery 2000), and indeed the intracellular GFP level did not respond rapidly to dynamic changes in external conditions. This implies that rapid changes in NAD(H) cofactor balancing may not be monitored with the developed system. However alternative GFP variants with shorter half-life, such as the GFP fused with a fragment of the CLN2 (yeast G1 cyclin) protein having a half-life of 30 minutes (Mateus and Avery 2000), could then be used instead.

Lately, much attention has been drawn in using FACS for screening large genetic space (Dietrich et al. 2010) and the technique has already been applied for the screening of NADPH-dependent alcohol dehydrogenases in *E. coli* (Siedler et al. 2014). However, the combination of a NADH biosensor and single-cell analysis has, to our knowledge, not yet been implemented in *S. cerevisiae*. Our tool offers the first single-cell high-throughput NAD(H) cofactor balance sensor in *S. cerevisiae*. In contrast to previously reported methods where extraction of intracellular NAD(H) metabolites was performed (Canelas et al. 2008), the method presented in this work is faster and allows for analysis with single-cell level resolution. Therefore the constructed tool offers possible applications in the selection of new NADH-dependent enzymes based on single cell sorting. Additionally it could be applied in monitoring the impact of various metabolic engineering strategies on the cytosolic NAD(H) cofactor balance of the cell.

Competing interests

The authors declare that they have no competing interests.

Authors' contribution

JDK, MC and MGG contributed equally in the design of the study. JDK performed the experimental work and wrote the manuscript. MC and MGG helped in drafting the manuscript. All authors have read and approved the submission of the manuscript.

Acknowledgements

This work was financially supported by the Swedish Research Council (Vetenskapsrådet). Camille Robert is acknowledged for assistance in creating some of the yeast strains. Stefan Morberg is acknowledged for technical assistance with the flow cytometry analysis. The pYGFP3 plasmid was kindly provided by Alistair Brown (University of Aberdeen).

References

Andreasen AA, Stier TJB (1953) Anaerobic nutrition of *Saccharomyces cerevisiae*. I. Ergosterol requirement for growth in a defined medium. J Cell Physiol 41(1):23–36

Andreasen AA, Stier TJB (1954) Anaerobic nutrition of *Saccharomyces cerevisiae*. II. Unsaturated fatty and requirement for growth in a defined medium. J Cell Physiol 43(3):271–281

Ansell R, Granath K, Hohmann S, Thevelein JM, Adler L (1997) The two isoenzymes for yeast NAD$^+$-dependent glycerol 3-phosphate dehydrogenase encoded by *GPD1* and *GPD2* have distinct roles in osmoadaptation and redox regulation. EMBO J 16(9):2179–2187

Bengtsson O, Hahn-Hägerdal B, Gorwa-Grauslund MF (2009) Xylose reductase from *Pichia stipitis* with altered coenzyme preference improves ethanolic xylose fermentation by recombinant *Saccharomyces cerevisiae*. Biotechnol Biofuels 2:9

Björkqvist S, Ansell R, Adler L, Lidén G (1997) Physiological response to anaerobicity of glycerol-3-phosphate dehydrogenase mutants of *Saccharomyces cerevisiae*. Appl Environ Microbiol 63(1):128–132

Blake WJ MK, Cantor CR, Collins JJ (2003) Noise in eukaryotic gene expression. Nature 422(6932):633–637

Blomberg A, Adler L (1989) Roles of glycerol and glycerol-3-phosphate dehydrogenase (NAD$^+$) in acquired osmotolerance of *Saccharomyces cerevisiae*. J Bacteriol 171(2):1087–1092

Bro C, Knudsen S, Regenberg B, Olsson L, Nielsen J (2005) Improvement of galactose uptake in *Saccharomyces cerevisiae* through overexpression of phosphoglucomutase: example of transcript analysis as a tool in inverse metabolic engineering. Appl Environ Microbiol 71(11):6465–6472

Bruinenberg P, de Bot PM, van Dijken J, Scheffers WA (1983) The role of redox balances in the anaerobic fermentation of xylose by yeasts. Eur J Appl Microbiol Biotechnol 18(5):287–292

Bücher T, Brauser B, Conze A, Klein F, Langguth O, Sies H (1972) State of oxidation-reduction and state of binding in the cytosolic NADH-system as disclosed by equilibration with extracellular lactate-pyruvate in hemoglobin-free perfused rat liver. Eur J Biochem 27(2):301–317

Canelas AB, van Gulik WM, Heijnen JJ (2008) Determination of the cytosolic free NAD/NADH ratio in *Saccharomyces cerevisiae* under steady-state and highly dynamic conditions. Biotechnol Bioeng 100(4):734–743

Carlquist M, Fernandes RL, Helmark S, Heins AL, Lundin L, Sorensen SJ, Gernaey KV, Lantz AE (2012) Physiological heterogeneities in microbial populations and implications for physical stress tolerance. Microb Cell Fact 11:94

Cirillo VP (1968) Galactose transport in *Saccharomyces cerevisiae*. I. Nonmetabolized sugars as substrates and inducers of the galactose transport system. J Bacteriol 95(5):1727–1731

Cormack BP, Valdivia RH, Falkow S (1996) FACS-optimized mutants of the green fluorescent protein (GFP). Gene 173(1):33–38

Cormack BP, Bertram G, Egerton M, Gow NA, Falkow S, Brown AJ (1997) Yeast-enhanced green fluorescent protein (yEGFP): a reporter of gene expression in *Candida albicans*. Microbiology 143(Pt 2):303–311

Dardalhon M, Kumar C, Iraqui I, Vernis L, Kienda G, Banach-Latapy A, He T, Chanet R, Faye G, Outten CE, Huang ME (2012) Redox-sensitive YFP sensors monitor dynamic nuclear and cytosolic glutathione redox changes. Free Radic Biol Med 52(11–12):2254–2265

Dietrich JA, McKee AE, Keasling JD (2010) High-throughput metabolic engineering: advances in small-molecule screening and selection. Annu Rev Biochem 79:563–590

Eriksson P, Andre L, Ansell R, Blomberg A, Adler L (1995) Cloning and characterization of *GPD2*, a second gene encoding sn-glycerol 3-phosphate dehydrogenase (NAD+) in *Saccharomyces cerevisiae*, and its comparison with *GPD1*. Mol Microbiol 17(1):95–107

Förster J, Famili I, Fu P, Palsson BO, Nielsen J (2003) Genome-scale reconstruction of the *Saccharomyces cerevisiae* metabolic network. Genome Res 13(2):244–253

Garcia Sanchez R, Hahn-Hägerdal B, Gorwa-Grauslund MF (2010) Cross-reactions between engineered xylose and galactose pathways in recombinant *Saccharomyces cerevisiae*. Biotechnol Biofuels 3:19

Gietz RD, Sugino A (1988) New yeast-*Escherichia coli* shuttle vectors constructed with *in vitro* mutagenized yeast genes lacking six-base pair restriction sites. Gene 74(2):527–534

Gietz RD, Woods RA (2002) Transformation of yeast by lithium acetate/single-stranded carrier DNA/polyethylene glycol method. Methods Enzymol 350:87–96

Gonzalez E, Fernandez MR, Larroy C, Sola L, Pericas MA, Pares X, Biosca JA (2000) Characterization of a (2R,3R)-2,3-butanediol dehydrogenase as the *Saccharomyces cerevisiae YAL060W* gene product. Disruption and induction of the gene. J Biol Chem 275(46):35876–35885

Hamacher T, Becker J, Gardonyi M, Hahn-Hägerdal B, Boles E (2002) Characterization of the xylose-transporting properties of yeast hexose transporters and their influence on xylose utilization. Microbiology 148(Pt 9):2783–2788

Hebisch E, Knebel J, Landsberg J, Frey E, Leisner M (2013) High variation of fluorescence protein maturation times in closely related *Escherichia coli* strains. PLoS One 8(10):e75991

Inoue H, Nojima H, Okayama H (1990) High efficiency transformation of *Escherichia coli* with plasmids. Gene 96(1):23–28

Ishii J, Izawa K, Matsumura S, Wakamura K, Tanino T, Tanaka T, Ogino C, Fukuda H, Kondo A (2009) A simple and immediate method for simultaneously evaluating expression level and plasmid maintenance in yeast. J Biochem 145(6):701–708

Jeppsson M, Träff K, Johansson B, Hahn-Hägerdal B, Gorwa-Grauslund MF (2003) Effect of enhanced xylose reductase activity on xylose consumption and product distribution in xylose-fermenting recombinant *Saccharomyces cerevisiae*. FEMS Yeast Res 3(2):167–175

Johnston M (1987) A model fungal gene regulatory mechanism: the *GAL* genes of *Saccharomyces cerevisiae*. Microbiol Rev 51(4):458–476

Karhumaa K, Hahn-Hägerdal B, Gorwa-Grauslund MF (2005) Investigation of limiting metabolic steps in the utilization of xylose by recombinant *Saccharomyces cerevisiae* using metabolic engineering. Yeast 22(5):359–368

Lidén G, Walfridsson M, Ansell R, Anderlund M, Adler L, Hahn-Hägerdal B (1996) A glycerol-3-phosphate dehydrogenase-deficient mutant of *Saccharomyces cerevisiae* expressing the heterologous *XYL1* gene. Appl Environ Microbiol 62(10):3894–3896

Mateus C, Avery SV (2000) Destabilized green fluorescent protein for monitoring dynamic changes in yeast gene expression with flow cytometry. Yeast 16(14):1313–1323

Mumberg D, Müller R, Funk M (1995) Yeast vectors for the controlled expression of heterologous proteins in different genetic backgrounds. Gene 156(1):119–122

Murray DB, Haynes K, Tomita M (2011) Redox regulation in respiring *Saccharomyces cerevisiae*. Biochim Biophys Acta 1810(10):945–958

Nissen TL, Schulze U, Nielsen J, Villadsen J (1997) Flux distributions in anaerobic, glucose-limited continuous cultures of *Saccharomyces cerevisiae*. Microbiology 143(Pt 1):203–218

Rizzi M, Erlemann P, Bui-Thanh N-A, Dellweg H (1988) Xylose fermentation by yeasts. Appl Microbiol Biotechnol 29(2–3):148–154

Romanos MA, Scorer CA, Clare JJ (1992) Foreign gene expression in yeast: a review. Yeast 8(6):423–488

Runquist D, Hahn-Hägerdal B, Bettiga M (2010) Increased ethanol productivity in xylose-utilizing *Saccharomyces cerevisiae* via a randomly mutagenized xylose reductase. Appl Environ Microbiol 76(23):7796–7802

Sambrock J, Russell DW (2001) Molecular Cloning, a Laboratory Manual. Press Cold Spring Harbor Laboratory, New York, USA

Sandström AG, Wikmark Y, Engström K, Nyhlen J, Bäckvall JE (2012) Combinatorial reshaping of the *Candida antarctica* lipase A substrate pocket for enantioselectivity using an extremely condensed library. Proc Natl Acad Sci U S A 109(1):78–83

Siedler S, Schendzielorz G, Binder S, Eggeling L, Bringer S, Bott M (2014) SoxR as a single-cell biosensor for NADPH-consuming enzymes in *Escherichia coli*. ACS Synth Biol 3(1):41–47

Theobald U, Mailinger W, Baltes M, Rizzi M, Reuss M (1997) In vivo analysis of metabolic dynamics in *Saccharomyces cerevisiae* : I. Experimental observations. Biotechnol Bioeng 55(2):305–316

Träff KL, Jönsson LJ, Hahn-Hägerdal B (2002) Putative xylose and arabinose reductases in *Saccharomyces cerevisiae*. Yeast 19(14):1233–1241

Valadi A, Granath K, Gustafsson L, Adler L (2004) Distinct intracellular localization of Gpd1p and Gpd2p, the two yeast isoforms of NAD^+-dependent glycerol-3-phosphate dehydrogenase, explains their different contributions to redox-driven glycerol production. J Biol Chem 279(38):39677–39685

Valkonen M, Mojzita D, Penttila M, Bencina M (2013) Noninvasive high-throughput single-cell analysis of the intracellular pH of *Saccharomyces cerevisiae* by ratiometric flow cytometry. Appl Environ Microbiol 79(23):7179–7187

van Dijken JP, Scheffers WA (1986) Redox balances in the metabolism of sugars by yeasts. FEMS Microbiol Lett 32(3–4):199–224

Veech RL, Eggleston LV, Krebs HA (1969) The redox state of free nicotinamide-adenine dinucleotide phosphate in the cytoplasm of rat liver. Biochem J 115(4):609–619

Williamson DH, Lund P, Krebs HA (1967) The redox state of free nicotinamide-adenine dinucleotide in the cytoplasm and mitochondria of rat liver. Biochem J 103(2):514–527

Zacharias DA, Tsien RY (2005) Molecular Biology and Mutation of Green Fluorescent Protein. Green Fluorescent Protein. John Wiley & Sons, Inc., In, pp 83–120

Zhang Q, Piston DW, Goodman RH (2002) Regulation of corepressor function by nuclear NADH. Science 295(5561):1895–1897

14

Trametes meyenii possesses elevated dye degradation abilities under normal nutritional conditions compared to other white rot fungi

Peter R Chenaux, Narisa Lalji and Daniel D Lefebvre*

Abstract

Several species of white-rot fungi were investigated for their utility in prolonged decolouration of the recalcitrant sulfonated azo dye, amaranth. *Trametes pubescens, T. multicolor, T. meyenii and T. versicolor* decoloured amaranth azo-dye best on low-nitrogen agar-solidified media whereas *Bjerkandera adusta* and *Phlebia radiata* were most effective in low nitrogen medium supplemented with manganese. *Trametes cotonea* did not decolour effectively under any condition. The decolouring *Trametes* species were also effective in liquid culture whereas *B. adusta* and *P. radiata* were not. *Trametes meyenii, T. pubescens* and *T. multicolor* were equal to or better than commonly employed *T. versicolor* at decolouring amaranth. This is the first study to show the dye decolouration potential of *T. meyenii*, *T. pubescens*, and *T. multicolor*. Supplementing with Mn(II) increased assayable manganese peroxidase activity, but not long-term decolouration, indicating that laccase is the main decolourizing enzyme in these *Trametes* species. This appears to be because of inadequate Mn^{3+} chelation required by manganese peroxidase because adding relatively low amounts of malonate enhanced decolouration rates. The ability of *Trametes meyenii* to simultaneously decolour dye over prolonged periods of time while growing in relatively nutrient-rich medium appears to be unique amongst white-rot fungi, indicating its potential in wastewater bioremediation.

Keywords: White-rot fungi; Dye decolouration; *Trametes*; Laccase; Manganese peroxidase; Manganic chelation

Introduction

One of the largest sources of environmental pollutants is the textile industry, which can produce over 800 kilo tonnes of dye annually, with 90% of the waste discharged into the environment (Nigam et al. 2000; Hessel et al. 2007; Martin et al. 2012). The strong structural integrity and toxic nature of dyes poses risk to flora, fauna and human populations (Chagas and Durrant 2001). The four main chromophoric groups include azo, anthraquinone, triarylmethane and phthalocyanine, with azo dyes accounting for over 50% used in the textile industry (Heinfling et al. 1998a; Reddy and Mathew 2001). Azo dyes are generally also the most difficult to degrade (Toh et al. 2003; Dafale et al. 2010). Amaranth, a sulfonated compound, was employed as a recalcitrant azo dye in this study.

White-rot fungi are able to degrade lignin, a complex plant biopolymer (Evans and Hedger 2001). They employ relatively non-specific enzymes that are also able to attack a wide range of pollutants including textile dyes (Archibald et al. 1997; Van Aken et al. 1999; Pointing 2001; Ramsay and Goode 2004). Even though *Trametes versicolor* (Swamy and Ramsay 1999b) and some other white-rot fungi (Heinfling et al. 1998b; Levin et al. 2002) do produce these enzymes under nutrient-rich conditions, it is generally accepted that effective degradation occurs during the induction of secondary metabolism only when carbon or nitrogen supplies are low (Archibald et al. 1997; Kaal et al. 1995; Swamy and Ramsay 1999a; Hatvani and Mécs 2002). It would, however, be more desirable from a bioremediation perspective to utilize well nourished growing organisms to decolor dyes for prolonged periods of time.

Manganese peroxidase (MnP) is the most common ligninolytic peroxidase as it is produced by almost all white-rot basidiomycetes (Morgenstern et al. 2008; Tomsovsky et al. 2009), and laccase occurs in almost all wood- and litter-transforming basidiomycetes (Wesenberg et al. 2003). These enzymes play major roles in decolorization processes

* Correspondence: lefebvre@queensu.ca
Department of Biology, Queen's University, Kingston, ON K7L 3 N6, Canada

in the fungal genus *Trametes* where they can be expressed to some degree under primary as well as secondary metabolism (Libra et al. 2003). They are also the main lignin modifying enzymes produced by *T. versicolor* during decoloration of amaranth dye (Swamy and Ramsay 1999b; Champagne and Ramsay 2005). Although laccase activity predominates under well nourished conditions, it may not be able to decolorize dye in the absence of MnP (Wesenberg et al. 2003; Viswanath et al. 2014). Because of their common occurence and enzyme efficiencies (Morgenstern et al. 2008; Wesenberg et al. 2003; Tomsovsky et al. 2009) and the fact that normal culture conditions for fungi do not induce lignin peroxidase activity (Swamy and Ramsay 1999b), enzyme investigations were limited to MnP and laccase in this study.

Although there have been several studies on dye degradation by *T. versicolor*, many of the estimated fifty *Trametes* species (Kirk et al. 2009) have not been investigated. This provides a large resource within which to search for species with desireable bioremediation properties. In the current study we investigate white-rot fungal species, including five species of *Trametes*, to determine how well they degrade amaranth and produce MnP and laccase in response to different nutritional treatments. Notable species of the genus *Trametes* were distinguished by their ability to effectively degrade dye over extended periods of time, and to do so, *Trametes meyenii* in particular, did not require nutrient deprivation.

Materials and methods

Culture maintenance and media

Seven species of fungi (Table 1) were maintained as stocks in 100 x 15 mm petri plates containing 15 mL of modified Kirk's medium (Kirk and Fenn 1982) with 3% (w:v) malt agar at 4°C and pH 5.0. The modified Kirk's medium consisted of 10 g L^{-1} glucose, 1.2 g L^{-1} ammonium tartrate, 0.05 g L^{-1} $MgSO_4.7H_2O$, 0.01 g L^{-1} $CaCl_2.2H_2O$, 0.20 g L^{-1} K_2HPO_4, 1 µg L^{-1} thiamine, 1 mL L^{-1} trace mineral solution and 15 g L^{-1} agar. The trace mineral solution contained 1 g L^{-1} NaCl, 0.5 g L^{-1} $MnSO_4.H_20$, 0.1 g L^{-1} $CoSO_4$, 0.1 g L^{-1} $FeSO_4.7H_2O$, 0.1 g L^{-1} $ZnSO_4.7H_2O$, 82 mg L^{-1} $CaCl_2$, 10 mg L^{-1} $CuSO_4.5H_2O$, 10 mg L^{-1} $NaMoO_4.2H_2O$, 10 mg L^{-1} H_3BO_3, and 0.1 g L^{-1} ethylene diamine tetraacetic acid at pH 5.0.

Table 1 White-rot fungal strains

Species	Strain code
Bjerkandera adusta	ATCC[a] MYA-264
Phlebia radiata	ATCC 64658
Trametes cotonea	CBS[b] 352.80
Trametes meyenii	CBS 453.76
Trametes multicolor	VIAM[c] MB 49
Trametes pubescens	CBS 396.90
Trametes versicolor	ATCC 20869

[a]ATCC, American Type Culture Collection, Manassas, USA;
[b]CBS, Centraalbureau voor Schimmelcultures, Utrecht, The Netherlands;
[c]VIAM, Culture Collection of the Institute of Applied Microbiology, University of Agricultural Sciences, Vienna, Austria.

Influence of culture parameters on growth and amaranth decoloration on agar-solidified media

Circular plugs measuring 0.5 cm diam. were taken from stock plates using the wide end of a sterilized 200 µL pipet tip (Advantech AD200Y-K, Diamed Lab Supplies Inc., Mississauga, Canada) and placed in the center of fresh plates and grown at 28°C. After 4 days fresh plugs were placed on experimental media shown in Table 2. Amaranth was added at 50 ppm (83 µM) unless otherwise stated. Plates were kept at 28°C and fungal growth and dye decoloration, where appropriate, were measured daily until either fungal growth or decoloration reached the edge of the plate. Growth and decolored zones were determined by measuring their areas on the plates. Experiments were performed in octuplicate.

Influence of culture parameters on amaranth decoloration and enzyme activity in liquid media

Ten 0.5 cm diameter circular agar plugs from 4 day old petri plate cultures were added to each 500 mL Erlenmeyer flask containing 100 mL of Kirk's medium. These were grown at 28°C with rotary shaking at 100 rpm. The Kirk's medium used followed the same recipe as for the agar-solidified medium without the agar and supplementation with 20 mM 2,2-dimethylsuccinate to act as a buffer (Swamy and Ramsay 1999a). Amaranth treatment was performed as follows. After 5 days the medium was decanted and the fungal pellets were resuspended in 100 mL of fresh Kirk's medium containing 0.12 mM ammonia tartrate (low-N medium) and 83 µM (50 ppm) amaranth. After 4 days, this medium was replaced with that of the same composition and samples were taken at increasing time intervals to determine decoloration of amaranth and enzyme activities. Where appropriate, the effect of Mn was by the addition of 200 µM $MnSO_4.H_2O$. All

Table 2 Modifications made to standard Kirk's medium

Medium	Details
Low-N Kirk's (1.2 mM)	Basic Kirk's containing 0.22 g L^{-1} ammonium tartrate.
High-N Kirk's (12 mM)	Basic Kirk's containing 2.2 g L^{-1} ammonium tartrate.
Low-Glu Kirk's	Basic Kirk's containing 1 g L^{-1} glucose.
Low-N/Low-Glu Kirk's	Basic Kirk's containing 1 g L^{-1} glucose & 0.22 g L^{-1} ammonium tartrate.
Low-N Kirk's + 200 µM Mn(II)	Basic Kirk's containing 0.22 g L^{-1} ammonium tartrate & 0.034 g L^{-1} $MnSO_4.H_2O$.

experiments were performed in duplicate with four replicates for each treatment.

Long term decoloration experiments were initiated by adding 5 mL wet volume of fungal biomass from 5 day old culture in liquid Kirk's medium without amaranth to 100 mL culture flasks containing various media. Experiments were started at 83 μM amaranth which was replenished after complete decolorations occurred as often as required.

Assays of decoloration and enzyme activities

One mL samples were taken at each time point and immediately replaced with Kirk's medium containing no glucose or ammonia tartrate. Amaranth dye concentrations were determined spectrophotometrically at 523 nm. An $O.D._{523}$ of 1.0 per cm corresponded to an amaranth concentration of 41.5 μM. MnP activity was measured by monitoring specific oxidation of Mn^{2+} to chelated Mn^{3+} (Wariishi et al. 1992). One unit (U) of activity equals 1 μmol Mn^{2+} oxidized per minute at 25°C and pH 4.5. Laccase activity assays were performed by measuring the oxidation of 2,2'-azino-bis(3-ethylbenzthiazoline-6-sulfonic) acid (ABTS) at 420 nm ($\varepsilon_{420} = 36{,}000\ M^{-1}\ cm^{-1}$) (Johannes and Majcherczyk 2000). One unit (U) of activity equals 1 μmol ABTS oxidized per minute at 25°C and pH 5.0. These assays contained 50 mM sodium acetate buffer and 0.2 mM ABTS. All assays were performed in quadruplicate using 96-well microtiter plates and measured with a Spectra Max Plus Spectrophotometer (Molecular Devices, Sunnyvale, CA). One direction ANOVA with a Tukey-Kramer HSD post hoc test was performed using JMP 10.0 software (SAS Incorporated, Toronto, Canada).

Results

Effect of nutrition on decoloration on agar-solidified media

The effect of medium composition on fungal abilities to decolor the azo-dye, amaranth in seven different species of white-rot fungi (Table 1) was tested on different variations of nutrient media (Table 2).

Decoloration abilities after 6 days growth on agar plates are presented in Table 3. After a lag of 3-4 days all species except for *Trametes cotonea* decolored amaranth. Complete dye decoloration occurred for *Phlebia radiata* on low N Kirk's with Mn, and for *Trametes meyenii, Trametes pubescens and Trametes multicolor* on low N/low glucose Kirk's. Significantly, *T. meyenii* also decolored completely on full nutrient Kirk's medium.

Decoloration studies in liquid cultures

Seven species of white-rot fungi were grown in liquid medium containing amaranth for four days. Low-N Kirk's was chosen because most species (6 of 7) had substantial decoloration capabilities on agar-solidified medium of this composition (Table 3). All species began decoloration immediately (Figure 1). *Phlebia radiata, Bjerkandera adusta,* and *T. cotonea* decolored amaranth only slightly which may be attributed to sorption to biomass rather than by enzymatic degradation. This is corroborated by the lack of detectable MnP and laccase activities in these species (Figure 2). The remaining species did not absorb dye. *Trametes multicolor* and *T. pubescens* decolored the best and both were able to decolor a second addition of dye within a 24 hour period, with the rates of the 2nd decoloration being approximately double that of the first exposures. *Trametes versicolor* and *T. meyenii* were able to achieve approx. 94 and 77 percent decoloration within 24 hours, respectively.

Secreted laccase and MnP were detected for the four species that showed the highest decoloring abilities, with laccase as the main lignin modifying enzyme (Figure 2). *Trametes pubescens* had the most laccase activity, which was three times that of the next best laccase producing species, *T. multicolor. Trametes pubescens* also had the highest MnP production, which was twice that of *T. multicolor*. Enzyme activities for lignin peroxidase, versatile peroxidase and dye-decolorizing peroxidase were not detected under the culturing conditions employed.

Table 3 Decoloration of amaranth by white-rot fungi grown on agar-solidified media

Species	Kirk's	LG Kirk's	HN Kirk's	LN Kirk's	LN/LG Kirk's	LN Kirk's + Mn
B.adusta	+	++	-	++	++	+++
P.radiata	+	+	+	++	-	++++
T.cotonea	-	-	-	-	-	-
T.meyenii	++++	+++	+++	+++	++++	++
T.pubescens	++	+++	+++	+++	++++	+++
T.multicolor	+	-	+	+++	++++	++
T.versicolor	+	-	+	+++	+++	+++

L, low; H, high; G, glucose; N, nitrogen; -, no decolouration; +, low decolouration; ++, intermediate decolouration; +++, strong decolouration; ++++, complete decolouration. Results were consistent over four independent experiments.

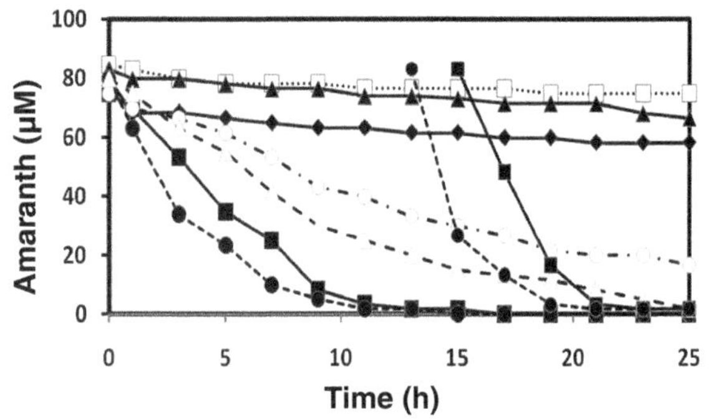

Figure 1 Amaranth decolouration by 9-day-old cultures in low-N Kirk's liquid medium. *B. adusta* (♦), *P. radiata* (□), *T. cotonea* (▲), *T. meyenii* (○), *T. pubescens* (●), *T. multicolor* (■), *T. versicolor* (Δ). S.D. always less than 4% (n = 5).

Sequential decoloration experiments

The four *Trametes* species that efficiently decolored amaranth dye in liquid medium were investigated to determine their respective most appropriate nutrient conditions for decoloration. A summary of these decoloration and enzyme activity characteristics are given in Table 4.

The greatest number of repeat decolorations occurred with *T. pubescens* grown in low-N Kirk's, which was able to decolor 26 sequential daily additions of 83 μM amaranth over a 29 day period. This species also had the highest level of laccase and MnP activities in this medium. The addition of Mn(II) or sufficient N resulted in 20 and 19 sequential decolorations, respectively, with reduced decoloration rates by day 20 in both. This decrease in decoloration rate was not reflected in measured enzyme activities which remained elevated (data not shown).

Trametes multicolor grown in low-N Kirk's was able to decolor 24 sequential additions of amaranth over 29 days. As with *T. pubescens*, this also corresponded to the highest levels of MnP and laccase activity in that species. The addition of Mn(II) decreased these enzymes' activities, and this was even more pronounced in normal Kirk's medium. After day 5 both laccase and MnP activities declined (data not presented). Despite this, decoloration rate was maintained until day 23 with most daily pigment additions being completely removed between 5 and 24 hours. This trend was most notable in low-N Kirk's, which was still able to decolor somewhat at day 29 despite there being very little measurable laccase or MnP activity.

Trametes versicolor also had its highest number of decolorations under low-N conditions, decoloring 20 sequential additions in 29 days. This also did not correspond to the highest activity of laccase and MnP. The highest enzyme activity was found in normal Kirk's medium, where only 9 sequential decolorations occurred. The addition of 200 μM Mn(II) to the medium doubled MnP but decreased laccase activity.

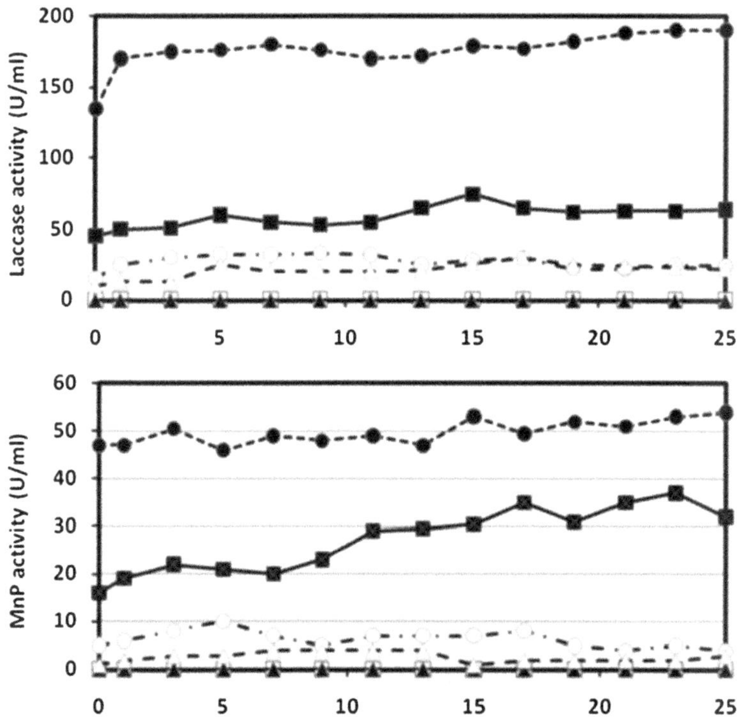

Figure 2 Laccase and MnP activities by 9-day-old cultures in low-N Kirk's liquid medium. *B. adusta* (♦), *P. radiata* (□), *T. cotonea* (▲), *T. meyenii* (○), *T. pubescens* (●), *T. multicolor* (■), *T. versicolor* (Δ). S.D. always less than 5% (n = 5). *B. adusta, P. radiata* and *T. cotonea* all displayed negligible enzyme activities.

Table 4 Decolouration and enzyme characteristics of *Trametes* species cultured in liquid media for 29 days

	Decoloration				Enzyme activities					
					Mn Peroxidase			Laccase		
	Initial decoloration period (days)	Number of decolorations	Rate of decoloration at 29 days	Average time for decoloration (days)	Maximum (U/ml)	Day at maximum	Average (U/ml)	Maximum (U/ml)	Day at maximum	Average (U/ml)
Trametes versicolor										
Kirks	4	8	Low	2.75	18.2	15	8.0	90.5	8	55.0
LN Kirks	3	19	Inter	1.50	8.3	6	5.3	91.8	6	34.1
LN Mn Kirks	3	16	Low	1.38	25	4	11.0	72.2	5	40.4
Trametes multicolor										
Kirks	3	21	Inter	1.29	22.5	4	11.1	84.1	6	42.7
LN Kirks	3	24	High	1.25	38.9	5	6.5	87.4	5	40.2
LN Mn Kirks	3	20	Low	1.10	50.0	5	18.3	84.6	6	35.0
Trametes pubescens										
Kirks	3	19	Inter	1.39	10.4	17	4.5	78.6	21	26.1
LN Kirks	3	26	High	1.08	11.3	21	5.0	80.3	22	29.8
LN Mn Kirks	3	20	Low	1.30	19.1	14	5.2	77.2	17	26.3
Trametes meyenii										
Kirks	4	23	High	1.26	20.5	15	7.3	88.1	17	31.5
LN Kirks	4	10	inter	2.91	20.8	6	4.8	86.7	8	39.6
LN Mn Kirks	6	8	Inter	3.49	46.3	11	15.1	90.3	29	36.1

Normal (Kirk's), low nitrogen (LN Kirk's), and low nitrogen supplemented with Mn (LN Mn Kirk's); for more details see Table 2. Each value is an average of five separate experiments. S.D. always less than 4% (n = 5).

Trametes meyenii was the only species that had higher decoloration rates in the higher-N containing Kirk's medium. It decolored over twice as many sequential additions of amaranth in contrast to the low-N treatment. Both laccase and MnP activities were also highest for this condition. Even though the addition of Mn(II) to the low-N culture medium doubled MnP activity in *T. meyenii*, there was no concomitant increase in decoloration rate. In fact, the initial lag phase took the longest in this condition, with 7 days to complete the first decoloration as opposed to 5 days for the Kirk's treatments.

Decoloration in the presence of malonate

Rates of decoloration increased significantly for all species of *Trametes* when 50 μM sodium malonate was added to cultures that had been pretreated in low-N Kirk's (Figure 3). However, a comparison between species found no significant differences among the resultant rates of decoloration on a per fresh wt basis.

Discussion

The bioremediation of chemical contaminants requires the identification of appropriate species that are easily maintained in active condition. It is commonly accepted that stress conditions such as nutrient deprivation with their associated growth inhibition must elicit secondary metabolism before appropriate enzymes are activated in white-rot fungi (Archibald et al. 1997; Hatvani and Mécs 2002; Swamy and Ramsay 1999b; Kaal et al. 1995; Tatarko and Bumpus 1998; Tenuta and Lazarovits 2002). However, the provision of highly effective degradative processes by healthy and actively growing organisms would be more practical and efficaceous. This could also help to overcome the limited utility of fungal cultures for dye decolorization (Kokol et al. 2007).

Screening for effectiveness of dye decoloration on agar-solidified media indicated that *T. cotonea* had little activity and that of both *B. adusta* and *P. radiata* were much less effective than the four other tested *Trametes* species (Table 3). This might be attributed to the strategies of the

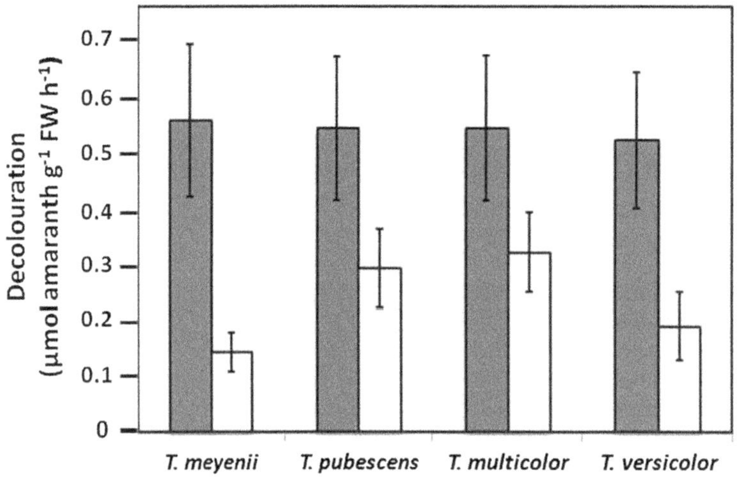

Figure 3 Rate of amaranth decolouration in the presence and absence of malonate by four *Trametes* species grown in low-N Kirk's. Presence and absence of 50 µM sodium malonate. All values are means ± S.D. (n = 5).

individual species with respect to natural substrate degradation (Barrasa et al. 2014). Furthermore, *P. radiata*, *B. adusta*, and *T. cotonea* decolored amaranth only slightly in liquid media through a process that involved sorption to their biomass (Figure 1). This was characterized by darkening of their mycelia with little or no secreted enzyme production (Figure 2). Results on agar-solidified media indicated that *P. radiata* and *B. adusta* are able to decolor amaranth and studies by other researchers using liquid culture have shown decoloration (Kaal et al. 1995; Robinson et al. 2001; Arora and Gill 2001). However, submerged cultures can display reduced decoloration because of low enzyme levels, and peroxidase activities are generally only optimal at high O_2 tension. Laccase, on the other hand, can be enhanced at relatively low O_2 (Wesenberg et al. 2003).

Trametes meyenii, T. pubescens and *T. multicolor* were equal to or better at decoloring amaranth than commonly employed *T. versicolor*. Of these, *T. pubescens* has been used to a limited extent in bioremediation of phenols (Ryan et al. 2005) and wine wastewater (Strong and Burgess 2007), and *T. multicolor* has been studied in the context of PCB degradation (Köller et al. 2000). Direct comparisons in decoloration and enzyme activites between these *Trametes* species has not been previously made.

Trametes meyenii had the highest rate of growth at 8% over the next best grower, *T. multicolor*, and was the only species to decolor best on normal Kirk's medium where it possessed prolonged activity. This supports an important role for laccase in the decoloration of amaranth by *T. meyenii* (Table 4). Laccase is produced during growth in *T. versicolor*, while peroxidases are produced during secondary metabolism (Wong and Yu 1999). Laccase production is also stimulated by low glucose (Tavares et al. 2005) and the presence of nitrogenous dye in this species (Casas et al. 2013), and dye degradation has been attributed solely to laccase in the white-rot fungus *Datronis* sp. (Vaithanomsat et al. 2010). The present study is the first to test dye decoloration and associated enzyme production in *T. meyenii*.

Liquid culturing more accurately reflects real wastewater treatment conditions. Low-N medium was tested initially because previous studies (Swamy and Ramsay 1999a; Swamy and Ramsay 1999b; de Jong et al. 1994) and our investigations with agar-solidified media showed that this enabled all species to decolor dye, with the exception of *T.* cotonea. Unlike the situations for the remaining *Trametes* species, *P. radiata* and *B. adusta* were severely affected by oxygen limiting conditions of liquid media. *Trametes pubescens* and *T. multicolor* were the most efficient and displayed the highest levels of secreted laccase and MnP activity in low-N media. However, there was not always a correlation between measured enzyme activities and a species' ability to decolor amaranth. High nitrogen has been previously shown to increase MnP (Kaal et al. 1995) and laccase production without increasing decoloration (Robinson et al. 2001). However, these other studies used very high ammonia that could cause stress in fungi (Tenuta and Lazarovits 2002), thereby invoking secondary metabolism.

The most efficient process from a bioremediation perspective would be if white-rot fungi were able to decolor consecutive additions of dye over an extended period of time. Therefore the four efficient *Trametes* species in this study were investigated further to determine their ability to decolor successive additions of the dye. The activities of laccase and MnP enzymes were monitored to determine their contributions to the degradative process because of their common occurence in white-rot fungi (Morgenstern et al. 2008; Wesenberg et al. 2003; Tomsovsky et al. 2009) and active roles in dye degradation under normal culture conditions, which does not appear to be the case for lignin peroxidase (Swamy and Ramsay 1999b).

Laccase and MnP were the only detected decolorizing enzymes in the culture media. With respect to lignin peroxidase, it is known that the culturing conditions used in the present study for *Trametes* fungi do not result in lignin peroxidase activity (Swamy and Ramsay 1999b). Furthermore, although DNA sequences for putative versatile and dye-decolorizing peroxidases exist in at least one of the fungal species of the presented study (Hofrichter et al. 2010; Ruiz-Dueñas et al. 2007), respective enzyme activities have yet to be demonstrated in this genus. For example, manganese-independent peroxidase (versatile peroxidase) was not detected in *T. versicolor* by Gavril et al. (2007). In addition, assaying for this enzyme (de Jong et al. 1992) and dye-decolorizing peroxidase (Sugano et al. 2006) did not detected any activity in the present study. The roles of laccase and MnP in the decoloration process is further supported by lag periods prior to decoloration corresponding to enzyme production in the culture media (Table 4). Once induced, degradation rates improved dramatically, with complete decoloration occurring in as little as 15 hours (Figure 1). However, dye decoloration by *T. versicolor* has been shown to be initially low regardless of the levels of measurable MnP and laccase (Swamy and Ramsay 1999b). This is supported by our findings that enzyme activities did not closely correspond to rate of decoloration (Table 4), thus indicating that other important factors are required for prolonged decoloration.

Despite *T. versicolor* having the highest measurable maximum and average laccase activities in long-term experiments, it was out performed by *T. multicolor* and *T. pubescens* in decoloration in low nitrogen. This could be in part because MnP activity was higher in *T. multicolor*, but that of *T. pubescens* was lower despite other research to the contrary (Strong and Burgess 2007). *Trametes meyenii* was exceptional because it possessed the longest period of rapid decoloration under higher nitrogen nutrition even though its enzyme activities were similar to those of *T. versicolor* and T. *multicolor*.

Supplying white-rot fungi with Mn(II) in low N Kirk's led to increases in measurable MnP production, but no improvement in long-term decoloration in all four *Trametes* species. This may indicate that either MnP is not the main decolorizing enzyme, or other necessary factors are more readily produced at low nitrogen.

It has been shown that despite high levels of extracellular enzymes, decoloration diminishs as glucose is depleted and is restored by its replenishment (Swamy and Ramsay 1999a). This could act as a general fixed carbon source required for the production of factors needed by the enzymes or more specifically, it could be a direct precursor of H_2O_2 via glucose-2-oxidase (Champagne and Ramsay 2005; Sen et al. 2012). We found that a relatively low concentration of malonate that would not be a significant source of carbon nutrition enhanced decoloration rates to the same magnitude on a per g FW basis in all four *Trametes* species when they were grown in low-N Kirk's medium (Figure 3). Manganese peroxidase oxidizes Mn^{2+} to Mn^{3+} which because of its instability, must chelate with an organic compound such as malonate (Schlosser and Hofer 2002; Kersten and Cullen 2007; Lundell et al. 2010; Liu et al. 2011). These chelators can act as low-molecular mass mediators (Higuchi 2004) that react with each other to eventually form peroxyl radicals (Gianfreda et al. 1999) able to attack a wide range of compounds non-specifically (Wariishi et al. 1988; Kuan and Tien 1993; Watanabe et al. 2000).

White-rot fungal species respond to nutrient conditions indicative of their specific ecological niches (Lundell et al. 2010; Tuor et al. 1995; Hiscox et al. 2010). This is the first study to show the dye decoloration potential of *T. meyenii*, *T. pubescens*, and *T. multicolor*. All three of these species possessed a strong capacity for prolonged dye decoloration by comparison with the more commonly studied *T. versicolor*. Enzyme activity for both laccase and MnP occurred over the decolorizing periods in all of these species. However, it appears that manganic chelation limited the degradative process and that the most efficient conditions for any given species may be that which enables the organisms to adequately produce chelating compounds. Studies with *T. meyenii* may help provide insight into the chelation phenomena because unlike other species it functions best when provided with full nutrient medium, a condition conducive to high metabolite production. In addition, more needs to be understood about the relative importance of secreted laccase and MnP enzymes. In some species of fungi, laccase and MnP have been shown to have an indirect interactive effect that enhances lignin depolymerization many times over that of individual enzymes, because of the production of H_2O_2 by laccase as a result of the oxidation of Mn^{2+} (Galliano et al. 1991; Leonowicz et al. 1999; 2001). However, the contribution of laccase and MnP to amaranth decoloration appeared to be only additive in *T. versicolor* with MnP providing considerably more to the decoloration process (Champagne and Ramsay 2005). It remains to be seen if this is the case for the other species of *Trametes* utilized in this study.

Of all the species tested, *T. meyenii* showed the most promise for industrial application purposes even though *T. multicolor* and *T. pubescens* functioned well in low-N medium. This is because *T. meyenii* possesses an elevated ability to decolor in higher-N environments, and most textile wastewaters contain relatively high levels of nitrogen, in part because of the nitrogenous structure of many dyes. This capacity to continue to simultaneously decolor dye over prolonged periods while growing in relatively nutrient-rich medium appears to be unique amongst white-rot fungi. The ecological significance of an ability to maintain lignolytic activity during higher nutritional states is also noteworthy.

Competing interests

The authors declare that they have no competing interests.

Authors' contributions

PRC designed and performed decoloration and enzyme assays and helped draft the manuscript. NL designed and undertook decoloration studies in the presence of chelators. DDL conceived and designed the study, supervised the research group, provided funding support, drafted and revised the manuscript. All authors approved the final manuscript.

Acknowledgements

This work was supported by the Natural Sciences and Engineering Research Council of Canada and the Advisory Committee of Queen's University, Canada

References

Archibald F, Bourbonnais R, Jurasek L, Paice M, Reid I (1997) Kraft pulp bleaching and delignification by *Trametes versicolor*. J Biotechnol 53:215–236

Arora D, Gill P (2001) Effects of various media and supplements on laccase production by some white rot fungi. Bioresour Technol 77:89–91

Barrasa JM, Blanco MN, Esteve-Raventós F, Altés A, Checa J, Martínez AT, Ruiz-Dueñas FJ (2014) Wood and humus decay strategies by white-rot basidiomycetes correlate with two different dye decolourization and enzyme secretion patterns on agar plates. Fungal Genet Biol, doi: 10.1016/j.fgb.2014.03.007

Casas N, Blánquez P, Vicent T, Sarrà M (2013) Laccase production by *Trametes versicolor* under limited-growth conditions using dyes as inducers. Environ Technol 34:113–119

Chagas EP, Durrant LR (2001) Decolourization of azo dyes by *Phanerochaete chrysosporium* and *Pleurotus sajorcaju*. Enzyme Microb Technol 29:473–477

Champagne P, Ramsay J (2005) Contribution of manganese peroxidase and laccase to dye decolouration by *Trametes versicolor*. Appl Microbiol Biotechnol 69:276–285

Dafale N, Wate S, Meshram S, Neti N (2010) Bioremediation of wastewater containing azo dyes through sequential anaerobic-aerobic bioreactor system and its biodiversity. Env Rev 18:21

de Jong E, Field J, de Bont J (1992) Evidence for a new extracellular peroxidase. Manganese-inhibited peroxidase from white-rot fungus *Bjerkandera* sp. BOS 55. FEBS Lett 299:107–110

de Jong E, Field J, de Bont J (1994) Aryl alcohols in the physiology of ligninolytic fungi. FEMS Microbiol Rev 13:153–187

Evans CS, Hedger JN (2001) Degradation of plant cell wall polymers. In: Gadd GM (ed) Fungi in bioremediation. Cambridge University Press, Cambridge, UK, pp 1–25

Galliano H, Gas G, Seris J, Boudet A (1991) Lignin degradation by *Rigidoporus lignosus* involves synergistic action of two oxidizing enzymes: Mn peroxidase and laccase. Enzyme Microb Technol 13:478–482

Gavril M, Hodson PV, McLellan J (2007) Decoloration of Amaranth by the white-rot fungus *Trametes versicolor*. Part 1. Statistical analysis. Can J Microbiol 53:313–326

Gianfreda L, Xu F, Bollag J (1999) Laccases: A useful group of oxidoreductive enzymes. Biorem J 3:1–26

Hatvani N, Mécs I (2002) Effect of the nutrient composition on dye decolourisation and extracellular enzyme production by *Lentinus edodes* on solid medium. Enzyme Microb Technol 30:381–386

Heinfling A, Martinez MJ, Martinez AT, Bergbauer M, Szewzyk U (1998a) Transformation of industrial dyes by manganese peroxidases from *Bjerkandera adusta* and *Pleurotus eryngii* in a manganese-independent reaction. Appl Environ Microbiol 64:2788–2793

Heinfling A, Martínez MJ, Martínez AT, Bergbauer M, Szewzyk U (1998b) Purification and characterization of peroxidases from the dye-decolourizing fungus *Bjerkandera adusta*. FEMS Microbiol Lett 165:43–50

Hessel C, Allegre C, Maisseu M, Charbit F, Moulin P (2007) Guidelines and legislation for dye house effluents. J Environ Manage 83:171–180

Higuchi T (2004) Microbial degradation of lignin: Role of lignin peroxidase, manganese peroxidase, and laccase. Proc Japan Acad Ser B Phys Biol Sci 80:204–214

Hiscox J, Baldrian P, Rogers H, Boddy L (2010) Changes in oxidative enzyme activity during interspecific mycelial interactions involving the white-rot fungus *Trametes versicolor*. Fungal Genet Biol 47:562–571

Hofrichter M, Ullrich R, Pecyna MJ, Liers C, Lundell T (2010) New and classic families of secreted fungal heme peroxidases. Applied Microbiol Biotechnol 87:871–897

Johannes C, Majcherczyk A (2000) Laccase activity tests and laccase inhibitors. J Biotechnol 78:193–199

Kaal E, Field J, Joyce T (1995) Increasing ligninolytic enzyme activities in several white-rot basidiomycetes by nitrogen-sufficient media. Bioresour Technol 53:133–139

Kersten P, Cullen D (2007) Extracellular oxidative systems of the lignin-degrading Basidiomycete *Phanerochaete chrysosporium*. Fungal Genet Biol 44:77–87

Kirk T, Fenn P (1982) Formation and action of the ligninolytic system in Basidiomycetes. In: Frankland JC, Hedger JN, Swift MJ (eds) Decomposer Basidiomycetes: their Biology and Ecology. Cambridge University Press, Cambridge, UK, pp 67–90

Kirk P, Cannon P, Minter D, Stalpers J (2009) Dictionary of the Fungi. Mycol Res 113:908–910

Kokol V, Doliska A, Eichlerova I, Baldrian P, Nerud F (2007) Decolourization of textile dyes by whole cultures of *Ischnoderma resinosum* and by purified laccase and Mn-peroxidase. Enzyme Microb Technol 40:1673–1677

Köller G, Möder M, Czihal K (2000) Peroxidative degradation of selected PCB: a mechanistic study. Chemosphere 41:1827–1834

Kuan I, Tien M (1993) Stimulation of Mn peroxidase activity: a possible role for oxalate in lignin biodegradation. Proc Natl Acad Sci U S A 90:1242–1246

Leonowicz A, Matuszewska A, Luterek J, Ziegenhagen D, Wojtas-Wasilewska M, Cho N, Hofrichter M, Rogalski J (1999) Biodegradation of lignin by white rot fungi. Fungal Genet Biol 27:175–185

Leonowicz A, Cho N, Luterek J, Wilkolazka A, Wojtas-Wasilewska M, Matuszewska A, Hofrichter M, Wesenberg D, Rogalski J (2001) Fungal laccase: properties and activity on lignin. J Basic Microbiol 41:185–227

Levin L, Forchiassin F, Ramos A (2002) Copper induction of lignin-modifying enzymes in the white-rot fungus *Trametes trogii*. Mycologia 94:377–383

Libra J, Borchert M, Banit S (2003) Competition strategies for the decolourization of a textile-reactive dye with the white-rot fungi *Trametes versicolor* under non-sterile conditions. Biotechnol Bioeng 82:736–744

Liu G, Zhou J, Wang J, Wang X, Jin R, Lv H (2011) Decolourization of azo dyes by *Shewanella oneidensis* MR-1 in the presence of humic acids. Appl Microbiol Biotechnol 91:417–424

Lundell T, Mäkelä M, Hildén K (2010) Lignin-modifying enzymes in filamentous basidiomycetes – ecological, functional and phylogenetic review. J Basic Microbiol 50:5–20

Martin G, Alibart O, Hierle R, Josse D, Baldi P, Zyss J (2012) Chemical safety in textile and apparel markets of emerging countries. Indian Textile J 122:118

Morgenstern I, Klopman S, Hibbett D (2008) Molecular evolution and diversity of lignin degrading heme peroxidases in the agaricomycetes. J Mol Evol 66:243–257

Nigam P, Armour G, Banat IM, Singh D, Marchant R (2000) Physical removal of textile dyes from effluents and solid-state fermentation of dye-adsorbed agricultural residues. Bioresour Technol 72:219–226

Pointing SB (2001) Feasibility of bioremediation by white-rot fungi. Appl Microbiol Biotechnol 57:20–33

Ramsay JA, Goode C (2004) Decolouration of a carpet dye effluent using *Trametes versicolor*. Biotechnol Lett 26:197–201

Reddy C, Mathew Z (2001) Bioremediation potential of white-rot fungi. In: Gadd GM (ed) Fungi in Bioremediation. Cambridge University Press, Cambridge, UK, pp 52–78

Robinson T, Chandran B, Nigam P (2001) Studies on the production of enzymes by white-rot fungi for the decolourisation of textile dyes. Enzyme Microb Technol 29:575–579

Ruiz-Dueñas FJ, Aguilar A, Martinez MJ, Zorn H, Martinez AT (2007) Gene cloning, heterologous expression, *in vitro* reconstitution and catalytic properties of a versatile peroxidase. Biocatal Biotransform 25:276–285

Ryan D, Leukes W, Burton S (2005) Fungal bioremediation of phenolic wastewaters in an airlift reactor. Biotechnol Prog 21:1068–1074

Schlosser D, Hofer C (2002) Laccase-catalyzed oxidation of Mn^{2+} in the presence of natural Mn^{3+} chelators as a novel source of extracellular H_2O_2 production and its impact on manganese peroxidase. Appl Environ Microbiol 68:3514–3521

Sen K, Pakshirajan K, Santra S (2012) Modelling the biomass growth and enzyme secretion by the white rot fungus *Phanerochaete chrysosporium* in presence of a toxic pollutant. J Environ Protection 3:114–119

Strong P, Burgess J (2007) Bioremediation of a wine distillery wastewater using white rot fungi and the subsequent production of laccase. Water Sci Technol 56:179–186

Sugano Y, Matsushima Y, Shoda M (2006) Complete decolorization of the anthraquinone dye Reactive blue 5 by the concerted action of two peroxidases from *Thanatephorus cucumeris* Dec 1. Appl Microbiol Biotechnol 73:862–871

Swamy J, Ramsay JA (1999a) Effects of glucose and NH^{4+} concentrations on sequential dye decolouration by *Trametes versicolor*. Enzyme Microb Technol 25:278–284

Swamy J, Ramsay JA (1999b) Effects of Mn^{2+} and NH^{4+} concentrations on laccase and manganese peroxidase production and Amaranth decolouration by *Trametes versicolor*. Appl Microbiol Biotechnol 51:391–396

Tatarko M, Bumpus JA (1998) Biodegradation of congo red by *Phanerochaete chrysosporium*. Water Res 32:1713–1717

Tavares A, Coelho M, Coutinho J, Xavier A (2005) Laccase improvement in submerged cultivation: induced production and kinetic modelling. J Chem Technol Biotechnol 80:669–676

Tenuta M, Lazarovits G (2002) Ammonia and nitrous acid from nitrogenous amendments kill the microsclerotia of *Verticillium dahliae*. Phytopathol 92:255–264

Toh Y, Yen JJL, Obbard JP, Ting Y (2003) Decolourisation of azo dyes by white-rot fungi (WRF) isolated in Singapore. Enzyme Microb Technol 33:569–575

Tomsovsky M, Popelarova P, Baldrian P (2009) Production and regulation of lignocellulose-degrading enzymes of poria-like wood-inhabiting basidiomycetes. Folia Microbiol (Praha) 54:74–80

Tuor U, Winterhalter K, Fiechter A (1995) Enzymes of white-rot fungi involved in lignin degradation and ecological determinants for wood decay. J Biotechnol 41:1–17

Vaithanomsat P, Apiwatanapiwat W, Petchoy O, Chedchant J (2010) Production of ligninolytic enzymes by white-rot fungus *Datronia* sp. KAPI0039 and their application for reactive dye removal. Int J Chem Eng 2010:6, Article ID 162504

Van Aken B, Hofrichter M, Scheibner K, Hatakka A, Naveau H, Agathos S (1999) Transformation and mineralization of 2,4,6-trinitrotoluene (TNT) by manganese peroxidase from the white-rot basidiomycete *Phlebia radiata*. Biodegradation 10:83–91

Viswanath B, Rajesh B, Janardhan A, Kumar AP, Narasimha G (2014) Fungal laccases and their applications in bioremediation. Enzyme Res 2014:21, Article ID 163242

Wariishi H, Akileswaran L, Gold MH (1988) Manganese peroxidase from the basidiomycete *Phanerochaete chrysosporium*: spectral characterization of the oxidized states and the catalytic cycle. Biochem 27:5365–5370

Wariishi H, Valli K, Gold MH (1992) Manganese oxidation by manganese peroxidase from the basidiomycete *Phanerochaete chrysosporium*. J Biol Chem 267:23689–23695

Watanabe T, Katayama S, Enoki M, Honda Y, Kuwahara M (2000) Formation of acyl radical in lipid peroxidation of linoleic acid by manganese-dependent peroxidase from *Ceriporiopsis subvermispora* and *Bjerkandera adusta*. Eur J Biochem 267:4222–4231

Wesenberg D, Kyriakides I, Agathos S (2003) White-rot fungi and their enzymes for the treatment of industrial dye effluents. Biotechnol Adv 22:161–187

Wong Y, Yu J (1999) Laccase-catalyzed decolourization of synthetic dyes. Water Res 33:3512–3520

15

L-citrulline production by metabolically engineered *Corynebacterium glutamicum* from glucose and alternative carbon sources

Dorit Eberhardt, Jaide V K Jensen and Volker F Wendisch*

Abstract

L-citrulline plays an important role in human health and nutrition and is an intermediate of the L-arginine biosynthetic pathway. L-citrulline is a by-product of L-arginine production by *Corynebacterium glutamicum*. In this study, *C. glutamicum* was engineered for overproduction of L-citrulline as major product without L-arginine being produced as by-product. To this end, L-arginine biosynthesis was derepressed by deletion of the arginine repressor gene *argR* and conversion of L-citrulline towards L-arginine was avoided by deletion of the argininosuccinate synthetase gene *argG*. Moreover, to facilitate L-citrulline production the gene encoding a feedback resistant N-acetyl L-glutamate kinase $argB^{fbr}$ as well as the gene encoding L-ornithine carbamoylphosphate transferase *argF* were overexpressed. The resulting strain accumulated 44.1 ± 0.5 mM L-citrulline from glucose minimal medium with a yield of 0.38 ± 0.01 $g \cdot g^{-1}$ and a volumetric productivity of 0.32 ± 0.01 $g \cdot l^{-1} \cdot h^{-1}$. In addition, production of L-citrulline from the alternative carbon sources starch, xylose, and glucosamine could be demonstrated.

Keywords: *Corynebacterium glutamicum*; L-citrulline; L-arginine; Alternative carbon sources; Starch; Xylose; Glucosamine; Metabolic engineering

Introduction

L-citrulline is a natural non-proteinogenic amino acid whose name is derived from watermelon *Citrullus lanatus* (Wada 1930). In mammalians it serves as a precursor for L-arginine. In contrast to the proteinogenic L-arginine, which is not transferred to the blood stream, when ingested, L-citrulline can be converted to L-arginine, which is then released by the kidney into the blood stream. It is applied in several medical approaches e.g. as a pharmaconutrient (Rimando and Perkins-Veazie 2005; Curis et al. 2005).

Currently, biocatalytic and fermentative methods to produce L-citrulline using *Pseudomonas putida* (Kakimoto et al. 1971; Yamamoto et al. 1974) or *Bacillus subtilis* strains exist (Okumura et al. 1966). Additionally, extraction processes from watermelon have been established (Fish 2012). L-citrulline is an intermediate of L-arginine biosynthesis and accumulates as a by-product of engineered L-arginine producing *Corynebacterium glutamicum* strains (Ikeda et al. 2009; Schneider et al. 2011).

C. glutamicum is a workhorse for amino acid production and is employed for the annual production of several million tons of L-glutamate and L-lysine (Wendisch 2014). *C. glutamicum* has been engineered to produce a wide range of bioproducts, such as diamines, carotenoids, terpenes, proteins (Schneider and Wendisch 2010; Schneider et al. 2012; Heider et al. 2014a, b; Frohwitter et al. 2014; Kikuchi et al. 2009; Teramoto et al. 2011; An et al. 2013) and the L-glutamate family amino acids L-arginine, L-ornithine, and L-proline (Schneider et al. 2011; Ikeda et al. 2009; Georgi et al. 2005; Blombach et al. 2009; Jensen and Wendisch 2013). However, the production of L-citrulline as the only or major product has not been published yet.

Due to its natural ability to produce L-glutamate under several eliciting conditions, *C. glutamicum* is a suitable producer of L-glutamate-derived products (Sato et al. 2008; Radmacher et al. 2005; Kim et al. 2009, 2010; Delaunay et al. 1999; Wendisch et al. 2014). L-ornithine is a non-proteinogenic glutamate-family amino acid and

* Correspondence: volker.wendisch@uni-bielefeld.de
Genetics of Prokaryotes, Faculty of Biology & Center for Biotechnology, Bielefeld University, Universitätsstraße 25, 33615 Bielefeld, Germany

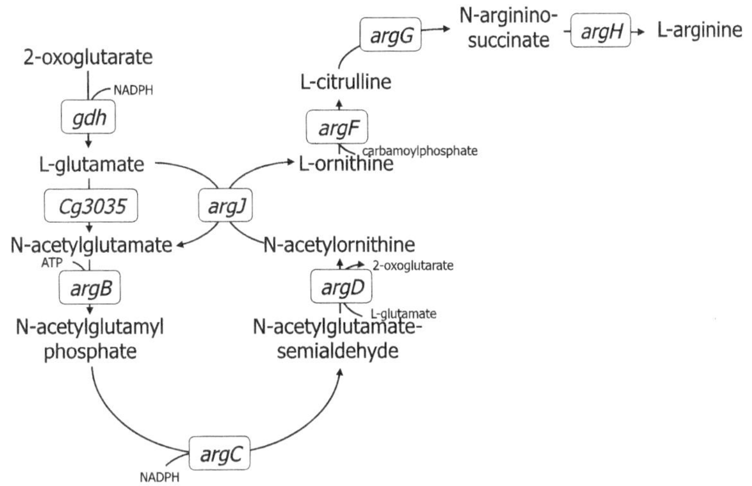

Figure 1 L-arginine pathway in *C. glutamicum* (modified from (Wendisch et al. 2014)). *gdh*: L-glutamate dehydrogenase, *cg3035*: anaplerotic N-acetylL-glutamate synthase, *argJ*: L-ornithine N-acetyltransferase, *argB*: N-acetylL-glutamate kinase; *argC*: N-acetyl-gamma-glutamyl-phosphate reductase; *argD*: acetylL-ornithine aminotransferase; *argE*: acetylL-ornithine deacetylase; *argF*: L-ornithine carbamoyltransferase; *argG*: argininosuccinate synthetase; *argH*: argininosuccinate lyase. Oxoglutarate is an intermediate of the central carbon metabolism.

an intermediate of L-arginine biosynthesis (Figure 1). An ornithine producer was obtained by deletion of *argR*, the gene encoding the genetic repressor of the arginine biosynthesis operon, and *argF* to prevent further processing of ornithine (Schneider et al. 2011). The production of L-proline from L-ornithine is possible by the heterologous overexpression of *ocd* from *Pseudomonas putida*, encoding ornithine cyclodeaminase (Jensen and Wendisch 2013). The diamine putrescine can be produced by overexpression of the *Escherichia coli* gene *speC*, which encodes ornithine decarboxylase (Schneider et al. 2012; Schneider and Wendisch 2010). As the arginine biosynthetic pathway is naturally regulated by feedback inhibition of N-acetylglutamate kinase (encoded by *argB*) by arginine, the use of feedback resistant enzyme variants in combination with deletion of *argR* has been described to overproduce L-arginine (Sakanyan et al. 1996; Ikeda et al. 2009; Schneider et al. 2011).

C. glutamicum can utilize a variety of carbon sources. In contrast to many other microorganisms used in biotechnology, simultaneous utilization of carbon sources e.g. present in mixtures such as lignocellulosic hydrolysates is a hall mark of *C. glutamicum* (Blombach and Seibold 2010; Meiswinkel et al. 2013a, b). The natural substrate spectrum of *C. glutamicum* includes monosaccharides, disaccharides, and organic acids as well as alcohols (Blombach and Seibold 2010; Arndt and Eikmanns 2008; Peters-Wendisch et al. 1998; Jolkver et al. 2009; Sasaki et al. 2011). To allow access to alternative carbon sources, *C. glutamicum* has also been engineered for utilization of glycerol, pentoses, and amino sugars as well as polysaccharides (Schneider et al. 2011; Rittmann et al. 2008; Seibold et al. 2006; Uhde et al. 2013; Gopinath et al. 2011; Matano et al. 2014).

One aim to reduce production cost is the use of complex sugar substrates for the production of biotechnological products. As an example of using a polymeric raw material without decomposition to its monomeric compounds e.g. by enzyme treatment, soluble starch could be used as a carbon source for the production of L-lysine and organic acids by engineered *C. glutamicum*

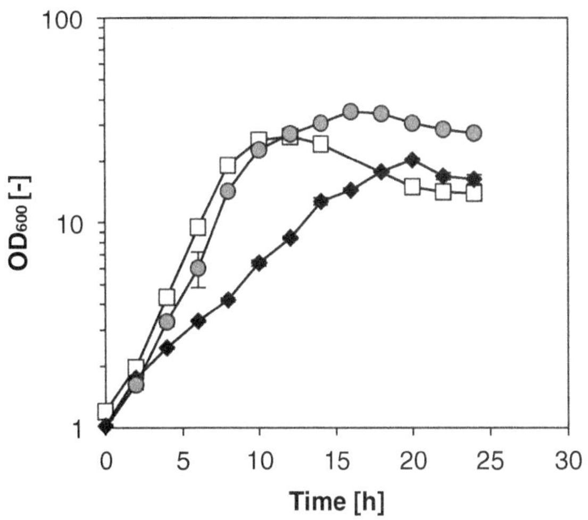

Figure 2 Biomass formation by various *C. glutamicum* strains. The cultivation was performed in CGXII minimal medium containing 20 g L-1 glucose, 1 mM IPTG, 750 µM L-arginine and 25 µg L-1 kanamycin. OD_{600} was determined of CIT0(pVWEx1) (open squares), CIT0(pVWEx1-*argF*) (gray circles) and CIT0 (pVWEx1-*argFB*^{fbr}) (black diamonds). Values and error bars represent the mean and the standard error of triplicates.

(Seibold et al. 2006; Tateno et al. 2007; Tsuge et al. 2013). However, due to the growing world population and a correlating higher demand for food, biotechnological processes based on non-food derived carbon sources are sought. Xylose is a pentose sugar compound present in the hemicellulosic fraction of agricultural wastes as for example rice straw. Glucosamine, on the other hand, is a constituent of chitin, the second most abundant biopolymer in Nature, which is accessible e.g. from shrimp shell waste accumulating in the food industry. *C. glutamicum* has been engineered to efficiently utilize both xylose and glucosamine as alternative carbon sources for growth and amino acid production (Gopinath et al. 2011; Meiswinkel et al. 2013a; Uhde et al. 2013; Matano et al. 2014).

In this study, the rational engineering of L-citrulline production by *C. glutamicum* is reported and the concept was extended to production of L-citrulline from the alternative carbon sources glucosamine, xylose, and starch.

Materials and methods

Microorganisms and growth conditions

Microorganisms and plasmids used in this study are listed in Table 1. *E. coli* DH5α was used for gene cloning. *C. glutamicum* and *E. coli* strains were routinely grown in lysogeny broth (LB) (10 g L^{-1} tryptone, 5 g L^{-1} yeast extract, 10 g L^{-1} sodium chloride) in 500-mL baffled flasks on a rotary shaker (120 rpm) at 30°C or 37°C. For growth experiments, CGXII minimal medium (Eggeling and Reyes 2005) was used for *C. glutamicum*. Growth was followed by measuring the optical density at 600 nm using a V-1200 Spectrophotometer (VWR, Radnor, PA, USA). An OD_{600} of 1 corresponds approximately to an estimated cell dry weight of 0.25 g/L.

When necessary, the growth medium was supplemented kanamycin (25 μg mL^{-1}), spectinomycin (100 μg mL^{-1}), tetracycline (10 μg mL^{-1}), isopropyl β-D-1-thiogalactopyranoside (IPTG) (1 mM) and L-arginine (750 μM). The growth behavior and L-citrulline production of recombinant *C. glutamicum* strains were analyzed in 500 ml baffled flasks. Briefly, a 50 mL BHI (37 g L^{-1}) seed culture was inoculated from an agar plate and grown overnight. The cells were harvested by centrifugation (4,000 × g, 10 min) and washed twice with CGXII minimal medium lacking the carbon source. Subsequently, 50 mL CGXII medium, containing a given concentration of carbon source and necessary supplements, was inoculated to an optical density of 1.0. Detailed information on the carbon source concentrations employed are given in the Results chapter.

Molecular genetic techniques

Standard methods such as restriction digestions, and ligation were carried out as described elsewhere (Sambrook and Russell 2012). Digested DNA was purified by using the QIAquick Gel Extraction Kit (Qiagen, Hilden, Germany).

Table 1 Strains and plasmids used in this study

E. coli		
DH5α	F^- *thi-1 endA1 hsdr17*(r^-, m^-) *supE44* Δ*lacU169* (Φ80*lacZ*ΔM15) *recA1 gyrA96 relA1*	(Hanahan 1983)
C. glutamicum		
MB001	ATCC 13032 with in-frame deletion of prophages CGP1 (cg1507-cg1524), CGP2 (cg1746-cg1752), and CGP3 (cg1890-cg2071)	(Baumgart et al. 2013)
CIT0	MB001 with Δ*argF*, Δ*argG*, Δ*argR*	This study
CIT1	CIT0 carrying the pVWEx-*argFB*^{fbr} vector	This study
Plasmids		
pEKEx3	$Spec^R$, P_{tac}, $lacI^q$	(Stansen et al. 2005)
pVWEx1	Kan^R, P_{tac}, $lacI^q$	(Peters-Wendisch et al. 2001)
pEC-XT99A	Tet^R, P_{trc}, $lacI^q$	(Kirchner and Tauch 2003)
pK19Δ*argFR*	Kan^R, pk19mobsacB with the deletion construct of genes *argFR*	(Schneider et al. 2011)
pK19Δ*argG*	Kan^R, pk19mobsacB with the deletion construct of genes *argG*	This study
pEKEx3- *argB*^{fbr}	$Spec^R$, pEKEx3 carrying *argB* from *C. glutamicum* ATCC 13032 with amino acid exchanges A49VM54V	(Schneider et al. 2011)
pVWEx1-*argF*	Kan^R, pVWEx1 carrying *argF* from *C. glutamicum* ATCC 13032	This study
pVWEx1-*argFG*	Kan^R, pVWEx1 carrying *argF* and *argG* from *C. glutamicum* ATCC 13032	This study
pVWEx1-*argFB*^{fbr}	Kan^R, pVWEx1 carrying *argF* from *C. glutamicum* ATCC 13032 and *argB*^{fbr} from pEKEx3- *argB*^{fbr}	This study
pEKEx3-*nagB*	$Spec^R$, pEKEx3 carrying *nagB* from *C. glutamicum* ATCC 13032	(Uhde et al. 2013)
pEKEx3-*xylAB*	$Spec^R$, pEKEx3 carrying *xylA* from *Xanthomonas campestris* XCC1758 and *xylB* from *C. glutamicum* ATCC 13032	(Meiswinkel et al. 2013a)
pAMY	Tet^R, pEC-XT99A carrying *amy* from *Streptomyces griseus* IMRU 3570	(Seibold et al. 2006)

E. coli cells were transformed by heat shock (Sambrook and Russell 2012) and *C. glutamicum* cells were transformed by electroporation (Eggeling and Reyes 2005). Isolation of genomic DNA was performed as previously described (Jensen and Wendisch 2013). Chromosomal changes in *C. glutamicum* were performed as described elsewhere (Eggeling and Reyes 2005).

Construction of strains and plasmids

The deletion of Δ*argFR* in MB001 was performed by using pK19mob*sacB*Δ*argFR*. Afterwards *argG* was deleted by using pK19mob*sacB*Δ*argG* to obtain CIT0. pK19mob*sacB*Δ*argG* contains the up- and downstream regions of *argG* in the Δ*argFR* strain. The plasmid was constructed by amplifying the upstream region with *argG*_up_f (CTT*gaattc*AGAAGCTGCGCCGCATG) and *argG*_up_r (agagacgacctaagccagtctAACGATGCGGTTAGTCATGAGG) and the downstream region with *argG*_down_f (agactggct taggtcgtctctGCTAACAAGCGCGATCGC) and *argG*_down_r (CCT*ctgcag*AACGACCAGCGCGCAGA). The two fragments were combined by crossover PCR using *argG*_up_f and *argG*_down_r and finally cloned into pK19mob*sacB* with *Pst*I and *Eco*RI.

pVWEx1-*argF* was constructed by amplifying *argF* with primers argF_f (CTT*gtcgac*AAGGAGATATAGATATGAC TTCACAACCACAGGTTCG) and argF_r (CCT*ggatcc*TT ACCTCGGCTGGTTGGC). The PCR product was treated with *Sal*I and *Bam*HI and ligated with similarly treated pVWEx1. pVWEx1-*argFG* was constructed by amplifying *argG* with primers argG_f (GGG*gtcgac*GAAAGG AGGCCCTTCAGATGACTAACCGCATCGTTCTTG) and argG_r (GGG*gtcgac*TTAGTTGTTGCCAGCTTCG CGA). The PCR product was treated with *Sal*I and ligated with similarly treated pVWEx1-*argF*.

The plasmid vector pEKEx-$argB^{fbr}$ ($argB_{A49VM54V}$ (Schneider et al. 2011)) was digested with *Bam*HI and *Kpn*I and the DNA fragment with a size of 0.9 kb harboring the $argB^{fbr}$ gene was cloned into the *Bam*HI/*Kpn*I digested vector pVWEx1-*argF*.

Determination of amino acid and carbohydrate concentrations

For the quantification of extracellular amino acids and carbohydrates, a high-performance liquid chromatography system was used (1200 series, Agilent Technologies Deutschland GmbH, Böblingen, Germany). Samples were withdrawn from the cultures, centrifuged (13,000 × g, 10 min), and the supernatant used for analysis.

Glucose and xylose were analyzed on a normal phase column (organic acid resin 300 × 8 mm, 10 μm particle size, 25 Å pore diameter; Chromatographie Service GmbH, Langerwehe, Germany) using 5 mM sulfuric acid as the mobile phase at a flow rate of 1 mL min^{-1} and were detected with a refractive index detector (RID G1362A, 1200 series, Agilent Technologies). Amino acids were automatically modified by precolumn derivatisation with ortho-phthalaldehyde and separated as described previously (Georgi et al. 2005). L-ornithine was quantified using a pre-column (LiChrospher 100 RP18 EC-5 μ (40 × 4 mm), CS-Chromatographie Service GmbH, Langerwehe, Germany) and a reversed phase column (LiChrospher 100 RP18 EC-5 μ (125 × 4 mm), CS Chromatographie) as a main coulumn and detected with a fluorescence detector at excitation at 230 nm and 450 nm emission (FLD G1321A, 1200 series, Agilent Technologies). For the determination of L-citrulline, a reverse-phase (RP) LiChrospher 100 RP8 EC-5 μ precolumn (40 × 4.6 mm) and a RP8 EC-5 μ (125 × 4.6 mm) main column (CS Chromatographie, Langerwehe, Germany) were used. 100 μM L-Asparagine was used as an internal standard. The mobile phases used were in case of RP8 A: 0.25% Na-acetate pH 6, B: methanol. The gradient used was: 0 min 30% B, 1 min 30% B, 6 min, 70% B, 11 min 90% B, 14 min 70% B, 16 min 30% B. In case of RP18, the mobile phases used were A:0.1 M Na-acetate pH 7.2, B: methanol. The gradient used was: 0 min 20% B, 0.5 min 38% B, 2.5 min 46% B, 3.7 min 65% B, 5.5 min 70% B, 6 min 75% B, 6.2 min 85% B, 6.7 min 20% B.

Results

Engineering a prophage-free *C. glutamicum* strain for L-citrulline production

C. glutamicum has recently been cured of prophage sequences to yield MB001 (Baumgart et al. 2013). This strain was used as the parental strain because it can be transformed easily and plasmid-based gene overexpression is more efficient (Baumgart et al. 2013). As *C. glutamicum* ATCC 13032, this strain does not accumulate L-citrulline, an intermediate of L-arginine biosynthesis (Figure 1). The deletion of three genes of the L-arginine operon (L-ornithine carbamoyltransferase (EC 2.1.3.3) *argF*, argininosuccinate synthetase (EC 6.3.4.5) *argG*, and L-arginine biosynthesis operon repressor gene *argR*) in *C. glutamicum* MB001 yielded the L-arginine auxotrophic strain CIT0 (Table 1). When supplemented with 0.75 mM L-arginine, *C. glutamicum* CIT0 accumulated 25.2 ± 2.6 mM L-ornithine from 2% glucose (Table 2). The deletion of *argF* and *argG* could be complemented by plasmid-borne expression of these genes since the complemented strain CIT0(pVWEx1-*argFG*) grew without L-arginine supplement while the empty vector carrying control CIT0(pVWEx1) did not (data not shown). Comparable growth rates and biomass concentrations were observed.

To enable L-citrulline accumulation, two plasmids were constructed and used to transform *C. glutamicum* CIT0. While pVWEx1-*argF* only carries *argF* encoding L-ornithine carbamoyltransferase, pVWEx1-$argFB^{fbr}$ in addition carries

Table 2 Growth on different carbon sources

C. glutamicum strain	Carbon source concentration	Maximum OD_{600}	Growth rate (h^{-1})
CIT1(pEKEx3-*xylAB*)	Xylose: 15 g/L	6 ± 1	0.03 ± 0.01
CIT1(pEKEx3-*nagB*)	Glucosamine: 10 g/L	3 ± 1	0.02 ± 0.01
CIT1(pAMY)	Soluble starch: 10 g/L Glucose: 2.5 g/L	9 ± 1	0.21 ± 0.01
CIT1(pEC-XT99A)	Soluble starch: 10 g/L Glucose: 2.5 g/L	3 ± 1	0.10 ± 0.01

Fermentations were performed in CGXII minimal medium containing the respective carbon source and were supplemented by 750 μM L-arginine. 1 mM IPTG and 25 μg/ml kanamycin and spectinomycin were added. Values and error bars represent the mean and the standard error of triplicates.

argB^{fbr} encoding feedback-resistant N-acetyl L-glutamate kinase (NAGK, EC 2.7.2.8). When grown in minimal medium with 2% glucose and 0.75 mM L-arginine *C. glutamicum* CIT0(pVWEx1-*argF*) grew to a higher OD than CIT0(pVWEx1) (Figure 2) and did not accumulate notable concentrations of L-citrulline. As opposed to CIT0 (pVWEx1), CIT0(pVWEx1-*argF*) did not produce L-ornithine (Figure 3). By contrast, the combined overexpression of *argF* and *argB*^{fbr} entailed L-citrulline production and the respective strain was named CIT1. *C. glutamicum* CIT1 accumulated 44.1 ± 0.5 mM L-citrulline in minimal medium with 2% glucose (Figure 4).

When comparing the growth of *C. glutamicum* CITO (pVWEx1) to that of CIT0(pVWEx1-*argF*), similar growth rates (0.37 ± 0.01 h^{-1} and 0.35 ± 0.04 h^{-1}, respectively) were obtained, whereas L-citrulline formation by CITO (pVWEx1-*argFB*^{fbr}) was accompanied by a reduced growth rate (0.15 ± 0.01 h^{-1}) (Figure 1). Moreover, the final OD_{600} of CIT0(pVWEx1-*argFB*^{fbr}) was 20 ± 1 as compared to an OD_{600} of 26 ± 1 of CIT0(pVWEx1). By contrast, *C. glutamicum* CIT0(pVWEx1-*argF*) grew to a higher biomass concentration with a final OD_{600} of 35 ± 1. As shown in Figure 3, the lower growth rates of CIT0(pVWEx1) and CIT0(pVWEx1-*argFB*^{fbr}) correlated inversely with the formation of the respective amino acids L-ornithine and L-citrulline, whereas *C. glutamicum* CIT0(pVWEx1-*argF*) reaches a higher final biomass and neither produces L-ornithine nor L-citrulline.

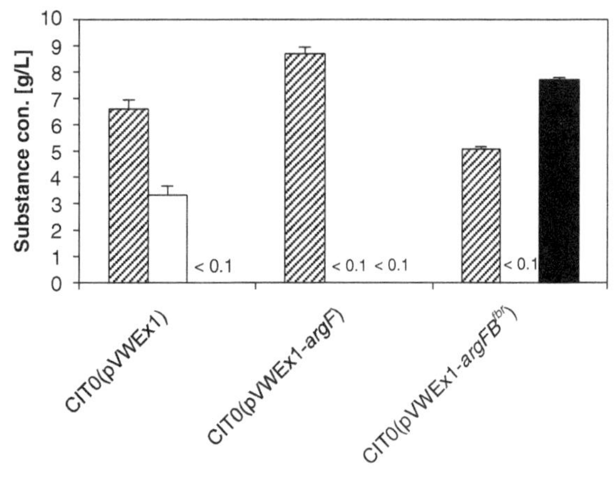

Figure 3 Biomass formation and production of ornithine and citrulline on glucose by various *C. glutamicum* strains: cell dry weight (hatched bars), L-ornithine concentration (open bars) and L-citrulline concentration (filled bars). The cultivation was performed in CGXII minimal medium containing 20 g/L glucose, 1 mM IPTG, 750 μM L-arginine and 25 μg/L kanamycin. The amino acid concentrations in the supernatant were determined after the consumption of glucose. Values and error bars represent the mean and the standard error of triplicates.

Production of L-citrulline from alternative carbon sources

Due to the high demand of biotechnological processes of using complex sugar substrates derived from raw materials and industrial wastes, the L-citrulline producer strain CIT1 was enabled to utilize the alternative carbon sources starch (as an example of a high molecular weight carbohydrate), xylose, and glucosamine (as an example of a carbohydrates, derived from forestry and food industrial wastes).

To enable *C. glutamicum* CIT1 to consume starch, the gene *amyA* from *Streptomyces griseus* was overexpressed. The combined overexpression of *xylA* from *Xanthomonas campestris* and endogenous *xylB* allowed the utilization of xylose by *C. glutamicum* CIT1. The endogenous *nagB* was overpressed ectopically to facilitate the consumption of glucosamine. The resulting strains were tested for growth and L-citrulline production.

When cultured in CGXII medium supplemented 0.75 mM L-arginine all strains engineered for alternative carbon source consumption grew with their respective substrate (Table 1). The empty vector carrying strain CIT1(pEKEx3) neither grew in xylose or glucosamine minimal medium nor consumed these substrates. By contrast, the recombinant strain CIT1(pEKEx3-*xylAB*) grew in xylose minimal medium with a growth rate of 0.03 ± 0.01 h^{-1} and reached a final OD_{600} of 6 ± 1. In glucosamine minimal medium, *C. glutamicum* CIT1(pEKEx3-*nagB*) grew to a final OD_{600} of 3 ± 1 with a growth rate of 0.02 ± 0.01 h^{-1}. In minimal medium containing 1% starch and 0.25% glucose as carbon sources, the empty vector harbouring strain CIT1(pEC-XT99A) formed roughly one third of the biomass as compared to *C. glutamicum* CIT1(pAmy). Growth of CIT1(pEC-XT99A) was slower (growth rate of 0.10 ± 0.01 h^{-1}) than that of CIT1(pAmy) (growth rate of 0.21 ± 0.01 h^{-1}). While strain CIT1(pEC-

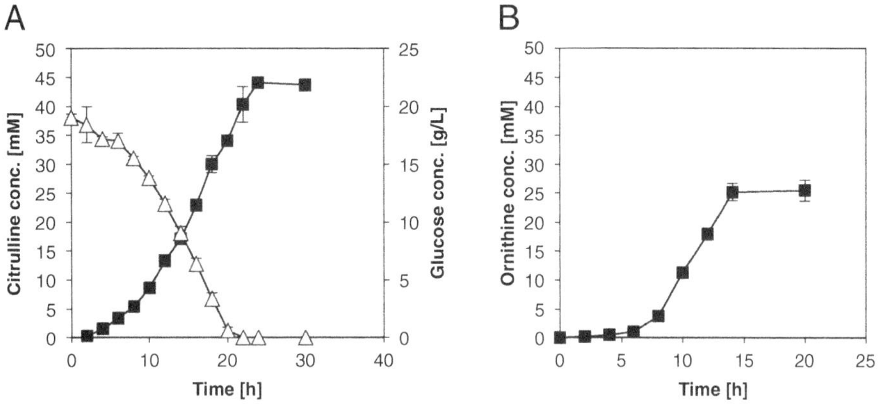

Figure 4 Amino acid production by various *C. glutamicum strains.* L-ornithine production by *C. glutamicum* CIT0(pVWEx1) (filled squares) **(A)** and L-citrulline accumulation (filled squares) and glucose consumption (open triangles) by strain CIT0(pVWEx1-*argFB*^{fbr}) **(B).** The experiments were performed in CGXII minimal medium with 20 g/L glucose, 1 mM IPTG, 25 µg/L kanamycin and supplemented with 750 µM L-arginine. Values and error bars represent the mean and the standard error of triplicates.

XT99A) only utilized glucose, but not starch, CIT1(pAmy) was able to consume both, glucose and starch.

The strains engineered for utilization of xylose and glucosamine, respectively, also produced L-citrulline from these carbon sources (Figure 5). *C. glutamicum* CIT1(pEKEx3-*nagB*) accumulated 2.6 ± 0.3 mM L-citrulline which corresponds to a yield of 0.045 ± 0.002 g/g since glucosamine was utilized completely. Similarly, after complete utilization of xylose by *C. glutamicum* CIT1(pEKEx3-*xylAB*) 6.4 ± 0.1 mM L-citrulline accumulated corresponding to a yield of 0.075 ± 0.001 g per g xylose.

As the determination of the starch concentration by HPLC was not possible, residual starch content was assayed by the use of Lugols solution. However, as it is known that overexpression of *amyA* in *C. glutamicum* results in high molecular mass degradation products of starch, which remain in the medium and are not detectable by Lugols solution (Seibold et al. 2006), the L-citrulline concentration was measured until no change in OD_{600}, starch content and L-citrulline concentration was observed. The starch utilizing strain CIT1(pAmy) was able to produce 11.9 ± 0.5 mM L-citrulline which corresponds to a yield of 0.167 g/g.

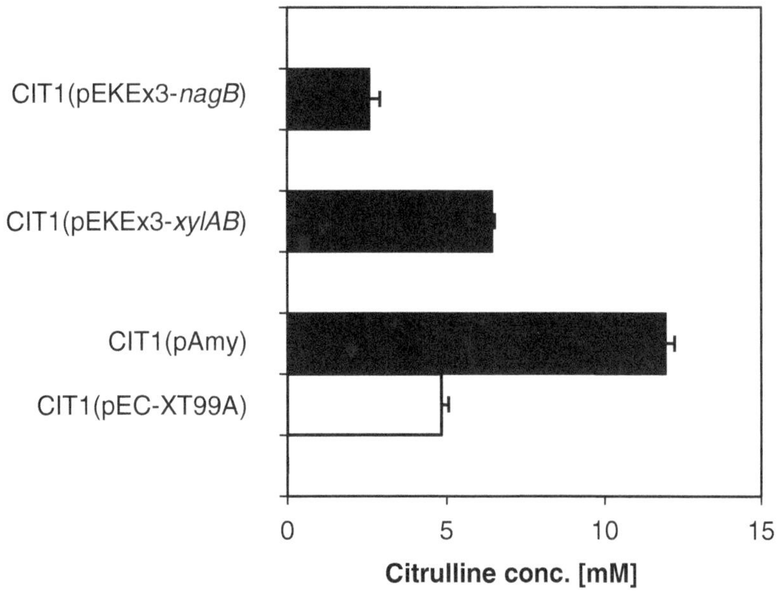

Figure 5 L-citrulline concentration in the engineered strains after the consumption of the respective carbon source. CIT1(pEC-XT99A), CIT1(pAmy) with 10 g/L soluble starch, 2,5 g/L glucose after 31 h. CIT1(pEKEx2-*xylAB*) with 15 g/L xylose after xylose consumption. CIT1(pEKEx3-*nagB*) with 10 g/L glucosamine after glucosamine consumption. Values and error bars represent the mean and the standard error of triplicates.

Discussion

C. glutamicum was engineered to accumulate L-citrulline as major product, both from glucose as well as from the alternative carbon sources starch, glucosamine and xylose.

Feedback insensitive N-acetyl L-glutamate kinase (encoded by *argB*^{fbr}; (A49VM54V)) was required for production of L-citrulline since CIT0(pVWEx1-*argF*) did not produce L-citrulline, while CIT0(pVWEx1-*argFB*^{fbr}) produced L-citrulline. It is unlikely that addition of L-arginine to CIT0(pVWEx1-*argF*) inhibited generation of L-ornithine, a precursor of L-citrulline, because strain CIT0(pVWEx1) produced L-ornithine when supplemented with L-arginine. However, it is possible that intracellular L-citrulline affects arginine biosynthesis. As overexpression of *argB*^{fbr} entailed L-citrulline formation, we assume that L-citrulline inhibits the NAGK of *C. glutamicum*, but this has not yet been described. As expected due to its structural similarity to L-arginine, L-citrulline inhibits NAGK of other microorganisms (Farago and Denes 1967; Haas and Leisinger 1975). In *Chlamydomonas reinhardtii*, NAGK is inhibited by several L-arginine structure analogs, including L-citrulline, however, inhibition was less pronounced than L-arginine inhibition (Farago and Denes 1967). NAGK from *Pseudomonas aeruginosa* lost two thirds of its activity in the presence of 2.5 mM L-citrulline which was claimed to be too weak under physiologic conditions (Haas and Leisinger 1975). However, it is conceivable that inhibition of NAGK by L-citrulline may play a role in recombinant *C. glutamicum* strains engineered for L-citrulline production, thus, possibly explaining the finding that L-citrulline production required overexpression *argB*^{fbr} encoding NAGK feedback resistant to L-arginine. Commensurate with this notion, simultaneous production of L-arginine and L-citrulline resulted from *argB*^{fbr} overexpression in a Δ*argR* background (Ikeda et al. 2009). In this *argB*^{fbr} overexpressing strain, the ratio of L-citrulline to L-arginine was higher than by classically obtained strains, which solely contain native *argB* (Ikeda et al. 2009). Currently, it remains to be studied if L-citrulline inhibits NAGK from *C. glutamicum* and if (some) variants feed-back resistant to L-arginine are also desensitized to L-citrulline.

Notably, about two fold more L-citrulline (about 7.7 g/L) was produced by strain CIT1 than L-ornithine was produced (about 3.3 g/L) by the isogenic strain CIT0 (pVWEx1). Both, overexpression of *argF* and *argB*^{fbr} may have contributed to this effect. It is more likely that *argB*^{fbr} is responsible as L-arginine supplementation may have limited flux in the arginine biosynthesis pathway of strain CIT0(pVWEx1) especially in the beginning of the cultivation. In *C. glutamicum* CIT1, only feedback-resistant NAGK is present and additionally a gene dosage effect due ectopic overexpression of *argB*^{fbr} might have contributed to increase L-citrulline production.

Glucose, glucosamine, xylose, and starch were shown to be suitable substrates for the production of L-citrulline. Strain construction was based on previously established engineering strategies (Seibold et al. 2006; Uhde et al. 2013; Meiswinkel et al. 2013a; Gopinath et al. 2011). The achieved L-citrulline concentrations on these substrates were lower than with glucose as carbon source. However, L-citrulline production from xylose (6.44 ± 0.12 mM) by CIT1(pEKEx3-*xylAB*) was lower, but in a similar range as production of L-ornithine (19.6 ± 1.9 mM) and putrescine (15.1 ± 1.2 mM), respectively, from the same xylose concentration by the respective recombinant *C. glutamicum* strains (Meiswinkel et al. 2013a). Similarly, product yields with glucosamine as carbon source were lower for L-citrulline (0.067 g/g) than for putrescine (0.112 g/g) (Uhde et al. 2013). Unexpectedly and hitherto not understood, the growth rate (0.02 ± 0.01 h^{-1}) and, thus, productivity by CIT1(pEKEx3-*nagB*) were very low. By contrast, a putrescine producing strain carrying pEKEx3-*nagB* showed only a slightly decreased growth rate (Uhde et al. 2013).

C. glutamicum strains carrying pAMY co-utilized starch with glucose (Seibold et al. 2006). Substrate co-utilization is observed with *C. glutamicum* WT as well as recombinant strains for almost all mixtures of carbon sources (Blombach and Seibold 2010). A L-lysine producing strain carrying pAMY showed increased biomass formation by addition of 10 g/L starch to 10 g/L glucose, whereas L-lysine production increased only upon addition of higher starch concentrations (Seibold et al. 2006).

In this study, the additional presence of starch increased the growth rate of CIT1 (from 0.15 to 0.21 h^{-1}) as well as L-citrulline production. Production of L-citrulline by CIT1(pAMY) from a starch glucose mixture was higher (11.95 ± 0.48 mM) than that by the empty vector carrying control strain (4.83 ± 0.4 mM) demonstrating that starch contributed to production of L-citrulline. It has to be noted that starch cannot be utilized completely by *C. glutamicum* strains overexpressing the α-amylase gene *amyA* because high-molecular-weight carbohydrates are generated from starch and remain unutilized in the medium (Seibold et al. 2006).

Taken together, production of L-citrulline as major product from glucose, starch, glucosamine, and xylose by recombinant *C. glutamicum* strains was achieved.

Competing interests
The authors declare that they have no competing interests.

Authors' contributions
DE designed experiments, performed experiments, analysed results and drafted the manuscript. JVKJ designed experiments, performed experiments and analysed results. VFW coordinated the study, designed experiments, analysed results and wrote the manuscript. All authors read and approved the final manuscript.

Acknowledgment
This work was partially supported by the Bundesministerium für Bildung und Forschung (BMBF, grant. no. 0316017) and by the program ZIM (grant. no.

KF2969003SB2). We acknowledge support for the Article Processing Charge by the Deutsche Forschungsgemeinschaft and the Open Access Publication Fund of Bielefeld University.

References

An SJ, Yim SS, Jeong KJ (2013) Development of a secretion system for the production of heterologous proteins in *Corynebacterium glutamicum* using the Porin B signal peptide. Protein Expr Purif 89(2):251–257, doi:10.1016/j.pep.2013.04.003

Arndt A, Eikmanns BJ (2008) Regulation of carbon metabolism in *Corynebacterium glutamicum*. In: Burkovski A (ed) Corynebacteria: genomics and molecular biology. Caister Academic Press, Wymondham, UK, pp 155–182

Baumgart M, Unthan S, Ruckert C, Sivalingam J, Grunberger A, Kalinowski J, Bott M, Noack S, Frunzke J (2013) Construction of a prophage-free variant of *Corynebacterium glutamicum* ATCC 13032 for use as a platform strain for basic research and industrial biotechnology. Appl Environ Microbiol 79(19):6006–6015, doi:10.1128/AEM. 01634-13

Blombach B, Seibold GM (2010) Carbohydrate metabolism in *Corynebacterium glutamicum* and applications for the metabolic engineering of L-lysine production strains. Appl Microbiol Biotechnol 86(5):1313–1322, doi:10.1007/s00253-010-2537-z

Blombach B, Hans S, Bathe B, Eikmanns BJ (2009) Acetohydroxyacid synthase, a novel target for improvement of L-lysine production by *Corynebacterium glutamicum*. Appl Environ Microbiol 75(2):419–427, doi:10.1128/AEM. 01844-08

Curis E, Nicolis I, Moinard C, Osowska S, Zerrouk N, Benazeth S, Cynober L (2005) Almost all about citrulline in mammals. Amino Acids 29(3):177–205, doi:10.1007/s00726-005-0235-4

Delaunay S, Gourdon P, Lapujade P, Mailly E, Oriol E, Engasser JM, Lindley NL, Goergen JL (1999) An improved temperature triggered process for glutamate production with *Corynebacterium glutamicum*. Enz Microb Biotechnol 25:762–768

Eggeling L, Reyes O (2005) Experiments. In: Eggeling L, Bott M (eds) Handbook of *Corynebacterium glutamicum*. CRC Press, Boca Raton, USA, pp 3535–3566

Farago A, Denes G (1967) Mechanism of arginine biosynthesis in *Chlamydomonas reinhardti*. II. Purification and properties of N-acetylglutamate 5-phosphotransferase, the allosteric enzyme of the pathway. Biochim Biophys Acta 136(1):6–18

Fish WW (2012) Process for the production of L-citrulline from watermelon flesh and rind. US 8173837 B1

Frohwitter J, Heider SA, Peters-Wendisch P, Beekwilder J, Wendisch VF (2014) Production of the sesquiterpene (+)-valencene by metabolically engineered *Corynebacterium glutamicum*. J Biotechnol. doi:10.1016/j.jbiotec.2014.05.032

Georgi T, Rittmann D, Wendisch VF (2005) Lysine and glutamate production by *Corynebacterium glutamicum* on glucose, fructose and sucrose: roles of malic enzyme and fructose-1,6-bisphosphatase. Metab Eng 7(4):291–301

Gopinath V, Meiswinkel TM, Wendisch VF, Nampoothiri KM (2011) Amino acid production from rice straw and wheat bran hydrolysates by recombinant pentose-utilizing Corynebacterium glutamicum. Appl Microbiol Biotechnol. doi:10.1007/s00253-011-3478-x

Haas D, Leisinger T (1975) N-acetylglutamate 5-phosphotransferase of *Pseudomonas aeruginosa*. Catalytic and regulatory properties. Eur J Biochem 52(2):377–393

Hanahan D (1983) Studies on transformation of *Escherichia coli* with plasmids. J Mol Biol 166(4):557–580

Heider SA, Peters-Wendisch P, Netzer R, Stafnes M, Brautaset T, Wendisch VF (2014a) Production and glucosylation of C50 and C 40 carotenoids by metabolically engineered *Corynebacterium glutamicum*. Appl Microbiol Biotechnol 98(3):1223–1235, doi:10.1007/s00253-013-5359-y

Heider SA, Peters-Wendisch P, Wendisch VF, Beekwilder J, Brautaset T (2014b) Metabolic engineering for the microbial production of carotenoids and related products with a focus on the rare C50 carotenoids. Appl Microbiol Biotechnol 98(10):4355–4368, doi:10.1007/s00253-014-5693-8

Ikeda M, Mitsuhashi S, Tanaka K, Hayashi M (2009) Reengineering of a *Corynebacterium glutamicum* L-arginine and L-citrulline producer. Appl Environ Microbiol 75(6):1635–1641, doi:10.1128/AEM. 02027-08

Jensen JV, Wendisch VF (2013) Ornithine cyclodeaminase-based proline production by *Corynebacterium glutamicum*. Microb Cell Factories 12:63, doi:10.1186/1475-2859-12-63

Jolkver E, Emer D, Ballan S, Kramer R, Eikmanns BJ, Marin K (2009) Identification and characterization of a bacterial transport system for the uptake of pyruvate, propionate, and acetate in *Corynebacterium glutamicum*. J Bacteriol 191(3):940–948, doi:10.1128/JB.01155-08

Kakimoto T, Shibatani T, Nishimura N, Chibata I (1971) Enzymatic production of L-citrulline by Pseudomonas putida. Appl Microbiol 22(6):992–999

Kikuchi Y, Itaya H, Date M, Matsui K, Wu LF (2009) TatABC overexpression improves *Corynebacterium glutamicum* Tat-dependent protein secretion. Appl Environ Microbiol 75(3):603–607, doi:10.1128/AEM. 01874-08

Kim J, Hirasawa T, Sato Y, Nagahisa K, Furusawa C, Shimizu H (2009) Effect of *odhA* overexpression and *odhA* antisense RNA expression on Tween-40-triggered glutamate production by *Corynebacterium glutamicum*. Appl Microbiol Biotechnol 81(6):1097–1106, doi:10.1007/s00253-008-1743-4

Kim J, Fukuda H, Hirasawa T, Nagahisa K, Nagai K, Wachi M, Shimizu H (2010) Requirement of de novo synthesis of the OdhI protein in penicillin-induced glutamate production by *Corynebacterium glutamicum*. Appl Microbiol Biotechnol 86(3):911–920, doi:10.1007/s00253-009-2360-6

Kirchner O, Tauch A (2003) Tools for genetic engineering in the amino acid-producing bacterium *Corynebacterium glutamicum*. J Biotechnol 104(1–3):287–299

Matano C, Uhde A, Youn JW, Maeda T, Clermont L, Marin K, Kramer R, Wendisch VF, Seibold GM (2014) Engineering of *Corynebacterium glutamicum* for growth and L-lysine and lycopene production from N-acetyl-glucosamine. Appl Microbiol Biotechnol 98(12):5633–5643, doi:10.1007/s00253-014-5676-9

Meiswinkel TM, Gopinath V, Lindner SN, Nampoothiri KM, Wendisch VF (2013a) Accelerated pentose utilization by Corynebacterium glutamicum for accelerated production of lysine, glutamate, ornithine and putrescine. Microb Biotechnol 6(2):131–140, doi:10.1111/1751-7915.12001

Meiswinkel TM, Rittmann D, Lindner SN, Wendisch VF (2013b) Crude glycerol-based production of amino acids and putrescine by *Corynebacterium glutamicum*. Bioresour Technol 145:254–258, doi:10.1016/j.biortech.2013.02.053

Okumura S, Shibuya M, Shimpachi K, Teruo S, Noboru K (1966) Method of producing citrulline by bacterial fermentation. US 3282794 A

Peters-Wendisch PG, Kreutzer C, Kalinowski J, Patek M, Sahm H, Eikmanns BJ (1998) Pyruvate carboxylase from *Corynebacterium glutamicum*: characterization, expression and inactivation of the *pyc* gene. Microbiology 144(Pt 4):915–927

Peters-Wendisch PG, Schiel B, Wendisch VF, Katsoulidis E, Mockel B, Sahm H, Eikmanns BJ (2001) Pyruvate carboxylase is a major bottleneck for glutamate and lysine production by *Corynebacterium glutamicum*. J Mol Microbiol Biotechnol 3(2):295–300

Radmacher E, Stansen KC, Besra GS, Alderwick LJ, Maughan WN, Hollweg G, Sahm H, Wendisch VF, Eggeling L (2005) Ethambutol, a cell wall inhibitor of *Mycobacterium tuberculosis*, elicits L-glutamate efflux of *Corynebacterium glutamicum*. Microbiology 151(Pt 5):1359–1368

Rimando AM, Perkins-Veazie PM (2005) Determination of citrulline in watermelon rind. J Chromatogr A 1078(1–2):196–200

Rittmann D, Lindner SN, Wendisch VF (2008) Engineering of a glycerol utilization pathway for amino acid production by *Corynebacterium glutamicum*. Appl Environ Microbiol 74(20):6216–6222, doi:10.1128/AEM. 00963-08

Sakanyan V, Petrosyan P, Lecocq M, Boyen A, Legrain C, Demarez M, Hallet JN, Glansdorff N (1996) Genes and enzymes of the acetyl cycle of arginine biosynthesis in *Corynebacterium glutamicum*: enzyme evolution in the early steps of the arginine pathway. Microbiology 142(Pt 1):99–108

Sambrook J, Russell D (2012) Molecular cloning. A laboratory manual, 4th edn. Cold Spring Harbor Laboratoy Press, Cold Spring Harbor, NY

Sasaki M, Teramoto H, Inui M, Yukawa H (2011) Identification of mannose uptake and catabolism genes in *Corynebacterium glutamicum* and genetic engineering for simultaneous utilization of mannose and glucose. Appl Microbiol Biotechnol 89(6):1905–1916, doi:10.1007/s00253-010-3002-8

Sato H, Orishimo K, Shirai T, Hirasawa T, Nagahisa K, Shimizu H, Wachi M (2008) Distinct roles of two anaplerotic pathways in glutamate production induced by biotin limitation in *Corynebacterium glutamicum*. J Biosci Bioeng 106(1):51–58

Schneider J, Wendisch VF (2010) Putrescine production by engineered *Corynebacterium glutamicum*. Appl Microbiol Biotechnol 88(4):859–868, doi:10.1007/s00253-010-2778-x

Schneider J, Niermann K, Wendisch VF (2011) Production of the amino acids L-glutamate, L-lysine, L-ornithine and L-arginine from arabinose by recombinant *Corynebacterium glutamicum*. J Biotechnol 154(2–3):191–198, doi:10.1016/j.jbiotec.2010.07.009

Schneider J, Eberhardt D, Wendisch VF (2012) Improving putrescine production by *Corynebacterium glutamicum* by fine-tuning ornithine transcarbamoylase activity using a plasmid addiction system. Appl Microbiol Biotechnol 95(1):169–178, doi:10.1007/s00253-012-3956-9

Seibold G, Auchter M, Berens S, Kalinowski J, Eikmanns BJ (2006) Utilization of soluble starch by a recombinant *Corynebacterium glutamicum* strain: growth and lysine production. J Biotechnol 124(2):381–391

Stansen C, Uy D, Delaunay S, Eggeling L, Goergen JL, Wendisch VF (2005) Characterization of a *Corynebacterium glutamicum* lactate utilization operon induced during temperature-triggered glutamate production. Appl Environ Microbiol 71(10):5920–5928

Tateno T, Fukuda H, Kondo A (2007) Production of L-Lysine from starch by *Corynebacterium glutamicum* displaying alpha-amylase on its cell surface. Appl Microbiol Biotechnol 74(6):1213–1220, doi:10.1007/s00253-006-0766-y

Teramoto H, Watanabe K, Suzuki N, Inui M, Yukawa H (2011) High yield secretion of heterologous proteins in *Corynebacterium glutamicum* using its own Tat-type signal sequence. Appl Microbiol Biotechnol 91(3):677–687, doi:10.1007/s00253-011-3281-8

Tsuge Y, Tateno T, Sasaki K, Hasunuma T, Tanaka T, Kondo A (2013) Direct production of organic acids from starch by cell surface-engineered *Corynebacterium glutamicum* in anaerobic conditions. AMB Express 3(1):72, doi:10.1186/2191-0855-3-72

Uhde A, Youn JW, Maeda T, Clermont L, Matano C, Kramer R, Wendisch VF, Seibold GM, Marin K (2013) Glucosamine as carbon source for amino acid-producing *Corynebacterium glutamicum*. Appl Microbiol Biotechnol 97(4):1679–1687, doi:10.1007/s00253-012-4313-8

Wada M (1930) Über Citrullin, eine neue Aminosäure im Presssaft der Wassermelone, *Citrullus vulgaris* Schrad. Biochem Z 224:420–429

Wendisch VF (2014) Microbial production of amino acids and derived chemicals: synthetic biology approaches to strain development. Curr Opin Biotechnol 30C:51–58, doi:10.1016/j.copbio.2014.05.004

Wendisch VF, Eberhardt D, Herbst M, Jensen JVK (2014) Amino acids and nucleotides. In: Bicas J (ed) Biotechnological production of natural ingredients for food industry. Bentham eBooks.

Yamamoto K, Sato T, Tosa T, Chibata I (1974) Continuous production of L-citrulline by immobilized *Pseudomonas putida* cells. Biotechnol Bioeng 16(12):1589–1599, doi:10.1002/bit.260161203

16

Assessing medium constituents for optimal heterologous production of anhydromevalonolactone in recombinant *Aspergillus oryzae*

Songsak Wattanachaisaereekul[1*], Anuwat Tachaleat[1], Juntira Punya[2], Rachada Haritakun[2], Chollaratt Boonlarppradab[2] and Supapon Cheevadhanarak[1,3]

Abstract

Anhydromevalonolactone (AMVL) is a bioactive natural product that arises from a molecular biology technique using *Aspergillus oryzae* as a heterologous host. AMVL has been used as a precursor for the synthesis of insect pest control reagents and has numerous applications in the biotechnological and medical industries. In this study, the Plackett-Burman Design and the Central Composite Design, which offer efficient and feasible approaches, were complemented to screen significant parameters and identify the optimal values for maximum AMVL production. The results suggested that sucrose, $NaNO_3$, yeast extract and K_2HPO_4 were the key factors affecting AMVL production in a complex medium, whereas the major components required for a defined medium were $NaNO_3$, K_2HPO_4, KH_2PO_4 and trace elements. These factors were subsequently optimized using the response surface methodology. Under optimal conditions, a maximum AMVL production of 250 mg/L in the complex medium and 200 mg/L in the defined medium was achieved, which represents an increase of approximately 3-4-fold compared to the commonly used malt extract medium.

Keywords: Anhydromevalonolactone; *Aspergillus oryzae*; Plackett-Burman design; Central composite design; Response surface methodology

Introduction

Aspergillus oryzae is an asexual, ascomycete filamentous fungus that plays an essential role in the production of oriental fermented foods and beverages such as soy sauce, miso, sake, and rice vinegar (Yokotsuka 1961; Yong and Wood 1974; Sakaguchi et al. 1992). This fungus is generally regarded as safe (GRAS) by the U.S. Food and Drug Administration (FDA) and the U.S. Environmental Protection Agency because it does not appear to be pathogenic for plants or animals (U.S. Environmental Protection Agency 1997). *A. oryzae* has long been used for large-scale industrial enzyme production: e.g., amylases, proteases, lipases and other hydrolytic enzymes (Ohnishi et al. 1994; Carlsen and Nielsen 2001; Chutmanop et al. 2008). More recently, it has been used as a heterologous host for the production of anhydromevalonolactone (AMVL) (Figure 1A) and other novel pyrones based on ectopic expression of the gene encoding polyketide synthase (*pksmt*) from *Xylaria* sp. BCC 1067 under the control of the *gpdA* promoter (Punya et al. 2013).

AMVL has been used as a precursor for the synthesis of lineatin (Figure 1B), which is a pheromone produced by the female striped ambrosia beetle (McKay et al. 1982). This type of beetle is responsible for severe economic damage to the timber and wood industries around the world (Orbay et al. 1994; Bumrungsri et al. 2008). Lineatin can act as a lure for mass trapping of the ambrosia beetle; therefore, AMVL is being studied as a possible precursor for a pest control reagent. Interestingly, it has been reported that AMVL can inhibit the growth of *Lactobacillus acidophilus* and *L. heterohiochii* in the presence of

* Correspondence: songsakwat@yahoo.com
[1]Pilot Plant Development and Training Institute, King Mongkut's University of Technology Thonburi, 49 Soi Thianthale 25, Bangkhunthian-Chaithale Rd., Thakham, Bangkhunthian, Bangkok 10150, Thailand
Full list of author information is available at the end of the article

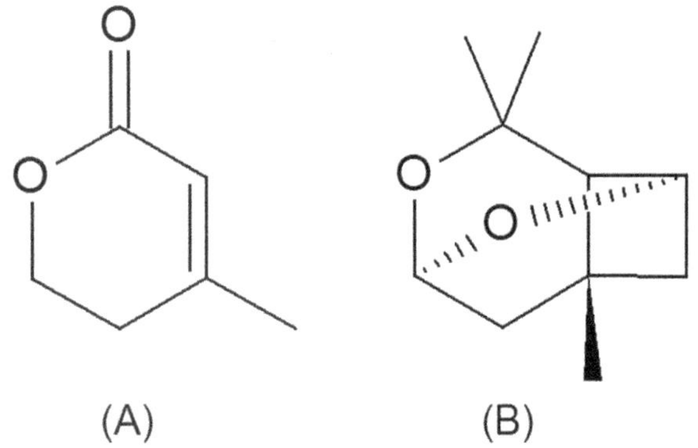

Figure 1 Structures of anhydromevalonolactone (A) and lineatin (B).

mevalonic acid, and it completely inhibits the growth of *Saccharomyces cerevisiae* and *S. carlsbergensis* (Kitamura et al. 1976). This finding indicates that AMVL can also be used as a compound to inhibit the incorporation of mevalonic acid into cholesterol in the mevalonate biosynthetic pathway. Furthermore, not only was AMVL found to be a precursor for two novel pyrones that exhibited anti-oral cavity cancer activity, but its derivatives also offer a wide range of biological activities (Punya et al. 2013; Busch and Hertweck 2009, Wilk et al. 2009). Therefore, AMVL has attracted a great deal of attention as a possible precursor for drugs and other valuable compounds that are of economic importance. Despite the interesting aspects of AMVL, scientific reports on its production are very limited. AMVL has previously been chemically synthesized and identified as a metabolite present in the culture broths of various fungi, but the production of AMVL has been hampered by low and unstable yields.

Malt extract broth (MEB) is the commonly used medium in our laboratory for fungal cultivation. MEB contains 17 g/L malt extract and 3 g/L mycological peptone. Malt extract is a diastase-free extract of malt and provides both nitrogenous and carbohydrate constituents; therefore, the use of this medium has been recommended for the cultivation of various molds and yeast. Because MEB was successfully used in our laboratory for the production of xyrrolin from *Xylaria* sp. BCC 1067 (phonghanpot et al. 2012), it was further used for the production of AMVL by *Aspergillus oryzae* MTG4. Unfortunately, a low AMVL yield was obtained using this medium. Additionally, there is no report on the statistical optimization of the nutritive conditions on AMVL production. For this reason, improvement of AMVL production using culture medium optimization is required.

Generally, the experimental procedure of medium optimization is carried out by varying a single factor to find the optimal solution while other factors are kept at a constant level, and this approach is referred to as one-factor-at-a-time. This technique obviously has serious disadvantages because interactions among the factors are not considered; therefore, it does not reflect all the potential effects on the optimization process (Lundstedt et al. 1998). More efficient approaches to determining the optimal conditions are based on the Plackett-Burman Design (PBD) and the Response Surface Methodology (RSM). The PBD is a method used to identify the components in a culture medium that have significant influences on the targeted response. Following the PBD, RSM is usually used for further optimization to detect the optimal concentration of the key components obtained from the PBD. RSM is a statistical technique for designing experiments, evaluating the relationships between independent variables and responses, searching for the optimum conditions and generating mathematical models that accurately describe the overall process (Myers 2002). This statistical optimization method has been used extensively for optimization in many areas of industrial research and process development in chemistry and biotechnology (Khamduang et al. 2009; Chen et al. 2010; Guo et al. 2010). The most popular statistical designs for RSM are the Central Composite Design (CCD) and the Box-Behnken Design (BBD). These designs differ based on the location of the experimental points in the studied region and the levels of one numeric factor (Bezerra et al. 2008).

The present study aims to optimize the medium components of complex and defined media to improve AMVL production by the recombinant *A. oryzae*. In this study, the significant medium components were identified by PBD, and the determining factors were optimized through CCD. Finally, the optimal conditions were experimentally validated, and AMVL production with optimized culture media was compared with the initial conditions. To our best knowledge, this is the first report illustrating a statistical medium optimization process for increased AMVL production in *A. oryzae*.

Materials and methods

Microorganisms and culture conditions

A. oryzae MTG4 (BIOTEC Molecular Genomics Collection, BMGC, no. 115) (Punya et al. 2013) harboring the gene encoding pksmt from *Xylaria* sp. BCC 1067 was cultivated on malt extract agar containing 20 g/L malt extract and 15 g/L agar, at 30°C for 1 week. The spores were harvested in 0.05% Tween 80 solution. The spores were transferred to 250-mL Erlenmeyer flasks containing 50 mL medium with different medium compositions designed by PBD and CCD. The pH of the medium was adjusted to 6.5 with 2 M HCl prior to autoclaving. With the initial spore concentration of 10^7 spores/50 mL medium, the shake-flask cultures were incubated at 25°C on a rotary shaker at 200 rpm for 23 days. All experiments were carried out in triplicate.

Biomass measurement and analytical methods for AMVL and sucrose

Biomass measurements were made by measuring the dry weight of biomass samples filtered with Whatman™ no. 1 filter paper (GE healthcare UK Limited) at 105°C for 48 hours. The fermentation broth was filtered through a 0.2-μm pore-sized nylon filter (VertiClean™, Vertical Chromatography Co., Ltd.) for later analysis of AMVL and sucrose. AMVL was analyzed by HPLC (Agilent Technologies 1200 series) on a device equipped with a UV detector (220 nm) and a C18 reversed-phase column (VertiSep™, UPS 4.6 x 250 mm, 5 μm, Vertical Chromatography Co., Ltd.), which was operated using a gradient of deionized water and acetonitrile at a flow rate of 0.5 mL/min. Sucrose was quantified by HPLC, which was operated at 60°C, with the flow rate of 0.6 mL/min for 5 mM H_2SO_4, using a refractive index detector and an ion exclusion column (Aminex HPX-87H, 300 mm × 7.8 mm, Bio-Rad Laboratories).

Experimental designs and optimization

The approach for medium optimization adopted for improving AMVL production was divided into the following 3 steps:

Screening of carbon and nitrogen sources

Because secondary metabolite production in filamentous fungi is strongly influenced by the carbon and nitrogen sources, the screening of carbon and nitrogen sources is crucial. A selection of carbon and nitrogen sources was therefore examined through a traditional non-statistical method. In this study, glucose, sucrose, and soluble starch were used to optimize the carbon source, and NH_4Cl, NH_4NO_3, $(NH_4)_2SO_4$, $NaNO_3$, KNO_3, urea, yeast extract and malt extract were used to optimize the nitrogen source. Based on Czapex Yeast broth (CYB), the concentrations of the carbon and nitrogen sources were set at 40 g/L and 5 g/L, respectively, and the rest of the medium components were 5 g/L K_2HPO_4, 0.5 g/L $MgSO_4.7H_2O$, 0.5 g/L KCl, and 10 mL/L trace elements (0.3575 g/L $ZnSO_4.7H_2O$, 0.0625 g/L $CuSO_4.5H_2O$, 0.25 g/L $MnSO_4.H_2O$, 0.345 g/L $FeSO_4.7H_2O$, and 0.075 g/L citric acid). *A. oryzae* MTG4 was cultivated in the media with different carbon and nitrogen sources on a rotary shaker at 25°C, 200 rpm for 23 days, and the fermentation broth was analyzed for AMVL production.

Plackett-Burman design

The Plackett-Burman design has been widely and successfully used in the screening of the major constituents of cultivation media. This experimental design was adopted in this study to identify the medium components that have significant effects on AMVL production. This technique is based on the first-polynomial model according to Eq. (1):

$$Y = \beta_0 + \Sigma\beta_i x_i \qquad (1)$$

where Y is the response, which is the concentration of AMVL, β_0 is the model intercept, β_i is the linear coefficient, and x_i is the level of independent variables. This model does not describe the interaction among factors, but it is used to evaluate and screen the factors that significantly impact the response (Y). The independent variables chosen to be screened for the complex medium were sucrose, $NaNO_3$, yeast extract, K_2HPO_4, KH_2PO_4, KCl, $MgSO_4.7H_2O$ and trace elements, while the nutrients selected for the defined medium were sucrose, $NaNO_3$, K_2HPO_4, KH_2PO_4, KCl, $MgSO_4.7H_2O$, $CaCl_2.2H_2O$, $Na_3C_6H_5O_7.2H_2O$, biotin and trace elements.

Based on the Plackett-Burman design, each factor was illustrated in two levels: −1 for the low level and +1 for the high level. A center point was run to evaluate the linear and curvature effects of the variables (Plackett and Burman, 1946). The experimental design for screening the significant variables in AMVL production with complex and defined media are provided in Table 1. However, the

Table 1 The levels of independent variables in the Plackett-Burman Design for the complex and defined media

Varible	Components	Levels of variable		
		−1 (Low)	0 (Central)	+1 (High)
	Complex medium			
A	Sucrose (g/L)	20	40	60
B	$NaNO_3$ (g/L)	1	3	5
C	Yeast extract (g/L)	0	2	4
D	KH_2PO_4 (g/L)	0	1	2
E	K_2HPO_4 (g/L)	1	3	5
F	KCl (g/L)	0.5	0.75	1
G	$MgSO_4.7H_2O$ (g/L)	0.5	0.75	1
H	Trace elements (mL/L)	0.5	0.75	1
	Defined medium			
I	Sucrose (g/L)	40	70	100
J	$NaNO_3$ (g/L)	3	5	7
K	K_2HPO_4 (g/L)	3	5	7
L	KH_2PO_4 (g/L)	0	1	2
M	KCl (g/L)	0.25	0.50	0.75
N	$MgSO_4.7H_2O$ (g/L)	0.25	0.50	0.75
O	$CaCl_2.2H_2O$ (g/L)	0	0.10	0.20
P	$Na_3C_6H_5O_7.2H_2O$ (g/L)	0	1	2
Q	Biotin (mL/L)	0	0.10	0.20
R	Trace elements (mL/L)	0.25	0.50	0.75

concentration levels in the defined medium were defined based on the optimum concentration of each component in the complex medium.

In this study, the variables were screened in 12 experimental runs with the foldover augmenting method to increase the resolution of the design in addition to 6 runs at the center points, giving a total of 30 experimental runs. All experiments were performed in triplicate, and the average concentration of AMVL was used as the response (the dependent variable). The factors significant at 95% ($p < 0.05$) were considered to have a significant effect on AMVL production and were used for further optimization by the response surface methodology.

Central composite design

Response surface methodology was carried out to illustrate the nature of the response surface in the experimental region and to optimize the medium components for enhanced AMVL production by using the central composite design. For the complex medium, 4 critical components identified based on the Plackett-Burman design, namely, sucrose, $NaNO_3$, yeast extract and K_2HPO_4, were optimized, whereas $NaNO_3$, K_2HPO_4, KH_2PO_4 and trace elements were optimized for the defined medium.

According to the experimental design, a full 2^4 factorial design, 8 axial points and 6 replications of center points were used, which led to 30 sets of experiments. Each factor optimized for the complex and defined media was studied at five different levels (−2, −1, 0, +1, +2). Other non-significant factors were held constant. The relationship between independent variables and responses was described by a second order polynomial equation, Eq. (2):

$$Y = \beta_0 + \sum \beta_i x_i + \sum \beta_{ii} x_i^2 + \sum \beta_{ij} x_i x_j, \quad i = 1, 2, \ldots k \tag{2}$$

where Y is the predicted response, x_i and x_j are the independent variables that influence the response Y, β_0 is the offset term, β_i is the linear coefficient of x_i, β_{ii} is the quadratic coefficient of x_i, and β_{ij} is the interaction coefficient between x_i and x_j. All experiments were carried out in triplicate. The results were analyzed and interpreted using Design Expert Version 8 (Design-Expert®, Stat-Ease, Inc. Minneapolis, MN, USA). The statistical significance of the second-order model was determined using Fischer's test, and the significances of all terms in the polynomial were assessed according to their p-values. The quality of the regression model was assessed statistically by the coefficient of determination R^2. Response surface plots were generated, and the optimum concentration of each variable was calculated by the differential equation of the quadratic model. Finally, an experiment on the optimal medium was conducted to validate the central composite design model developed.

Results

Screening of optimal carbon and nitrogen sources

To choose the most efficient nutrients, glucose, sucrose, and soluble starch were used as carbon sources, and various compounds containing nitrogen, namely NH_4Cl, NH_4NO_3, $(NH_4)_2SO_4$, $NaNO_3$, KNO_3, urea, yeast extract and malt extract, were used as nitrogen sources for AMVL production by *A. oryzae* MTG4. Based on CYB medium, the concentrations of carbon and nitrogen sources were initially set as 40 g/L and 5 g/L, respectively, in addition to 5 g/L K_2HPO_4, 0.5 g/L KCl, 0.5 g/L $MgSO_4.7H_2O$ and 10 mL/L trace elements.

As illustrated in Figure 2, AMVL production varied with the different carbon and nitrogen sources. Sucrose, which provides an energy source, was a more suitable carbon source for AMVL production than glucose and soluble starch, while $NaNO_3$ was the most preferable nitrogen source. The combined use of sucrose and $NaNO_3$ led to the highest AMVL production at 82 mg/L. Yeast extract, which is an extract of autolyzed brewer's yeast and provides essential nutrients such as vitamins, amino acids and other nitrogenous compounds required for fungal growth, is also beneficial for AMVL production. AMVL produced approximately 65 mg/L with the combination of sucrose and yeast extract. Additionally, it was observed that AMVL production was low in the media containing ammonium ions such as NH_4Cl, NH_4NO_3 and $(NH_4)_2SO_4$ because after the ammonium ions were used up, the media became acidic and unfavorable for AMVL production. Therefore, in this study, sucrose and $NaNO_3$ were selected as carbon and inorganic nitrogen sources for the defined medium, respectively, whereas yeast extract were chosen as an organic nitrogen source for the complex medium.

Plackett-Burman design

The Plackett-Burman design is the screening design that was used to identify important variables that had significant effects on AMVL production. In this experimental design, the exact quantity of each variable is not determined, and the interactions between the variables are considered negligible (Plackett and Burman 1946). The design matrix and its corresponding AMVL concentration with the complex and defined media are illustrated in Tables 2 and 3, respectively.

For the complex medium, the p-value, which indicates the statistical confidence of an estimated factor, and the main effects of each medium constituent are summarized in Table 4. According to the statistical analysis, sucrose, $NaNO_3$, yeast extract, K_2HPO_4, KCl and $MgSO_4.7H_2O$ had positive effects. However, KH_2PO_4 and trace elements showed negative effects on AMVL

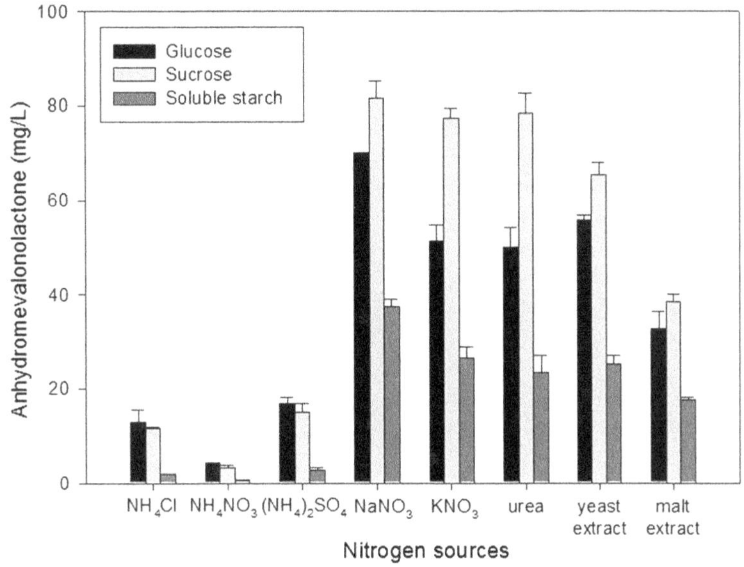

Figure 2 Effects of different carbon and nitrogen sources on AMVL production.

production. A maximum p-value of 0.05 was used as a cutoff point for selecting significant variables. The result suggested that only four variables, sucrose, $NaNO_3$, yeast extract and K_2HPO_4, were significant factors because their p-values were 0.0201, <0.001, 0.0025 and 0.0049, respectively. Because these factors showed positive effects as previously mentioned, a +1 level would be expected to improve AMVL production. Other variables had low confidence levels because the p-values were greater than 0.05, and consequently, these variables were considered insignificant and were not included in the next optimization experiment.

For the defined medium, ten media constituents were screened by PBD. The main effects, including the p-values, are presented in Table 4. Based on the results, $NaNO_3$, K_2HPO_4, KH_2PO_4 and trace elements had confidence levels above 95%, indicating that they significantly influenced AMVL production. Therefore, the optimal level of these factors was further determined by central composite design. The rest of the factors were considered insignificant because their confidence levels were below 95%. As a result, the insignificant factors were not included in the next round of optimization.

Response surface methodology

The response surface methodology (RSM) is a mathematical and statistical technique used for developing and optimizing processes. This technique is also used for evaluating the significance of relevant factors in experiments. Central composite design (CCD), which is a popular experimental design in RSM, has been successfully used for medium optimization. CCD has three sets of experimental runs as follows: (1) full factorial runs in which each factor is studied at +1 and −1 levels, (2) replication at the center points of all factors to help in understanding the curvature and to estimate pure error, and (3) axial points with the values outside the median of two factorial levels. The factors optimized for the complex and defined media are listed in Table 5.

Optimization of the complex medium for AMVL production

According to the screening results obtained by the Plackett-Burman design, sucrose, $NaNO_3$, yeast extract and K_2HPO_4 were chosen as independent input variables, while AMVL production was used as an output variable. A 2^4-factorial central composite experimental design, with six replications at the center points and a 2×4 axial point ($\alpha = 2$) leading to a total number of 30 experimental runs, was employed for the optimization. Table 6 shows the central composite design along with the predicted and observed response for each individual experiment. The relationships between sucrose, $NaNO_3$, yeast extract and K_2HPO_4 were identified by RSM. The analysis of variance (ANOVA) for the quadratic response surface model is presented in Table 7, and the full second-order polynomial equation for AMVL production was given in coded values as Eq. (3):

$$\begin{aligned} Y = {} & 242.18 + 3.80A + 7.53B + 5.31C + 17.19D \\ & + 7.96AB + 1.81AC - 5.76AD - 11.90BC - 0.28BD \\ & - 6.16CD - 15.62A^2 - 19.60B^2 - 14.73C^2 - 13.41D^2 \end{aligned} \quad (3)$$

where Y is the predicted AMVL production (mg/L), A is sucrose, B is $NaNO_3$, C is yeast extract, D is

Table 2 Experimental design using the Plackett-Burman method with AMVL production for the complex medium

Run	Coded values								AMVL (mg/L)
	A	B	C	D	E	F	G	H	
1	1	1	−1	1	1	1	−1	−1	161.1 ± 7.2
2	−1	1	1	−1	1	1	1	−1	139.1 ± 5.3
3	1	−1	1	1	−1	1	1	1	17.6 ± 0.3
4	−1	1	−1	1	1	−1	1	1	68.3 ± 4.7
5	−1	−1	1	−1	1	1	−1	1	50.7 ± 3.1
6	−1	−1	−1	1	−1	1	1	−1	1.4 ± 0.1
7	1	−1	−1	−1	1	−1	1	1	12.4 ± 0.5
8	1	1	−1	−1	−1	1	−1	1	81.0 ± 3.6
9	1	1	1	−1	−1	−1	1	−1	123.8 ± 5.8
10	−1	1	1	1	−1	−1	−1	1	55.0 ± 3.2
11	1	−1	1	1	1	−1	−1	−1	39.2 ± 1.6
12	−1	−1	−1	−1	−1	−1	−1	−1	1.1 ± 0.1
13	0	0	0	0	0	0	0	0	80.5 ± 4.4
14	0	0	0	0	0	0	0	0	86.2 ± 4.7
15	0	0	0	0	0	0	0	0	81.2 ± 3.5
16	−1	−1	1	−1	−1	−1	1	1	34.4 ± 2.0
17	1	−1	−1	1	−1	−1	−1	1	1.6 ± 0.2
18	−1	1	−1	−1	1	−1	−1	−1	74.8 ± 3.4
19	1	−1	1	−1	−1	1	−1	−1	48.1 ± 2.2
20	1	1	−1	1	−1	−1	1	−1	86.0 ± 4.1
21	1	1	1	−1	1	−1	−1	1	136.1 ± 5.7
22	−1	1	1	1	−1	1	−1	−1	80.9 ± 3.8
23	−1	−1	1	1	1	−1	1	−1	57.2 ± 2.3
24	−1	−1	−1	1	1	1	−1	1	2.4 ± 0.3
25	1	−1	−1	−1	1	1	1	−1	25.0 ± 0.9
26	−1	1	−1	−1	−1	1	1	1	44.5 ± 2.3
27	1	1	1	1	1	1	1	1	142.9 ± 6.8
28	0	0	0	0	0	0	0	0	92.8 ± 3.7
29	0	0	0	0	0	0	0	0	76.9 ± 2.9
30	0	0	0	0	0	0	0	0	87.1 ± 3.3

Note: A, Sucrose; B, $NaNO_3$; C, Yeast extract; D, KH_2PO_4; E, K_2HPO_4; F, KCl; G, $MgSO_4.7H_2O$; H, Trace elements.

Table 3 Experimental design using the Plackett-Burman method with AMVL production for the defined medium

Run	Coded values										AMVL (mg/L)
	I	J	K	L	M	N	O	P	Q	R	
1	1	1	−1	1	1	1	−1	−1	−1	1	44.5 ± 2.2
2	−1	1	1	−1	1	1	1	−1	−1	−1	57.9 ± 3.1
3	1	−1	1	1	−1	1	1	1	−1	−1	32.6 ± 1.4
4	−1	1	−1	1	1	−1	1	1	1	−1	54.7 ± 2.6
5	−1	−1	1	−1	1	1	−1	1	1	1	17.7 ± 0.6
6	−1	−1	−1	1	−1	1	1	−1	1	1	11.7 ± 0.4
7	1	−1	−1	−1	1	−1	1	1	−1	1	24.2 ± 1.3
8	1	1	−1	−1	−1	1	−1	1	1	−1	89.9 ± 3.8
9	1	1	1	−1	−1	−1	1	−1	1	1	38.4 ± 1.5
10	−1	1	1	1	−1	−1	−1	1	−1	1	23.8 ± 1.1
11	1	−1	1	1	1	−1	−1	−1	1	−1	19.5 ± 0.8
12	−1	−1	−1	−1	−1	−1	−1	−1	−1	−1	56.2 ± 3.3
13	0	0	0	0	0	0	0	0	0	0	45.4 ± 1.7
14	0	0	0	0	0	0	0	0	0	0	41.8 ± 1.9
15	0	0	0	0	0	0	0	0	0	0	39.9 ± 1.6
16	−1	−1	1	−1	−1	−1	1	1	1	−1	41.5 ± 2.3
17	1	−1	−1	1	−1	−1	−1	1	1	1	20.1 ± 0.7
18	−1	1	−1	−1	1	−1	−1	−1	1	1	56.4 ± 3.0
19	1	−1	1	−1	−1	1	−1	−1	−1	1	21.8 ± 1.1
20	1	1	−1	1	−1	−1	1	−1	−1	−1	58.4 ± 2.9
21	1	1	1	−1	1	−1	−1	1	−1	−1	71.4 ± 3.7
22	−1	1	1	1	−1	1	−1	−1	1	−1	71.8 ± 4.3
23	−1	−1	1	1	1	−1	1	−1	−1	1	8.8 ± 0.2
24	−1	−1	−1	1	1	1	−1	1	−1	−1	45.3 ± 1.9
25	1	−1	−1	−1	1	1	1	−1	1	−1	42.0 ± 2.2
26	−1	1	−1	−1	−1	1	1	1	−1	1	52.1 ± 3.5
27	1	1	1	1	1	1	1	1	1	1	47.5 ± 2.4
28	0	0	0	0	0	0	0	0	0	0	47.0 ± 3.2
29	0	0	0	0	0	0	0	0	0	0	41.6 ± 1.8
30	0	0	0	0	0	0	0	0	0	0	40.5 ± 1.5

Note: I, Sucrose; J, $NaNO_3$; K, K_2HPO_4; L, KH_2PO_4; M, KCl; N, $MgSO_4.7H_2O$; O, $CaCl_2.2H_2O$; P, $Na_3C_6H_5O_7.2H_2O$; Q, Biotin; R, Trace elements.

K_2HPO_4, and AB represents the interaction term between A and B.

The model determination coefficient (R^2) of the above equation was 0.9766, indicating a good agreement between the experimental and predicted values. The sample variation of 97.66% was attributed to the variable, and only less than 2.34% of the total variance could not be explained by the model. The adjusted R^2 of 0.9548 was in reasonable agreement with the predicted R^2 of 0.8860. The model of adequate precision (signal-to-noise ratio) of 21.513, which is very high compared to the desirable value (greater than 4), indicates an adequate signal and that the model can be used to navigate the design space. Finally, the lower value of the coefficient of variation (C.V.) was 3.84, showing that the experiments were precise and reliable (Box et al. 1978).

To measure how well the factors describe the variation of the mean data, the F-test was used in this study. A greater F-value indicates that the factors adequately explain the variation in the data. A p-value < 0.05 indicates the significant model term. It was observed that the quadratic regression model (Eq. [3]) was highly significant because of a very low probability value ($p_{model} > F = 0.0001$). Additionally, the lack of fit (LOF), which is a variation of

Table 4 Statistical analysis of the Plackett-Burman design for the complex and defined media

Variable	Effect	Coefficient	p-value
Complex medium			
constant		65.04	< 0.0001
A, Sucrose	22.1004	11.05	0.0201
B, $NaNO_3$	75.1828	37.59	< 0.0001
C, Yeast extract	30.4603	15.23	0.0025
D, KH_2PO_4	−4.7852	−2.39	0.5877
E, K_2HPO_4	27.8186	13.91	0.0049
F, KCl	8.7311	4.37	0.3272
G, $MgSO_4.7H_2O$	1.7222	0.86	0.8448
H, Trace elements	−15.9140	−7.96	0.0830
Defined medium			
constant		42.14	< 0.0001
I, Sucrose	1.0233	0.51	0.7089
J, $NaNO_3$	27.1197	13.56	< 0.0001
K, K_2HPO_4	−8.5673	−4.28	0.0052
L, KH_2PO_4	−10.8947	−5.45	0.0008
M, KCl	−2.3609	−1.18	0.3930
N, $MgSO_4.7H_2O$	5.1297	2.56	0.0733
O, $CaCl_2.2H_2O$	−5.7316	−2.87	0.0542
P, $Na_3C_6H_5O_7.2H_2O$	2.7889	1.39	0.3149
Q, Biotin	1.1972	0.6	0.6625
R, Trace elements	−22.8260	−11.41	< 0.0001

the data around the fitted model, was also used to determine the adequacy of the model fit. If the model does not fit the data well, this term is significant. As shown in Table 7, LOF was not significant at the F-value and p-value of 1.93 and 0.2428, respectively. This result indicated that the model fitted the response well and that there is a 24.28% chance that a lack of fit F-value could occur due to noise.

Table 5 Coded and uncoded values of independent variables used in the central composite design for the complex and defined media

Independent variables	Coded value				
	-α(−2)	−1	0	+1	+α(+2)
Complex medium					
A, Sucrose (g/L)	10	40	70	100	130
B, $NaNO_3$ (g/L)	1	3	5	7	9
C, Yeast extract (g/L)	1	3	5	7	9
D, K_2HPO_4 (g/L)	1	3	5	7	9
Defined medium					
A, $NaNO_3$ (g/L)	3	5	7	9	11
B, K_2HPO_4 (g/L)	1	2	3	4	5
C, KH_2PO_4 (g/L)	0	0.5	1	1.5	2
D, Trace elements (mL/L)	0	0.15	0.30	0.45	0.60

Table 6 Central composite design for the complex medium

Run	Coded factors				AMVL (mg/L)	
	A	B	C	D	Actual	Predicted
1	−1	−1	−1	−1	137.9 ± 5.3	129.1
2	1	−1	−1	−1	134.5 ± 6.2	132.3
3	−1	1	−1	−1	157.9 ± 7.1	152.1
4	1	1	−1	−1	183.0 ± 7.7	187.1
5	−1	−1	1	−1	176.8 ± 8.4	175.9
6	1	−1	1	−1	184.9 ± 8.0	179.1
7	−1	1	1	−1	154.0 ± 6.5	151.2
8	1	1	1	−1	189.8 ± 8.3	186.3
9	−1	−1	−1	1	195.0 ± 8.6	187.4
10	1	−1	−1	1	165.8 ± 7.5	167.5
11	−1	1	−1	1	211.6 ± 9.7	210.3
12	1	1	−1	1	224.1 ± 9.3	222.3
13	−1	−1	1	1	214.6 ± 9.1	209.5
14	1	−1	1	1	194.9 ± 6.8	189.6
15	−1	1	1	1	185.4 ± 5.7	184.8
16	1	1	1	1	195.2 ± 7.9	196.8
17	−2	0	0	0	161.0 ± 7.6	172.1
18	2	0	0	0	187.2 ± 6.4	187.3
19	0	−2	0	0	137.2 ± 4.1	148.7
20	0	2	0	0	179.2 ± 6.8	178.8
21	0	0	−2	0	167.3 ± 6.7	172.6
22	0	0	2	0	188.1 ± 7.3	193.9
23	0	0	0	−2	146.8 ± 5.2	154.2
24	0	0	0	2	219.2 ± 8.9	222.9
25	0	0	0	0	238.9 ± 9.5	242.2
26	0	0	0	0	248.8 ± 8.0	242.2
27	0	0	0	0	241.5 ± 8.2	242.2
28	0	0	0	0	243.3 ± 9.4	242.2
29	0	0	0	0	233.1 ± 6.7	242.2
30	0	0	0	0	247.5 ± 7.2	242.2

Note: A, Sucrose; B, $NaNO_3$; C, Yeast extract; D, K_2HPO_4.

According to Table 7, the variables that were highly significant ($P < 0.0001$) were the square terms of sucrose (A^2), $NaNO_3$ (B^2), yeast extract (C^2), and K_2HPO_4 (D^2). Moreover, the linear effect of all variables was significant. The significance of the linear and quadratic terms indicated that sucrose, $NaNO_3$, yeast extract and K_2HPO_4 can act as limiting nutrients, and little variation in their concentrations would alter AMVL production to a considerable extent. Based on the interaction terms, the results clearly showed that the interaction terms between sucrose and yeast extract (AC) and $NaNO_3$ and K_2HPO_4 (BD)

Table 7 ANOVA for response surface quadratic model of the complex and defined media

Source	Sum of Squares	df	Mean Square	F Value	p-value Prob > F	
Complex medium						
Model	33907.83	14	2421.99	44.71	< 0.0001	Significant
A-Sucrose	345.73	1	345.73	6.38	0.0233	
B-$NaNO_3$	1359.75	1	1359.75	25.1	0.0002	
C-Yeast extract	675.49	1	675.49	12.47	0.003	
D-K_2HPO_4	7094.29	1	7094.29	130.97	< 0.0001	
AB	1015	1	1015	18.74	0.0006	
AC	52.56	1	52.56	0.97	0.3402	
AD	530.92	1	530.92	9.8	0.0069	
BC	2264.16	1	2264.16	41.8	< 0.0001	
BD	1.27	1	1.27	0.023	0.8806	
CD	606.84	1	606.84	11.2	0.0044	
A^2	6695.69	1	6695.69	123.61	< 0.0001	
B^2	10541.5	1	10541.5	194.61	< 0.0001	
C^2	5955.23	1	5955.23	109.94	< 0.0001	
D^2	4934.49	1	4934.49	91.1	< 0.0001	
Residual	812.53	15	54.17			
Lack of Fit	645.21	10	64.52	1.93	0.2428	Not significant
Pure Error	167.32	5	33.46			
Cor Total	34720.36	29				
Defined medium						
Model	61950.92	14	4425.07	23.27	< 0.0001	Significant
A-$NaNO_3$	1466.37	1	1466.37	7.71	0.0141	
B-K_2HPO_4	7429.48	1	7429.48	39.07	< 0.0001	
C-KH_2PO_4	1619.47	1	1619.47	8.52	0.0106	
D-Trace elements	3979	1	3979	20.92	0.0004	
AB	240.85	1	240.85	1.27	0.2781	
AC	3034.41	1	3034.41	15.96	0.0012	
AD	12478.18	1	12478.18	65.61	< 0.0001	
BC	1745.83	1	1745.83	9.18	0.0084	
BD	1394.94	1	1394.94	7.33	0.0162	
CD	1498.72	1	1498.72	7.88	0.0133	
A^2	4217.35	1	4217.35	22.18	0.0003	
B^2	11914.43	1	11914.43	62.65	< 0.0001	
C^2	42.42	1	42.42	0.22	0.6435	
D^2	15571	1	15571	81.88	< 0.0001	
Residual	2852.69	15	190.18			
Lack of Fit	2166.36	10	216.64	1.58	0.3207	Not significant
Pure Error	686.33	5	137.27			
Cor Total	64803.62	29				

Note: For the complex medium: $R^2 = 0.9766$, R^2 (adj) = 0.9548, R^2 (pred) = 0.8860, CV = 3.84%, Adeq Precision = 21.513.
For the defined medium: $R^2 = 0.956$, R^2 (adj) = 0.9149, R^2 (pred) = 0.7922, CV = 12.93%, Adeq Precision = 17.282.

were not significant because their p-values were higher than 0.05. Although the full model was significantly fitted to the data, it was necessary to remove those non-significant interaction terms from the full model. The polynomial model for AMVL production considered only the significant term that was expressed as a coded value by Eq. (4):

$$Y = 242.18 + 3.80A + 7.53B + 5.31C + 17.19D + 7.96AB - 5.76AD - 11.90BC - 6.16CD - 15.62A^2 - 19.60B^2 - 14.73C^2 - 13.41D^2 \quad (4)$$

where Y is the predicted AMVL production (mg/L), A is sucrose, B is $NaNO_3$, C is yeast extract, and D is K_2HPO_4.

According to Eq. 4, the effects of the interactions of sucrose, $NaNO_3$, yeast extract and K_2HPO_4 on AMVL production were studied by observing the interactions of two of these variables while maintaining the other two independent variables at constant levels. The 3D response surface graphs can be used to determine the optimum values of the variables within the range considered. The 3D plots for interactions between two variables were presented in Figure 3. The result showed that an elliptical response surface was derived from the second order quadratic equation for AMVL production with the interactions between sucrose and $NaNO_3$, sucrose and K_2HPO_4, $NaNO_3$ and yeast extract, and yeast extract and K_2HPO_4. The predicted AMVL production decreased at the higher and lower values of these independent parameters. The maximum

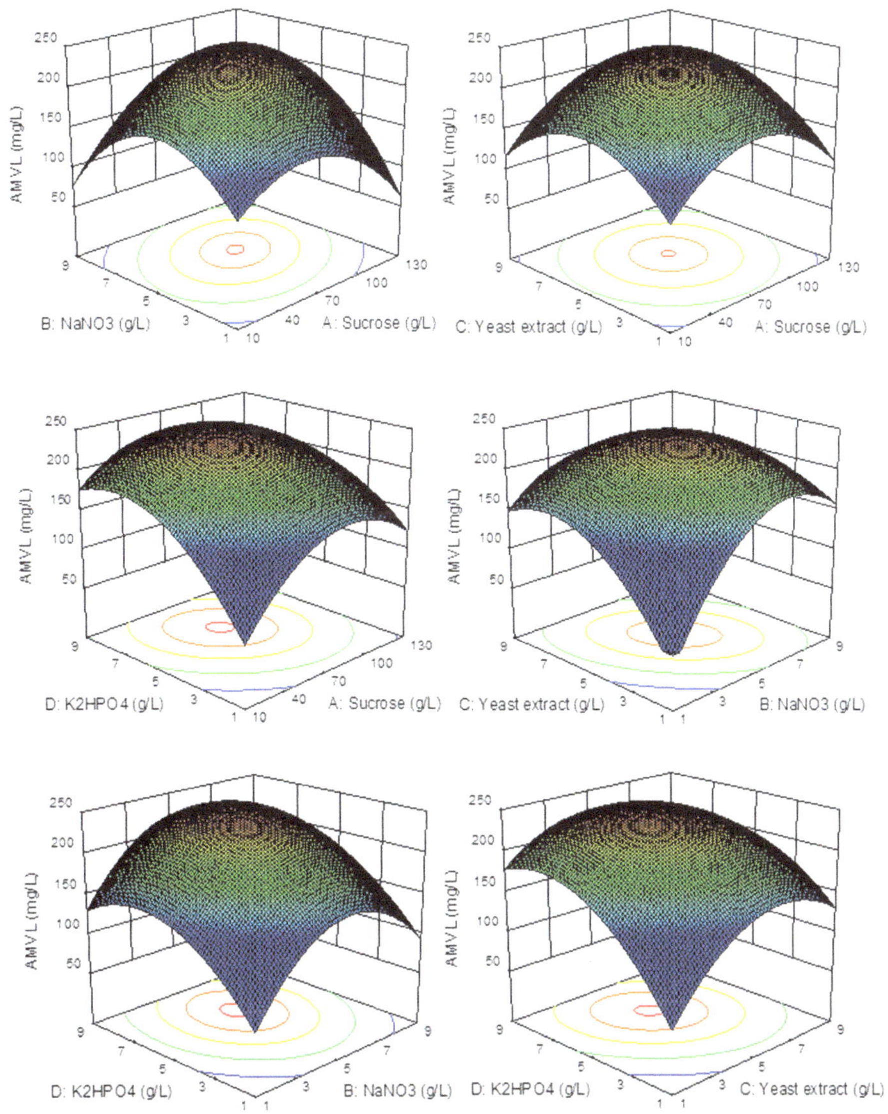

Figure 3 Response surface plots of AMVL production in the complex medium as a function of sucrose (A), $NaNO_3$ (B), yeast extract (C) and K_2HPO_4 (D) based on the results of the central composite design.

predicted value, identified by the surface confined in the smallest ellipse in the contour plot, was located near the center points of the response surface. The study of response surface graphs revealed that the maximum AMVL production of 248.478 mg/L was predicted with the media containing 71.651 g/L sucrose (A = 0.055), 5.418 g/L $NaNO_3$ (B = 0.209), 4.931 g/L yeast extract (C = -0.0345) and 6.271 g/L K_2HPO_4 (D = 0.6355).

Optimization of the defined medium for AMVL production

The defined medium, known as chemically defined medium or synthetic medium, is the medium in which all the chemicals used are known and does not contain any yeast, animal or plant tissue. The media constituents, namely, $NaNO_3$, K_2HPO_4, KH_2PO_4 and trace elements as independent variables screened by PBD, were further studied by the CCD. The CCD consisted of a 2^4 full factorial design, 2 × 4 axial designs and 6 center points. The design matrix of CCD, the independent variables and the corresponding responses are presented in Table 8.

To determine the optimal value, the ANOVA for the full second-order polynomial quadratic model was carried out as shown in Table 7. By applying multiple regression analysis to the experimental data, the second-order polynomial equation was developed as a coded value as Eq. (5):

$$\begin{aligned} Y = {} & 151.33 + 7.82A - 17.59B - 8.21C \\ & + 12.88D - 3.88AB - 13.77AC + 27.93AD \\ & + 10.45BC + 9.34BD \\ & + 9.68CD - 12.40A^2 - 20.84B^2 \\ & + 1.24C^2 - 23.83D^2 \end{aligned} \qquad (5)$$

where Y is the predicted AMVL production (mg/L) and A, B, C and D represent the coded values of $NaNO_3$, K_2HPO_4, KH_2PO_4 and trace elements, respectively. Eq. (5) is a full model because it includes all terms regardless of their significance. The fitness of the full model was estimated by the determination coefficient (R^2), which was 0.956, indicating a good correlation between the experimental and predicted value of AMVL production, and above 95% of the variability in the response could be explained by the mode. Additionally, the predicted determination coefficient (predicted $R^2 = 0.7922$) was in reasonable agreement with the adjusted determination coefficient (adjusted $R^2 = 0.9149$). The model F-values of 23.27 with p-values less than 0.0001 also implied that the model was significant. Moreover, the lack of fit for an F-value of 1.58 implied that the LOF is not significant relative to the pure error, indicating that the model fitted the data well. Although the full model showed a significant fit, there were non-significant terms ($p \geq 0.05$) in the model, which were the square terms of KH_2PO_4 and the interaction terms between $NaNO_3$ and K_2HPO_4. After excluding the non-significant terms, the model was modified as Eq. (6):

$$\begin{aligned} Y = {} & 151.33 + 7.82A - 17.59B - 8.21C \\ & + 12.88D - 13.77AC + 27.93AD \\ & + 10.45BC + 9.34BD \\ & + 9.68CD - 12.40A^2 - 20.84B^2 - 23.83D^2 \end{aligned} \qquad (6)$$

The response surface plots were used to explain the interactions of $NaNO_3$, K_2HPO_4, KH_2PO_4 and trace elements as illustrated in Figure 4. Each figure showed the effect of two variables, while others were held at level

Table 8 Central composite design for the defined medium

Run	Coded factors				AMVL (mg/L)	
	A	B	C	D	Actual	Predicted
1	−1	−1	−1	−1	157.6 ± 7.6	143.0
2	1	−1	−1	−1	141.9 ± 6.3	130.3
3	−1	1	−1	−1	71.0 ± 3.7	68.2
4	1	1	−1	−1	43.0 ± 2.4	55.6
5	−1	−1	1	−1	117.3 ± 5.9	113.9
6	1	−1	1	−1	49.1 ± 2.2	46.1
7	−1	1	1	−1	81.2 ± 3.5	80.9
8	1	1	1	−1	21.8 ± 2.6	13.1
9	−1	−1	−1	1	68.2 ± 3.2	74.9
10	1	−1	−1	1	177.4 ± 9.1	173.9
11	−1	1	−1	1	38.2 ± 3.3	37.5
12	1	1	−1	1	135.1 ± 7.4	136.5
13	−1	−1	1	1	85.3 ± 4.0	84.4
14	1	−1	1	1	143.3 ± 7.5	128.4
15	−1	1	1	1	94.8 ± 4.3	88.8
16	1	1	1	1	106.4 ± 5.7	132.8
17	−2	0	0	0	80.5 ± 3.6	86.1
18	2	0	0	0	122.1 ± 5.2	117.4
19	0	−2	0	0	86.0 ± 4.4	103.2
20	0	2	0	0	49.1 ± 3.5	32.8
21	0	0	−2	0	171.9 ± 8.6	167.8
22	0	0	2	0	139.9 ± 7.4	134.9
23	0	0	0	−2	19.8 ± 2.6	30.3
24	0	0	0	2	91.4 ± 5.8	81.8
25	0	0	0	0	141.9 ± 7.4	151.3
26	0	0	0	0	135.8 ± 6.1	151.3
27	0	0	0	0	154.2 ± 7.3	151.3
28	0	0	0	0	168.5 ± 8.7	151.3
29	0	0	0	0	149.4 ± 7.0	151.3
30	0	0	0	0	158.3 ± 8.2	151.3

Note: A, $NaNO_3$; B, K_2HPO_4; C, KH_2PO_4; D, Trace elements.

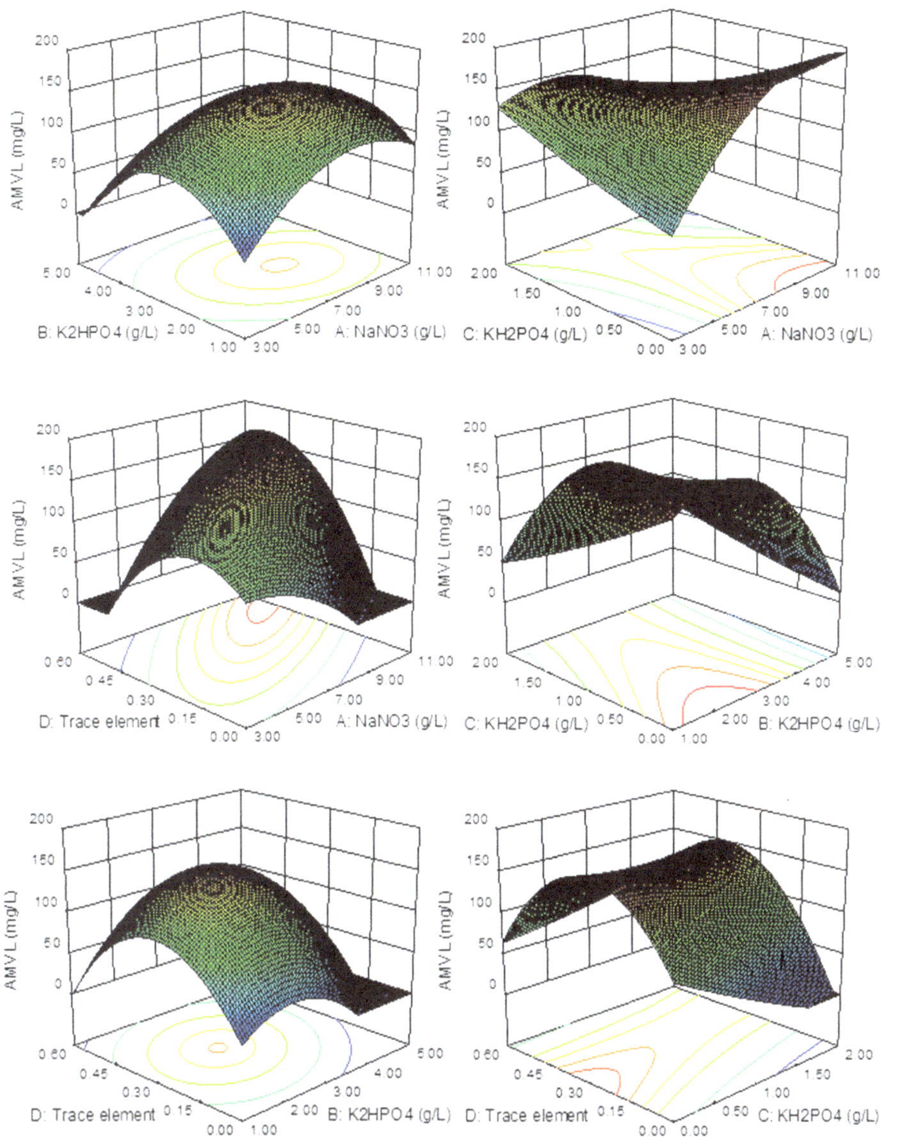

Figure 4 Response surface plots of AMVL production in the defined medium as a function of $NaNO_3$ (A), K_2HPO_4 (B), KH_2PO_4 (C) and trace elements (D) based on the results of the central composite design.

zero. According to Figure 4, AMVL production was affected by varying the concentrations of $NaNO_3$, K_2HPO_4, KH_2PO_4 and trace elements. The predicted AMVL production decreased at the higher and lower value ranges for both K_2HPO_4 and trace elements, but it increased with the increasing concentration of $NaNO_3$. In contrast, the AMVL production increased with decreasing KH_2PO_4 concentrations because KH_2PO_4 showed a negative influence on AMVL production. With Eq. (6), it can be shown that the optimum point is located at A = 1.562, B = −0.568, C = −1.706 and D = 0.647 with the corresponding actual concentrations of 9.687 g/L $NaNO_3$, 2.432 g/L K_2HPO_4, 0.147 g/L KH_2PO_4, and 0.397 mL/flask trace elements with a maximum predicted AMVL production of 209.85 mg/L.

Validation of the experimental design

According to the results obtained from the PBD and CCD, the optimized complex medium for AMVL production was prepared as follows: 71.65 g/L sucrose, 5.42 g/L $NaNO_3$, 4.93 g/L yeast extract, 6.27 g/L K_2HPO_4, 0.5 g/L KCl, 0.5 g/L $MgSO_4.7H_2O$, and 0.25 mL/flask trace elements. In contrast, the optimal concentration of the defined medium components were 70 g/L sucrose, 9.69 g/L $NaNO_3$, 2.43 K_2HPO_4, 0.15 g/L KH_2PO_4, 0.4 mL/flask trace elements, 2 g/L $Na_3C_6H_5O_7.2H_2O$, 0.25 g/L KCl, and 0.75 g/L $MgSO_4.7H_2O$. The initial pH of both media was adjusted to 6.5. To validate the prediction of the model, *A. oryzae* MTG4 was characterized using two optimized media and a non-optimized medium (MEB) in triplicate. The cultivations were carried out for 23 days to investigate the influence of the media on the growth of *A. oryzae* MTG4 and AMVL production. Samples were taken for the analysis of biomass dry weight, sucrose concentration and the amount of AMVL produced. The results from cultivations using three different media are illustrated in Figure 5 and showed that exponential growth was observed in all cultivations. During this phase, the cells consumed carbon sources at their maximum rate until the carbon was exhausted. A maximum specific growth rate (μ_{max}) of 0.039 h^{-1} was observed in MEB medium, whereas the μ_{max} of *A. oryzae* MTG4 in the optimized complex and defined media were 0.080 and 0.052 hr^{-1}, respectively (Table 9). At the end of the cultivation, the cultures reached a phase where AMVL was constantly produced. The maximum titer of AMVL was approximately 60 mg/L in the MEB medium, whereas AMVL was approximately 200 and 250 mg/L in the optimized defined and complex media, respectively. The yields of AMVL per biomass in the optimized defined

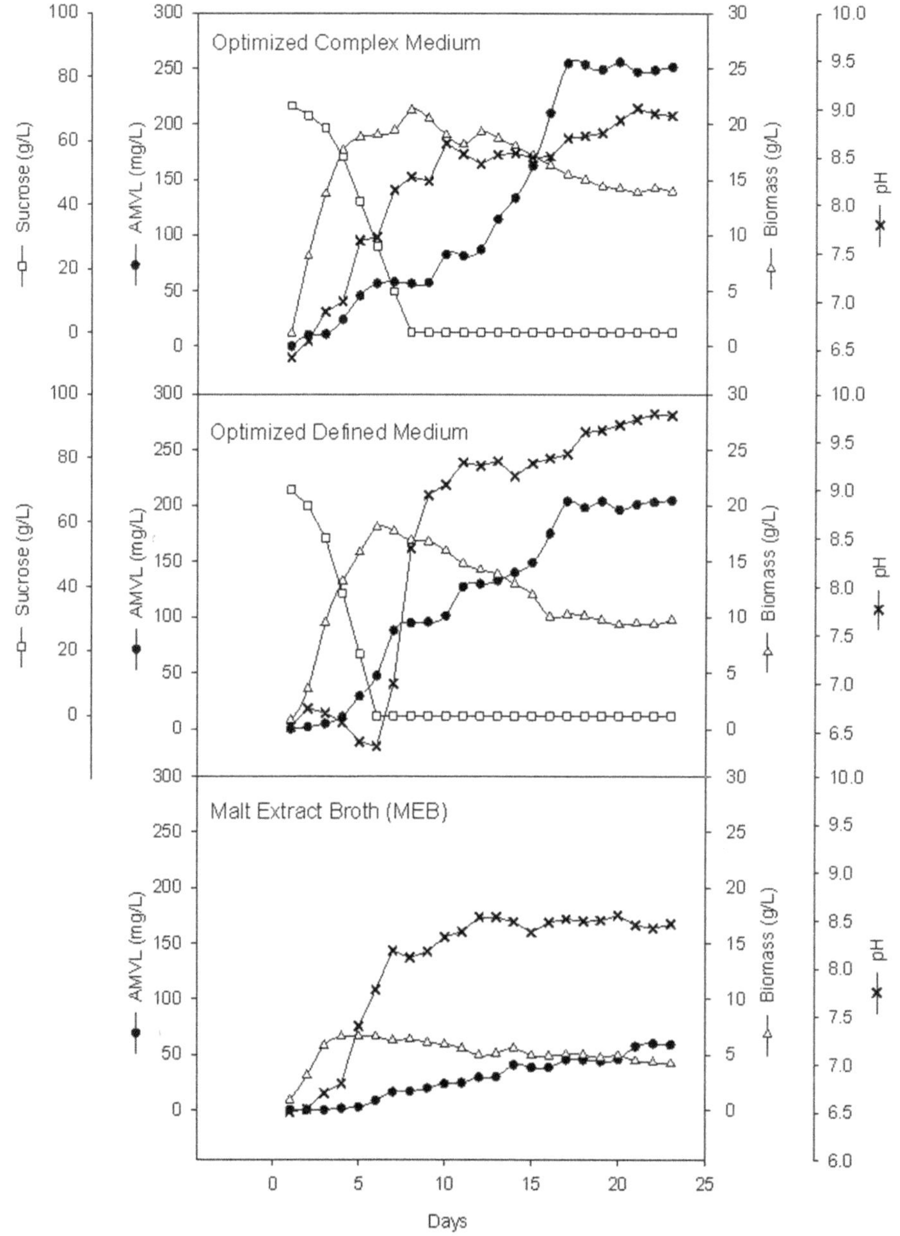

Figure 5 AMVL production, changes in pH, sucrose concentration and growth curves of *A. oryzae* MTG4 in optimized media and MEB.

Table 9 Maximum specific growth rate, AMVL production and yields during the cultivations of *A. oryzae* MTG4 using three different media

	Medium		
	Optimized complex medium	Optimized defined medium	MEB
μ_{max} (h^{-1})	0.080 ± 0.002	0.052 ± 0.003	0.039 ± 0.001
Maximum titer of AMVL (mg/L)	250.9 ± 7.1	201.3 ± 13.9	58.8 ± 5.4
Y_{xp} (mg AMVL/g dw)	17.4 ± 0.5	20.8 ± 1.4	13.6 ± 1.3
r_p (mg AMVL/g dw/h)	1.02 ± 0.03	1.22 ± 0.08	0.62 ± 0.06

and complex media were also higher compared to the yields of AMVL in the MEB medium (Table 9). Thus, the yield and production of AMVL clearly improved when the optimal media were used for the cultivation of *A. oryzae* MTG4. However, under the calculated optimal culture conditions, the maximum AMVL concentrations produced from the defined and complex media were 209.9 and 248.5 mg/L, respectively, which agreed well with the experimental results and suggested that the models (Eq. [4] and Eq. [6]) were valid for predicting the AMVL production in the complex and defined media.

Discussion

During the past decade, the technology of microbial secondary metabolite or bioactive compound production has undergone significant advances in the biotechnological and pharmaceutical industries. Recombinant products in particular have drawn much interest due to their great bioactivity and increasing applications. This includes the substances produced by non-native producers or heterologous products. Many efforts have been made regarding the efficient production of these compounds in well-characterized and fast-growing hosts using medium optimization and strain re-engineering to achieve the maximum yields. In this study, a recombinant strain, *A. oryzae* MTG4, was applied. This strain carries the gene encoding polyketide synthase from *Xylaria* sp. BCC 1067 and is capable of producing the product identified as a 2-pyrone containing compound, known as AMVL. The present work is the first report to assess medium components for the efficient heterologous production of AMVL.

Carbon and nitrogen sources are crucial factors affecting AMVL production. According to the results of the screening of carbon and nitrogen sources, the strain was able to utilize all carbon sources, including glucose, sucrose and soluble starch, and all nitrogen sources, including NH_4^+, NO_3^-, and yeast and malt extracts, suggesting that the recombinant *A. oryzae* MTG4 can utilize a broad spectrum of nutrients.

Several reports have illustrated the use of sucrose instead of glucose for the growth of microorganisms and secondary metabolite production. Glucose is found to be the strongest catabolite repressor, and many examples of secondary metabolites have been reported to be repressed by the presence of glucose in the cultivation media (Ruijter and Visser 1997; Jonsbu et al. 2002). Soluble starch had also been used as a carbon source in various types of microorganisms. It had been reported that enzyme synthesis in *Aspergillus terreus* and *Trichoderma viride* was induced in the presence of soluble starch (Ghosh et al. 1990; Schellart et al. 1976). Additionally, the majority of *Streptomyces* prefer starch as a carbon source for the production of secondary metabolites (Jia et al. 2008; Gao et al. 2009). Although, there are no observations of soluble starch acting as a repressor, soluble starch was not a suitable carbon source for AMVL production by *A. oryzae* MTG4 due to the low amount of amylase produced by this strain; thus, this strain cannot efficiently utilize starch as a substrate (Punya et al., 2013; Kalayanamitr et al. 1987).

Sucrose is the predominant sugar produced from sugarcane, which is one of the largest crops in Thailand. Because it is relatively cheap and abundant, and a higher product yield was observed in the media containing sucrose, sucrose is considered to be the optimal carbon source for the cultivation of *A. oryzae* MTG4. Several studies have shown that sucrose was the main nutritional factor for the production of secondary metabolites in fungi. The amount of sucrose used in the media varied depending on the strain. Only 3 g/L of sucrose was used as the main carbon source in Tannic acid-nitrate broth for the cultivation of *Fusarium*. To support growth and metabolite production by *Aspergillus* and *Penicillium*, 40 g/L of sucrose was supplemented in Czapek yeast extract broth and 150 g/L in yeast extract sucrose broth (YES). However, the concentration of sucrose could be as high as 200 g/L or 400 g/L in CZ20S (Czapek sucrose broth) and M40Y (Malt yeast broth), respectively, for the cultivation of xerophilic fungi such as *Eurotium* and *Wallemia* (Samson and Pitt, 1985; Samson et al. 2004).

Yeast extract and $NaNO_3$ were considered as efficient nitrogen sources in this study. Several studies reported the use of yeast extract and $NaNO_3$ for metabolite production in fungi (Barratt et al. 1965; Frisvad 1981; Klich and Pitt 1988; Filtenborg et al. 1990; Srinubabu et al. 2007). Previous work suggested that NO_3^- was superior to NH_4^+ as a nitrogen source for the biomass and the production of extracellular enzymes in *A. oryzae* (Kundu et al. 1973). These results contradict the findings of Pedersen and Nielsen (2000), who reported that the a mixture of NH_4^+ and complex nitrogen sources, such as yeast extract, gave better biomass yield and enzyme productivity for *A. oryzae* than NH_4^+ or NO_3^- as the sole nitrogen source. However, these experiments were carried out in chemostat cultures with pH control. Carlsen et al. (1996) showed that when *A. oryzae* was cultivated using NH_4^+ as the sole nitrogen source with no pH control, ammonium was taken up with a proton, resulting in a substantial acidification of the medium. The acidification had a negative effect on growth and led to the inactivation of enzyme production in *A. oryzae*, but this did not happen when NO_3^- was used as a nitrogen source. This result indicates that the carbon and nitrogen sources used in the cultivation media completely depended on the strain and the cultivation process.

After sucrose, yeast extract and $NaNO_3$ were identified as the important carbon, organic nitrogen and inorganic nitrogen sources, respectively, other components such as KH_2PO_4, K_2HPO_4, KCl, $MgSO_4.7H_2O$, $CaCl_2.2H_2O$, $Na_3C_6H_5O_7.2H_2O$, biotin and trace elements were investigated by the PBD to determine whether these factors have significant effects on AMVL production. Because the use of yeast extract in the complex medium was not economical, the production of AMVL in the optimal defined medium without yeast extract was established for a comparison. The advantage of using the defined medium is that this medium produces more consistent titers, which allows easier process control and simplifies downstream recovery of AMVL.

KH_2PO_4 and K_2HPO_4 were used in the media because they act as buffering agents and provide the phosphate necessary for the production of sugar phosphates, nucleic acids, ATP and membrane phospholipids (Deacon 2006). Sodium citrate or $Na_3C_6H_5O_7.2H_2O$ can also act as buffering agents or acidity regulators, resisting changes in pH. KCl was added to maintain the osmotic equilibrium. The inorganic salt, $MgSO_4.7H_2O$, stimulates fungal growth and enhances sporulation. $CaCl_2.2H_2O$ provides Ca^{2+}, which induces a loose pelleted form of growth and plays an important role in signal transduction in fungi (Pera and Callieri 1999; Miller et al. 1990). Biotin is a cofactor for acetyl-CoA carboxylase, which catalyzes the carboxylation of acetyl-CoA to form malonyl-CoA and is required for the de novo biosynthesis of long-chain fatty acids and polyketide biosynthesis (Hasslacher et al. 1993; Wattanachaisaereekul et al. 2008). Trace elements supply the Zn^{2+}, Cu^{2+}, Mn^{2+}, and Fe^{2+} required for the functional components of enzymes produced by fungi. Fe^{2+} is particularly important in cytochrome P450 oxidases, which are involved in the biosynthesis of secondary metabolites. Zn^{2+} plays a crucial role in stabilizing protein structures (Hanson 2008).

For the complex medium, sucrose, $NaNO_3$, yeast extract and K_2HPO_4 were selected by the PBD as the major factors that influence AMVL production, while $NaNO_3$, K_2HPO_4, KH_2PO_4 and trace elements were the important components affecting AMVL production in the defined medium. The results from the CCD shown in Figures 3 and 4 revealed that the increase of carbon source increased AMVL production until the optimal point was reached. A further increase of the sucrose concentration reversed the trend, most likely because the very high sucrose concentration in the media caused hyperosmotic stress, which may be detrimental to growth, cell survival and secondary metabolism in various fungi (Hafnawy 2001; Duran et al. 2010).

The amount of the nitrogen source in the media has been shown to be a critical factor in metabolite production. In this study, it was found that increased yeast extract or $NaNO_3$ resulted in increased AMVL production. However, AMVL production was decreased at high concentrations of nitrogen sources. Previous work reported that excess nitrogen sources had been implicated in the repression of secondary metabolism such as penicillin biosynthesis in *Penicillium* sp. (Sanchez et al. 1981). For secondary metabolite production, the concentrations of both carbon and nitrogen sources and their balance, or the C:N ratio, are very important. According to the results, higher AMVL production was observed in the N-limited medium with the optimal molar C:N ratio in the defined medium of 22:1 (70 g/L sucrose and 9.69 g/L $NaNO_3$). The result from this study is similar to the findings of Mao et al. (2005), who reported that the highest cordycepin production in *Cordyceps militaris* was attained in the culture with a high C:N ratio. Additionally, López et al. (2003) illustrated that the maximum productivity of lovastatin by *A. terreus* was obtained in the nitrogen-limited media or the media with a high C:N ratio. This could be because secondary metabolite production is generally associated with nitrogen-limited growth when excess carbon can be channeled into secondary metabolism (Demain 1986).

It was not surprising that KH_2PO_4 and K_2HPO_4 were important nutrients in the defined media in this study because phosphate salts provide a buffering capacity against pH fluctuations that could adversely affect normal metabolic activity. Phosphate is also known to be important for attaining high biomass densities of various fungi. It is also the crucial growth-limiting nutrient in many secondary metabolite fermentations. Thus, supplementing the media with phosphate enhanced AMVL production in *A. oryzae* MTG4. However, AMVL production decreased when a high concentration of phosphate was applied. This result is consistent with the report of Demain (1986), who showed that phosphate in the range of 0.3-300 mM generally supports extensive cell growth, whereas concentrations of 10 mM and above suppress the biosynthesis of many secondary metabolites.

The optimized complex and defined media were subsequently validated and compared with the original malt extract broth or MEB. Three different media cultivations in Figure 5 illustrated that the strain grew at different growth rates. The high maximum specific growth rates (μ_{max}) of *A. oryzae* MTG4 in the optimized media were accomplished. After the exponential phase, the biomass had a decreasing trend in all cultivations, but AMVL production continued to increase until 16–17 days. Because AMVL is synthesized using the *pksmt* gene, it is classified as a secondary metabolite. Although a constitutive *gpdA* promoter was used to control the gene, there was a delay in the production of AMVL because it requires the enzymatic reaction to convert the intermediates or end products of primary metabolism to AMVL. In batch cultivation, cultures that exhibit a distinct growth phase are often followed by the secondary metabolite production phase. However, the timing of the primary and secondary phase may overlap depending on the strain and the nutrients in the culture.

During the cultivation of *A. oryzae* MTG4, the pH of the optimized complex and the MEB media increased with fungal growth. In contrast, the pH of the defined medium during the exponential phase did not increase but was almost constant at 6.3-6.7. This result might be due to the buffering effect of KH_2PO_4 and K_2HPO_4 present in the defined media. After that, the pH of all cultures dramatically increased, especially in the defined medium. The pH was maintained at over 8.0 for MEB and the complex medium and at 9.0 for the defined medium after 10 days of cultivations. It has been reported that the pH requirements for the growth of fungi is broad, ranging from 2.5 to 9.0. Generally, *Penicillium* species are tolerant of acidic pH, while *Aspergillus* species appear to be more tolerant of alkaline pH (Wheeler et al. 1991). However, previous reports have shown that the growth of fungi could be affected by the pH of the culture medium either directly by the action of pH on cell surfaces or indirectly by the effect of substrate availability (Ahmed and Naresh 2009). In our case, the decrease in biomass, along with the increase in pH, was caused by the limiting of media components because sucrose was completely exhausted at approximately 6–7 days of cultivation.

Additionally, the pH level may affect secondary metabolite production by the regulation of the gene encoding PacC (Espeso et al. 1993; Denison 2000). Under alkaline conditions, PacC is a transcriptional activator of alkaline-expressed genes and a repressor of acid-expressed genes. Several genes in secondary metabolism have been identified as alkaline-expressed genes, such as the genes required for penicillin production in *Penicillium chrysogenum* and *A. nidulans*. Thus, when growth was at an alkaline pH, higher levels of penicillins were observed (Shah et al. 1991; Espeso et al. 1993). Other examples of alkaline-expressed genes in secondary metabolism include genes in sterigmatocystin biosynthesis in *A. nidulans* (Keller et al. 1997). According to the results, a high AMVL production was observed with the optimized media, even though the pH of the media was high. Therefore, it is possible that genes involved in AMVL biosynthesis may be controlled by a pH regulatory system.

Although the AMVL yields and productivities from the optimized media were improved compared to the original medium, this experiment was a preliminary study that requires further studies for the improvement of production. Additional process parameters such as aeration and pH should be investigated, and physiological fermentation in a controlled bioreactor should be studied in detail. The present study certainly promotes the development of biotechnological applications of AMVL and provides a basis for future studies with large-scale fermentation for the production of AMVL and other novel fungal secondary metabolites using *A. oryzae* as a heterologous host.

Competing interests
The authors declare that they have no conflict of interests.

Acknowledgements
This research was supported by National Research Council of Thailand and King Mongkut's University of Technology Thonburi, Thailand.

Author details
[1]Pilot Plant Development and Training Institute, King Mongkut's University of Technology Thonburi, 49 Soi Thianthale 25, Bangkhunthian-Chaithale Rd., Thakham, Bangkhunthian, Bangkok 10150, Thailand. [2]Bioresource Research Unit, National Center for Genetic Engineering and Biotechnology, 113 Thailand Science Park, Phahonyothin Rd., Khlong Nueng, Klong Luang, Pathum Thani 12120, Thailand. [3]School of Bioresources and Technology, King Mongkut's University of Technology Thonburi, 49 Soi Thianthale 25, Bangkhunthian-Chaithale Rd., Thakham, Bangkhunthian, Bangkok 10150, Thailand.

References
Ahmed A, Naresh M (2009) Influence of physiological factors on growth, sporulation and ochratoxin A/B production of new *Aspergillus ochraceus* grouping. World Mycotoxin J 2(4):429–434

Barratt RW, Johnson GB, Ogata WN (1965) Wild-type and mutant stocks of *Aspergillus nidulans*. Genetics 52(1):233–246

Bezerra MA, Santelli RE, Oliveira EP, Villar LS, Escaleira LA (2008) Response surface methodology (RSM) as a tool for optimization in analytical chemistry. Talanta 76:965–977

Box GEP, Hunte WG, Hunte JS (1978) Statistics for experiments: an introduction to design. Data Analysis and Model Building. John Wiley, New York, pp 291–334

Bumrungsri S, Beaver R, Phongpaichit S, Sittichaya W (2008) The infestation by an exotic ambrosia beetle, *Euplatypus parallelus* (F.) (Coleoptera: Curculionidae: Platypodinae) of Angsana trees (*Pterocarpus indicus* Willd.) in southern Thailand. Songklanakarin J Sci Technol 30(5):579–582

Busch B, Hertweck C (2009) Evolution of metabolic diversity in polyketide-derived pyrones: Using the non-colinear aureothin assembly line as a model system. Phytochemistry 70(15–16):1833–1840

Carlsen M, Nielsen J (2001) Influence of carbon source on alpha-amylase production by *Aspergillus oryzae*. Appl Microbiol Biotechnol 57(3):346–349

Carlsen M, Nielsen J, Villadsen J (1996) Kinetic studies of the acid-inactivation of α-amylase from *Aspergillus oryzae*. Chem Eng Sci 51:37–43

Chen PT, Chiang CJ, Chao YP (2010) Medium optimization and production of secreted *Renilla* luciferase in *Bacillus subtilis* by fed-batch fermentation. Biochem Eng J 49:395–400

Chutmanop J, Chuichulcherm S, Christi Y, Srinophakun P (2008) Protease production by *Aspergillus oryzae* in solid-state fermentation using agroindustiral substrates. J Chem Technol Biotechnol 83:1012–1018
Deacon JW (2006) Fungal Biology 4th ed. Blackwell Publishing Ltd., MA, USA
Demain AL (1986) Regulation of secondary metabolism in fungi. Pure Appl Chem 58(2):219–226
Denison SH (2000) pH regulation of gene expression in fungi. Fungal Genet Biol 29:61–71
Duran R, Cary JW, Calvo AM (2010) Role of the osmotic stress regulation pathway in morphogenesis and secondary metabolism in filamentous fungi. Toxins 2:367–381
Espeso EA, Tilburn J, Arst HN Jr, Peñalva MA (1993) pH regulation is a major determinant in expression of a fungal penicillin biosynthetic gene. EMBO J 12(10):3947–3956
Filtenborg O, Frisvad JC, Thrane U (1990) The significant of yeast extract composition on metabolite production in *Penicillium*. In: Samson RA, Pitt JI (eds) *Penicillium* and *Aspergillus* classification. Plenum Press, New York, USA, pp 433–440
Frisvad JC (1981) Physiological criteria and mycotoxin production as aids in identification of common asymmetric *Penicillia*. Appl Environ Microbiol 41:568–579
Gao H, Liu M, Liu J, Dai H, Zhou X, Liu X, Zhuo Y, Zhang W, Zhang L (2009) Medium optimization for the production of avermectin B1a by *Streptomyces avermitilis* 14-12A using response surface methodology. Bioresour Technol 100(17):4012–4016
Ghosh A, Chatterjee B, Das A (1990) Induction and catabolite repression of high-affinity glucoamylase in *Aspergillus terreus* strain 4. J Gen Microbiol 136:1307–1311
Guo Y, Xu J, Zhang Y, Xu H, Yuan Z, Li D (2010) Medium optimization for ethanol production with *Clostridium autoethanogenum* with carbon monoxide as sole carbon source. Bioresour Technol 101:8784–8789
Hafnawy MA (2001) Effect of osmotic stress on *Aspergillus chevalieri* respiratory system. Folia Microbiol 46(5):407–412
Hanson JR (2008) The chemistry of fungi. The Royal Society of Chemistry, Cambridge, UK, 221 p
Hasslacher M, Ivessa AS, Paltauf F, Kohlwein SD (1993) Acetyl-CoA carboxylase from yeast is an essential enzyme and is regulated by factors that control phospholipid metabolism. J Biol Chem 268(15):10946–10952
Jia B, Jin ZH, Mei LH (2008) Medium optimization based on statistical methodologies for pristinamycins production *Streptomyces pristinaespiralis*. Appl Biochem Biotechnol 144:133–143
Jonsbu E, McIntyre M, Nielsen J (2002) The influence of carbon sources and morphology on nystatin production by *Streptomyces noursei*. J Biotechnol 95:133–144
Kalayanamitr A, Bhumiratana A, Flegel TW, Glinsukon T, Shinmyo A (1987) Occurrence of toxicity among protease, amylase, and color mutants of a nontoxic soy sauce koji mold. Appl Environ Microbiol 53:1980–1982
Keller NP, Nesbit C, Sarr B, Phillips TD, Burrow GB (1997) pH regulation of sterigmatocystin and aflatoxin biosynthesis in *Aspergillus* spp. Phytopathology 87:643–648
Khamduang M, Chutmanop J, Packdibamrung K, Srinophakun P (2009) Optimization of medium composition for L-phenylalanine production from glycerol using response surface methodology. Kasetsart J (Nat Sci) 43:727–737
Kitamura J, Shima M, Hiratuka K, Asano S (1976) Studies on the antimetabolites of mevalonic acid I: Microbiological activities of 2,3-anhydro-mevalonic acid. Yakugaku Zasshi 96(6):732–736
Klich MA, Pitt JI (1988) A laboratory guide to the common *Aspergillus* species and their teleomorphs. Commonwealth Scientific and Industrial Research Organization, Division of Food Processing, North Ryde, N.S.W, Australia
Kundu AK, Das D, Gupta TK (1973) Influence of culture and nutritional conditions on the production of amylase by submerged culture of *Aspergillus oryzae*. J Ferment Technol 51:142–150
López JL, Pérez JA, Sevilla JM, Fernández FG, Grima EM, Chisti Y (2003) Production of lovastatin by *Aspergillus terreus*: Effects of the C: N ratio and the principal nutrients on growth and metabolite production. Enzyme Microb Technol 33:270–277
Lundstedt T, Seifert E, Abramo L, Thelin B, Nyström Å, Petterson J, Bergman R (1998) Experimental design and optimization. Chemom Intell Lab Syst 42:3–40
Mao XB, Eksriwong T, Chauvatcharin S, Zhong JJ (2005) Optimization of carbon source and carbon/nitrogen ratio for cordycepin production by submerged cultivation of medicinal mushroom *Cordyceps militaris*. Proc Biochem 40:1667–1672
McKay WR, Ounsworth J, Sum PE, Weiler L (1982) Synthesis of (±)-lineatin by the photochemical cycloaddition of allene to anhydromevalonolactone. Can J Chem 60:872–880
Miller AJ, Vogg G, Sanders D (1990) Cytosolic calcium homeostasis in fungi: Roles of plasma membrane transport and intracellular sequestration of calcium. Proc Natl Acad Sci 87:9348–9352
Myers RH (2002) Response surface methodology: Process and product optimization using design experiments, 2nd edn. John Wiley & Sons Press, NY., USA
Ohnishi K, Yoshida Y, Sekiguchi J (1994) Lipase production of *Aspergillus oryzae*. J Ferment Bioeng 77(5):490–495
Orbay L, McLean JA, Sauder BJ, Cottell PL (1994) Economic losses resulting from ambrosia beetle infestation of sawlogs in coastal British Columbia, Canada. Can J Forest Res 24:1266–1276
Pedersen H, Nielsen J (2000) The influence of nitrogen sources on the α-amylase productivity of *Aspergillus oryzae* in continuous cultures. Appl Microbiol Biotechnol 53:278–281
Pera LM, Callieri DA (1999) Influence of calcium on fungal growth and citric acid production during fermentation of a sugarcane molasses-based medium by a strain of *Aspergillus niger*. World J Microbiol Biotechnol 15:647–649
Phonghanpot S, Punya J, Tachaleat A, Laoteng K, Bhavakul V, Tanticharoen M, Cheevadhanarak S (2012) Biosynthesis of xyrrolin, a new cytotoxic hybrid polyketide/non-ribosomal peptide pyrroline with anticancer potential, in *Xylaria* sp. BCC 1067. ChemBioChem 13(6):895–903
Plackett RL, Burman JP (1946) The design of optimum multifactorial experiments. Biometrika 33(4):305–325
Punya J, Tachaleat A, Wattanachaisaereekul S, Haritakun R, Boonlarppradab C, Cheevadhanarak C (2013) Functional expression of a foreign gene in *Aspergillus oryzae* producing new pyrone compounds. Fungal Genet Biol 50:55–62
Ruijter GJG, Visser J (1997) Carbon repression in *Aspergilli*. FEMS Microbiol Lett 151(2):103–114
Sakaguchi K, Takagi M, Horiuchi H, Gomi K (1992) Fungal enzymes used in oriental food and beverage industries. In: Kinghorn JR, Turner G (eds) Applied molecular genetics of filamentous fungi. Blackie Academic & Professional, Glasgow, UK, pp 54–99
Samson RA, Pitt JI (1985) Advances in *Penicillium* and *Aspergillus* systematics. Plenum Publishers, London & New York, p 483
Samson RA, Hoekstra ES, Frisvad JC (2004) Introduction to food-and airborne fungi, 7th edn. ASM press, Utrecht, The Netherlands, p 389
Sanchez S, Paniagua L, Mateos RC, Lara F, Mora J (1981) Nitrogen regulation of penicillin G biosynthesis in *Penicillium chrysogenum*. In: Vezina C, Singh K (eds) Advances in Biotechnology, Vol. III Fermentation products. Pergamon Press, London, UK, p 147
Schellart JA, Van Arem EJF, Boekel MHJS, Middelhoven WJ (1976) Starch degradation by the mould *Trichoderma viride*. II. Regulation of enzyme synthesis. Antonie Van Leeuwenhoek 42:239–244
Shah AJ, Tilburn JL, Adlard MW, Arst HN Jr (1991) pH regulation of penicillin production in *Aspergillus nidulans*. FEMS Microbiol Lett 77:209–212
Srinubabu G, Lokeswari N, Jayaraju K (2007) Screening of nutritional parameters for the production of protease from *Aspergillus oryzae*. E J Chem 4(2):208–215
U.S. Environmental Protection Agency (1997) *Aspergillus oryzae* final risk assessment (attachment I) biotechnology program under the toxic substances control act (TSCA). http://www.epa.gov/biotech_rule/pubs/fra/fra007.htm
Wattanachaisaereekul S, Lantz AE, Nielsen ML, Nielsen J (2008) Production of the polyketide 6-MSA in yeast engineered for increased malonyl-CoA supply. Metab Eng 10(5):246–254
Wheeler KA, Hurdman BF, Pitt JI (1991) Influence of pH on the growth of some toxigenic species of *Aspergillus*, *Penicillium* and *Fusarium*. Int J Food Microbiol 12:141–149
Wilk W, Waldmann H, Kaiser M (2009) γ-Pyrone natural products - A privileged compound class provided by nature. Bioorg Med Chem 17(6):2304–2309
Yokotsuka T (1961) Aroma and flavor of Japanese soy sauce. Adv Food Res 10:75–134
Yong FM, Wood BJB (1974) Microbiology and biochemistry of soy sauce fermentation. Adv Appl Microbiol 17:157–194

17

Fe(III) mineral reduction followed by partial dissolution and reactive oxygen species generation during 2,4,6-trinitrotoluene transformation by the aerobic yeast *Yarrowia lipolytica*

Ayrat M Ziganshin[1*], Elvira E Ziganshina[1], James Byrne[2], Robin Gerlach[3], Ellen Struve[2], Timur Biktagirov[4], Alexander Rodionov[4] and Andreas Kappler[2]

Abstract

Understanding the factors that influence pollutant transformation in the presence of ferric (oxyhydr)oxides is crucial to the efficient application of different remediation strategies. In this study we determined the effect of goethite, hematite, magnetite and ferrihydrite on the transformation of 2,4,6-trinitrotoluene (TNT) by *Yarrowia lipolytica* AN-L15. The presence of ferric (oxyhydr)oxides led to a small decrease in the rate of TNT removal. In all cases, a significant release of NO_2^- from TNT and further NO_2^- oxidation to NO_3^- was observed. A fraction of the released NO_2^- was abiotically decomposed to NO and NO_2, and then NO was likely oxidized abiotically to NO_2 by O_2. ESR analysis revealed the generation of superoxide in the culture medium; its further protonation at low pH resulted in the formation of hydroperoxyl radical. Presumably, a fraction of NO released during TNT degradation reacted with superoxide and formed peroxynitrite, which was further rearranged to NO_3^- at the acidic pH values observed in this study. A transformation and reduction of ferric (oxyhydr)oxides followed by partial dissolution (in the range of 7–86% of the initial Fe(III)) were observed in the presence of cells and TNT. Mössbauer spectroscopy showed some minor changes for goethite, magnetite and ferrihydrite samples during their incubation with *Y. lipolytica* and TNT. This study shows that i) reactive oxygen and nitrogen species generated during TNT transformation by *Y. lipolytica* participate in the abiotic conversion of TNT and ii) the presence of iron(III) minerals leads to a minor decrease in TNT transformation.

Keywords: 2,4,6-Trinitrotoluene; Biodegradation; Ferric (oxyhydr)oxides; *Yarrowia lipolytica*; Reactive oxygen species; Reactive nitrogen species

Introduction

Nitroaromatic compounds have multiple applications in the chemical industry, amongst which special attention should be paid to 2,4,6-trinitrotoluene (TNT) due to its extensive use throughout the world (Harter 1985). A significant number of land and water territories are contaminated with this explosive and its conversion products as a result of the manufacture, transportation, storage and testing of munitions. These compounds are classified as relatively recalcitrant to biological degradation and their fate is affected by biogeochemical processes (Smets et al. 2007; Singh et al. 2012; Ziganshin and Gerlach 2014; Chien et al. 2014).

On the one hand, bioremediation strategies could offer attractive solutions for soil and water decontamination, although, TNT is exceedingly recalcitrant to complete biological mineralization. On the other hand, TNT nitro groups can be fairly easily reduced to hydroxylamino and amino derivatives by bacteria and fungi. However, these metabolites are also known to be toxic, potentially mutagenic and persistent contaminants (Smets et al. 2007; Singh et al. 2012; Khan et al. 2013). The most promising

* Correspondence: a.ziganshin06@fulbrightmail.org
[1]Department of Microbiology, Kazan (Volga Region) Federal University, ul. Kremlyovskaya 18, Kazan 420008, The Republic of Tatarstan, Russia
Full list of author information is available at the end of the article

and efficient pathway of nitrogen release from TNT is the transformation of parent compound through the addition of hydride ions resulting in the formation of Meisenheimer hydride complexes followed by their subsequent destruction (ring fission) (Pak et al. 2000; Jain et al. 2004; Wittich et al. 2009; Ziganshin et al. 2010a). Therefore, the application of microorganisms, which are able to degrade TNT and its metabolites, can be useful for bioremediation of TNT-contaminated soil and water.

Iron is one of the most abundant elements in the biosphere, naturally occurring in the form of various Fe(II)- and Fe(III)-bearing minerals and present at many TNT-contaminated field sites. The biogeochemical cycling of iron also has an influence on nitroaromatic compounds, including TNT, mostly through the stimulation of nitro group reduction process in different soils, sediments and waters. In almost all cases TNT transformation in the presence of iron species is accompanied by the enhanced formation and accumulation of undesirable metabolites, such as hydroxylamino-dinitrotoluenes (HADNTs), amino-dinitrotoluenes (ADNTs) and/or diamino-nitrotoluenes (DANTs) (Hofstetter et al. 1999; Borch et al. 2005; Oh et al. 2008; Boparai et al. 2010), which are fairly persistent and approximately as toxic as TNT itself (Smets et al. 2007; Khan et al. 2013). Only a few studies have reported denitration of TNT in the presence of ferrihydrite (Eyers et al. 2008; Khilyas et al. 2013). In terms of the efficiency of remediation strategies, TNT denitration processes resulting in nitro group elimination would be advantageous if the nitro group reduction pathways could be minimized. Therefore, for the development of effective TNT-bioremediation technologies in the presence of iron-containing minerals and because of the distinct solubilities and reactivities of various Fe (III) minerals (Borch et al. 2010; Posth et al. 2014), it is important to investigate TNT transformation processes in the presence of different ferric (oxyhydr)oxides. In addition, very little is known about the reduction of Fe (III) minerals by strictly aerobic bacteria and yeasts in the presence of polynitroaromatic compounds.

Systematic data showing the transformation of TNT by distinct microorganisms in the presence of iron (oxyhydr) oxides under oxic conditions is lacking, despite the fact that oxygen can have a significant effect on Fe-transformation processes (Morgan and Lahav 2007). Only a few studies have so far demonstrated the generation of the superoxide radical anion ($O_2^{\cdot-}$) during TNT reduction by purified enzymes (Kumagai et al. 2000; Kumagai et al. 2004) and in the presence of bacterial cells (Naumenko et al. 2008) via the formation of TNT nitroanion radicals. In some studies the involvement of abiotically and biologically produced $O_2^{\cdot-}$ in the denitration of TNT and its metabolites was also reported (Fritsche et al. 2000; Van Aken and Agathos 2002; Stenuit et al. 2012). However, little is known regarding the biological production of $O_2^{\cdot-}$ and other reactive oxygen species in culture medium during the microbial transformation of polynitroaromatic xenobiotics.

Given the importance of iron minerals for the fate of TNT, the main goal of this study was to investigate the effect of different ferric (oxyhydr)oxides on the TNT transformation processes by the yeast strain *Yarrowia lipolytica* AN-L15 and to identify iron mineral transformation in the presence of yeast cells and mineral additions of goethite (α-FeOOH), hematite (α-Fe_2O_3), magnetite (Fe_3O_4) or ferrihydrite ($5Fe_2O_3{\bullet}9H_2O$). Finally, reactive oxygen and nitrogen species generated over the course of TNT transformation were characterized.

Materials and methods

Yeast growth media and growth conditions

The yeast strain *Yarrowia lipolytica* AN-L15 deposited in the Russian National Collection of Industrial Microorganisms under a collection number of VKPM Y-3492 was cultivated aerobically at 30°C for 24 hours in Petri dishes with Sabouraud agar medium containing (per liter) 10 g of glucose, 10 g of peptone, 5 g of yeast extract, 0.25 g of NaCl and 20 g of agar (Ziganshin et al. 2007a). Yeast cells were harvested, washed twice with 16 mM phosphate buffer (pH 6.0) and added into 250 mL Erlenmeyer flasks containing 50 mL of a synthetic medium of the following content (mM): glucose, 28; $(NH_4)_2SO_4$, 7.6; $MgSO_4$, 2; Na_2HPO_4, 1.94; KH_2PO_4, 14.06 (pH 6.0). In ferric (oxyhydr)oxide containing treatments, goethite, hematite, magnetite or ferrihydrite were added to the synthetic medium to a final concentration of 0.15 g L^{-1} or 0.3 g L^{-1} (measured as Fe). As ferric (oxyhydr)oxides, commercially available goethite, hematite and magnetite were used (Lanxess). In the case of ferrihydrite-containing experiments, 2-line ferrihydrite was synthesized by dissolving of 4.0 g of $Fe(NO_3)_3{\bullet}9H_2O$ in 50 mL of Millipore water following Amstaetter et al. (2012). 1 M KOH was then added dropwise over 3 min until a pH of 7.3 was reached. After 2 h, the pH was adjusted to a final value of 7.5. After centrifugation at 5,000 × g for 10 min and washing of the solid phase with Millipore water, 2-line ferrihydrite was obtained and used for the experiments. Cell growth was monitored photometrically by measuring the optical density at 600 nm (OD_{600}) using a SPECOL 1300 spectrophotometer (Analytik Jena, Germany) with cell-free supernatant as a reference; the initial OD_{600} was adjusted to 0.05. TNT was added to a final concentration of 0.44 mM from an ethanolic stock solution (0.8 ml of 96% ethanol into 50 ml of medium), and the culture was incubated in the absence of light at 30°C on a rotary shaker at 150 rpm. Experiments conducted in the absence of TNT contained the same amount of 96% ethanol. All experiments were set up in triplicate.

Analytical methods

TNT and biotransformation products were detected and quantified with a Shimadzu high-performance liquid chromatograph equipped with an autoinjector, a diode array detector, a column oven, a Supelcosil LC-8 guard column and a Supelcosil octyl (C-8) column (150 by 4.6 mm; particle size, 5 μm) as described previously (Ziganshin et al. 2007b; Ziganshin et al. 2010b).

The nitrite and nitrate ions in the culture fluid were determined by using an AA3 SEAL Analyzer equipped with a XY2 Sampler (SEAL Analytical, Germany).

Dissolved Fe(II) concentrations were quantified after sample centrifugation at 15,000 × g for 10 min and subsequent filtration through 0.2 μm filters (Spartan 13/0.2 RC; Whatman). Filtered sample (100 μL) was added to 900 μL of 1 M HCl. The received solution (20 μL) was added to 80 μL of 1 M HCl and to 100 μL of 0.1% ferrozine solution in 96-well microtiter plates and incubated for 5 min at room temperature. The resulting absorbance (562 nm) in the plates was measured using a Flashscan 550 plate reader (Analytik Jena, Germany) (Stookey 1970).

For ^{57}Fe Mössbauer spectroscopy, biologically produced mineral precipitates were centrifuged at 10,000 × g for 10 min and then dried in an anoxic glovebox (100% N_2). Samples were prepared by loading dried powders into Plexiglas holders (area 1 cm^2). In order to ensure a homogeneous sample with ideal thickness, each sample was mixed and ground using a pestle and mortar with 80 mg glucose monohydrate. The samples were transferred to the Mössbauer spectrograph and inserted into a closed-cycle exchange gas cryostat (Janis cryogenics). Spectra were collected at 295 K, 77 K and 5 K by using a constant acceleration drive system (WissEL) in transmission mode with a ^{57}Co/Rh source and calibrated against a 7 μm thick α-^{57}Fe foil measured at room temperature. Spectra were analyzed using Recoil (University of Ottawa) using the Voigt Based Fitting (VBF) routine. The HWHM was fixed to 0.122 mm/s, as determined from the minimum line width of the calibration foil, measured at 295 K.

Superoxide anion radical (O_2^-) generation during TNT transformation was determined by electron spin resonance (ESR) spectroscopy through the formation of 5-(Diethoxyphosphoryl)-5-methyl-1-pyrroline-N-oxide (DEPMPO) spin adducts (Frejaville et al. 1995). Firstly DEPMPO (50 mM) was added to a solution to be analyzed and after 30 sec incubation process at room temperature the mixture was transferred to glass tubes with an inner diameter of 1 mm (Sigma-Aldrich, Germany). ESR spectra were then recorded at room temperature with an ESP 300 Electron Spin Resonance spectrometer (Bruker, Germany) and a super-high Q microwave cavity. The experimental parameters were as follows: microwave power 2 mW, microwave frequency 9.73 GHz, modulation amplitude 0.2 mT, time constant 82 ms and scan rate 2.1 G/s. The received ESR spectra were analyzed using EasySpin toolbox for MATLAB. TNT-free or cell-free experiments were used as a control. The xanthine-xanthine oxidase reaction was used as the standard system for O_2^- generation; superoxide dismutase was used to remove superoxide anions in the samples (Stenuit et al. 2012).

Chemicals

2,4,6-Trinitrotoluene and 2,4-dinitrotoluene (2,4-DNT) are the chemicals of ChemService (West Chester, USA); 2-amino-4,6-dinitrotoluene (2-ADNT) and 4-amino-2,6-dinitrotoluene (4-ADNT) were received from Supelco (Bellefonte, USA); 2-hydroxylamino-4,6-dinitrotoluene (2-HADNT) and 4-hydroxylamino-2,6-dinitrotoluene (4-HADNT) as well as some other potential metabolites of TNT nitro group reduction pathway were purchased from AccuStandard (New Haven, USA). DEPMPO as a spin trapping reagent for reactive oxygen species was received from Enzo Life Sciences (Farmingdale, USA). Hypoxanthine, xanthine oxidase, superoxide dismutase as well as all other chemicals and reagents were purchased from Sigma-Aldrich (Germany).

Results

Influence of ferric (oxyhydr)oxides on TNT biotransformation processes

The effect of various ferric (oxyhydr)oxides with a final concentration of 0.3 g L^{-1} (measured as Fe) on TNT (440 μM) transformation and denitration by *Yarrowia lipolytica* AN-L15 in the presence of glucose and ethanol as carbon and electron sources was evaluated. Independent of the presence of ferric (oxyhydr)oxides, yeast growth in the presence of TNT occurred with a pH decrease of the synthetic medium from an initial pH of 6.0 to below 3.0 because of production of organic acids (Additional file 1: Table S1, Supporting Information, SI). In the absence of TNT, the pH change of the medium was observed to decrease to almost the same values (data not shown). HPLC analysis demonstrated that TNT was efficiently removed from the medium during yeast growth (Figure 1A). As shown in Figure 1A, complete TNT removal by *Y. lipolytica* AN-L15 occurred within 5 days in the presence or absence of the various ferric (oxyhydr)oxides. The highest rate of TNT biotransformation was observed in the absence of iron minerals followed by the treatments in the presence of goethite. The presence of ferrihydrite, magnetite and hematite led to a small decrease in the rate of TNT removal. *Y. lipolytica* removed TNT in the range of 63–82% of TNT during the first day of incubation, while the remaining 18–37% were further transformed in the following 3 to 4 days of the experiments. The TNT biotransformation by *Y. lipolytica* was slower compared to previous

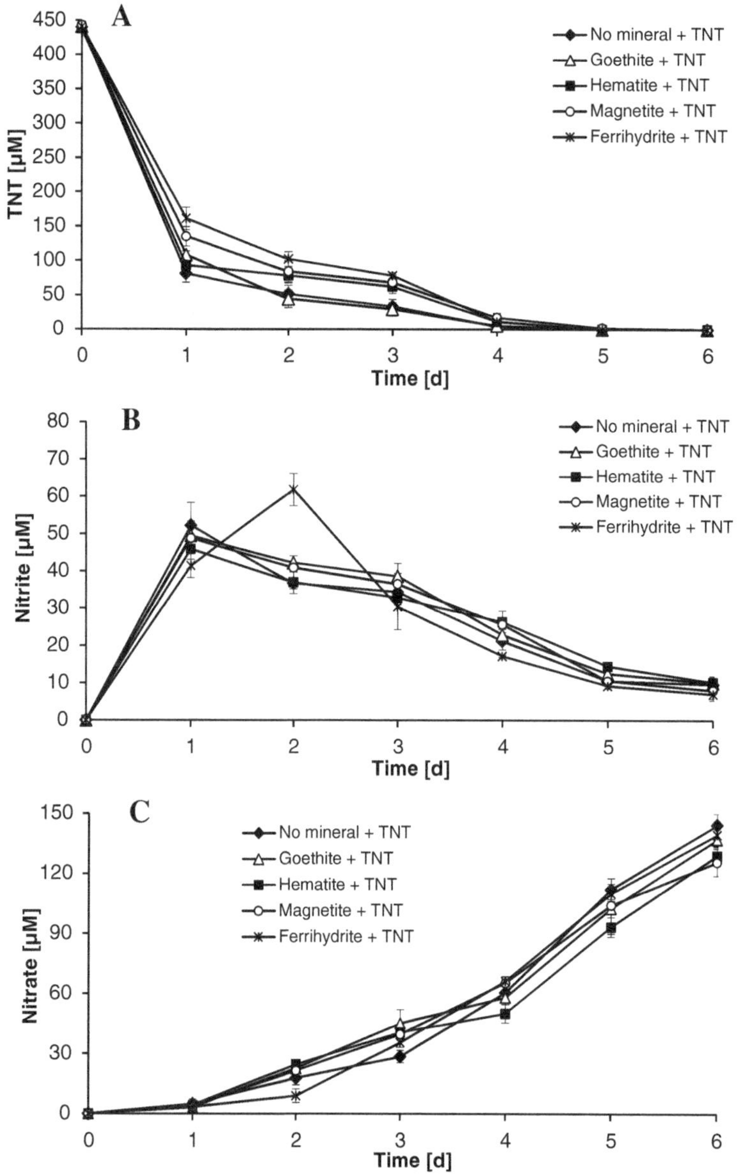

Figure 1 TNT removal (A), nitrite accumulation and removal (B) and nitrate accumulation during transformation of TNT (C) by yeast cells of *Y. lipolytica* AN-L15 in the absence and presence of various ferric (oxyhydr)oxides. Error bars represent the standard deviation of triplicate experiments.

studies (Ziganshin et al. 2010b; Khilyas et al. 2013), which can be explained by the lower initial yeast cell concentration used for these experiments (initial OD_{600} of 0.05). When an initial OD_{600} of 0.2 or 1.0 was used in Ziganshin et al. (2010b) and Khilyas et al. (2013), TNT was completely removed in the absence of ferric (oxyhydr)oxides within 24 and 10 hours, respectively.

During the conversion of TNT by yeast cells of *Y. lipolytica* AN-L15, different metabolites successively appeared in the culture medium. Among the aromatic ring hydrogenation products, eight hydride-Meisenheimer complexes of TNT were detected, which were previously described in detail (Ziganshin et al. 2007b; Ziganshin et al. 2010b). Independent of the presence of ferric (oxyhydr)oxides, the main intermediate metabolite detected during the first day of incubation was a monohydride-Meisenheimer complex of TNT (3-H^--TNT). The maximum concentration of 3-H^--TNT detected after 1 day ranged from 255–294 μM (data not shown); its concentration was estimated as described previously (Ziganshin et al. 2007b). Its subsequent biotransformation into 2,4-DNT and other dihydride-Meisenheimer complexes of TNT (3,5-2H^--TNT and 3 isomers of 3,5-2H^--TNT · H^+) as well as degradation of TNT-dihydride complexes over the next 3–4 days occurred with the liberation of nitrite into the medium in the absence as well as in the presence of the different ferric (oxyhydr)oxides (Figure 1B). As can be seen in Figures 1A and 1B, the maximum amount of nitrite detected during TNT removal was in the range of 45–61 μM. In the presence of ferrihydrite, TNT transformation was slightly less significant combined with delayed nitrite ion liberation, which might be due to the increase in buffer capacity of the medium in the presence of ferrihydrite as asserted by Khilyas et al. (2013). After 1 day, nitrate was also detected indicating nitrite oxidation to nitrate (Figure 1C). The highest concentration of nitrate was observed in iron mineral-free systems (145 ± 7 μM), whereas the lowest nitrate accumulation was found in the culture medium containing magnetite (126 ± 7 μM) and hematite (129 ± 5 μM). Interestingly, 2,4-DNT as a product of TNT degradation was found at high levels only in iron mineral-free treatments (72 ± 7 μM), while the addition of ferric (oxyhydr)oxides resulted in lower concentrations of this metabolite (15–46 μM; SI, Additional file 1: Table S1). TNT nitro group reduction metabolites, such as HADNTs and ADNTs, were also detected and their concentrations are summarized in Additional file 1: Table S1 (SI). Additional abiotic experiments conducted in the presence of iron minerals at pH 6.0 or 7.0 showed that nitrite, nitrate as well as TNT were relatively stable and did not undergo any transformation reactions in the absence of yeast cells (data not shown). Also, we were unable to detect other potential conversion products of TNT such as diarylamines, tetranitro-azoxytoluenes, tetranitro-azotoluenes, amino-dimethyl-tetranitrobiphenyls or Bamberger rearrangement products during aerobic growth of *Y. lipolytica* AN-L15 in the presence of Fe(III) minerals.

Superoxide generation during TNT biotransformation

ESR spectra of DEPMPO radical adducts obtained during TNT biotransformation clearly showed the generation of $O_2^{\cdot-}$ radical species in the culture medium (Figure 2A). This ESR spectrum was obtained 30 minutes after the start of the experiment when the culture medium pH was around 6.0. Superoxide anion presents in equilibrium with its protonated form, hydroperoxyl radical ($HO_2^{\cdot}$), in an aqueous solution, and this equilibrium is shifted towards $HO_2^{\cdot}$ under acidic conditions (Bielski et al. 1985). Therefore, a fraction of superoxide anions is expected to be protonated at the low pH values observed in the present work. Both species, $O_2^{\cdot-}$ and $HO_2^{\cdot}$, can react with DEPMPO yielding the same spin adduct (DEPMPO-OOH) (Lauricella et al. 2004). Therefore, the nitroxide radical associated with the superoxide/hydroperoxyl conjugate pair will be referred

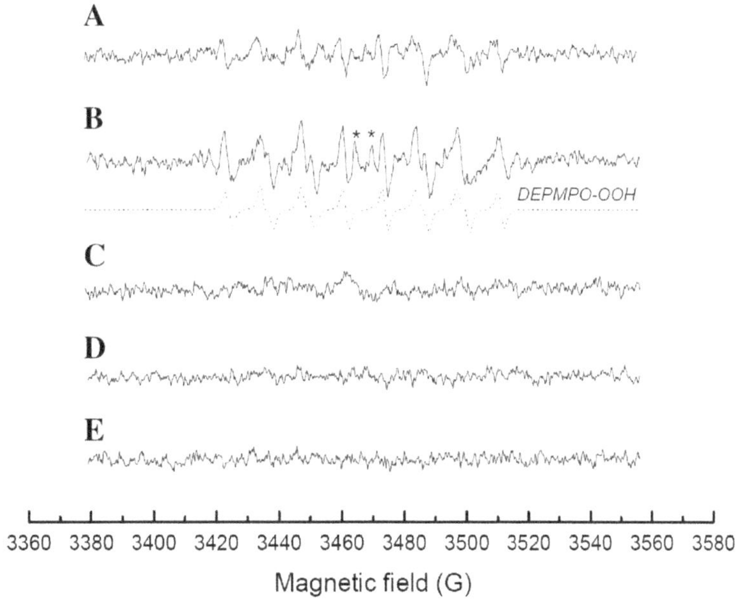

Figure 2 ESR spectra of DEPMPO spin adducts generated in the presence of yeast cells of *Y. lipolytica* AN-L15 and TNT in the absence of Fe(III) (oxyhydr)oxides (measured 30 minutes after the start of the experiment) (A) and in a standard xanthine/xanthine oxidase system (reference spectrum for DEPMPO-OOH). (B) Simulated spectrum of DEPMPO-OOH is presented as a reference (dashed line); the typical set of hyperfine coupling constants was used for simulation (Frejaville et al. 1995). The spectral lines attributed to DEPMPO-OH spin adduct are marked with asterisks; their appearance is the result of DEPMPO-OOH decomposition to DEPMPO-OH (Stenuit et al. 2012). **(C)** Same as in **(A)** with the exception that superoxide dismutase (200 U mL^{-1}) was added. Control experiments in the presence of *Y. lipolytica* AN-L15 and absence of TNT **(D)** and in the absence of *Y. lipolytica* AN-L15 and presence of TNT **(E)**. The spectra were recorded for 10 min after the DEPMPO was added. The experimental parameters were as follows: microwave power 2 mW, microwave frequency 9.73 GHz, modulation amplitude 0.2 mT, time constant 82 ms and scan rate 2.1 G/s.

to as DEPMPO-superoxide spin adduct hereafter. For confirmation of the lines in the spectra obtained during the TNT biotransformation namely to DEPMPO-OOH, a standard xanthine/xanthine oxidase reaction was used that generates $O_2^{\cdot-}$ (Figure 2B). The standard spectrum was compared with those received during TNT transformation by *Y. lipolytica* AN-L15 and associated with the generation of $O_2^{\cdot-}$. The addition of superoxide dismutase (SOD) removed the original signal, indicating that this mechanism was superoxide-dependent and did not lead to hydroxyl radical generation (Figure 2C). Since SOD was able to remove the original signal, the appearance of spectral lines attributed to the hydroxyl radical spin adduct (DEPMPO-OH) is the result of DEPMPO-OOH decomposition to DEPMPO-OH (Stenuit et al. 2012). This suggests that TNT transformation in the presence of the yeast strain results in the formation of $O_2^{\cdot-}/HO_2^{\cdot}$ in the culture medium. However, we cannot exclude the partial generation of hydroxyl radicals ($HO^{\cdot}$) as well as other similarly reactive species in these systems. Control experiments in the presence of *Y. lipolytica* AN-L15 but absence of TNT, as well as those in the absence of *Y. lipolytica* AN-L15 and presence of TNT did not result in the generation of $O_2^{\cdot-}$ (Figure 2D and 2E). Additional experiments conducted in the presence of yeast cells of *Y. lipolytica*, different ferric (oxyhydr)oxides and TNT led to formation of the same ESR spectra, indicating the formation of superoxide anions as well (results not shown).

Reduction of ferric (oxyhydr)oxides during incubation with *Y. lipolytica* in the presence of TNT

Ferric (oxyhydr)oxides dissolution by aerobically growing *Y. lipolytica* AN-L15 cells was observed in the presence of TNT with the accumulation of dissolved Fe(II) (Figure 3). All ferric (oxyhydr)oxides (0.3 g L^{-1} Fe) tested in this research appeared to undergo partial reduction to Fe(II) in TNT-containing synthetic medium in the presence of *Y. lipolytica* AN-L15. The greatest extent of Fe(III) reduction was observed when ferrihydrite was used for the experiments, while some dissolved Fe(II) production was also observed in magnetite-containing treatments. Even less dissolved Fe(II) production was observed in the presence of hematite and goethite. After 6 days, the concentrations of dissolved Fe(II) in TNT- and yeast-containing medium in the presence of ferrihydrite, magnetite, hematite and goethite reached 4.62 mM, 2.05 mM, 1.02 mM and 0.39 mM, respectively (Figure 3). Approximately 86% of the Fe(III) added as ferrihydrite had been converted into Fe(II), while for the other ferric minerals, the extent of Fe(III) conversion to Fe(II) was less significant. Approximately 38, 19 and 7% of Fe(III) were liberated through dissolution when magnetite, hematite and goethite were used, respectively.

In contrast, significantly lower amounts of ferrous iron production were observed in TNT-free setups and in the presence of *Y. lipolytica* AN-L15 when compared to TNT-containing treatments. The maximum concentrations of dissolved Fe(II) in the absence of TNT after 6 days of incubation with strain AN-L15 were 0.62 mM, 0.26 mM, 0.2 mM and 0.2 mM (approximately 12, 5, 4 and 4% of the added ferrihydrite, magnetite, hematite and goethite, respectively; Figure 3). When lower concentrations of ferric (oxyhydr)oxides (0.15 g L^{-1} Fe) were tested, lower amounts of Fe(II) were detected (SI, Additional file 1: Figure S1); however, the trends remained the same. In control experiments, using cell-free medium, no Fe(II) accumulation was observed regardless of the presence of TNT (data not shown).

Transformation of ferric (oxyhydr)oxides during incubation with *Y. lipolytica* in the absence/presence of TNT

^{57}Fe Mössbauer spectroscopy was performed on all samples to determine the iron mineral phases present and to

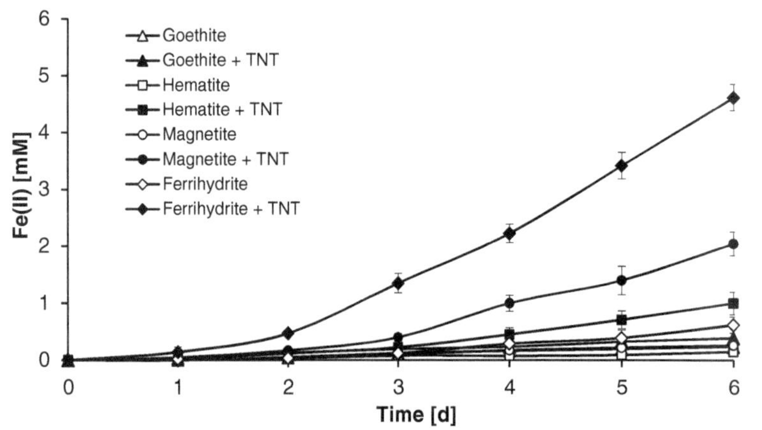

Figure 3 Formation of dissolved Fe(II) during aerobic growth of *Y. lipolytica* AN-L15 in the presence of different ferric (oxyhydr)oxides (0.3 g L^{-1} Fe). Error bars represent the standard deviation of triplicate experiments.

identify potential differences between the materials after a 6 day incubation period in the presence of *Y. lipolytica* AN-L15 cells alone, and also in the presence of yeast cells with TNT (Figures 4 and 5). Room temperature spectra for magnetite, hematite and goethite are shown in Figure 4, whilst Additional file 1: Table S2 (SI) shows the parameters obtained through spectral fitting of the different ferric (oxyhydr)oxides after aerobic incubation with *Y. lipolytica* AN-L15 cells in the absence or the presence of TNT.

Goethite is usually expected to have a narrow hyperfine field at room temperature (38.0 T) and a small center shift (0.37 mm/s). The spectra obtained for both goethite samples have a narrower hyperfine field than expected (33.1–33.9 T); however, they also both exhibit superparamagnetic relaxation. This occurs close to the blocking temperature of a material and is likely the result of small average crystallite size or poor crystallinity of the sample. The paramagnetic background decreases in area relative to the sextet corresponding to the goethite from 39.5% to 24.8% for goethite and goethite + TNT, respectively (Figure 4). A small increase in center shift (CS) from 0.37 mm/s (goethite) to 0.43 mm/s (goethite + TNT) was also observed.

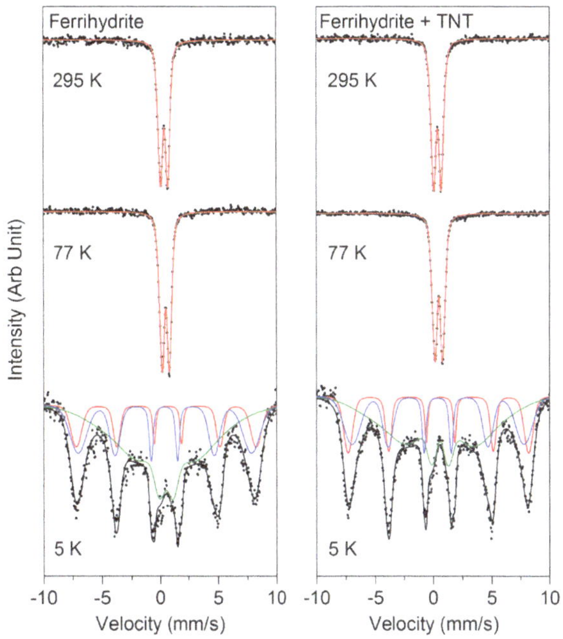

Figure 5 Mössbauer spectra of ferrihydrite after 6 days of incubation in the presence of *Y. lipolytica* AN-L15 with and without TNT (spectra collected at 295 K, 77 K and 5 K).

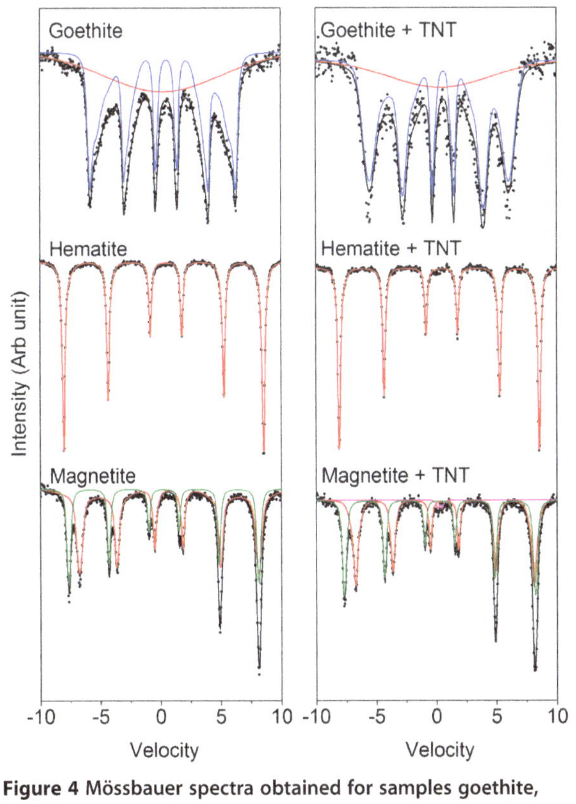

Figure 4 Mössbauer spectra obtained for samples goethite, hematite and magnetite at room temperature (295 K) after 6 days of incubation in the presence of *Y. lipolytica* AN-L15 with and without TNT.

Both hematite samples (+/− TNT) possess a single sextet with very wide magnetic splitting (51.4 T) and low center shift (0.37–0.38 mm/s) that is characteristic of natural and synthetic hematite (Cornell and Schwertmann 2003). Its incubation in the presence of *Y. lipolytica* cells alone and with yeast cells and TNT did not yield any additional mineral phases or differences from the parent mineral (Figure 4).

Magnetite samples exhibit two overlapping sextets which are characteristic of the iron mineral. One sextet corresponds to Fe^{3+} in tetrahedral coordination, denoted the A-site, or $^{Tet}Fe^{3+}$. The second sextet is observed above the Verwey temperature of magnetite (~120 K) and is formed due to electron hopping within the octahedral lattice and corresponds to both Fe^{2+} and Fe^{3+}. This second sextet is commonly denoted the B site, or $^{Oct}Fe^{2.5+}$. The results obtained here indicate that the spectrum of a magnetite + cells + TNT sample can only be fully fitted with an additional doublet which contributes to 1.3% of the total area (Figure 4). This doublet has a very low CS of 0.28 mm/s and low quadrupole splitting (ΔE_Q) of 0.35 mm/s, which is consistent with a poorly crystalline Fe(III) mineral phase; however, it is not clear what the mineral identity is.

Finally, Mössbauer spectra were collected for ferrihydrite samples at three different temperatures, 295 K, 77 K and 5 K (Figure 5). Both spectra collected at 295 K have the same center shift (0.38 mm/s) and very similar ΔE_Q (0.68 and 0.70 mm/s for samples of ferrihydrite incubated with *Y. lipolytica* AN-L15 and with *Y. lipolytica* AN-L15 and TNT, accordingly). These parameters are characteristic for ferrihydrite at room temperature (Cornell and Schwertmann 2003; Mikutta et al. 2008). Spectra obtained at 77 K maintain the paramagnetic doublet observed at room temperature, albeit with an increase in CS to 0.49 mm/s for both samples (Figure 5). Additional analyses were carried out at 5 K, with both samples requiring the fitting of two magnetically ordered (S1 and S2) and one poorly ordered (S3) sextets (Figure 5, Additional file 1: Table S2). The hyperfine fields of the ordered sextets are at close to the expected values of ferrihydrite (Eusterhues et al. 2008), although it is clear that neither sample is fully magnetically ordered, as seen by the requirement of a poorly ordered sextet. The total contribution of this background phase decreases from 55.0% (ferrihydrite + cells) to 49.9% (ferrihydrite + cells + TNT). Pure 2-line or 6-line ferrihydrite is usually expected to be fully magnetically split at 5 K and does not require this background feature to fit the data. This suggests that both samples have a very low blocking temperature, which is potentially much lower than the value expected for typical ferrihydrite (Murad 1996).

Discussion

In this study we investigated the impact of various ferric (oxyhydr)oxides on TNT transformation by *Y. lipolytica* AN-L15. The fastest rate of TNT removal was seen in the absence of iron minerals followed by experiments in the presence of goethite. The presence of ferrihydrite, magnetite and hematite led to a small decrease in the rate of TNT transformation. However, significant TNT denitration by *Y. lipolytica* occurred in the absence and presence of ferric (oxyhydr)oxides, indicating the possibility of TNT aromatic ring cleavage independent of the presence of iron minerals. The significant liberation of NO_2^- from TNT and further NO_2^- oxidation to NO_3^- was observed with yeast cells of *Y. lipolytica*. Also, the presence of ferric (oxyhydr)oxides did not stimulate the nitro group reduction. It has previously been suggested that the increase of nitro group reduction metabolites during TNT bioconversion is possible in the presence of iron minerals without any denitration processes (Hofstetter et al. 1999; Borch et al. 2005). However, our results indicate that TNT denitration by aerobically grown *Y. lipolytica* AN-L15 can take place in iron rich environments, and that the presence of ferric (oxyhydr)oxides does not lead to an increase of HADNT production. Given that Fe (II)- and Fe(III)-containing minerals are found at many TNT-contaminated field sites, these results are of significance for TNT bioremediation in such environments.

ESR spectra of DEPMPO radical adducts obtained during TNT biotransformation clearly demonstrated the generation of $O_2^{\cdot-}/HO_2^{\cdot}$ radical species in the culture medium. Presumably, the generation of reactive oxygen species here is the result of the interaction of *Y. lipolytica* AN-L15 cells with TNT and its transformation products, resulting in TNT-induced oxidative stress. It is known that TNT can produce superoxide radical anions through TNT nitroanion radical formation (Kumagai et al. 2000; Kumagai et al. 2004; Naumenko et al. 2008). Recently, an involvement of abiotically produced superoxide radical anions in the TNT denitration process with the accumulation of nitrite ions in the reaction mixture was reported (Stenuit et al. 2012). The authors suggested that denitration occurred through a nucleophilic attack on the parent trinitroaromatic compound; however, the exact mechanism is yet unknown. The generation of superoxide anions in our treatments also indicates their possible participation in the TNT denitration process. However, the main pathway of nitrite/nitrate liberation from TNT by *Y. lipolytica* AN-L15 occurred via biodegradation of TNT-hydride complexes resulting in the formation of 2,4-DNT and other unidentified, carbon-containing metabolites.

In recent work we detected nitric oxide (NO) as a metabolite of TNT biodegradation by *Y. lipolytica* AN-L15 (Khilyas et al. 2013). A potential pathway for its formation is the abiotic conversion of nitrous acid (formed through nitrite release during Meisenheimer hydride complexes destruction) to NO and nitric acid (nitrate) via a disproportionation reaction at low pH (Van Cleemput and Baert 1984). It is also likely that a fraction of the released NO_2^- was decomposed to NO and nitrogen dioxide (NO_2) and the produced NO could have been oxidized to NO_2 in the presence of molecular oxygen (O_2) as ascertained by Klueglein and Kappler (2013). In addition, the produced NO_2 may potentially be scavenged by natural organic matter (NOM) through nitration reactions. Another fraction of the released NO_2^- could have reacted with amino group-containing cell components in the low pH systems with the formation of nitrosoamines (Fischer and Warneck 1996). All of these reactions might have lead to the decrease of the total amount of NO_2^-/NO_3^- detected in our experimental systems.

Furthermore, NO released during TNT biodegradation by *Y. lipolytica* AN-L15 might have reacted with $O_2^{\cdot-}$ to form peroxynitrite ($ONOO^-$) which is a stronger oxidant than both NO and $O_2^{\cdot-}$ (Pfeiffer et al. 1997; Vásquez-Vivar et al. 1997). Kumagai et al. (2004) demonstrated the production of NO and $O_2^{\cdot-}$ in the presence of TNT catalyzed by nitric oxide synthase and also proposed a mechanism of peroxynitrite formation. Peroxynitrite is not stable under acidic conditions resulting in rearrangement to nitrate (Pryor and Squadrito 1995; Pfeiffer et al. 1997). Thus, together with the biological and abiotic oxidation of released NO_2^- into NO_3^-, NO_3^- accumulation during TNT biodegradation by *Y. lipolytica* AN-L15 at low pH could also occur through rearrangement of $ONOO^-$. Some other oxidative reactions are also possible as $ONOO^-$ reacts rapidly with CO_2, producing highly oxidizing species such as NO_2 and $CO_3^{\cdot-}$ (Denicola et al. 1996). Since *Y. lipolytica* AN-L15 is able to synthesize and excrete different organic acids, mostly pyruvate and citrate, into the growth medium (Ziganshin et al. 2010b), these metabolites would probably provide the cells with some protection against oxidative damage. Pyruvic acid, like other α-keto acids, acts as a hydrogen peroxide (H_2O_2) scavenger and is able to react with peroxynitrite and peroxynitrous acid via pyruvate decarboxylation to acetate (Vásquez-Vivar et al. 1997). Furthermore, some protective properties of citric acid against superoxide anions (Van den Berg et al. 2003) and peroxynitrite (Han et al. 2005) are also known.

The results of our experiments also indicate a greater likelihood of ferrihydrite and magnetite to be reduced to ferrous iron in the presence of aerobic yeast cells and TNT than hematite or goethite.

Possible mechanisms for the reduction of the ferric (oxyhydr)oxides here could be the induction of yeast enzymes with Fe(III) reducing capacity by TNT or its transformation products as well as the participation of TNT degradation products and/or some unidentified extracellular metabolites in the abiotic reduction of Fe (III). Kerem et al. (1999) demonstrated that the brown rot fungus *Gloeophyllum trabeum* was able to degrade another recalcitrant compound, polyethylene glycol. *G. trabeum* excreted an extracellular metabolite, 2,5-dimethoxy-1,4-benzoquinone, into the medium and after its reduction to 2,5-dimethoxyhydroquinone, this compound led to the non-enzymatic reduction of dissolved Fe(III) to Fe(II). The production of electron shuttling compounds by microbes can also enhance metal reduction (Gerlach et al. 2011; Field et al. 2013). Similar mechanisms during TNT biodegradation by *Y. lipolytica* AN-L15 could potentially also occur.

Another possible reason for Fe(II) accumulation and changing ferric (oxyhydr)oxide mineralogy in the presence of dissolved O_2 is the participation of reactive oxygen species in Fe(III) reduction (Melton et al. 2014). The reactivity of biologically produced $O_2^{\cdot-}$ toward amorphous ferric (oxyhydr)oxides leading to Fe(II) formation was previously investigated (Kustka et al. 2005; Fujii et al. 2006). It is known that $O_2^{\cdot-}/HO_2^{\cdot}$ can reduce dissolved ferric iron at a range of pH values (Rush and Bielski 1985), and organically associated Fe(III) has also been reported to be reduced by $O_2^{\cdot-}/HO_2^{\cdot}$ (Rose and Waite 2005). In addition, a partial re-oxidation of the released Fe(II) to Fe(III) might occur by either molecular oxygen (Morgan and Lahav 2007), nitric oxide or nitrogen dioxide (Klueglein and Kappler 2013) generated during TNT biodegradation. We also cannot exclude that a fraction of the produced Fe(II) reacted with $O_2^{\cdot-}/HO_2^{\cdot}$ with the subsequent formation of Fe(III) and H_2O_2. However, the rate of ferric ion reduction by $O_2^{\cdot-}$ is usually faster than that of ferrous iron oxidation by $O_2^{\cdot-}$ (Rush and Bielski 1985). In addition, $O_2^{\cdot-}$ might have reacted with NOM to form H_2O_2 in our systems (Fischer and Warneck 1996). Yet another possible mechanism of partial Fe(II) oxidation is related to the Fenton chemistry that involves H_2O_2 produced either via reaction of Fe(II) with $O_2^{\cdot-}$ (Rush and Bielski 1985), via reaction with NOM (Fischer and Warneck 1996), or via superoxide, hydroperoxyl and superoxide/hydroperoxyl disproportionation reactions (Rose and Waite 2005). Fe(III) can be also reduced back to Fe(II) by H_2O_2 with the generation of a hydroperoxyl radical. Additionally, $O_2^{\cdot-}$ could have also interacted with H_2O_2 to produce hydroxyl radicals (Haber and Weiss 1934). Due to the high reactivity of the different radical and peroxide species, a large number of reactions might have occurred in these systems and it is quite possible that reactions other than the described ones occurred in the experimental systems. Table 1 lists a number of possible reactions and products that might occur during TNT transformation in the presence of *Y. lipolytica* AN-L15 and Fe(III) minerals. Due to the complexity of radical-based reactions, however, the list of reactions is very likely incomplete.

Mössbauer spectroscopy revealed only relatively minor differences between ferrihydrite, goethite and magnetite minerals incubated with and without TNT; however, hematite exhibited no differences at all. No secondary Fe (II) mineral phases were detected despite the release of a significant proportion of aqueous Fe(II) as detected by the ferrozine assay. These results suggest that all of the Fe(II) which was released in the presence of only cells, or cells + TNT remained in solution and was not reprecipitated as Fe(II) minerals. This is likely due to the low pH of the growth medium which dropped to pH 3 over the course of the incubation. Incubation of the magnetite sample in the presence of *Y. lipolytica* AN-L15 and TNT resulted in formation of a minor mineral phase which is consistent with the Mössbauer spectrum of a poorly crystalline Fe(III) mineral phase; however, it is not clear what the mineral identity is. A decrease in the relative area of the paramagnetic component in the goethite sample was detected during incubation in the presence of *Y. lipolytica* AN-L15 and TNT. Analyses performed on both ferrihydrite samples showed that a significant fraction of the mineral in both of them is still not fully magnetically ordered even at 5 K. This would imply the samples have a very low blocking temperature, (i.e. temperature below which a mineral is magnetically ordered) likely much lower than the value which is typically expected for ferrihydrite (Murad 1996). Decreases in the blocking temperature can be caused by poor crystallinity (e.g. due to decreasing particle size) or by association with organic complexes. Mikutta et al. (2008) observed a decrease in blocking temperature from 68 K for pure ferrihydrite to 50 K for ferrihydrite synthesized in the presence of polysaccharides. Thus, it is possible

Table 1 Possible reactions that can take place during TNT transformation in the presence of *Y. lipolytica* AN-L15 and Fe(III) minerals

Biological reactions					
TNT	→	3-H⁻-TNT isomers; 3,5-2H⁻-TNT; 3,5-2H⁻-TNT · H^+ isomers; 1-H⁻-TNT			
TNT	→	2-HADNT; 4-HADNT; 2-ADNT; 4-ADNT			
3-H⁻-TNT	→	2,4-DNT + NO_2^-			
3,5-2H⁻-TNT · H^+	→	NO_2^- + ?			
1-H⁻-TNT	→	NO_2^- + ?			
NO_2^-	→	NO_3^-			
Abiotic reactions					
3-H⁻-TNT	⇄	3-H⁻-TNT isomers	$HO_2^{\cdot} + HO_2^{\cdot}$	→	$O_2 + H_2O_2$
3-H⁻-TNT	⇄	3,5-2H⁻-TNT	$HO_2^{\cdot} + O_2^{\cdot-}$	→	$O_2 + HOO^-$
3,5-2H⁻-TNT	→	3,5-2H⁻-TNT · H^+ isomers	$O_2^{\cdot-} + O_2^{\cdot-} + 2H^+$	→	$O_2 + H_2O_2$
3-H⁻-TNT	→	TNT	$O_2^{\cdot-}$ + NOM	→	H_2O_2 + ?
$NO_2^- + H^+$	⇆	HNO_2	$O_2^{\cdot-} + H_2O_2$	→	$HO^{\cdot} + OH^- + O_2$
$3HNO_2$	→	$HNO_3 + 2NO + H_2O$	Fe(III) + $O_2^{\cdot-}$	→	Fe(II) + O_2
$2HNO_2$	→	$NO_2 + NO + H_2O$	Fe(III) + H_2O_2	→	Fe(II) + $HO_2^{\cdot} + H^+$
$2NO + O_2$	→	$2NO_2$	Fe(II) + $HO_2^{\cdot} + H^+$	→	Fe(III) + H_2O_2
NO_2 + NOM*	→	?	Fe(II) + $O_2^{\cdot-} + 2H^+$	→	Fe(III) + H_2O_2
$O_2^{\cdot-}$ + NO	→	$ONOO^-$	Fe(II) + H_2O_2	→	Fe(III) + $HO^{\cdot} + OH^-$
$ONOO^- + H^+$	=	ONOOH	Fe(II) + O_2	→	Fe(III) + $O_2^{\cdot-}$
ONOOH	⇆	$NO_3^- + H^+$	2Fe(II) + $NO_2 + 2H^+$	→	2Fe(III) + NO + H_2O
$ONOO^- + CO_2$	→	$NO_2 + CO_3^{\cdot-}$	Fe(II) + NO + H^+	→	Fe(III) + HNO
$O_2^{\cdot-} + H^+$	⇆	$HO_2^{\cdot}$	2HNO	→	$N_2O + H_2O$

*Natural organic matter.

that our samples were not fully magnetically ordered because they were associated with organic material; however, this effect is usually only observed when ferrihydrite is synthesized in the presence of organic matter. The decrease in the relative area of the paramagnetic background for samples containing both goethite and ferrihydrite in the presence of cells and TNT could be primarily attributed to an increase in particle crystallinity. However, this somewhat contradicts the idea that Fe is being released into solution (as Fe(II)) which would require dissolution of the mineral phase, thus leading to a decrease in the particle crystallinity. An alternative explanation could be due to a shift in the particle size distribution which may be present in the goethite and ferrihydrite minerals with a mixture of some large crystalline phases and some smaller (i.e. more poorly crystalline) phases. The smaller particles would have a larger surface area to volume ratio, could thus be more accessible to the yeast cells, and therefore more likely to be the phases which are dissolved, leading to the release of Fe which was detected as Fe(II) in solution by ferrozine. ^{57}Fe Mössbauer only accounts for the solid mineral fraction; therefore as the more poorly crystalline phases are dissolved, the relative amount of more crystalline phase increases which could be interpreted as an increase in crystallinity, as seen in our data.

Our results clearly show the complexity of TNT transformation by *Y. lipolytica* AN-L15 in the presence of ferric (oxyhydr)oxides. In addition to biotic processes mediated by the yeast strain *Y. lipolytica* AN-L15, oxygen and nitrogen reactive species generated during TNT transformation process also participated in the abiotic transformation of iron and TNT. Since Fe(II)- and Fe(III)-bearing minerals are found at many TNT-contaminated field areas, the received results are significant for further understanding of TNT degradation and detoxification in the environment.

Additional file

Additional file 1: Table S1. Change of medium pH during aerobic growth of *Y. lipolytica* AN-L15 and maximum amounts of some metabolites (μM) detected over the course of TNT biotransformation. **Table S2.** Mössbauer spectroscopic analysis of various ferric (oxyhydr) oxides after their aerobic incubation with *Y. lipolytica* AN-L15 cells in the absence or presence of TNT. **Figure S1.** Formation of dissolved Fe(II) during aerobic growth of *Y. lipolytica* AN-L15 in the presence of different ferric (oxyhydr)oxides (0.15 g L^{-1} Fe Error bars represent the standard deviation of triplicate experiments.

Competing interests
The authors declare that they have no competing interests.

Authors' contributions
AMZ and EEZ designed and conducted experiments, performed statistical data treatments and interpretation of the data and drafted the manuscript. JB carried out Mössbauer spectroscopy and drafted and revised the manuscript. RG participated in the design of the study and drafted and revised the manuscript. ES helped to perform experiments and revised the manuscript. TB and AR helped to conduct ESR spectroscopic studies and participated in the ESR data analysis. AK supervised the research group, participated in the design of the study, drafted and revised the manuscript. All authors read and approved the final manuscript.

Acknowledgments
This work is performed according to the Russian Government Program of Competitive Growth of Kazan Federal University. Partial financial support was provided by the Russian Foundation for Basic Research, Grant No. 14-04-31694.

Author details
[1]Department of Microbiology, Kazan (Volga Region) Federal University, ul. Kremlyovskaya 18, Kazan 420008, The Republic of Tatarstan, Russia. [2]Geomicrobiology, Center for Applied Geosciences, University of Tübingen, Tübingen D-72076, Germany. [3]Department of Chemical and Biological Engineering, Center for Biofilm Engineering, Montana State University, Bozeman, Montana 59717, USA. [4]Institute of Physics, Kazan (Volga Region) Federal University, Kazan 420008, The Republic of Tatarstan, Russia.

References
Amstaetter K, Borch T, Kappler A (2012) Influence of humic substance imposed changes of ferrihydrite aggregation on microbial Fe(III) reduction. Geochimica et Cosmochimia Acta 85:326–341

Bielski BHJ, Cabelli DE, Arudi RL, Ross AB (1985) Reactivity of HO_2/O_2^- radicals in aqueous solution. J Phys Chem Ref Data 14:1041–1100

Boparai HK, Comfort SD, Satapanajaru T, Szecsody JE, Grossl PR, Shea PJ (2010) Abiotic transformation of high explosives by freshly precipitated iron minerals in aqueous Fe^{II} solutions. Chemosphere 79:865–872

Borch T, Inskeep WP, Harwood JA, Gerlach R (2005) Impact of ferrihydrite and anthraquinone-2,6-disulfonate on the reductive transformation of 2,4,6-trinitrotoluene by a gram-positive fermenting bacterium. Environ Sci Technol 39:7126–7133

Borch T, Kretzschmar R, Kappler A, Van Cappellen P, Ginder-Vogel M, Voegelin A, Campbell K (2010) Biogeochemical redox processes and their impact on contaminant dynamics. Environ Sci Technol 44:15–23

Chien CC, Kao CM, Chen DY, Chen SC, Chen CC (2014) Biotransformation of trinitrotoluene (TNT) by *Pseudomonas* spp. isolated from a TNT-contaminated environment. Environ Toxicol Chem 33:1059–1063

Cornell RM, Schwertmann U (2003) The iron oxides: structure, properties, reactions, occurances and uses. Wiley-VCH, Weinheim

Denicola A, Freeman BA, Trujillo M, Radi R (1996) Peroxynitrite reaction with carbon dioxide/bicarbonate: kinetics and influence on peroxynitrite-mediated oxidations. Arch Biochem Biophys 333:49–58

Eusterhues K, Wagner FE, Häusler W, Hanzlik M, Knicker H, Totsche KU, Kögel-Knabner I, Schwertmann U (2008) Characterization of ferrihydrite-soil organic matter coprecipitates by X-ray diffraction and Mössbauer spectroscopy. Environ Sci Technol 42:7891–7897

Eyers L, Stenuit B, Agathos SN (2008) Denitration of 2,4,6-trinitrotoluene by *Pseudomonas aeruginosa* ESA-5 in the presence of ferrihydrite. Appl Microbiol Biotechnol 79:489–497

Field EK, Gerlach R, Viamajala S, Jennings LK, Peyton BM, Apel WA (2013) Hexavalent chromium reduction by *Cellulomonas* sp. strain ES6: the influence of carbon source, iron minerals, and electron shuttling compounds. Biodegradation 24:437–450

Fischer M, Warneck P (1996) Photodecomposition of nitrite and undissociated nitrous acid in aqueous solution. J Phys Chem 100:18749–18756

Frejaville C, Karoui H, Tuccio B, Le Moigne F, Culcasi M, Pietri S, Lauricella R, Tordo P (1995) 5-(Diethoxyphosphoryl)-5-methyl-1-pyrroline N-oxide: a new efficient phosphorylated nitrone for the *in vitro* and *in vivo* spin trapping of oxygen-centered radicals. J Med Chem 38:258–265

Fritsche W, Scheibner K, Herre A, Hofrichter M (2000) Fungal degradation of explosives: TNT and related nitroaromatic compounds. In: Spain JC, Hughes JB, Knackmuss HJ (eds) Biodegradation of nitroaromatic compounds and explosives. Lewis Publishers, New York, pp 213–238

Fujii M, Rose AL, Waite TD, Omura T (2006) Superoxide-mediated dissolution of amorphous ferric oxyhydroxide in seawater. Environ Sci Technol 40:880–887

Gerlach R, Field EK, Viamajala S, Peyton BM, Apel WA, Cunningham AB (2011) Influence of carbon sources and electron shuttles on ferric iron reduction by *Cellulomonas* sp. strain ES6. Biodegradation 22:983–995

Haber F, Weiss J (1934) The catalytic decomposition of hydrogen peroxide by iron salts. Proc R Soc London Ser A 147:332–351

Han D, Canali R, Garcia J, Aguilera R, Gallaher TK, Cadenas E (2005) Sites and mechanisms of aconitase inactivation by peroxynitrite: modulation by citrate and glutathione. Biochemistry 44:11986–11996

Harter DR (1985) The use and importance of nitroaromatic chemicals in the chemical industry. In: Rickert DE (ed) Toxicity of nitroaromatic compounds. Hemisphere Publishing Corporation, Washington, DC, pp 1–14

Hofstetter TB, Heijman CG, Haderlein SB, Holliger C, Schwarzenbach RP (1999) Complete reduction of TNT and other (poly)nitroaromatic compounds under iron-reducing subsurface conditions. Environ Sci Technol 33:1479–1487

Jain MR, Zinjarde SS, Deobagkar DD, Deobagkar DN (2004) 2,4,6-Trinitrotoluene transformation by a tropical marine yeast, *Yarrowia lipolytica* NCIM. Mar Pollut Bull 49:783–788

Kerem Z, Jensen KA, Hammel KE (1999) Biodegradative mechanism of the brown-rot basidiomycete *Gloeophyllum trabeum*: evidence for an extracellular hydroquinone-driven Fenton reaction. FEBS Lett 446:49–54

Khan MI, Lee J, Park J (2013) A toxicological review on potential microbial degradation intermediates of 2,4,6-trinitrotoluene, and its implications in bioremediation. KSCE J Civ Eng 17:1223–1231

Khilyas IV, Ziganshin AM, Pannier AJ, Gerlach R (2013) Effect of ferrihydrite on 2,4,6-trinitrotoluene biotransformation by an aerobic yeast. Biodegradation 24:631–644

Klueglein N, Kappler A (2013) Abiotic oxidation of Fe(II) by reactive nitrogen species in cultures of the nitrate-reducing Fe(II)-oxidizer *Acidovorax* sp. BoFeN1 – questioning the existence of enzymatic Fe(II) oxidation. Geobiology 11:180–190

Kumagai Y, Wakayama T, Li S, Shinohara A, Iwamatsu A, Sun G, Shimojo N (2000) Zeta-crystallin catalyzes the reductive activation of 2,4,6-trinitrotoluene to generate reactive oxygen species: a proposed mechanism for the induction of cataracts. FEBS Lett 478:295–298

Kumagai Y, Kikushima M, Nakai Y, Shimojo N, Kunimoto M (2004) Neuronal nitric oxide synthase (nNOS) catalyzes one-electron reduction of 2,4,6-trinitrotoluene, resulting in decreased nitric oxide production and increased nNOS gene expression: implication for oxidative stress. Free Radic Biol Med 37:350–357

Kustka AB, Shaked Y, Milligan AJ, King DW, Morel FMM (2005) Extracellular production of superoxide by marine diatoms: contrasting effects on iron redox chemistry and bioavailability. Limnol Oceanogr 50:1172–1180

Lauricella R, Allouch A, Roubaud V, Bouteiller JC, Tuccio B (2004) A new kinetic approach to the evaluation of rate constants for the spin trapping of superoxide/hydroperoxyl radical by nitrones in aqueous media. Org Biomol Chem 2:1304–1309

Melton ED, Swanner ED, Behrens S, Schmidt C, Kappler A (2014) The interplay of microbially mediated and abiotic reactions in the biogeochemical Fe cycle. Nature Rev Microbiol 12:797–808

Mikutta C, Mikutta R, Bonneville S, Wagner F, Voegelin A, Christl I, Kretzschmar R (2008) Synthetic coprecipitates of exopolysaccharides and ferrihydrite. Part I: Characterization. Geochim Cosmochim Acta 72:1111–1127

Morgan B, Lahav O (2007) The effect of pH on the kinetics of spontaneous Fe(II) oxidation by O_2 in aqueous solution – basic principles and a simple heuristic description. Chemosphere 68:2080–2084

Murad E (1996) Magnetic properties of microcrystalline iron(III) oxides and related materials as reflected in their Mössbauer spectra. Phys Chem Minerals 23:248–262

Naumenko EA, Naumov AV, Suvorova ES, Gerlach R, Ziganshin AM, Lozhkin AP, Silkin NI, Naumova RP (2008) Participation of oxygen in the bacterial transformation of 2,4,6-trinitrotoluene. Biochemistry Moscow 73:463–469

Oh SY, Chiu PC, Cha DK (2008) Reductive transformation of 2,4,6-trinitrotoluene, hexahydro-1,3,5-trinitro-1,3,5-triazine, and nitroglycerin by pyrite and magnetite. J Hazard Mater 158:652–655

Pak JW, Knoke KL, Noguera DR, Fox BG, Chambliss GH (2000) Transformation of 2,4,6-trinitrotoluene by purified xenobiotic reductase B from *Pseudomonas fluorescens* I-C. Appl Environ Microbiol 66:4742–4750

Pfeiffer S, Gorren AC, Schmidt K, Werner ER, Hansert B, Bohle DS, Mayer B (1997) Metabolic fate of peroxynitrite in aqueous solution. Reaction with nitric oxide and pH-dependent decomposition to nitrite and oxygen in a 2:1 stoichiometry. J Biol Chem 272:3465–3470

Posth N, Canfield DE, Kappler A (2014) Biogenic Fe(III) minerals: from formation to diagenesis and preservation in the rock record. Earth-Sci Rev 135:103–121

Pryor WA, Squadrito GL (1995) The chemistry of peroxynitrite: a product from the reaction of nitric oxide with superoxide. Am J Physiol 268:699–722

Rose AL, Waite TD (2005) Reduction of organically complexed ferric iron by superoxide in a simulated natural water. Environ Sci Technol 39:2645–2650

Rush JD, Bielski BHJ (1985) Pulse radiolytic studies of the reactions of $HO_2^{\cdot}/O_2^{-}$ with Fe(II)/Fe(III) ions. The reactivity of $HO_2^{\cdot}/O_2^{-}$ with ferric ions and its implication on the occurrence of the Haber-Weiss reaction. J Phys Chem 89:5062–5066

Singh B, Kaur J, Singh K (2012) Microbial remediation of explosive waste. Crit Rev Microbiol 38:152–167

Smets BF, Yin H, Esteve-Núñez A (2007) TNT biotransformation: when chemistry confronts mineralization. Appl Microbiol Biotechnol 76:267–277

Stenuit B, Lamblin G, Cornelis P, Agathos SN (2012) Aerobic denitration of 2,4,6-trinitrotoluene in the presence of phenazine compounds and reduced pyridine nucleotides. Environ Sci Technol 46:10605–10613

Stookey L (1970) Ferrozine – a new spectrometric reagent for iron. Anal Chem 42:779–781

Van Aken B, Agathos SN (2002) Implication of manganese (III), oxalate, and oxygen in the degradation of nitroaromatic compounds by manganese peroxidase (MnP). Appl Microbiol Biotechnol 58:345–351

Van Cleemput O, Baert L (1984) Nitrite: a key compound in N loss processes under acid conditions? Plant Soil 76:233–241

Van den Berg AJ, Halkes SB, Van Ufford HC, Hoekstra MJ, Beukelman CJ (2003) A novel formulation of metal ions and citric acid reduces reactive oxygen species *in vitro*. J Wound Care 12:413–418

Vásquez-Vivar J, Denicola A, Radi R, Augusto O (1997) Peroxynitrite-mediated decarboxylation of pyruvate to both carbon dioxide and carbon dioxide radical anion. Chem Res Toxicol 10:786–794

Wittich RM, Ramos JL, Van Dillewijn P (2009) Microorganisms and explosives: mechanisms of nitrogen release from TNT for use as an N-source for growth. Environ Sci Technol 43:2773–2776

Ziganshin AM, Gerlach R (2014) Pathways of 2,4,6-trinitrotoluene transformation by aerobic yeasts. In: Singh SN (ed) Biological remediation of explosive residues. Environmental science and engineering. Springer International Publishing AG, Switzerland, pp 301–311

Ziganshin AM, Naumov AV, Suvorova ES, Naumenko EA, Naumova RP (2007a) Hydride-mediated reduction of 2,4,6-trinitrotoluene by yeasts as the way to its deep degradation. Microbiology 76:676–682

Ziganshin AM, Gerlach R, Borch T, Naumov AV, Naumova RP (2007b) Production of eight different hydride complexes and nitrite release from 2,4,6-trinitrotoluene by *Yarrowia lipolytica*. Appl Environ Microbiol 73:7898–7905

Ziganshin AM, Gerlach R, Naumenko EA, Naumova RP (2010a) Aerobic degradation of 2,4,6-trinitrotoluene by the yeast strain *Geotrichum candidum* AN-Z4. Microbiology 79:178–183

Ziganshin AM, Naumova RP, Pannier AJ, Gerlach R (2010b) Influence of pH on 2,4,6-trinitrotoluene degradation by *Yarrowia lipolytica*. Chemosphere 79:426–433

18

Alkyl hydroperoxide reductase enhances the growth of *Leuconostoc mesenteroides* lactic acid bacteria at low temperatures

Seitaro Goto[1], Jun Kawamoto[2], Satoshi B Sato[3], Takashi Iki[1], Itaru Watanabe[1], Kazuyuki Kudo[1], Nobuyoshi Esaki[2] and Tatsuo Kurihara[2*]

Abstract

Lactic acid bacteria (LAB) can cause deterioration of food quality even at low temperatures. In this study, we investigated the cold-adaptation mechanism of a novel food spoilage LAB, *Leuconostoc mesenteroides* NH04 (NH04). *L. mesenteroides* was isolated from several spoiled cooked meat products at a high frequency in our factories. NH04 grew rapidly at low temperatures within the shelf-life period and resulted in heavy financial losses. NH04 grew more rapidly than related strains such as *Leuconostoc mesenteroides* NBRC3832 (NBRC3832) at 10°C. Proteome analysis of NH04 demonstrated that this strain produces a homolog of alkyl hydroperoxide reductase--AhpC--the expression of which can be induced at low temperatures. The expression level of AhpC in NH04 was approximately 6-fold higher than that in NBRC3832, which was grown under the same conditions. Although AhpC is known to have an anti-oxidative role in various bacteria by catalyzing the reduction of alkyl hydroperoxide and hydrogen peroxide, the involvement of AhpC in cold adaptation of food spoilage bacteria was unclear. We introduced an expression plasmid containing *ahpC* into NBRC3832, which grows slower than NH04 at 10°C, and found that expression of AhpC enhanced growth. These results demonstrated that AhpC, which likely increases anti-oxidative capacity of LAB, plays an important role in their rapid growth at low temperatures.

Keywords: Lactic acid bacteria; Food spoilage; Cold adaptation; AhpC

Introduction

Lactic acid bacteria (LAB) are useful for the production of fermented foods such as soy sauce and cheese (Caplice and Fitzgerald 1999; McKay and Baldwin 1990; Murooka and Yamshita 2008). Although the importance of LAB is widely recognized, LAB sometimes cause quality defects in fresh foods and fermented meat products stored at low temperatures (Borch et al. 1996). LAB that cause food spoilage can grow and cause quality defects by producing an unpleasant taste, white liquid, and a slimy substance (Asano et al. 2009; Hamasaki et al. 2003; Kondo and Ikeda 2000) even in cooked meat products such as ham and sausage stored at 10°C. Therefore, to ensure the safety of stored food and reduce such risks, elucidation of the cold-adaptation mechanism of LAB in food products is important. LAB that are closely related to *Leuconostoc mesenteroides*, *Lactococcus lactis*, *Leuconostoc citreum*, and *Weissella viridescens* grow rapidly at temperatures below 10°C (Borch et al. 1996; Chenoll et al. 2007; Diez et al. 2009; Hamasaki et al. 2003; Metaxopoulos et al. 2002; Samelis et al. 2006; Samelis et al. 1998), and are frequently isolated from spoiled cooked meat products stored at 10°C. These findings suggest a unique system of adaptation to low temperatures, causing rapid food spoilage in the refrigerator.

At low temperatures, microbes must overcome unfavorable conditions such as decreased membrane fluidity, molecular dynamics, and enzymatic activities. They can adapt to such an extreme environment via the induction of cold-shock and cold-acclimation proteins responsible for the maintenance of membrane fluidity, protein synthesis and folding, and metabolism (Feller and Gerday 2003; Graumann and Marahiel 1998; Marceau et al. 2004; Salotra et al. 1995; Wang et al. 2005; Wouters et al. 2000; Yamanaka et al. 1998). Identification

* Correspondence: kurihara@scl.kyoto-u.ac.jp
[2]Institute for Chemical Research, Kyoto University, Uji, Kyoto 611-0011, Japan
Full list of author information is available at the end of the article

and characterization of these proteins is necessary to understand the environmental adaptations of psychrotrophic LAB. Many studies on cold-inducible proteins have been reported, but few have focused on the relationship between these proteins and the growth of food spoilage bacteria. In order to prevent the contamination of foods by psychrotrophic LAB and to develop a method for their detection, we focused on the cold-adaptation mechanism of LAB at the molecular level. In this study, we characterized a novel psychrotrophic LAB, *Leuconostoc mesenteroides* NH04 (NH04), isolated from cooked meat stored at 10°C. *L. mesenteroides* was isolated from several spoiled cooked meat products at high frequency in our factories. NH04 grows rapidly at low temperatures within the shelf-life period and is responsible for the significant financial losses incurred. We analyzed the proteins that were inducibly expressed by this strain at low temperatures and identified a protein that facilitates the growth of *L. mesenteroides* at low temperatures.

Materials and methods

Bacterial strains, plasmids, and culture conditions

The strains used in this study are NH04 isolated from spoiled sausage and the related LAB, *Leuconostoc mesenteroides* NBRC3832 (NBRC3832) (NITE Biological Resource Center; Kisarazu, Japan). NH04 was deposited in NITE Biological Resource Center with accession number NBRC110676. For isolation of NH04, 25 g of spoiled product was suspended in 225 mL saline, and the suspension was subjected to 10-fold serial dilution. The dilutions (1 mL) were mixed with 15–20 mL BCP medium containing agar (Nissui Pharmaceutical Co., Ltd.; Tokyo, Japan), and the plates were incubated at 25°C for 72 h. The dominant species in the plates were isolated, and identified by sequencing their 16S rRNA genes. The nucleotide sequence of the 16S rRNA gene of NH04 was deposited in DDBJ with accession number LC005518. NH04 thus isolated and NBRC3832 were grown in 5 mL GAM broth (Nissui Pharmaceutical Co., Ltd.), with 1.0% D-glucose added to improve LAB growth. GAM broth consisted of peptone (10.0 g), soy peptone (3.0 g), proteose peptone (10.0 g), digested serum (13.5 g), yeast extract (5.0 g), meat extract (2.2 g), liver extract (1.3 g), dextrose (3.0 g), potassium dihydrogen phosphate (2.5 g), sodium chloride (3.0 g), soluble starch (5.0 g), L-cysteine hydrochloride (2.5 g), and sodium thioglycolate (0.3 g) dissolved in 1 L of deionized water (final pH of 7.1). The cells were grown at 10°C or 25°C in a compact rocking incubator (TVS062CA; ADVANTEC Toyo; Tokyo, Japan) by shaking at 70 rpm, and the growth was simultaneously monitored by measuring the turbidity at 660 nm with a spectrophotometer installed in the incubator. The kinetics of the growth was determined from three independent experiments. The LAB shuttle vector pGK::*nuc*MCS (Le Loir et al. 1994) was used for the transformation of NBRC3832. Transformants harboring pGK::*nuc*MCS and pGK*ahpC*, which was constructed as described below, were cultivated in the presence of erythromycin (5 μg/mL).

Identification of cold-inducible proteins by two-dimensional electrophoresis (2DE)

Cells (10 mL) grown to stationary phase ($1.5 < OD_{600} < 2.5$) at 10°C and 25°C were harvested by centrifugation and resuspended in 100 μl of 50 mM Tris–HCl (pH 7.0). Cell suspensions were sonicated for 50 min at 10% amplitude (3 mm microtip, 400 W; Digital Sonifier; BRANSON Ultrasonics Corporation; Danbury, CT) while they were chilled on ice, and then centrifuged at 2,200 × *g* for 10 min at 4°C. Proteins in the supernatants were purified using the ReadyPrep 2-D Cleanup Kit (Bio-Rad Laboratories, Inc.; Hercules, CA) before an initial isoelectric focusing. Proteins (150 μg) were loaded onto Ready Strip IPG strips (17 cm, pH 4–7; Bio-Rad Laboratories, Inc.), and isoelectric focusing was performed using the PROTEAN IEF Cell (Bio-Rad Laboratories, Inc.), according to the manufacturer's protocol. Treatment of the gel strips for 2DE was carried out as described previously (Kawamoto et al. 2007). After fixation and staining with SYPRO Ruby (Invitrogen Corp.; Carlsbad, CA), gels were scanned using Typhoon 9400 image analyzer (GE Healthcare Ltd.; Buckinghamshire, UK). Each experiment was performed three times to ensure reproducibility.

N-Terminal amino acid sequence of the protein inducibly produced at 10°C was determined using a PPSQ-21 protein sequencer (Shimadzu; Kyoto, Japan). A BLAST search (http://blast.ncbi.nlm.nih.gov/Blast.cgi) with the identified N-terminal sequence suggested that the cold-inducible protein was a homolog of AhpC. Degenerate primers for amplification of the *ahpC* gene of NH04 [peroxi130_For and peroxi569_Rev (Table 1)] were designed from conserved regions of the genes coding for AhpC from several LAB, including *Leuconostoc, Lactobacillus*, and

Table 1 Primers used in this study

Primer	Sequence
peroxi130_For	GACTTCTCATTTGTTTGYCC
peroxi569_Rev	TAMAKYTTRCCRACTARRTCWARRCT
peroxi232_inv_For	GTATCCGTAGCTTCTGCC
peroxi407_inv_Rev	CATATACCATCAACAACATGGG
peroxi_start_For	ATGACTACAAATTTTATTGATTCAGAAATAACAGA
peroxi_end_SalI_Rev	ACGCGTCGACGTCCTAAATTTTACCAACTAAATCT AAGCTCG
thr_prm_XhoI_For	CCGCTCGAGCGGATCATCTGATAGATATCGATCA TAAGAG
thr_prm_peroxi_Rev	TCAATAAAATTTGTAGTCATGATTAATTCTCCTTTTT TGTGACAAAAGTA

Enterococcus species, and the internal nucleotide fragment was amplified. Based on the internal sequence of the gene coding for AhpC in NH04, an inverse PCR was performed with two primers (peroxi232_inv_For and peroxi407_inv_Rev) listed in Table 1. The sequences of the PCR products were determined and assembled to obtain the complete sequence of *ahpC*, which was deposited in DDBJ with accession number AB819067.

RNA extraction and quantitative real-time RT-PCR

Total RNA was extracted from cells cultivated at 10°C and 25°C using the RNeasy Kit (QIAGEN Inc., Valencia, CA). RNA pellets were dissolved in 0.1% diethyl pyrocarbonate-treated water and stored at −80°C until use. Quantitative real-time RT-PCR was performed with SuperScript III Platinum SYBR Green One-Step qRT-PCR Kit (Invitrogen Corp.) and an Mx3000P Multiple Quantitative RT-PCR system (Stratagene; La Jolla, CA). The amount of mRNA in each sample was normalized with the amount of 16S rRNA. Each experiment was performed three times to ensure reproducibility.

Construction of AhpC-overexpressing strain

A gene fragment coding for AhpC was amplified using the primers peroxi_start_For and peroxi_end_SalI_Rev (Table 1), and the PCR product was fused with a DNA fragment containing a putative promoter region of the NH04 mannitol dehydrogenase gene (*mdh*) amplified by PCR with the primers thr_prm_XhoI_For and thr_prm_peroxi_Rev. The gene fragment was ligated with the pGK::*nuc*MCS plasmid after digestion with XhoI and SalI. The constructed plasmid, pGK*ahp*C, was introduced into NBRC3832 and NH04 by electroporation. Electroporation was performed using a previously described method with slight modifications (Leathers et al. 2004). Briefly, the cells from 50 mL culture were harvested at an OD_{600} of 0.6, washed twice with 75 mL of ice-cold deionized water, then washed once with 5 mL of ice-cold sterile electroporation buffer (1 mM of potassium phosphate buffer (pH 7.4) containing 1 mM of $MgCl_2$ and 0.5 M of sucrose), and suspended in 1 mL of electroporation buffer. Plasmids (1 μg) were mixed with 40 μL of the cell suspensions and placed into a pre-chilled electroporation cuvette (0.2 cm). Electroporation was performed with the Gene Pulser Xcell PC system (Bio-Rad Laboratories, Inc.) at a capacitance of 25 μF. After electroporation, the cell suspensions were immediately diluted with 1 mL of GAM broth containing 1% D-glucose and incubated at 25°C for 2 h. Transformants were selected on GAM plates containing 1% D-glucose and 5 μg/mL of erythromycin.

Production of AhpC from NH04 was analyzed by SDS-PAGE. SDS-PAGE was performed on a 12.5% polyacrylamide gel (ePAGEL; ATTO Corp.; Tokyo, Japan). Soluble proteins (5 μg) were loaded and run on the gel. After fixation and staining with SYPRO Ruby, the gels were scanned using a Typhoon 9400 imager.

Results

Growth of NH04

The growth of NH04 and NBRC3832 at 10°C and 25°C was monitored. NH04 grew faster than NBRC3832 at both 10°C and 25°C (Figure 1a, b). The doubling times of NH04 were 9 ± 0.6 h and 1.8 ± 0.1 h at 10°C and 25°C, respectively, whereas those of NBRC3832 were 13 ± 0.7 h and 2.5 ± 0.3 h at 10°C and 25°C, respectively. These results indicate that there were significant differences ($p < 0.05$, n = 3) between the doubling times of the two strains at both 10°C and 25°C.

Identification of the cold-inducible protein of cold-adapted *Leuconostoc mesenteroides* NH04

To identify the proteins contributing to the cold adaptation of NH04, we performed 2DE of the soluble proteins of NH04 grown at 10°C and 25°C. 2DE analysis showed that the amount of a specific protein, indicated by arrowhead in

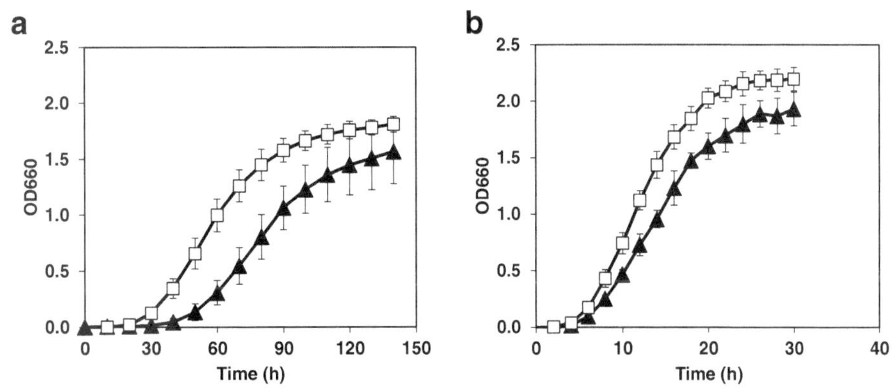

Figure 1 Growth of NH04 and NBRC3832. NH04 (square) and NBRC3832 (triangle) were cultured at 10°C (a) and 25°C (b). Each growth curve was plotted using data obtained from three experiments.

Figure 2a and b, was 2 ± 0.2 times higher ($p < 0.05$, n = 3) in the cells grown at 10°C than in the cells grown at 25°C. We further characterized this protein because it was one of the major proteins of NH04 and reproducibly more abundant at 10°C. The N-terminal amino acid sequence of this protein was MTTNFIDSEITDFKVNAYHD. The gene encoding this protein was cloned and sequenced as described in Materials and Methods, and it was found that this cold-inducible protein was a homolog of alkyl hydroperoxide reductase (AhpC). The *ahpC* gene (AB819067) coded for a protein of 189 amino acid residues, and the calculated molecular weight of the protein was 21,118.5. AhpC of NH04 shares a high sequence similarity (99.5% identity) with that of NBRC3832 (AB921977). Despite the high sequence similarity, the expression level of this protein was much higher in NH04 than in NBRC3832. 2DE analysis of the soluble proteins of NBRC3832 grown under the same conditions demonstrated that the fluorescence intensity of the spot for AhpC (arrowhead in Figure 2c) was approximately 17% ± 12% of that for NH04 (arrowhead in Figure 2a).

Transcriptional levels of *ahpC* in *Leuconostoc* species

The transcriptional level of *ahpC* in the NH04 cells grown at 10°C was approximately 3.8 ± 1.5-fold greater ($p < 0.05$, n = 3) than that in the cells grown at 25°C. The level of *ahpC* in NH04 grown at 10°C was approximately 2.8 ± 0.4-fold higher ($p < 0.05$, n = 3) than that in NBRC3832 grown under the same conditions.

AhpC facilitates the growth of LAB at low temperatures

To examine the effects of high-level expression of AhpC on the growth of LAB at low temperatures, we constructed an overexpression vector for *ahpC* (pGK*ahpC*) and introduced it into NBRC3832, which grows slower than NH04. The level of AhpC in NBRC3832 harboring pGK*ahpC* was approximately 6.5 ± 1.6-fold higher than that in the cells harboring a control plasmid, pGK::*nuc*MCS (Figure 3a). NBRC3832 overexpressing *ahpC* grew more rapidly than the cells containing a control plasmid at 10°C (Figure 3b). The doubling time of NBRC3832 harboring pGK*ahpC* was 15.4 ± 0.8 h, whereas that of the strain harboring pGK::*nuc*MCS was 17.6 ± 1.1 h. The difference between the growth rates of NBRC3832 harboring pGK*ahpC* and pGK::*nuc*MCS was statistically significant ($p < 0.05$, n = 3). In contrast to the case of NBRC3832, the growth rate of NH04 was not affected by introduction of pGK*ahpC* (Figure 3b). The doubling times of NH04 harboring pGK*ahpC* and pGK::*nuc*MCS were both 9.8 ± 0.3 h at 10°C. At 25°C, introduction of pGK*ahpC* did not significantly affect the growth rate of NBRC3832 and NH04 (data not shown). The doubling times of NBRC3832 harboring pGK*ahpC*

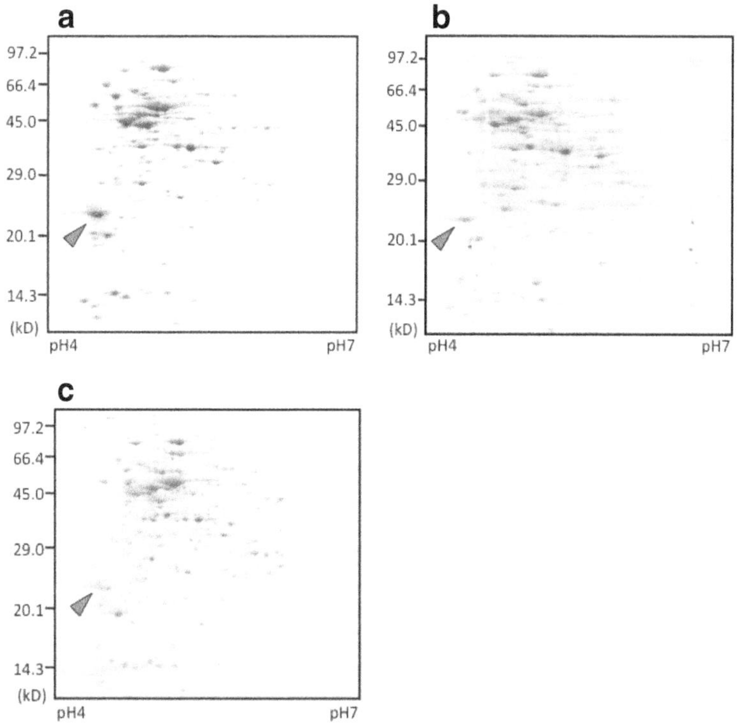

Figure 2 Two-dimensional polyacrylamide gel electrophoresis of soluble proteins extracted from NH04 grown at 10°C (a) and 25°C (b) and the related strain, NBRC3832, grown at 10°C (c). Arrowhead indicates the spot corresponding to AhpC.

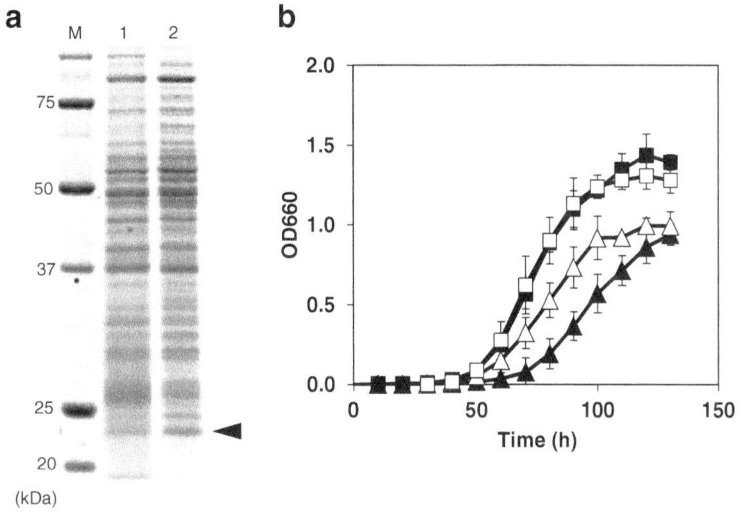

Figure 3 Effects of *ahpC* overexpression on the growth of NBRC3832. (a) Production of AhpC from NH04 in NBRC3832. Soluble proteins from NBRC3832 harboring a control plasmid (lane 1) and an *ahpC* overexpression plasmid (lane 2) were analyzed by SDS-PAGE. Arrowhead indicates the band corresponding to AhpC. Lane M represents the molecular weight marker. **(b)** Growth of NH04 (square) and NBRC3832 (triangle) containing a control plasmid (filled symbol) and an *ahpC* overexpression plasmid (open symbol) at 10°C. Each growth curve was plotted using data obtained from three experiments.

and pGK::*nuc*MCS were 3.0 ± 0.2 h and 3.2 ± 0.1 h, respectively, at 25°C, which were not significantly different from each other ($p < 0.05$, n = 3). The doubling times of NH04 harboring pGK*ahp*C and pGK::*nuc*MCS were both 2.0 ± 0.1 h at 25°C.

Discussion

We isolated a novel food-spoilage bacterium, NH04, from spoiled meat product at 10°C, which causes food quality defects at low temperatures. Although little is known regarding the effects of various stresses on *Leuconostoc* species, they can survive in unfavorable environments such as oxidative and low temperature conditions (Con and Seamus 1986; Salotra et al. 1995; van de Guchte et al. 2002). Most *Leuconostoc* strains grow even at 10°C, and NH04 grows faster than other related strains at both 10°C and 25°C (Figure 1).

A comprehensive analysis of proteins synthesized in NH04 demonstrated that this strain inducibly produces a homolog of alkyl hydroperoxide reductase, AhpC, which likely has a role in anti-oxidative functions at low temperatures (Figure 2). AhpC homologs are widely distributed among prokaryotes, and AhpC shares approximately 40% amino acid sequence identity with thioredoxin peroxidase from yeast, rat, plant, amoebae, nematodes, rodents, and humans (Chen et al. 1998). AhpC homologs define a large family of anti-oxidants present in organisms from all kingdoms, and they protect cells from reactive oxygen species (ROS). ROS such as O_2^-, •OH, and H_2O_2 cause oxidative damage to cells (Cabiscol et al. 2000). The O_2^- generated during cell growth is typically converted to hydrogen peroxide spontaneously or by the activity of superoxide dismutase (SOD), from which the hydroxyl radical is generated by a metal-catalyzed redox reaction (the Fenton reaction). Hydroxyl radicals react immediately with cellular components and generate organic hydroperoxide (ROOH) that causes cell damage. AhpC catalyzes the reduction of organic hydroperoxide and protects cell components, thus acting as an anti-oxidant against hydroxyl radicals. Since cold stress induces production of ROS (Gocheva et al. 2009), psychrotrophic LAB that induce the expression of anti-oxidative protein(s) may grow rapidly at low temperatures. We demonstrated that the production of AhpC from NH04 in the related strain, NBRC3832, promotes growth at low temperatures (Figure 3). This suggests an important role for AhpC in the growth of LAB at low temperatures. Reduction of organic hydroperoxide may be crucial for LAB to grow at low temperatures. NH04 grows faster than NBRC3832 not only at 10°C but also at 25°C. However, the faster growth of NH04 at 25°C is not likely due to AhpC, because the expression level of AhpC in NH04 at 25°C is much lower than at 10°C and a statistically significant growth improvement of NBRC3832 by overexpression of AhpC was not observed at 25°C.

In this study, due to the lack of a gene disruption system for NH04 and related strains, the involvement of AhpC in the cold adaptation of NH04 could not be shown directly by a loss-of-function analysis. Nevertheless, the results obtained by heterologous expression of AhpC in a related strain strongly support the view that this protein is involved in the growth of LAB at low temperatures.

Our results provide new insight into the molecular mechanism of cold adaptation of food-spoilage LAB that cause food quality defects during cold storage. In the future studies, we will examine whether AhpC is involved in the cold adaptation of other food-spoilage LAB, and whether other anti-oxidative substances play a role in cold adaptation. Our findings raise a possibility for the control of food-spoilage related LAB growing at low temperatures by developing an inhibitor against their anti-oxidation system. It may also be possible in the future for food manufacturers to detect food products and manufacturing equipment contaminated with food-spoilage LAB by developing a system to detect AhpC as a marker.

Competing interests
The authors declare that they have no competing interests.

Authors' contributions
SG designed the study, performed all the experiments, and wrote the manuscript. JK designed the study, helped the experiments, and wrote the manuscript. SBS helped experimental design and participated in the discussion on the interpretation of the results. TI, IW, and KK were committed to all the experiments. NE participated in the discussion on the interpretation of the results. TK designed the study and wrote the manuscript. All authors approved the final manuscript.

Acknowledgments
This work was supported in part by Grants-in-Aid for Scientific Research (B) from JSPS (22404021, 24380047, and 25303028 to T.K.) and a Grant from the Japan Foundation for Applied Enzymology (to T.K.).

Author details
[1]Product Development Laboratory, NH Foods Ltd., Chikusei, Ibaraki 308-0042, Japan. [2]Institute for Chemical Research, Kyoto University, Uji, Kyoto 611-0011, Japan. [3]Research Center for Low Temperature and Materials Sciences, Kyoto University, Kyoto 606-8501, Japan.

References
Asano S, Iijima K, Suzuki K, Motoyama Y, Ogata T, Kitagawa Y (2009) Rapid detection and identification of beer-spoilage lactic acid bacteria by microcolony method. J Biosci Bioeng 108(2):124–129, doi:10.1016/j.jbiosc.2009.02.016
Borch E, Kant-Muermans ML, Blixt Y (1996) Bacterial spoilage of meat and cured meat products. Int J Food Microbiol 33(1):103–120, doi:10.1016/0168-1605(96)01135-X
Cabiscol E, Tamarit J, Ros J (2000) Oxidative stress in bacteria and protein damage by reactive oxygen species. Int Microbiol 3(1):3–8
Caplice E, Fitzgerald GF (1999) Food fermentations: role of microorganisms in food production and preservation. Int J Food Microbiol 50(1–2):131–149, doi:10.1016/S0168-1605(99)00082-3
Chen L, Xie QW, Nathan C (1998) Alkyl hydroperoxide reductase subunit C (AhpC) protects bacterial and human cells against reactive nitrogen intermediates. Mol Cell 1(6):795–805, doi: 10.1016/S1097-2765(00)80079-9
Chenoll E, Macian MC, Elizaquivel P, Aznar R (2007) Lactic acid bacteria associated with vacuum-packed cooked meat product spoilage: population analysis by rDNA-based methods. J Appl Microbiol 102(2):498–508, doi:10.1111/j.1365-2672.2006.03081.x
Con AL, Seamus EC (1986) Active role of oxygen and NADH oxidase in growth and energy metabolism of *Leuconostoc*. J Gene Microbiol 132(7):1789–1796, doi:10.1099/00221287-132-7-1789
Diez AM, Bjorkroth J, Jaime I, Rovira J (2009) Microbial, sensory and volatile changes during the anaerobic cold storage of *morcilla de Burgos* previously inoculated with *Weissella viridescens* and *Leuconostoc mesenteroides*. Int J Food Microbiol 131(2–3):168–177, doi:10.1016/j.ijfoodmicro.2009.02.019
Feller G, Gerday C (2003) Psychrophilic enzymes: hot topics in cold adaptation. Nat Rev Microbiol 1(3):200–208, doi:10.1038/nrmicro773
Gocheva YG, Tosi S, Krumova ET, Slokoska LS, Miteva JG, Vassilev SV, Angelova MB (2009) Temperature downshift induces antioxidant response in fungi isolated from Antarctica. Extremophiles 13(2):273–281, doi:10.1007/s00792-008-0215-1
Graumann PL, Marahiel MA (1998) A superfamily of proteins that contain the cold-shock domain. Trends Biochem Sci 23(8):286–290, doi:10.1016/S0968-0004(98)01255-9
Hamasaki Y, Ayaki M, Fuchu H, Sugiyama M, Morita H (2003) Behavior of psychrotrophic lactic acid bacteria isolated from spoiling cooked meat products. Appl Environ Microbiol 69(6):3668–3671, doi:10.1128/AEM.69.6.3668-3671.2003
Kawamoto J, Kurihara T, Kitagawa M, Kato I, Esaki N (2007) Proteomic studies of an Antarctic cold-adapted bacterium, *Shewanella livingstonensis* Ac10, for global identification of cold-inducible proteins. Extremophiles 11(6):819–826, doi:10.1007/s00792-007-0098-6
Kondo T, Ikeda T (2000) Rapid detection of substrate-oxidizing activity of hiochi bacteria using benzoquinone-mediated amperometric method. J Biosci Bioeng 90(2):217–219, doi:10.1016/S1389-1723(00)80114-0
Le Loir Y, Gruss A, Ehrlich SD, Langella P (1994) Direct screening of recombinants in Gram-positive bacteria using the secreted staphylococcal nuclease as a reporter. J Bacteriol 176(16):5135–5139
Leathers TD, Jones JD, Wyckoff HA (2004) Transformation of alternan-producing strains of *Leuconostoc* by electroporation. Biotechnol Lett 26(14):1119–1124, doi:10.1023/B:BILE.0000035482.41491.7e
Marceau A, Zagorec M, Chaillou S, Mera T, Champomier-Verges MC (2004) Evidence for involvement of at least six proteins in adaptation of *Lactobacillus sakei* to cold temperatures and addition of NaCl. Appl Environ Microbiol 70 (12):7260–7268, doi:10.1128/AEM.70.12.7260-7268.2004
McKay LL, Baldwin KA (1990) Applications for biotechnology: present and future improvements in lactic acid bacteria. FEMS Microbiol Rev 7(1–2):3–14, doi:10.1016/0378-1097(90)90694-L
Metaxopoulos J, Mataragas M, Drosinos EH (2002) Microbial interaction in cooked cured meat products under vacuum or modified atmosphere at 4°C. J Appl Microbiol 93(3):363–373, doi:10.1046/j.1365-2672.2002.01701.x
Murooka Y, Yamshita M (2008) Traditional healthful fermented products of Japan. J Ind Microbiol Biotechnol 35(8):791–798, doi:10.1007/s10295-008-0362-5
Salotra P, Singh DK, Seal KP, Krishna N, Jaffe H, Bhatnagar R (1995) Expression of DnaK and GroEL homologs in *Leuconostoc esenteroides* in response to heat shock, cold shock or chemical stress. FEMS Microbiol Lett 131(1):57–62, doi:10.1111/j.1574-6968.1995.tb07754.x
Samelis J, Bjorkroth J, Kakouri A, Rementzis J (2006) *Leuconostoc carnosum* associated with spoilage of refrigerated whole cooked hams in Greece. J Food Prot 69(9):2268–2273
Samelis J, Kakouri A, Georgiadou KG, Metaxopoulos J (1998) Evaluation of the extent and type of bacterial contamination at different stages of processing of cooked ham. J Appl Microbiol 84(4):649–660, doi:10.1046/j.1365-2672.1998.00392.x
van de Guchte M, Serror P, Chervaux C, Smokvina T, Ehrlich SD, Maguin E (2002) Stress responses in lactic acid bacteria. Antonie Van Leeuwenhoek 82(1–4):187–216, doi:10.1023/A:1020631532202
Wang Y, Delettre J, Guillot A, Corrieu G, Beal C (2005) Influence of cooling temperature and duration on cold adaptation of *Lactobacillus acidophilus* RD758. Cryobiology 50(3):294–307, doi:10.1016/j.cryobiol.2005.03.001
Wouters JA, Mailhes M, Rombouts FM, de Vos WM, Kuipers OP, Abee T (2000) Physiological and regulatory effects of controlled overproduction of five cold shock proteins of *Lactococcus lactis* MG1363. Appl Environ Microbiol 66 (9):3756–3763, doi:10.1128/AEM.66.9.3756-3763.2000
Yamanaka K, Fang L, Inouye M (1998) The CspA family in *Escherichia coli*: multiple gene duplication for stress adaptation. Mol Microbiol 27(2):247–255, doi:10.1046/j.1365-2958.1998.00683.x

19

Recombinant glucagon: a differential biological activity

Angelina M M Basso[1,2], Patrícia B Pelegrini[2], Fernanda Mulinari[2,3], Michelle C Costa[2], Antonio B Viana[2,4], Luciano P Silva[5] and Maria Fatima Grossi-de-Sa[2,4*]

Abstract

In Brazil, there is a growing demand for specialised pharmaceuticals, and the high cost of their importation results in increasing costs, reaching US$ 1.34 billion in 2012 and US$ 1.61 billion in 2013. Worldwide expenses related to drugs could reach US$ 1.3 trillion in 2018, especially due to new treatments for hepatitis C and cancer. Specialised or high-cost pharmaceutical drugs used for the treatment of viral hepatitis, multiple sclerosis, HIV and diabetes are distributed free of charge by the Brazilian government. The glucagon peptide was included in this group of high-cost biopharmaceuticals in 2008. Although its main application is the treatment of hypoglycaemia in diabetic patients, it can also be used with patients in an alcoholic coma, for those patients with biliary tract pain, and as a bronchodilator. Therefore, in order to reduce biopharmaceutical production costs, the Brazilian government passed laws focusing on the development and increase of a National Pharmaceutical Industrial Centre, including the demand for the national production of glucagon. For that reason and given the importance and high cost of recombinant glucagon, the purpose of this study was to develop methods to improve production, purification and performance of the biological activity of recombinant glucagon. Glucagon was recombined into a plasmid vector containing a Glutathione S-transferase tag, and the peptide was expressed in a heterologous *Escherichia coli* system. After purification procedures and molecular analyses, the biological activity of this recombinant glucagon was examined using *in vivo* assays and showed a highly significant ($p < 0.00001$) and prolonged effect on glucose levels when compared with the standard glucagon. The experimental procedure described here facilitates the high level production of recombinant glucagon with an extended biological activity.

Keywords: Recombinant Glucagon; Peptide expression; Heterologous system; Enterokinase; GST tag; Biological activity

Introduction

The development of recombinant proteins has been extremely promising. In 2009, pharmaceutical industries spent US$ 90 billion on the activity evaluation of 400 different drugs. Moreover, in 2012, more than 150 recombinant drugs were approved by FDA (Food and Drug Administration) regulators and/or by the European Medicine Agency (Huang et al. 2012). By 2017, it is expected that over 300 products will be in the biopharmaceutical market, with a value of about US$ 109 billion and an expected growth of US$ 166 billion (Ibarra Cabrera et al. 2013). Worldwide expenses related to drugs could reach US$ 1.3 trillion in 2018, especially due to new treatments for hepatitis C and cancer (Chen, 2014).

According to the World Health Organization, one third of the world's population has no regular access to essential medicines (Nowbike 2006; Blatt et al. 2012). In Brazil, the Government provides free, high-cost drugs to the population, which are classified into three different categories: basic, strategic and specialised. The specialised components are expensive, which limits access of the population to important medicines. Hence, in 1993, the Program for Specialized Medicaments (SM) became a prospective solution to this problem. However, most of these medicines are imported from other countries, resulting in a large cost to the Brazilian government. Moreover, the expectation is that the amount spent on the importation of SM will continue to grow over the next several years.

* Correspondence: fatima.grossi@embrapa.br
[2]Laboratory of Plant-Pest Interaction, Embrapa – Genetic Resources and Biotechnology, Brasília, DF, Brazil
[4]Catholic University of Brasilia, Brasilia, DF, Brazil
Full list of author information is available at the end of the article

In 2009, US$ 884 million was spent on specialised drugs, which is four times the amount (US$ 198 million) spent in 2003 (Valadares 2009). In 2013, US$ 1.61 billion was spent, and in 2014, preliminary data were provided by the Brazilian Ministry of Health showing US$ 1.5 billion for the purchase of SM (Brazil Ministry Health, 2014). Therefore, to reduce high costs associated with imported pharmaceutical products, the Brazilian government released the Technological Innovation Law, which stimulates the development of new biotechnological companies for the production of national biopharmaceuticals (Brazil Ministry Health 2008; de Castro 2014). The Technological Innovation Law is part of a strategic plan to strengthen the pharmaceutical industry, due the increased demand for health public services and consumption (medicines, diagnostics kits, vaccines, etc.). The strategy aimed to replace the imported drugs that were causing a significant impact on the economy and encourage scientific training and public–private partnerships (Victora et al. 2011; Silveira 2012; Reddy 2013; Sundfeld and Souza 2013). In 2012, 34 public–private partnerships were developed to produce several molecules with therapeutic activity, such as anti-retroviral and anti-asthmatic drugs and drugs used to treat rheumatoid arthritis and Crohn and Gaucher diseases, among others (Paiva 2012). In 2013, 63 public-private partnerships were developed, resulting in the production of 63 drugs, vaccines and rapid diagnostic tests, involving 16 public and 41 private laboratories, with an estimated US$ 2.8 billion per year in budget savings (Oliveira 2013). Alongside Brazil, other countries, such as South Korea and India, also used national biosimilar production as an alternative to increase the access of high-cost biopharmaceuticals to the population (IMS 2011) through similar biotechnology. Furthermore, the large accumulation of expired patents related to biopharmaceutical products led to the development of a second generation of drugs: biosimilars (Niederwieser and Schmitz 2010; Utzig et al. 2010). The substitution of classic drugs with biosimilars depends on the criteria established by each country (Rosenberg et al. 2010).

The glucagon peptide could be used in diagnostics tests or for the treatment of alcoholic coma, biliary tract pain, hypoglycaemia (as a bronchodilator), etc. (Guimarães et al. 2011). Furthermore, glucagon has been on the Brazilian SM Program list since 2008. Currently, glucagon is imported from other countries, and initial production of new substances is restricted to universities and research centres (Massi et al. 2013). The main activity of glucagon involves the primary regulation of glucose production in the liver in situations of fasting, exercise and hypoglycaemia (Ramnanan et al. 2011). Hancock and colleagues (2010) indicated that the main glucagon activity was associated with glycemic control in fasting and postprandial situations. Therefore, glucagon utilisation is important since this hormone shows a fundamental role in glycemic control and, consequently, the maintenance of homeostasis.

In this work, the glucagon peptide was expressed and purified from *Escherichia coli* cells using the pGEX vector system. An efficient purification procedure at the laboratory scale for a biologically active Glutathione S-transferase (GST)-glucagon fusion molecule was also performed. The biological activity of the biosimilar molecule was evaluated in *in vivo* tests, showing a prolonged activity when compared with the control (Glucagen NovoNordisk®). This study is the first report in which glucagon is efficiently produced in Brazil using a bacterial system, showing advanced biological activity in *in vivo* assays.

Materials and methods

Materials

Glucagon PCR primers were purchased from Integrated DNA Technologies (Coralville, IA, USA) and the FUSION Taq High-Fidelity enzyme was purchased from Finnzymes, Thermo Scientific (Vantaa, Finland). The glucagon expression vector *pGEX4t-3* and the Glutathione Sepharose 4 Fast Flow resin were purchased from GE Healthcare (Inc, Sweden). *E. coli* BL21 (DE3) cells (Invitrogen) were kindly provided by Embrapa Cenargen, Brazil. The expression inductor Isopropyl β-D-1-thiogalactopyranoside (IPTG) was acquired from Invitrogen (Carlsbad, CA., USA). Glucagen®, used as positive control, was obtained from Novo Nordisk A/S (Denmark). Polyclonal anti-glucagon antibodies were purchased from Santa Cruz Biotechnology (Santa Cruz, CA., USA).

Cloning of the human glucagon gene

The nucleotide sequence of human glucagon (GenBank GI 30582464), containing 128 bp, was cloned into the synthetic vector *pGEX4t-3* (GE Healthcare). The glucagon coding sequence was then amplified using designed primers containing a *Sal* I site (forward - 5' GGG GGT CGA CGA TTA CAA AGA TGA TGA TG 3'), a *Not* I site (reverse – 5' TTT TGC GGC CGC TCA GGT GTT CAT CAG CCA C 3') and a cleavage site for the FLAG tag. The amplicon was purified, and a recombination reaction was performed to clone the glucagon fragment into the *pGEX4-t3* expression vector (GE Healthcare). The recombinant plasmid was used to transform *E. coli* XLI-Blue cells via electroporation, and transformed cells were initially selected using LB (Luria and Bertani) agar plates [1% (w/v) tryptone, 0.5% (w/v) yeast extract, 0.5% (w/v) sodium chloride, and 1.5% (w/v) agar, supplemented with 100 μg/mL ampicillin, pH 7.0].

Recombinant glucagon expression

The construct was inserted into *E. coli* BL21 (DE3) cells through thermal shock. To evaluate the glucagon expression profile, a single colony was inoculated into 10 mL of LB medium containing ampicillin (100 μg/mL) and grown overnight at 37°C (200 rpm). The seed culture was inoculated into 1 L (Litre) of LB medium supplemented with ampicillin, and cells were grown at 30°C during 2 hours. The expression induction was carried out by the addition of IPTG (Isopropyl β-D-1-thiogalactopyranosid; 0.25 mM). After 4 hours of induction, cells were harvested by centrifugation (1680 g for 5 min at 25°C). The supernatant was discarded, and the pellet was resuspended in 1X sample buffer [10% glycerol (w/v), 2-10% mercaptoethanol (v/v), 2.3% sodium dodecyl sulfate (w/v), 124 mM Tris-Base, 0.01% bromophenol blue (w/v)] and boiled for 5 min. After that, soluble and insoluble proteins were separated by centrifugation (13171 g for 5 min), and protein molecular masses were analysed by Coomassie-blue stained, 12.5% acrylamide SDS-PAGE gels (Sambrook and Russell 2006). Glucagon expression levels were determined by densitometry using the ImageScannerIII (GE Healthcare Inc., Sweden) and ImageMaster2D Plattinum 7.02 and LabScan 6.0 (GE Healthcare Inc., Sweden) software for data analyses.

Bioreactor glucagon expression was performed. Larger lab-scale peptide production was obtained using the Wave Bioreactor™ 20/50 EHT system (GE Healthcare, Brazil). A series of parameters were adjusted to optimise glucagon expression in the Wave Bioreactor, such as the swing angle (30°), speed (30 rpm), temperature (30°C) and airflow (0.1 ppm for 30 seconds every 1 hour). Peptide expression was induced with the addition of IPTG (0.25 mM), and cells were harvested after 4 hours.

Glucagon purification

Recombinant glucagon was purified by Glutathione Sepharose 4 Fast Flow (GE Healthcare Inc., Sweden) affinity chromatography according to the manufacturer's instructions. Purified samples were dialysed overnight in a tube with a cut off of 12000–14000 kDa (Cheshire Sciences, UK), lyophilised and stored at −20°C for quantification. Each sample concentration was previously estimated using Qubit methodology (Invitrogen, USA). Quantified samples were used for the digestion procedure with enterokinase.

Western blot assays

Purified GST-FLAG-Glucagon and non-purified protein concentrations were estimated using the Bradford method (1976). Protein molecular mass was estimated as previously described. Glucagon was detected using the polyclonal anti-glucagon antibody (Santa Cruz, CA., USA) and alkaline phosphatase-conjugated anti-goat IgG (GE Healthcare Inc., Sweden). The Alkaline Phosphatase Conjugate Substrate Kit (BioRad, CA., USA) was used according to the manufacturer's instructions.

Cleavage of the fusion protein

The cleavage reaction for the fusion protein was performed using the enzyme enterokinase EKMax (Invitrogen, USA) at a concentration of 1 U per 20 μg of purified recombinant peptide and incubated at 22°C overnight. A 7-16% Tris-Tricine gel (Schangger 2006) was used to analyse the cleavage reaction.

Protein sequencing

After cleavage, the GST-FLAG and Glucagon proteins were diluted in 0.1% trifluoroacetic acid and purified by reverse-phase ultra fast high performance liquid chromatography (RP-UFLC) (RP-UFLC LC-20 AD/T LPGE kit, Shimadzu Corporation Kyoto, Japan) using a Vydac C_{-18}TP column. Retained proteins were eluted with a linear gradient of acetonitrile (0-100%). Samples were then subjected to speed-vac concentration, and major fractions were mixed with a saturated matrix solution of alpha-cyano-4-hydroxycinnamic acid (1:3) and spotted (0.5 μl) onto an MTP AnchoChip var/384 matrix-assisted laser desorption ionisation (MALDI) target plate. The monoisotopic molecular mass of glucagon was determined by MALDI-time of flight TOF/MS tandem mass spectrometry (MALDI-TOF/MS) using an AutoFlex Speed (Bruker Daltonics, Germany) controlled by the FlexControl 3.0 software. The MS spectra were acquired in the positive ion reflector mode at a laser frequency of 100 Hz, and the MS/MS spectra were obtained by LIFT™ mode, both with external calibration using the peptide calibration standard I (Bruker Daltonics, Germany). Data were analysed using the Flex-Analysis 3.0 software.

In vivo bioassays

In vivo bioassays were performed following the methodology described by Hinke et al. (2000) with modifications. Briefly, recombinant glucagon (7.1 nmol/kg), commercial glucagon (positive control) and dH_2O (negative control) were injected into three groups of five male Wistar rats, respectively. After administration, glucose levels were monitored at 0, 10, 20 and 40 minutes. The differences in glucose levels (differences between 10, 20 or 40 minutes and the baseline time) over time were calculated and evaluated. When the recombinant glucagon was used, the blood glucose level was maintained at a 1.3-fold increased level 40 minutes after injection, whereas it decreased to similar levels as the negative control after 40 minutes when the commercial glucagon was used. The biological test was repeated three times.

In this study, to biological assay the initial samples contained recombinant glucagon with a FLAG-GST tag. The fusion between FLAG-GST and recombinant glucagon was removed from digestion with enterokinase. After this step, samples were evaluated for *in vivo* biological activity. The FLAG-GST tag and recombinant glucagon were in the same sample to *in vivo* biological activity, but not were fusioned.

The study was previously approved by the Pioneer Union for Social Insertion – UPIS Committee for Ethical Use of Animals (CEUA), and the proposal (#071-11) entitled "Biological activity evaluation of recombinant peptide glucagon" was approved on August 12, 2011. Therefore, the reported experimental procedures comply with ethical procedures for animal use in *in vivo* experiments (Ellery 1985).

Statistical analyses were based on the generalised linear mixed model test (R Development Core Team 2010) and were used to validate the differences between values obtained by the different groups. In addition, analyses were performed to assign a fold-difference value comparing the increased glycemic levels seen at 10, 20 and 40 minutes with the basal levels, considering the random variation of rats. $P < 0.05$ was considered statistically significant. R statistical packages were used to validate the data.

Results

Glucagon expression and purification

After expression, it was observed that 40 to 50% of total soluble product corresponded to the glucagon peptide, which, including the GST fusion protein, presented a molecular mass of 31.44 kDa (Figure 1A; Additional file 1: Figure S1). After purification using affinity chromatography, densitometry analyses indicated that the GST-glucagon corresponded to 90% of the purified product. Western blotting assays using both anti-GST and anti-glucagon antibodies confirmed the presence of the GST-fused glucagon (Figure 1B). Later, the cleavage of glucagon from the GST fusion protein using the enterokinase enzyme produced a 3.5 kDa-peptide, as revealed by a 16% Tris-tricine gel (Figure 1C).

Amino acid sequencing of glucagon

After hydrolysis, the glucagon peptide was purified using UFLC (ultra-high performance liquid chromatography), and fractions were collected. The ion corresponding to the peptide was present in a fraction that eluted at 26 minutes of separation, and it was further detected by MALDI-TOF (Matrix Assisted Laser Desorption Ionisation Time-of-Flight) MS (Additional file 2: Figure S2). Hence, 16 amino acid residues from the N-terminus of glucagon (H-S-Q-G-T-F-T-S-D-Y-S-K-Y-L-D-S) were determined by *de novo* sequencing, indicating that the predicted digestion was performed as expected.

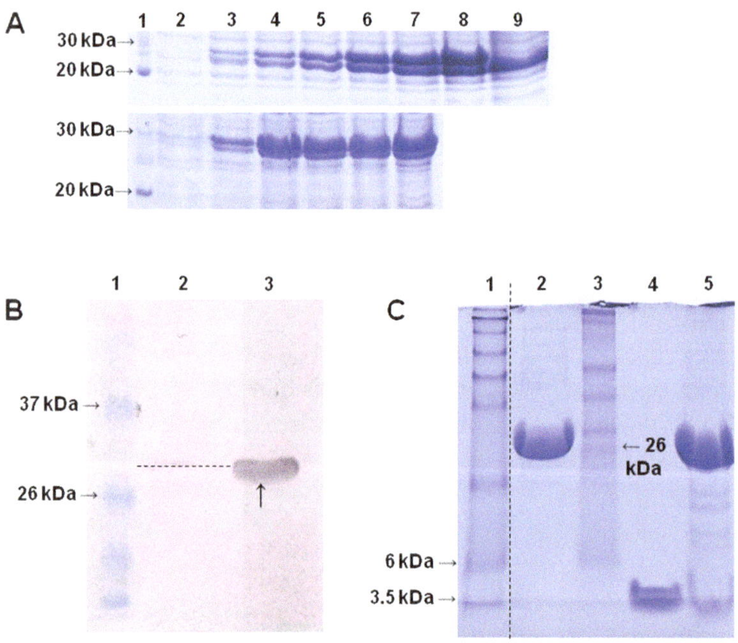

Figure 1 Recombinant glucagon peptide expression profiling and analysis. (A) Glucagon expression after different incubation times. Upper lane: protein expression using 1 L of culture media; Lower lane: protein expression in the bioreactor (5 L). Lane 1: BenchMark Protein Ladder (Invitrogen); Lanes 2 to 9 indicate expression protein patterns after 0 h, 1 h, 2 h, 3 h, 4 h, 5 h, 6 h and overnight incubation, respectively. **(B)** Western blotting of glucagon using anti-glucagon antibodies. Lane 1: Pre-Stained Protein Ladder (Invitrogen); Lane 2: purified GST protein (3 µg); Lane 3 purified non-digested glucagon fused with the GST protein (3 µg) arrows indicate recombinant glucagon fused with GST protein. **(C)** Glucagon digestion to separate the peptide from the GST tag in Tris-Tricine gel. Lane 1: Seeblue Plus2 Pre Stained Standard (Invitrogen); Lane 2: purified non-digested glucagon fused with the GST protein (5 µg); Lane 3: Pre-Stained Protein Ladder (Invitrogen); Lane 4: commercial glucagon (Glucagen, NovoNordisk, 5 µg); Lane 5: digested glucagon (40 µg total digestion).

In vivo bioassays

GST-Glucagon activity was evaluated by measuring glycemic levels in rats before and after exposure to the glucagon molecule. Both the recombinant peptide and the commercially available glucagon (NovoNordisk®) induced a similar blood glucose peak right after injection. Nevertheless, the recombinant glucagon maintained blood glucose levels at a 1.3-fold increased level 40 minutes after injection, whereas commercial glucagon decreased to the same levels of negative control after 40 minutes (Additional file 3: Figure S3). Statistical analyses demonstrated that the differential time profile of glucose levels between the recombinant glucagon and the commercial glucagon was significant (Figure 2A-C).

Glucagon expression using the WAVE Bioreactor

To evaluate the expression level of recombinant glucagon at a higher scale with controlled conditions, the WAVE bioreactor (GE Healthcare) was used at a scale up of 5 L (Figure 1A, lower lane). After expression and purification, the final yield of recombinant glucagon in the bioreactor reached 7 mg/L.

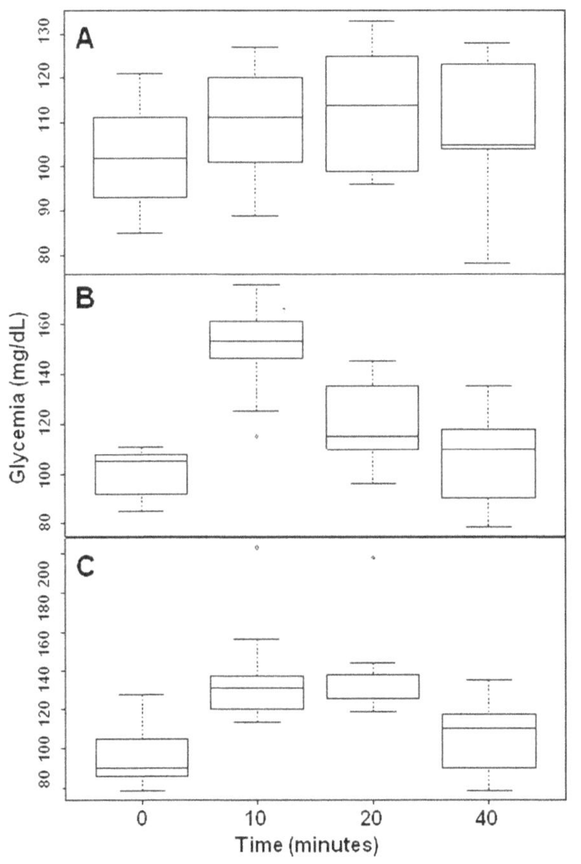

Figure 2 Statistical model analysis of the biological activity test based on glucose levels during specific times after of samples administration. (A) Negative control. **(B)** Positive control. **(C)** Recombinant glucagon. $p > 0,00001$.

Discussion

Among the seven emerging countries in the biopharmaceutical market (Russia, China, Turkey, Mexico, South Korea, India and Brazil), a growth from US$ 5.3 billion in 2010 to US$ 13.6 billion in 2015 is estimated (International Market Analysis Research & Consulting, 2010). Biosimilar sales are expected to reach between US$ 1.9 - 2.6 billion (IMS 2011) in 2015. In 2016, the medicines market will spend US$ 1.2 trillion, most of which are expenses focused on cancer, diabetes and asthma drugs (IMS Institute 2012).

The cost of drugs used for diabetes treatment (insulin, glucagon, and others) reached US$ 548 billion in 2013. In the United States, it is estimated that the diabetes drug market will expand by about US$ 125 million (Biodel 2013; Hassan et al. 2014). Statistical data suggest a worldwide growth from 382 million diabetic people in 2013 to 592 billion people in 2035 (Ledford 2013; Guariguata et al. 2014). There is an estimated 200.000 people/year that are hospitalised due to severe hypoglycaemia, and only 20% of this number use the glucagon emergency kit. One of the main causes for the lack of access to the kit is the elevated price (Biodel 2013). The first glucagon recombinant regimen was created by the NovoNordisk® Company in 1989 using a yeast heterologous system (*Saccharomyces cerevisiae*) (patent n°4.826.763). In 1994, the Eli Lilly® Company used an *E. coli* heterologous system to produce recombinant glucagon (patent n° 5.512.549). Today, recombinant glucagon is expressed in many countries for high scale production purposes, mainly own production (Agarwal et al. 2014) and such that is reported by When and colleagues (2001).

The glucagon production reached 40 to 50% of the total protein expression, with concentrations slightly above the 39% reported by Ohana and colleagues (2011), which evaluated different protein expression yields with differential tag systems. The tac promoter (de Boer et al. 1983) used in the expression vector contributed to the high level of glucagon production. It is already known that the tac promoter is strong and one of the most widely used in research and industry (Huang et al. 2012).

Bioreactors have been used in the recombinant peptide industry for more than a decade, mainly due to the flexibility of use reducing costs, labour and contaminations. However, the major limitations are the amount of oxygen used during expression and the low refrigeration capacity (Dreher et al. 2014). An initial peptide expression in bioreactors was also performed using up to 5 L of culture media. The ultimate goal of establishing a methodology for producing biopharmaceuticals is scale-up. Hence, the expression yielded 122 mg of semi-purified peptide, a 3.4-fold increase compared with the expected value observed earlier for the 1 L expression. This result suggested that the expression of glucagon in bioreactors using bacterial cells might provide better conditions for peptide production, indicating that the methodology used can easily be applied at industrial levels.

It is estimated that 77% of *E. coli* expression proteins are insoluble (Cuozzo and Soutter 2014). However, some studies showed high levels of soluble peptide production when fused with tag proteins, such as Thioredoxin (Trx) and GST (Gao et al. 2010). Hence, the glucagon peptide was expressed in the soluble phase partly due to the addition of a GST tag. At the end of the process, achieving satisfactory amounts of protein does not mean success (Rosano and Ceccarelli 2014). It is necessary to evaluate the biological activity to validate the process (Nausch et al. 2013). Here, we demonstrate that the recombinant GST-glucagon displayed high glycemic levels for a longer period of time than the commercial glucagon. For a biosimilar, it was expected that recombinant glucagon here produced and the positive control (commercial glucagon) should demonstrate similar glucose levels during the evaluated time, which was not observed in this study.

The analysis of recombinant proteins fused with GST in biological assays has been earlier described, with no interference of the targeted protein's activity. Recently, two antitumoral proteins, granzyme B (GrB) and perforin (PFP), earlier known by their ability to inhibit cell viability, were cloned into the *pGEX-4 T-1* vector and expressed in bacterial systems. The fusion proteins GST-GrB and GST-PFP were later challenged against Hep-2 cells, showing significant decreases in laryngeal cancer cell growth (Li et al. 2014). Using GST as a control, the assay also demonstrated the inability of the fusion protein to interfere with recombinant protein activities.

Therefore, the application of bacterial systems for the expression of human peptides fused with GST proteins is becoming a promising strategy in the production of recombinant molecules with potential for biopharmaceutical products. A previous report showed that a recombinant glucagon-like peptide-1 analogue (KGLP-1), also used on diabetes therapy, was expressed in *E. coli* cells as a fusion protein with GST (Liu et al. 2014). In this study, digestion reactions were performed to remove GST from glucagon and confirmed by Tris-Tricine gel (Figure 1C). Furthermore, in biological activity, the recombinant glucagon in the presence of GST tag, but not fused to it, showed prolonged high glycemic levels when compared with the positive control (Glucagen® NovoNordisk). The removal of tags from recombinant proteins is also critical, as it may interfere with the shape and activity of the protein of interest. Moreover, recombinant proteins with permanent tags are not allowed by the standards of therapeutic proteins (Walls et al. 2011; Einsfrldt et al. 2011). Therefore, the GST tag was removed from the fusion with the peptide glucagon using enterokinase. According to Hosfield and Lu (1999), enterokinase presents a cleavage efficiency, when the first amino acid residue of a protein fused with flag-tag is histidine, which is the case of recombinant glucagon, the cleavage efficiency of enterokinase is 74%. Moreover, according to Kahn (2009), of the 29 amino residues, the first amino acids close to the N-terminal portion are essential for biological activity, mainly histidine (position 1), asparagine (position 9) and serine (position 16). If glucagon was fused to GST, the histidine amino acid could not be connected to the glucagon receptor, and the biological activity would be partially impaired, which was not observed here.

All patents of technologies used in this work (Table 1) were deposited in 1997. Thereby, the patents expire in 2017, allowing the use of this protocol to obtain recombinant glucagon. The estimated value spent (Table 2, Additional file 4: Table S1) per mg of recombinant glucagon in this work was US$ 175.28. However, this value can be reduced to US$ 22.10 (see Table 2). In Brazil, the commercial value of recombinant glucagon is US$ 56.60, 2.5 times more expensive than recombinant glucagon. This value can be changed because the recombinant glucagon was semi-purified and total purification can increase the value.

In this work, the glucagon peptide was expressed in a soluble form using a pGEX vector in a bacterial system, yielding a high concentration of the purified product.

Table 1 Technologies used in the process to obtain recombinant glucagon and respective patents

Technology	Patent number	Patent year	Patent expiration
pGEX4t-3 t vector	US 5654176	1997	2017
GST Tag	US 5654176	1997	2017
FLAG Tag	US 4703004	1987	2007
BL21(DE3) strain	US 5693489	1997	2017
Enterokinase Enzyme	US 5665566	1997	2017
Glutathione Sepharore Resin	US 5654176	1997	2017
Glucagen®	US 5652216	1997*	2017

*The first patent filed was in 1989. This is an *evergreening* patent, a type of patent with novel technology is added to previous one. Basso (2012)

Table 2 Products used for recombinant glucagon expression and respective costs for 1 litre expression

Product	Price (US$)	Amount used	Final price spent (US$)
Ampicillin	134.33	100 mg	2.68
Benzamidine	207.92	180.225 mg	7.49
Enterokinase 1	781.50	1U/20 µg fusioned protein	1094.11*
Enterokinase 2	781.50	1U/1 mg fusion protein	21.87*
Yeast extract	56.60	5 g	1.14
Reduced glutathione	6135.00	3.07 g	14.21
IPTG	86.03	59.57 mg	5.12
KCl	47.54	3 mg	0.0002
KH_2PO_4	107.54	3.67 mg	0.007
NaCl	47.54	5.12 g	0.97
Na_2HPO_4	125.29	21.29 mg	0.005
PMSF	181.13	261.29 mg	9.46
Tips	16.98	15 units	0.25
Glutathione Sepharose	1864.90	4 mL	74.59
Tryptone	128.31	10 g	5.14
TrisHCl	93.59	6.05 g	2.26
Water, equipment and others			9.53
Total with enterokinase 1			**1226.96 (**/7 = 175.28)**
Total with enterokinase 2			**157.72 (**/7 = 20.10)**

Prices obtained from the companies Invitrogen, GE and Sigma. *Value may change depending on the concentration used for digestion. In this work, we used manufacturer's instructions (1U/20 µg), but literature reports (Ehrhardt et al. 2000; SUN et al. 2011) indicate 1U/mg recombinant protein to digestion. Thereby, this concentration was used to calculate this value in this condition. **Total value divided by 7, the recombinant protein total obtained in the 1 litre expression (7 mg in 1 litre) to indicate the value for 1 mg. Basso (2012)

Moreover, the recombinant glucagon showed a differentiated activity on glycemic levels when compared with the positive control, in the presence but not fused to the GST-tag. Further studies aimed to better understand the cause of this extended effect on glycemic levels and the role of GST are being performed.

Many recombinant biopharmaceuticals have been developed for use in biotechnology. However, after the expiration of several pharmaceutical patents, a variety of biosimilars have been produced. Some strategies are emerging for protein production at a large-scale to obtain higher levels of expression. Hence, in this report, we describe, for the first time, the production of the biosimilar glucagon hormone in a heterologous system by a Brazilian research group. Moreover, an improved expression and purification of the peptide is reported, which could support, in the future, the treatment of patients with several different diseases and will decrease Brazilian importation costs with medicaments. Furthermore, local production of the biosimilar glucagon will stimulate the development of national pharmaceutical industries. The peptide in the presence of GST, but not fused, showed a prolonged activity when compared with the positive control. This differential activity may be beneficial because it reduces the discomfort caused by the increase and decrease of glucose levels after commercial drug administration. However, further studies will indicate potential adverse effects of recombinant glucagon.

Additional files

Additional file 1: Figure S1. Recombinant glucagon peptide expression profiling. Soluble and insoluble expression of the glucagon samples. Lanes 1 and 6: Prestained SDS-PAGE Standards Low Range (Bio-Rad); Lane 2: Expression of the soluble glucagon; Lane 3: soluble fraction of expressed proteins in a non-induced sample; Lane 4: insoluble fraction of expressed glucagon; Lane 5: insoluble fraction of expressed proteins in a non-induced sample; Arrows indicate recombinant glucagon fused with the GST protein.

Additional file 2: Figure S2. Mass spectrometry sequencing using MALDI-TOF/TOF (LIFT^{TM} mode). Sequencing part of the fraction of the 1991.83 Da of the glucagon peptide showing the N-terminal portion (y) starting from amino acid H(y1) and reverse sequencing (b) starting from amino acid Q (b2) (despite the absence of b1, the same reveals its presence in y series). The calibrant used the peptide calibration standard II Mixture (Bruker Daltonics), detection range m/z 1000–4000. The analysed fraction retention time was 26 min.

Additional file 3: Figure S3. Analysis of glycemic levels using the *in vivo* biological activity test. This assay used four times (0, 10, 20 and 40 minutes after sample administration). Positive control (Glucagen® NovoNordisk), Negative control (dH_2O).

Additional file 4: Table S1. Apparatus used for recombinant glucagon and costs for 1litre expression.

Competing interests
The authors declare that they have no competing interests.

Authors' contributions

AM performed the experiments and developed the protocol for protein digestion and biological activity. PB developed the protocol for the bioreactor protein expression and purification. FM developed the protocol for protein expression, protein digestion, and biological activity and designed the study. MC performed the biological activity. AA developed the biological molecular experiments. LP performed protein sequencing by Mass Spectrometry analyses. MF designed the study. All authors read and approved the submission of the manuscript.

Acknowledgments

We acknowledge Dr. Gerard Murray for his contribution reading the manuscript. This study was supported by grants from the Brazilian Agricultural Research Corporation (Embrapa), the National Council for Science and Technology (CNPq) and the Brazilian Federal Agency for Support and Evaluation of Graduate Education (CAPES).

Author details

[1]Department of Molecular Pathology, University of Brasilia, Brasilia, DF, Brazil. [2]Laboratory of Plant-Pest Interaction, Embrapa – Genetic Resources and Biotechnology, Brasília, DF, Brazil. [3]Pioneer Union for Social Insertion – UPIS, Planaltina, DF, Brazil. [4]Catholic University of Brasilia, Brasilia, DF, Brazil. [5]Laboratory of Mass Spectrometry, Embrapa – Genetic Resources and Biotechnology, Brasília, DF, Brazil.

References

Agarwal DR, Chyung D, Carson BB, Yi L, Makaroun L, Kishore S, Reidenberg MM (2014) Application to add glucagon to the model list of essential medicines. http://www.who.int/selection_medicines/committees/expert/18/applications/Glucagon.pdf?ua=1. Acessed: 30 sept 2014

Basso AMM (2012) Recombinant glucagon production: expression, purification and biological activity evaluation. Dissertation, University of Brasilia

Blatt CR, Silvana CT, Schmidt EH, Marchesan S, Silva LM, Martins JL (2012) General awareness and use of generic medication among citizens of Tubarão, state of Santa Catarina, Brazil. Ciênc saúde coletiva 17:79–87, doi:10.1590/S1413-81232012000100011

Biodel (2013). Life sciences management access conference. http://files.shareholder.com/downloads/BIOD/0x0x684784/656e78a5-c336-4ad1-b985-dcbc7fdbf6a7/Biodel_at_Wedbush_Conference_13Aug_13_-_FINAL.pdf. Acessed: 18 junh 2014

Bradford MM (1976) A rapid and sensitive method for the quantification of microgram quantities of protein utilising the principle of protein-dye binding. Anal Biochem 72(1–2):248–54, doi:10.1016/0003-2697(76)90527-3

Brazil Ministery of Healthy (2008) National politics for science and technology innovation in health (in Portuguese). http://bvsms.saude.gov.br/bvs/publicacoes/Politica_Portugues.pdf. Accessed 24 Jan 2014

Brazil Ministery of Healthy (2014) http://portalsaude.saude.gov.br/images/pdf/2014/dezembro/16/livro-2-completo-para-site.pdf. Acessed 30 Dez 2014

Chen C (2014) World's drug bill will reach $1.3 trillion in 2018: study. http://www.bloomberg.com/news/2014-11-20/world-s-drug-bill-will-reach-1-3-billion-in-2018-study.html. Acessed: 30 dez 2014.

Cuozzo JW, Soutter HH (2014) Overview of recent progress in protein-expression technologies for small molecule screening. J Biomol Screen on line version p.1-14. doi:0.1177/1087057114520975

De Boer HA, Comstock LJ, Vasser M (1983) The tac promoter: a functional hybrid derived from the trp and lac promoters. Proc Natl Acad Sci U.S.A. 80(1):21–25. http://www.pnas.org/content/80/1/21.long

De Castro LAB (2014) Brazilian biotechnology network – the challenge to innovate in biotechnology in Brazil. BCM Proced 8(suppl 4):O40, doi:10.1186/1753-6561-8-S4-O40

Dreher T, Walcarius B, Husemann U, Klingenberg F, Zahnow C, Adams T, Wilde D, Casteels P, Greller G (2014) Microbial high cell density fermentations in a stirred single-use bioreactor. Adv Biochem Eng Biotechnol 138:127–47, doi:10.1007/10_2013_189

Ehrhardt RA, Slepetis RM, Siegal-Willott J, Van Amburgh ME, Bell AW, Boidclair YR (2000) Development of a specific radioimmunoassay to measure physiological changes of circulating lepting in cattle and sheep. J Endocrinol 166:519–28

Einsfrldt K, Severo-Junior JB, Correa AAP, Medeiros MA, Alves TL, Almeida RV, Larentis AL (2011) Cloning and expression of protease ClpP from *Streptococcus pneumoniae* in *Escherichia coli*: Study of influence of kanamycin and IPTG concentration on cell grown, recombinant protein production and plasmid stability. Vaccine 29:7136–43, doi:10.1016/j.vaccine.2011.05.073

Ellery AW (1985) Guidelines for specification of animals and husbandry methods when reporting the results of animal experiments. Working committee for the biological characterization of laboratory animals/GV-solas. Lab Anim 19:106–8, doi:10.1258/0023677857809422714

Gao M, Chen M, Liu W, Zhu J, Tian H, Gao X, Yao W (2010) Production and purification of an analog of glucagon-like peptide-1 by autoinduction and on-column cleavage in *Escherichia coli*. World J Microbiol Biotechnol 26:1675–82, doi:10.1007/s11274-010-0345-3

Guariguata L, Whiting DR, Hambleton I, Beagley J, Linnenkamp U, Shaw JE (2014) Global estimates of diabetes prevalence for 2013 and projections for 2035. Diabetes Res Clin Pract 103:137–149. http://dx.doi.org/10.1016/j.diabres.2013.11.002

Guimarães HP, Lopes RD, Lopes AC (2011) Treaty of urgent and emergency medicine: emergency room and UTI. Atheneu, São Paulo

Hancock AS, Du A, Liu J, Miller M, May CL (2010) Glucagon deficiency recues hepatic glucose production and improves glucose tolerance in adult mice. Mol Endocrinol 4(8):1605–14, doi:10.1210/me.2010-0120

Hassan A, Meo SA, Usmani AM, Shaikh TJ (2014) Diabetes during Ramadan – Pre-approach model: presentation, risk stratification, education. Eur Rev Med Pharmacol Sci 18:1798–1805. http://www.europeanreview.org/article/7534

Hinke SA, Pospisilik JA, Demuth HU, Mannhart S, Kuhn-Wache K, Hoffmann T, Nishimura E, Pederson RA, McIntosh HS (2000) Dipeptidyl peptidase IV (DPIV/CD26) degradation of glucagon. J Biol Chem 275:3827–34, doi:10.1074/jbc.275.6.3827

Huang CJ, Lin H, Yang X (2012) Industrial production of recombinante therapeutics in *Escherichia coli* and its recent advancements. J Ind Microbiol Biotechnol 39:383–99, doi:10.1007/s10295-011-1082-9

Hosfield T, Lu Q (1999) Influence of the amino acid residue downstream of $(Asp)_4$Lys on enterokinase cleavage of a fusion protein. Anal Biochem 269:10–6, doi:10.1006/abio.1998.3084

Ibarra-Cabrera R, Mena-Pérez SC, Bondani-Guasti A, Garcia-Arrazola R (2013) Review on the worldwide regulatory framework for bisimilar focusing on the Mexican case as an emerging in Latin America. Biotechnol Adv 31:1333–43, doi:10.1016/j.biotechadv.2013.05.005

IMS – Institute for Healthcare Informatics (2012) The Global Use of Medicines: Outlook through 2016., http://www.imshealth.com/deployedfiles/ims/Global/Content/Insights/IMS%20Institute%20for%20Healthcare%20Informatics/Global%20Use%20of%20Meds%202011/Medicines_Outlook_Through_2016_Report.pdf. Accessed: 06 May 2014

IMS – Institute for Healthcare Informatics (2013) The Global Use of Medicines: Outlook through 2017., http://www.imshealth.com/deployedfiles/imshealth/Global/Content/Corporate/IMS%20Health%20Institute/Reports/Global_Use_of_Meds_Outlook_2017/IIHI_Global_Use_of_Meds_Report_2013.pdf Accessed 13Set 2014

International Market Analysis Research & Consulting (2010) http://www.bioseeker.com/market-research-report/factsheet-pdf/world-pharmaceutical-and-biopharmaceutical-market-2010-2015-pipeline-analysis-of-the-top-50-companies.pdf. Accessed 27 Jan 2014

Kahn CR (2009) Joslin's Diabetes Melitus. Artmed, São Paulo

Ledford H (2013) Diabetes drug ride a bumpy road. Nat 504:198, doi:10.1038/504198a

Li X, Zhang G, An G, Liu S, Lai Y (2014) Expression, purification and anticancer analysis of GST-tagged human perforin and granzyme B proteins in human laryngeal cancer Hep-2 cells. Prot Exp Purif 95:38–43, doi:10.1016/j.pep.2013.11.009

Liu Y, Ren L, Ge L, Cui Q, Cao X, Hou Y, Bai F, Bai G (2014) A strategy for fusion expression and preparation of functional glucagon-like peptide-1 (GLP-1) analogue by introducing an enterokinase cleavage site. Biotechnol Lett 36:1675–8, doi:10.1007/s10529-014-1526-1

Massi L, Cerrutti BM, Queiroz SL (2013) Methodology Jigsaw in discipline of medicinal chemistry. (in Portuguese) Quím Nova 36:897–904. http://dx.doi.org/10.1590/S0100-40422013000600028

Nausch H, Huckauf J, Koslowski R, Meyer U, Broer I, Mikschofsky (2013) Recombinant Production of Human Interleukin 6 in Escherichia coli. PLOS ONE 8 (1). doi:10.1371/journal.pone.0054933

Niederwieser N, Schmitz S (2010) Biosimilar agents in oncology/hematology: from approval to practice. Eur J Haematolol 86:277–88, doi:10.1111/j.1600-0609.2010.01566.x

Nwobike JC (2006) Pharmaceutical corporations and access to drugs in developing countries: the way forward. Sur, Rev int direitos human 3:126–43, doi:10.1590/S1806-64452006000100008
Ohana RF, Hurst R, Vidugiriene J, Slater MR, Wood K, Urh M (2011) HaloTag-based purification of functional human kinases from mammalian cells. Protein Expr Purif 76:154–64, doi:10.1016/j.pep.2010.11.014
Oliveira EJV (2013) Heath ministry actions for promotion of national production. (in Portuguese) http://www.cimes.org.br/uploads/paginas/file/palestras/Eduardo%20Oliveira%20-%20SCTIE%20-%20Fomento%20%C3%A0%20Produ%C3%A7%C3%A3o.pdf. Accessed: 02 Nov 2014
Paiva, LB (2012) Health technology innovation and productive industrial complex in Brasil. (in Portuguese) http://www.saude.sp.gov.br/resources/instituto-de-saude/homepage/pdfs/seminario-de-inovacao-material/dr._leonardo_paiva.pdf?attach=true. Accessed: 02 Nov 2014
R Development Core Team (2010) (2010) R: A language and environment for statistical computing. R Foundation for Statistical Computing, Vienna, Austria, ISBN 3-900051-01-0. URL
Ramnanan CJ, Edgerton DS, Kraft G, Cherrington AD (2011) Physiologic action of glucagon in liver glucose metabolism. Diabetes Obes Metab 13(1):118–25, doi:10.1111/j.1463-1326.2011.01454.x
Reddy, GB (2013) Local working of patents – law and implementation in India. JIPR 18:15–27. http://nopr.niscair.res.in/handle/123456789/15742
Rosano G, Ceccarelli EA (2014) Recombinant protein expression in *Escherichia coli*: advances and challenges. Front Microbiol v 5:1–17, doi:10.3389/fmicb.2014.00172
Rosenberg G, Fonseca MGD, D'avila LA (2010) Comparative analysis of the industrial concentration and turnover of pharmaceutical industry in Brazil for the segments of brand and generic medications (in Portuguese) Econ Soc 19:107–134. doi:10.1590/S0104-06182010000100005
Sambrook J, Russell DW (2006) The condensed protocols. From molecular cloning: A laboratory manual. Cold Spring Harbor Laboratory, New York
Schangger H (2006) Tricine –SDS-PAGE. Nat Protoc 1:16–22, doi: 10.1038/nprot.2006.4
Silveira, E (2012) Nanoscale drugs (in portuguese). FAPESP. 201:76–79. http://revistapesquisa.fapesp.br/en/2013/01/18/medicine-in-nanoscale/. Acessed 02 Nov 2014.
Sun YL, Lin Y, Lin CS (2011) Soluble fusion expression and production of rabbit neutrophil peptide-1 in *Escherichia coli*. Rom Biotech Lett 16:6618–6629. http://www.rombio.eu/rbl5vol16/19%20Chih%20Seng%20Lin.pdf ISSN 1224–5984
Sundfeld CA, Souza RP (2013) Public-private partnerships for the development of medicines and the bidding process in Brazil. (in Portuguese) RDA 24:91–133 http://bibliotecadigital.fgv.br/ojs/index.php/rda/article/viewFile/14078/12945 Accessed: 02 Nov 2014
Utzig MFK, Virtuoso S, Oliveira SMM (2010) Verification of the adhesion of a community to generic medicines after seven years of implantation of the law. (in Portuguese) Visão Acad 10: 100–5. http://ojs.c3sl.ufpr.br/ojs/index.php/academica/article/download/21339/14067. Accessed: 02 Nov 2014
Valadares C (2009) Domensal. Bulletin of the Department of Pharmaceutical Care and Strategic Inputs. Department of Science, Technology and Strategic Inputs. Health Ministry. (in Portuguese) Available: http://www.sinfarmig.org.br/UserFiles/File/dose%20mensal%203.pdf. Accessed 19 Jan 2014.
Victora CG, Barreto ML, Leal MC, Monteiro CA, Schmidt MI, Paim J, Bastos FI, Almeida C, Bahia L, Travassos C, Reichenheim M, Barros FC (2011) Health conditions and health politics innovations in Brasil: way to go. (in Portuguese) Lancet Series 6:90–102. doi:10.1016/S0140-6736(11)60055-X
Walls D, Mcgrath R, Loughran ST (2011) A digest of protein purification. Protein Chromatography: Methods and Protocols. Methods Mol Biol 681:3–23, doi:10.1007/978-1-60761-913-0_1
When C, Wang Z, Du P, Gan R, Zhu S (2001) Secretion expression of recombinant glucagon in *Escherichia coli*. Sci China C Life Sci 44:233–40, doi:10.1007/BF02879330

20

Simple observation of *Streptococcus mutans* biofilm by scanning electron microscopy using ionic liquids

Yoko Asahi[1*], Jiro Miura[2], Tetsuya Tsuda[3], Susumu Kuwabata[3], Katsuhiko Tsunashima[4], Yuichiro Noiri[1], Takao Sakata[5], Shigeyuki Ebisu[1] and Mikako Hayashi[1]

Abstract

Scanning electron microscopy (SEM) has been successfully used to image biofilms because of its high resolution and magnification. However, conventional SEM requires dehydration and metal coating of biological samples before observation, and because biofilms consist mainly of water, sample dehydration may influence the biofilm structure. When coated with an ionic liquid, which is a kind of salt that exists in the liquid state at room temperature, biological samples for SEM observation do not require dehydration or metal coating because ionic liquids do not evaporate under vacuum conditions and are electrically conductive. This study investigates the ability of ionic liquids to allow SEM observation of *Streptococcus mutans* biofilms compared with conventional coating methods. Two hydrophilic and two hydrophobic ionic liquids, all of which are electronic conductors, are used. Compared with samples prepared by the conventional method, the ionic-liquid-treated samples do not exhibit a fibrous extracellular matrix structure and cracking on the biofilm surface. The hydrophilic ionic liquids give clearer images of the biofilm structure than those of the hydrophobic ionic liquids. This study finds that ionic liquids are useful for allowing the observation of biofilms by SEM without preparation by dehydration and metal coating.

Keywords: Biofilm; Extracellular polymeric substance; Ionic liquid; Scanning electron microscopy; *Streptococcus mutans*

Introduction

Over 90% of bacteria in nature exist in robust, sessile communities known as biofilms (Vu et al. 2009), which can cause problems in a number of fields such as agriculture, industry, and medicine. Biofilm bacteria attach to surfaces and are surrounded by an extracellular matrix called the extracellular polymeric substance (EPS) (Costerton et al. 1999). Ninety percent of the dry weight mass of biofilms is composed of EPS, and the other major component of the biofilm matrix (up to 97%) is water (Sutherland 2001). The EPS has many functions, such as promoting the microbial adhesion and structural stability of the biofilm (Flemming et al. 2007; Stoodley et al. 2002), and enhancing biofilm resistance to antimicrobial agents. The EPS also protects against environmental stresses such as desiccation, oxidizing or charged biocides, and host immune defenses (Flemming and Wingender 2010). *Streptococcus mutans* is a Gram-positive, facultatively anaerobic bacterium, and an important etiologic agent of dental caries (Loesche 1986). *S. mutans* produces exopolysaccharide, which is synthesized by glucosyltransferases and the main constituent in the EPS of cariogenic biofilms (Bowen and Koo 2011; Hamada and Slade 1980). The EPS is essential for the initial adherence of *S. mutans* to tooth surfaces and facilitates the formation of a mature dental biofilm (Hamada and Slade 1980; Koo et al. 2003).

Many early biofilm studies used scanning electron microscopy (SEM), which showed in detail the surface morphologies of microbial biofilms and their structure. Direct SEM observation clearly revealed that bacterial cells in a biofilm were extensively surrounded by fibrous or amorphous matrices, which represented the EPS (Marrie and Costerton 1984). However, when using a conventional scanning electron microscope, the sample chamber must be kept dry and under high vacuum during use, so the specimens are limited to nonvolatile, conductive materials

* Correspondence: yoko-a@dent.osaka-u.ac.jp
[1]Department of Restorative Dentistry and Endodontology, Osaka University Graduate School of Dentistry, 1-8 Yamadaoka, Suita, Osaka 565-0871, Japan
Full list of author information is available at the end of the article

(Tsuda et al. 2011b). If nonconducting samples are to be observed, such as biological specimens with a high water content, the samples must first be fixed with aldehyde, dehydrated with a graded alcohol series, dried, and then coated with a conductive film of a metal such as platinum prior to SEM observation (Figure 1). However, this dehydration process leads to considerable sample distortion and artifacts; for example, it causes the EPS to appear as fibers rather than as a thick gelatinous matrix surrounding the cells (Donlan and Costerton 2002). To avoid the draconian process of total sample dehydration required in conventional SEM, environmental SEM (ESEM), which can be used to visualize specimens in a wet or partially hydrated state, has been used to observe biofilm samples (Darkin et al. 2001; Priester et al. 2007). Other researchers have used variable-pressure SEM to observe biofilms (Weber et al. 2014).

Ionic liquids (ILs) are molten salts that remain in a liquid state even at room temperature (Torimoto et al. 2010). A large number of ILs have been synthesized and studied because of their many benefits such as negligible vapor pressure at room temperature and relatively high electronic conductivity (Tsuda et al. 2011a). It has been reported that ILs can behave as electronically conducting materials and enable SEM observation of samples without a metal or carbon coating (Kuwabata et al. 2006). Furthermore, replacing the water in samples with ILs allows the samples to remain wet during SEM observation because ILs resist evaporation even under vacuum conditions. The entire conventional biosample preparation process of dehydration, critical point drying and making the sample surface conductive can therefore be replaced by treatment with an IL. Such IL treatment has been demonstrated by the SEM observation of seaweed leaves swollen with water and subsequently soaked in ILs (Arimoto et al. 2008), as well as other biological samples such as fixed human culture cells (Ishigaki et al. 2011a), chromosomes (Dwiranti et al. 2012), and insects and pollen (Tsuda et al. 2011b). However, SEM observation of biofilm samples prepared by IL treatment has not yet been reported.

In this study, biofilms of *S. mutans* are treated with various kinds of ILs, and the conditions suitable for SEM observation of the IL-treated biofilms, including the optimum IL concentration and accelerating voltage used for SEM observations, are investigated.

Materials and methods

Bacterial strains and culture conditions

Streptococcus mutans MT8148 (JCM5175) was provided by Dr. Ooshima of Osaka University, Osaka, Japan. *S. mutans* MT8148 was grown in brain heart infusion (BHI) broth (Becton, Dickinson and Company, MD, USA) at 37°C under anaerobic conditions.

Ionic liquids

The ILs used in this study were choline lactate ([Ch][Lac]), 1-ethyl-3-methylimidazolium acetate ([C_2mim][AcO]), 1-butyl-3-methylimidazolium bis(trifluoromethanesulfonyl)amide ([C_4mim][Tf_2N]) and tri-*n*-butyldodecylphosphonium tetrafluoroborate ([$P_{4,4,4,12}$][BF_4]) (Figure 2). [Ch][Lac] and [C_2mim][AcO] are hydrophilic ILs, while [C_4mim][Tf_2N] and [$P_{4,4,4,12}$][BF_4] are hydrophobic ILs (Table 1). [Ch][Lac] and [$P_{4,4,4,12}$][BF_4] were prepared according to the methods described in previous papers (Tsuda et al. 2011a; Tsuda et al. 2011b; Tsunashima and Sugiya 2007). The other two ILs, [C_2mim][AcO] and [C_4mim][Tf_2N], were purchased from Kanto Chemical Co., Inc. (Tokyo, Japan). Prior to use, ILs were pretreated by vacuum drying at 80°C to remove impurities such as unreacted precursors and solvent.

Biofilm formation

The medium used was BHI broth containing 0.5% sucrose. Biofilm formation was accomplished with a modified Robbins device as previously described (Noiri et al. 2003). Briefly, plastic disks were placed face down in the modified Robbins device after being coated with a 0.22 μm-filter-sterilized human saliva. Culture medium (400 ml) containing *S. mutans* cells (10^7 cfu/ml) was perfused for 1 day using a peristaltic pump (SJ-1220; Atto Co., Tokyo, Japan) at a flow rate of 3.3 ml/min as described previously (Noiri

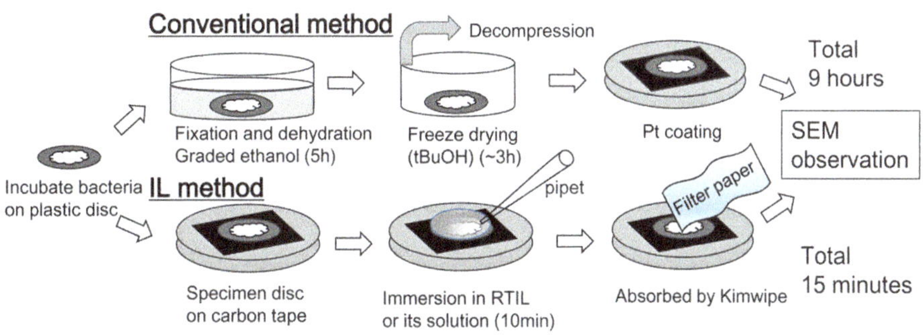

Figure 1 Pretreatment of samples for SEM observation using the conventional and IL methods.

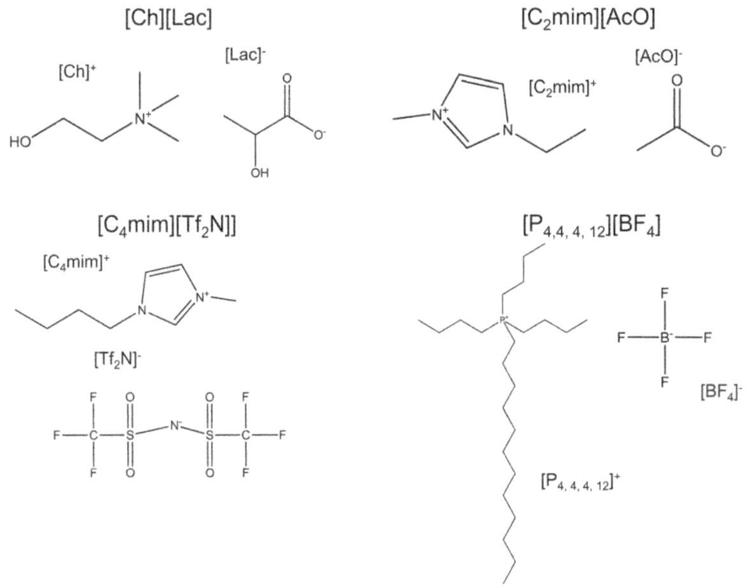

Figure 2 Chemical structures of the ILs used in this study.

et al. 2003). The culture medium was changed to fresh BHI medium without *S. mutans* and perfused for another 6 days. The medium was changed every 2 days.

Conventional treatment to prepare samples for SEM observation

The specimens were immersed in half-strength Karnovsky's solution (2% paraformaldehyde, 2.5% glutaraldehyde, pH 7.4) for 30 min. The fixed specimens were dehydrated using 50, 70, 80, 90, 95 and 100% (v/v) graded ethanol, and transferred into *t*-butyl alcohol. The specimens were then freeze-dried (JFD-320; JEOL, Tokyo, Japan). The samples were coated with platinum by sputtering with a plasma multicoater (PMC-5000; Meiwafosis, Tokyo, Japan) (Asahi et al. 2012). Samples that were treated with the conventional method were used as controls.

Ionic liquid treatment of samples for SEM observation

Hydrophilic ILs were diluted with distilled water and hydrophobic ILs were diluted with ethanol prior to use. Samples were prepared without a dehydration process by dropping 50 μl of diluted ILs (1, 10, and 20% (v/v)) on top of each biofilm sample and leaving the sample for 10 min at room temperature. Any excess IL was then absorbed using Kimwipes (Figure 1). To evaluate the effect of fixation, some samples were fixed with 2.5% glutaraldehyde for 30 min prior to IL treatment.

SEM observation

All specimens were examined with a scanning electron microscope (JSM-6390LV, JEOL, Tokyo, Japan) using the secondary electron emission mode with accelerating voltages of 1, 5, 10, and 20 kV. The magnifications used were × 1,500 and × 5,000.

Transmission electron microscopy (TEM) observation

TEM observation was performed by slightly modifying the previously described sample preparation method (Asahi et al. 2014). Biofilm samples were treated with 10% hydrophilic ILs and left for 10 min at room temperature. Samples that were not treated with ILs were used as controls. The specimens were fixed first with 2.5% glutaraldehyde in 0.2 M cacodylate buffer containing 0.15% ruthenium red for 1 h, and then with 2% osmium tetroxide and 0.15% ruthenium red in 0.2 M

Table 1 Physicochemical properties of the ILs used in this study

	Hydrophilicity	Viscosity/cP	Formula weight	Reference
[Ch][Lac]	+	895.0	193.24	Tsuda et al. 2014
[C_2mim][AcO]	+	143.61	170.20	Tsuda et al. 2014
[C_4mim][Tf_2N]	-	50.5	419.37	McHale et al. 2008, Gardas and Coutinho 2008
[$P_{4,4,4,12}$][BF_4]	-	1310	458.45	Tsunashima and Sugiya 2007

cacodylate buffer. The fixed samples were dehydrated in a graded ethanol series and embedded in epoxy resin (Quetol 812 NissinEM, Tokyo, Japan). Ultrathin sections (70 nm) were then cut with a diamond knife (Nanotome Thick, Sakai Advanced Electron Microscope Research Center, Saitama, Japan) in an ultramicrotome (Ultrotome V, LKB, Stockholm, Sweden). These sections were mounted on copper grids (#100), stained with 2% uranyl acetate and 0.4% Sato's lead stain, and examined using a transmission electron microscope (H800, Hitachi, Tokyo, Japan) with an accelerating voltage of 200 kV.

Confocal laser scanning microscope (CLSM) observation

Biofilm samples were treated with 10% hydrophilic ILs and left for 10 min at room temperature. Samples that were not treated with ILs were used as controls. The specimens were washed and then stained with Concanavalin A-tetramethylrhodamine conjugate (Invitrogen, Carlsbad, CA, USA) and SYTO® 9 (Invitrogen) for 30 min in the dark. After rinsing in water, images were obtained using a CLSM (LSM700, Carl Zeiss, Munchen-Hallbergmoos, Germany). The scanning images were analyzed three-dimensionally using imaging software (Imaris®, Bitplane AG, Zurich, Switzerland). Fluorescent images were also quantified and exopolysaccharide levels were calculated as the ratio of the numbers of exopolysaccharides and cells. Significant differences between experimental and control groups were analyzed using Dunnett's multiple comparison test ($P < 0.01$). Eight images per field per sample were acquired randomly. The experiment was independently repeated three times.

Results

Micromorphology of biofilms treated with platinum sputtering

The SEM images of platinum-coated biofilms obtained in this study are similar to those of *S. mutans* biofilms reported by other researchers (Li et al. 2002; Pandit et al. 2011). The biofilms were composed of aggregated cocci, and fibriform extracellular matrix-like structures

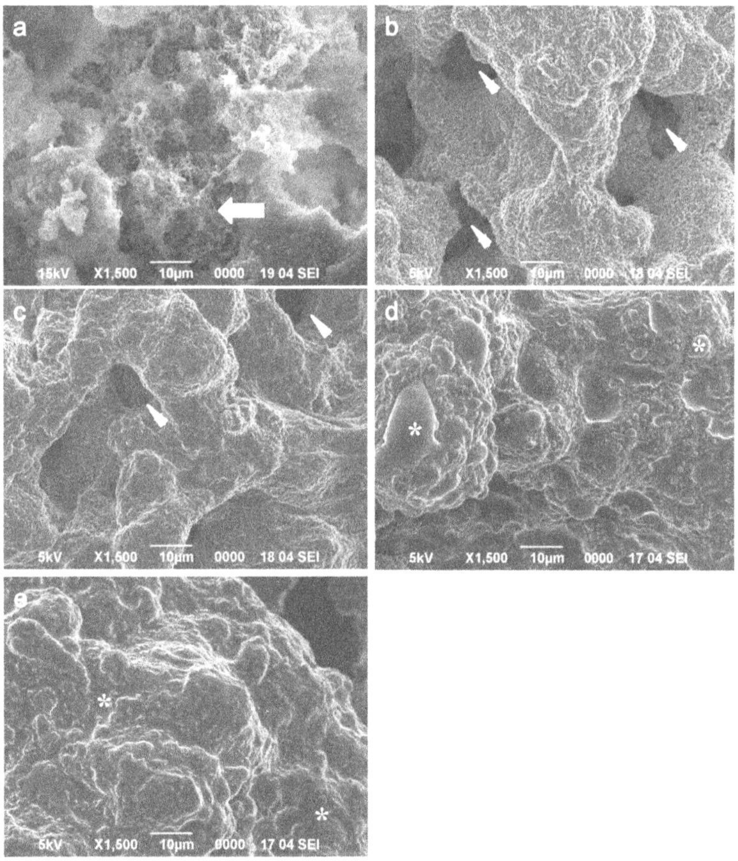

Figure 3 SEM images of biofilms prepared for observation using hydrophilic and hydrophobic ILs. (a): SEM images of platinum-coated *S. mutans* biofilms. Fibrous extracellular matrix-like structures (star) were partially observed. In some areas of the biofilm surface, the cells were separated from the biofilm (arrow). Images of biofilms prepared for observation using hydrophilic ILs **(b)** [Ch][Lac] and **(c)** [C_2mim][AcO], and hydrophobic ILs **(d)** [C_4mim][Tf_2N] and **(e)** [$P_{4,4,4,12}$][BF_4]. The concentration of all ILs was 10%. Some hydrophobic ILs were repelled by the biofilm surfaces and pooled on them. An asterisk indicates the accumulated IL. Arrowheads indicate the dark gaps thought to be water channels. Scale bars = 10 µm.

could be partially observed (Figure 3, Additional file 1: Figure S1). In some areas of the biofilm surface, the cells were almost separate from the biofilm (Figure 3a).

SEM observation of biofilms treated with ILs

To investigate the effectiveness of fixing the biofilm with glutaraldehyde, an *S. mutans* biofilm was treated with glutaraldehyde and 10% [Ch][Lac]. Fixation with glutaraldehyde did not change the image quality compared with that of the control (Additional file 1: Figure S2), so subsequent samples were not fixed with glutaraldehyde.

The ability of the two hydrophilic and two hydrophobic ILs to image the *S. mutans* biofilm using SEM was compared by treating the biofilms with 10% ILs. Using the hydrophilic ILs, the quality of the SEM images was as high as those of the platinum-coated samples (Figure 3b and c), although the IL-treated biofilms exhibited a smooth surface that was quite different from the rough surface of the platinum-coated biofilms (Figure 3a). In addition, the fibriform extracellular matrix-like structures observed at the surface of the platinum-coated biofilms did not appear on the surface of the IL-treated biofilms. Cracks were barely visible in the IL-treated biofilms at magnifications of ×1500 and ×5000, and the dark gaps thought to be water channels were more discernible in the IL-treated biofilms than in the platinum-coated ones. In contrast, the SEM images of the samples prepared using hydrophobic ILs showed that these ILs were repelled by the biofilm surfaces and formed pools on them (Figure 3d and e).

The images obtained were influenced by accelerating voltage (Figure 4). A lower accelerating voltage made it difficult to bring the image into focus, and limited the high-magnification observation of the IL-treated samples (Figure 4a). At a high voltage, however, the outlines of individual bacteria became obscured compared with those in the images obtained at lower voltage (Figure 4c and d). The optimum accelerating voltage for the IL-treated samples was around 5 kV (Figure 4b).

At some of the IL concentrations used in this study, the SEM images of the biofilm were clear (Additional file 1: Figure S3). However, for IL concentrations exceeding 20%, we observed that some hydrophilic IL accumulated on the surface of the biofilms (Additional file 1: Figure S3c).

TEM observation of biofilms treated with ILs

To investigate the micromorphological effects of ILs on *S. mutans* biofilms, the samples were stained with ruthenium red because this polyanionic stain helps to maintain the structural integrity of biofilms during TEM observation (Barber et al. 1993). TEM analysis showed that the bacterial cells were surrounded by electron-

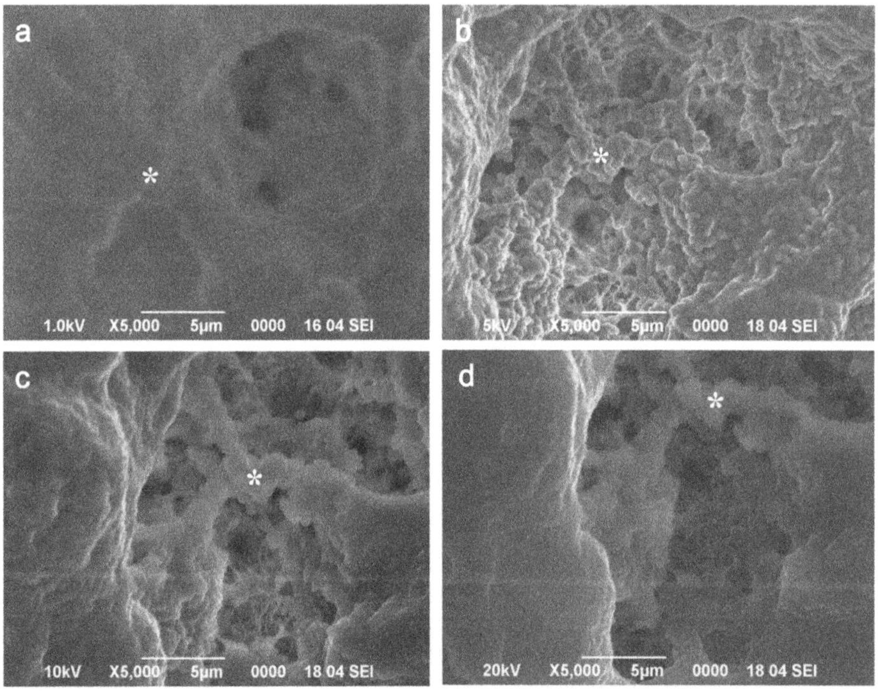

Figure 4 SEM images recorded using various accelerating voltages on the same area of an IL-treated biofilm specimen. The IL used was 1% [Ch][Lac]. The accelerating voltages were **(a)** 1 kV, **(b)** 5 kV, **(c)** 10 kV, and **(d)** 20 kV. At an accelerating voltage of 1 kV, it was difficult to focus the image. At 5 kV, the bacteria and biofilm outlines were observed clearly. Scale bars = 5 µm.

dense fibrous matrices that were stained with ruthenium red (Figure 5). There was no appreciable difference in the images of the bacterial cell membrane, cell wall and cytoplasm obtained for the IL-treated biofilm when compared with those of the control biofilm (Additional file 1: Figure S4).

CLSM observation of biofilms treated with ILs

In all samples, the three-dimensional structure of the biofilms was examined and the EPS was found to be spatially heterogeneous. The effect of ILs on the EPS of *S. mutans* was evaluated, and no significant difference in the exopolysaccharide level was found between corresponding experimental and control groups (Figure 6).

Discussion

To observe biological samples using SEM, wet specimens must be dehydrated before examination and coated by carbon or metal to provide electronic conductivity. The discovery that ILs can act as an electronically conducting material under vacuum conditions has led to the use of ILs to allow SEM observation of wet samples (Arimoto et al. 2008). In this study, we investigate the usefulness of the IL method for biofilm SEM observation. As shown by the results in the present

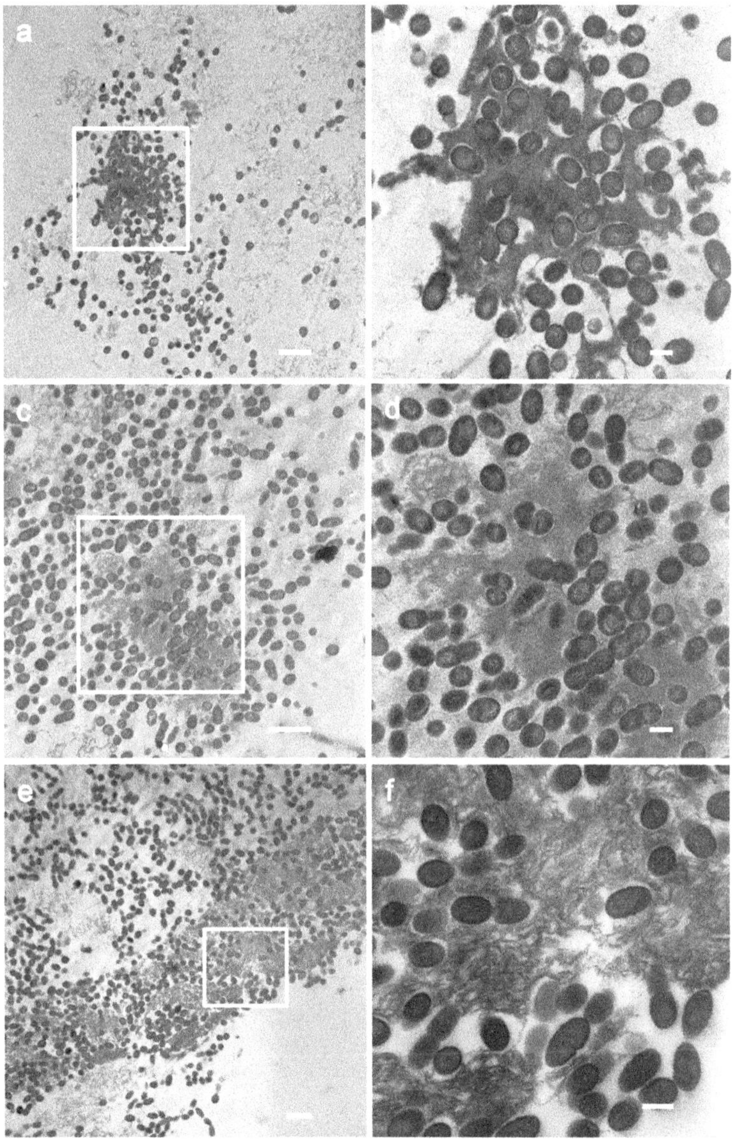

Figure 5 TEM images of IL-treated *S. mutans* biofilms. **(a)**, **(b)** control. Biofilms treated with **(c)**, **(d)** 10% [Ch][Lac] and **(e)**, **(f)** 10% [C_2mim][AcO]. The EPS appears as electron-dense materials around the bacterial cell. **(a)**, **(c)**, **(e)** Low-magnification images, scale bars = 2 μm. **(b)**, **(d)**, **(f)** High-magnification images, scale bars = 500 nm.

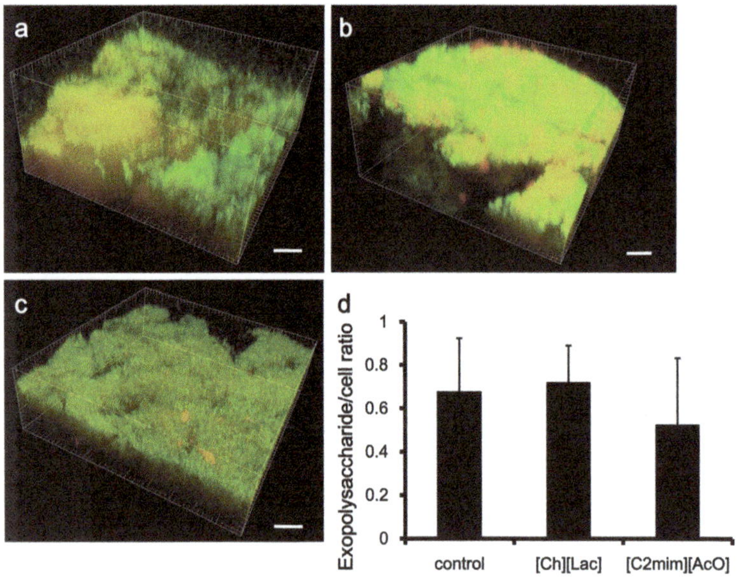

Figure 6 CLSM observation of *S. mutans* biofilms treated with ILs. Three-dimensional images of **(a)** control, **(b)** 10% [Ch][Lac]-treated and **(c)** 10% [C_2mim][AcO]-treated biofilms. Scale bars = 10 μm. Exopolysaccharide is shown in red and bacterial cells in green. **(d)** Levels of exopolysaccharide represented by the ratio of the fluorescence of the exopolysaccharide to that of the cells. There was no significant difference between corresponding experimental and control groups ($P > 0.01$). Data are presented as the mean ± standard error (n = 8).

study, clear SEM images of biofilms were obtained with a simple IL treatment that did not require a dehydration or metal deposition process. Some techniques like low-vacuum SEM and ESEM have been successfully used to observe wet samples, although problems have been reported when trying to obtain high-resolution ESEM images of biofilms. This is because of the lack of conductivity in the wet sample and because the focused electron beam appears to destroy the three-dimensional biofilm structure when the microscope is set to magnifications of ×10,000 and greater (Alhede et al. 2012). In this study, high-resolution images were obtained at ×10,000 magnification (Additional file 1: Figure S1) using biofilm samples prepared by the IL method.

The fibrous extracellular matrix structures observed on the biofilm surface of samples prepared by the conventional method were not detected on the surface of the IL-treated biofilms (Figure 3). It has been proposed that these fibrous structures are caused by the dehydration process in conventional sample preparation methods, so it appears that the biofilm structure observed after IL treatment has retained its form better than when the conventional method is used. Indeed, the SEM images of biofilms pretreated with ILs resemble those observed with low-vacuum SEM more closely than those obtained using samples prepared by the conventional technique (Darkin et al. 2001; Weber et al. 2014).

In the present study, we investigate the suitability of both hydrophilic and hydrophobic ILs for SEM observation of biofilms. Like other biological samples such as cultured human cells (Ishigaki et al. 2011b), hydrophilic ILs were found to be suitable for SEM observation of biofilms because of their affinity with the wet surface, but hydrophobic ILs were not (Figure 3). It is believed that the hydrophobic ILs are repelled by the water in the biofilm samples and are not able to replace it, causing the hydrophobic ILs to pool on the biofilm surface.

Higher concentrations of hydrophilic ILs also tended to pool on the biofilm surfaces (Additional file 1: Figure S3); this may arise from the higher viscosity of more concentrated solutions of ILs than that of more dilute ones (Table 1) (Gardas and Coutinho 2008; McHale et al. 2008). To prevent the ILs from accumulating on the surface of the biofilm, the ILs were therefore diluted. Using hydrous superabsorbent polymer particles for SEM observation after treatment with hydrophilic ILs, Tsuda et al. observed that more water in the particles was replaced with ILs upon using a higher concentration of ILs (Tsuda et al. 2014) and, when the substitution rate was higher, a decreased reduction of particle size was also observed. Therefore, a major factor to be considered when replacing the water in a biofilm is the IL concentration, and it is desirable to use a high concentration of ILs from the perspective of biofilm structure retention. Furthermore, considering that the major component of biofilms is water (Schmitt and Flemming 1999), the observation of an accumulation of ILs on the surface of the biofilm at high IL concentration may be the normal state of a biofilm that is covered with water (Additional file 1: Figure S3). It seems that the issue of efficiently replacing the water in the sample with ILs is an important one when using this method, so an IL that effectively replaces water is desired.

When using an IL as a conducting material for SEM observation, the accelerating voltage of the SEM is an important factor to consider (Ishigaki et al. 2011a). The images of the biofilms strongly depend on accelerating voltage (Figure 4), and it appears that the subsurface structure of the biofilm is observed better at higher accelerating voltages because the electrons are able to penetrate deeper into the specimen. This is an advantage of the IL method, and allows both the surface and subsurface structure to be imaged. When imaging IL-treated cultured human cells by SEM, the samples were easy to charge up at accelerating voltages of 10–15 kV, which are the optimum accelerating voltages for platinum-coated samples (Ishigaki et al. 2011a). In contrast, a high accelerating voltage did not charge up the biofilm samples treated with ILs in this study. To observe the outermost surface of the biofilm, 1 kV is believed to be the most suitable accelerating voltage, but such a low accelerating voltage gives poor image resolution.

Some ILs possess amphiphilic properties and behave as surfactants (Brown et al. 2013), especially surface-active ILs (SAILs) with long-chain alkyl groups (Galgano and El Seoud 2010). Rhamnolipid, a surfactant produced by *Pseudomonas aeruginosa*, is involved in the dispersion of biofilm cells by acting on the EPS (Boles et al. 2005). In this study, treatment with 10% IL for 10 min does not appear to disrupt the EPS within the biofilm (Figure 6). Among the ILs used in this study, [$P_{4,4,4,12}$][BF_4] is expected to behave as a SAIL because it has a long alkyl group. Conversely, [Ch][Lac] and [C_2mim][AcO] both have short side chains and are thought to have no surface activity, so these ILs are unlikely to considerably alter the EPS.

It has been reported that some kinds of ILs have antimicrobial activity (Pham et al. 2010), which is affected by the length of the substituent; activity against cocci, rods, and fungi is lost when short substituent groups are present (Pernak et al. 2003). It has been suggested that the mechanism of the antimicrobial activity of ILs is the disruption of bacterial cell membranes (Venkata et al. 2012). However, destruction of the cell membranes in the biofilm bacteria treated with ILs was not observed by TEM in this study (Figure 5). Because [Ch][Lac] and [C_2mim][AcO] possess short substituent groups and [Ch][Lac] contains carboxylate groups, it is believed that they will not act on the bacterial cell membrane. This is substantiated by the finding that [Ch][Lac] has a relatively high biocompatibility with fibroblast cells (Tsuda et al. 2011b).

Using ILs to prepare biofilm samples for SEM observation means that the biofilms do not require the extensive manipulation, fixation, dehydration, and coating that are used conventionally, and extreme sample dryness is also avoided. SEM observation using samples prepared with ILs does not require any special equipment and, importantly, high-resolution images can be obtained.

In conclusion, biofilm samples for SEM observation prepared by treatment with ILs can be readily visualized. Treatment with ILs can be added as an important sample preparation technique for the observation of biofilms by electron microscopy.

Additional file

Additional file 1: Figure S1. High-magnification images of *S. mutans* biofilm using IL. (a) Control prepared for observation using the conventional method at × 10,000, (b) 1% [Ch][Lac], (c) 10% [C_2mim][AcO] at × 10,000, and (d) 10% [Ch][Lac] at × 15, 000. Clear images were obtained at high magnification. Bars = 1 μm. Arrows indicate the fibriform extracellular matrix-like structures. **Figure S2.** SEM images of *S. mutans* biofilms prepared for observation using glutaraldehyde and IL. SEM images of *S. mutans* biofilms pretreated with (a) glutaraldehyde and 10% [Ch][Lac], and (b) 10% [Ch][Lac] (control). There were no differences detected between the SEM images of the glutaraldehyde-treated biofilm and control. Scale bars = 10 μm. **Figure S3.** SEM images of *S. mutans* biofilm using various concentrations of [C_2mim][AcO]. (a) 1%, (b) 10%, and (c) 20%. The asterisk indicates the accumulated IL. Scale bars = 5 μm. **Figure S4.** High-magnification TEM images of *S. mutans* biofilms prepared for observation using (a) the conventional method (control), (b) 10% [Ch][Lac], and (c) 10% [C_2mim][AcO]. Compared with those of the control biofilm, there was no difference incpf the images of the bacterial cell membrane, cell wall and cytoplasm for the IL-treated biofilms. Scale bars = 500 nm.

Competing interests
The authors declare that they have no competing interests.

Authors' contributions
KT and TT purified the ILs. YA participated in the SEM and CLSM observations and performed the statistical analysis. JM and TS carried out the TEM observation. YA, JM, and YN participated in the design of the study. YA, JM, and TT drafted the manuscript. SK, SE, and MH conceived the study, participated in its design and coordination, and helped to draft the manuscript. All authors read and approved the final manuscript.

Acknowledgements
This study was supported by the Japan Science and Technology Agency and a Grant-in-Aid for Scientific Research (No. 26462878) from the Japan Society for the Promotion of Science. Some of this work was supported by the "Nanotechnology Platform of the Ministry of Education, Culture, Sports, Science and Technology (MEXT), Japan" at the Research Center for Ultrahigh Voltage Electron Microscopy, Osaka University, Japan.

Author details
[1]Department of Restorative Dentistry and Endodontology, Osaka University Graduate School of Dentistry, 1-8 Yamadaoka, Suita, Osaka 565-0871, Japan. [2]Division for Interdisciplinary Dentistry, Osaka University Graduate School of Dentistry, 1-8 Yamadaoka, Suita, Osaka 565-0871, Japan. [3]Department of Applied Chemistry, Graduate School of Engineering, Osaka University, Suita, Osaka 565-0871, Japan. [4]Department of Material Science, Wakayama National College of Technology, 77 Noshima, Nada-cho, Gobo, Wakayama 644-0023, Japan. [5]Osaka University Research Center for Ultra-high-voltage Electron Microscopy, Suita, Osaka 565-0871, Japan.

References
Alhede M, Qvortrup K, Liebrechts R, Hoiby N, Givskov M, Bjarnsholt T (2012) Combination of microscopic techniques reveals a comprehensive visual

impression of biofilm structure and composition. FEMS Immunol Med Microbiol 65(2):335–342, doi:10.1111/j.1574-695X.2012.00956.x
Arimoto S, Sugimura M, Kageyama H, Torimoto T, Kuwabata S (2008) Development of new techniques for scanning electron microscope observation using ionic liquid. Electrochim Acta 53(21):6228–6234, doi:10.1016/j.electacta.2008.01.001
Asahi Y, Noiri Y, Igarashi J, Suga H, Azakami H, Ebisu S (2012) Synergistic effects of antibiotics and an N-acyl homoserine lactone analog on *Porphyromonas gingivalis* biofilms. J Appl Microbiol 112(2):404–411, doi:10.1111/j.1365-2672.2011.05194.x
Asahi Y, Noiri Y, Miura J, Maezono H, Yamaguchi M, Yamamoto R, Azakami H, Hayashi M, Ebisu S (2014) Effects of the tea catechin epigallocatechin gallate on *Porphyromonas gingivalis* biofilms. J Appl Microbiol 116(5):1164–1171, doi:10.1111/jam.12458
Barber PM, Fletcher JM, Newman HN (1993) Demonstration of the glycocalyces associated with three oral gram-negative bacterial species using a modern acrylic resin technique. J Struct Biol 110(3):180–187, doi:10.1006/jsbi.1993.1020
Boles BR, Thoendel M, Singh PK (2005) Rhamnolipids mediate detachment of *Pseudomonas aeruginosa* from biofilms. Mol Microbiol 57(5):1210–1223, doi:10.1111/j.1365-2958.2005.04743.x
Bowen WH, Koo H (2011) Biology of *Streptococcus mutans*-derived glucosyltransferases: role in extracellular matrix formation of cariogenic biofilms. Caries Res 45(1):69–86, doi:10.1159/000324598
Brown P, Butts CP, Eastoe J, Grillo I, James C, Khan A (2013) New catanionic surfactants with ionic liquid properties. J Colloid Interface Sci 395:185–189, doi:10.1016/j.jcis.2012.12.046
Costerton JW, Stewart PS, Greenberg EP (1999) Bacterial biofilms: a common cause of persistent infections. Science 284(5418):1318–1322
Darkin MG, Gilpin C, Williams JB, Sangha CM (2001) Direct wet surface imaging of an anaerobic biofilm by environmental scanning electron microscopy: application to landfill clay liner barriers. Scanning 23(5):346–350
Donlan RM, Costerton JW (2002) Biofilms: survival mechanisms of clinically relevant microorganisms. Clin Microbiol Rev 15(2):167–193
Dwiranti A, Lin L, Mochizuki E, Kuwabata S, Takaoka A, Uchiyama S, Fukui K (2012) Chromosome observation by scanning electron microscopy using ionic liquid. Microsc Res Tech 75(8):1113–1118, doi:10.1002/jemt.22038
Flemming HC, Wingender J (2010) The biofilm matrix. Nat Rev Microbiol 8(9):623–633, doi:10.1038/nrmicro2415
Flemming HC, Neu TR, Wozniak DJ (2007) The EPS matrix: the "house of biofilm cells". J Bacteriol 189(22):7945–7947, doi:10.1128/JB.00858-07
Galgano PD, El Seoud OA (2010) Micellar properties of surface active ionic liquids: a comparison of 1-hexadecyl-3-methylimidazolium chloride with structurally related cationic surfactants. J Colloid Interface Sci 345(1):1–11, doi:10.1016/j.jcis.2010.01.078
Gardas RL, Coutinho JAP (2008) A group contribution method for viscosity estimation of ionic liquids. Fluid Phase Equilib 266(1–2):195–201, doi:10.1016/j.fluid.2008.01.021
Hamada S, Slade HD (1980) Biology, immunology, and cariogenicity of *Streptococcus mutans*. Microbiol Rev 44(2):331–384
Ishigaki Y, Nakamura Y, Takehara T, Nemoto N, Kurihara T, Koga H, Nakagawa H, Takegami T, Tomosugi N, Miyazawa S, Kuwabata S (2011a) Ionic liquid enables simple and rapid sample preparation of human culturing cells for scanning electron microscope analysis. Microsc Res Tech 74(5):415–420, doi:10.1002/jemt.20924
Ishigaki Y, Nakamura Y, Takehara T, Shimasaki T, Tatsuno T, Takano F, Ueda Y, Motoo Y, Takegami T, Nakagawa H, Kuwabata S, Nemoto N, Tomosugi N, Miyazawa S (2011b) Scanning electron microscopy with an ionic liquid reveals the loss of mitotic protrusions of cells during the epithelial-mesenchymal transition. Microsc Res Tech 74(11):1024–1031, doi:10.1002/jemt.20989
Koo H, Hayacibara MF, Schobel BD, Cury JA, Rosalen PL, Park YK, Vacca-Smith AM, Bowen WH (2003) Inhibition of *Streptococcus mutans* biofilm accumulation and polysaccharide production by apigenin and tt-farnesol. J Antimicrob Chemother 52(5):782–789, doi:10.1093/jac/dkg449
Kuwabata S, Kongkanand A, Oyamatsu D, Torimoto T (2006) Observation of ionic liquid by scanning electron microscope. Chem Lett 35(6):600–601, doi:10.1246/cl.2006.600
Li YH, Tang N, Aspiras MB, Lau PC, Lee JH, Ellen RP, Cvitkovitch DG (2002) A quorum-sensing signaling system essential for genetic competence in *Streptococcus mutans* is involved in biofilm formation. J Bacteriol 184(10):2699–2708
Loesche WJ (1986) Role of *Streptococcus mutans* in human dental decay. Microbiol Rev 50(4):353–380
Marrie TJ, Costerton JW (1984) Scanning and transmission electron microscopy of in situ bacterial colonization of intravenous and intraarterial catheters. J Clin Microbiol 19(5):687–693
McHale G, Hardacre C, Ge R, Doy N, Allen RW, MacInnes JM, Bown MR, Newton MI (2008) Density-viscosity product of small-volume ionic liquid samples using quartz crystal impedance analysis. Anal Chem 80(15):5806–5811, doi:10.1021/ac800490q
Noiri Y, Okami Y, Narimatsu M, Takahashi Y, Kawahara T, Ebisu S (2003) Effects of chlorhexidine, minocycline, and metronidazole on *Porphyromonas gingivalis* strain 381 in biofilms. J Periodontol 74(11):1647–1651, doi:10.1902/jop.2003.74.11.1647
Pandit S, Kim JE, Jung KH, Chang KW, Jeon JG (2011) Effect of sodium fluoride on the virulence factors and composition of *Streptococcus mutans* biofilms. Arch Oral Biol 56(7):643–649, doi:10.1016/j.archoralbio.2010.12.012
Pernak J, Sobaszkiewicz K, Mirska I (2003) Anti-microbial activities of ionic liquids. Green Chem 5(1):52–56, doi:10.1039/b207543c
Pham TP, Cho CW, Yun YS (2010) Environmental fate and toxicity of ionic liquids: a review. Water Res 44(2):352–372, doi:10.1016/j.watres.2009.09.030
Priester JH, Horst AM, Van de Werfhorst LC, Saleta JL, Mertes LA, Holden PA (2007) Enhanced visualization of microbial biofilms by staining and environmental scanning electron microscopy. J Microbiol Methods 68 (3):577–587, doi:10.1016/j.mimet.2006.10.018
Schmitt J, Flemming HC (1999) Water binding in biofilms. Water Sci Technol 39(7):77–82
Stoodley P, Sauer K, Davies DG, Costerton JW (2002) Biofilms as complex differentiated communities. Annu Rev Microbiol 56:187–209, doi:10.1146/annurev.micro.56.012302.160705
Sutherland I (2001) Biofilm exopolysaccharides: a strong and sticky framework. Microbiology 147:3–9
Torimoto T, Tsuda T, Okazaki K, Kuwabata S (2010) New frontiers in materials science opened by ionic liquids. Adv Mater 22(11):1196–1221, doi:10.1002/adma.200902184
Tsuda T, Kondo K, Tomioka T, Takahashi Y, Matsumoto H, Kuwabata S, Hussey CL (2011a) Design, synthesis, and electrochemistry of room-temperature ionic liquids functionalized with propylene carbonate. Angew Chem 50(6):1310–1313, doi:10.1002/anie.201005208
Tsuda T, Nemoto N, Kawakami K, Mochizuki E, Kishida S, Tajiri T, Kushibiki T, Kuwabata S (2011b) SEM observation of wet biological specimens pretreated with room-temperature ionic liquid. Chembiochem 12(17):2547–2550, doi:10.1002/cbic.201100476
Tsuda T, Mochizuki E, Kishida S, Iwasaki K, Tsunashima K, Kuwabata S (2014) SEM observation of hydrous superabsorbent polymer pretreated with room-temperature ionic liquids. PLoS One 9(3):e91193, doi:10.1371/journal.pone.0091193
Tsunashima K, Sugiya M (2007) Physical and electrochemical properties of room-temperature dicyanamide ionic liquids based on quaternary phosphonium cations. Electrochemistry 75:734–736
Venkata NY, Reddy GK, Lalithamanasa P, Venugopalan VP (2012) The ionic liquid 1-alkyl-3-methylimidazolium demonstrates comparable antimicrobial and antibiofilm behavior to a cationic surfactant. Biofouling 28(10):1141–1149, doi:10.1080/08927014.2012.736966
Vu B, Chen M, Crawford RJ, Ivanova EP (2009) Bacterial extracellular polysaccharides involved in biofilm formation. Molecules 14(7):2535–2554, doi:10.3390/molecules14072535
Weber K, Delben J, Bromage TG, Duarte S (2014) Comparison of SEM and VPSEM imaging techniques with respect to *Streptococcus mutans* biofilm topography. FEMS Microbiol Lett 350(2):175–179, doi:10.1111/1574-6968.12334

21

Salmonella enterica biofilm-mediated dispersal by nitric oxide donors in association with cellulose nanocrystal hydrogels

Massimiliano Marvasi[1,2*], Ian A Durie[2], Eric S McLamore[3], Diana C Vanegas[3,4] and Prachee Chaturvedi[5]

Abstract

Protected by extracellular polymers, microbes within biofilms are significantly more resistant to disinfectants. Current research has been instrumental in identifying nitric oxide donors and hydrogels as potential disinfectant additives. Nitric oxide (NO) donors are considered a very promising molecule as biofilm dispersal agents and hydrogels have recently attracted a lot of interest due to their biocompatible properties and ability to form stable thin films. When the NO donor MAHMA NONOate was dissolved in phosphate saline buffer, it was able to reduce the biomass of well-established biofilms up to 15% for at least 24 h of contact time. Encapsulation of MAHMA NONOate and molsidomine within a hydrogel composed of cellulose nanocrystals (CNC) has shown a synergistic effect in dispersing well-established biofilms: after 2 h of exposure, moderate but significant dispersion was measured. After 6 h of exposure, the number of cells transitioning from the biofilm to the planktonic state was up to 0.6 log higher when compared with non-treated biofilms. To further explore the transport processes of NO donors within hydrogels, we measured the nitric oxide flux from gels, at 25°C for a composite of 0.1 μM MAHMA NONOate–CNC. Nitric oxide diffuses up to 500 μm from the hydrogel surface, with flux decreasing according to Fick's law. 60% of NO was released from the hydrogel composite during the first 23 min. These data suggest that the combined treatments with nitric oxide donor and hydrogels may allow for new sustainable cleaning strategies.

Keywords: *Salmonella enterica*, Biofilm dispersal, Nitric oxide donor, Microelectrodes, Cellulose nanocrystals hydrogel

Introduction

Biofilms formed on washing systems in produce production facilities can be recalcitrant reservoirs of human pathogens, which are difficult to control and can potentially cause costly outbreaks (Srey et al. 2013). Pathogens in biofilms are resistant to common disinfectants and contribute to an increased risk in contamination of produce by potentially contaminated water (Zhang and Mah 2008; Corcoran et al. 2014). Several disinfectants, antibiotics and messenger molecules have been studied for their ability to dislodge existing biofilms (Barraud et al. 2006, 2009a, 2012). Of these, nitric oxide (NO) appears to be very promising, and currently nitric oxide donors are used clinically. Nitric oxide is effective as a biofilm dispersant, functioning as a messenger rather than a generic poison (Barraud et al. 2006, 2009a, 2012; Marvasi et al. 2014). Nitric oxide can be delivered to biofilms by using donor molecules (Wang et al. 2005), and the application of nitric oxide donors has the same effect as the direct addition of nitric oxide gas (Barraud et al. 2009b). In bacteria, nitric oxide acts as an active signaling molecule at very low concentrations and is able to disperse preformed biofilms at nano- or picomolar concentration, indicating that truly minute quantities of the chemical are sufficient for dislodging much of the biofilm (Barraud et al. 2006, 2009a, 2012; Marvasi et al. 2014). Nitric oxide could have a universal effect on dispersal of bacteria biofilm including both Gram-positive and Gram-negative bacteria (Xiong and Liu 2010).

*Correspondence: m.marvasi@mdx.ac.uk
[1] Department of Natural Sciences, School of Science and Technology, Middlesex University, The Burroughs, London NW4 4BT, UK
[2] Soil and Water Science Department, University of Florida, Gainesville, FL, USA
Full list of author information is available at the end of the article

The mechanisms of biofilm dispersion are not clear, but it appears to function in the transition of sessile biofilm organisms to free-swimming bacteria. For example, microarray studies have revealed that genes involved in adherence are downregulated in *Pseudomonas aeruginosa* upon exposure to nitric oxide donors (Firoved et al. 2004).

Over 105 nitric oxide donors have been characterized, and 6 have been currently tested as biofilm dispersal. Among these, the main studied are the sodium nitroprusside, (Barraud et al. 2006; Marvasi et al. 2014; Charville et al. 2008; Barraud et al. 2009a), molsidomine, diethylamine NONOate diethylammonium, NONOate diethylammonium, MAHMA nonoate (Marvasi et al. 2014), NORS (sodium nitrite citric acid) (Regev-Shoshani et al. 2013) and NO-releasing silica nanoparticles (Hetrick et al. 2009). Among these, MAHMA NONOate is one of the best candidates in biofilm dispersion: it spontaneously dissociates in a pH/temperature dependent manner, in serum it has a half-life of 1 and 3 min at 37°C and 22–25°C respectively, at pH 7.4 to liberate 2 mol of NO per mole of parent compound (Wang et al. 2005; Keefer et al. 1996; Hrabie et al. 1993). Interesting, when MAHMA NONOate was used to disperse *Salmonella enterica* and *Escherichia coli* biofilms (24-h old), up to 50% of *Salmonella* 14028 and a cocktail of six *Salmonella* outbreak strains biofilms were dispersed when incubated for 6 and 24 h at 22°C (both $p < 0.0001$). About 40% of dispersion was measured also for the pathogenic *Escherichia coli* O157:H7 when exposed at room temperature for 6 and 24 h (Marvasi et al. 2014).

Nitric oxide donors can be associated with other disinfectants obtaining a synergistic effect in terms of biofilm dispersion. For example, a 20-fold increase in the efficiency was observed when nitric oxide was applied together with chlorine in removing multispecies biofilms (Barraud et al. 2009b). In addition, we recently showed that the nitric oxide donor MAHMA NONOate and molsidomine were also able to increase up to 15% *Salmonella* biofilm dispersion when associated with the industrial disinfectant SaniDate 12.0 (Marvasi et al. 2014).

We used this foundation to hypothesize that nitric oxide donors can also be applied to hydrogels in order to obtain a synergistic effect in terms of biofilm dispersal. Hydrogels have recently attracted a lot of interest due to high strength/stiffness, optical transparency, biocompatibility, biodegradability, and highly porous structural network (Klemm et al. 2011; Siro and Plackett 2011) and the ability to carry bactericidal molecules (Azizi et al. 2013; Drogat et al. 2011; Feese et al. 2011). Hydrogels are water-swollen, and cross-linked polymeric network produced by the simple reaction of one or more monomers (Ahmed 2013). The industrial applications of hydrogels are extremely heterogeneous, ranging from agriculture, drug delivery systems, food additives, biomedical applications such as implants, and separation of biomolecules (Thakur and Thakur 2014).

Among the hydrogels, cellulose nanocrystals (CNC) is one of the most promising polymer because it is an eco-friendly product, cheap and safe (George et al. 2011). CNC has high crystallinity, high water holding capacity and excellent mechanical and thermal properties (George et al. 2011). For our scope, CNC was procured from a process development center that focuses on production of wood-derived renewable nanomaterials. CNC was selected as the hydrogel base because this material has an ionic charged surface appropriate for binding biofilm outer extracellular polymeric substances (EPS) (Marvasi et al. 2014; George et al. 2011; Marvasi et al. 2010).

The main objective of this study was to measure the synergistic effect of the nitric oxide donor MAHMA NANOate in association with CNC during nitric oxide-mediated biofilm dispersion of *S. enterica* sv Typhimurium. We focused our experiments on well-structured biofilms and we measured the shortest effective exposure. The nitric oxide release profile, diffusion and flux rate were also measured during MAHMA NONOate–CNC association.

Materials and methods

Bacterial strains and culture media

Salmonella enterica serovar Typhimurium ATCC14028, was used in this study. pGFP-ON, a strongly fluorescent construct carrying GFP protein expressed from the Salmonella *dppA* promoter (Noel et al. 2010), was transformed into the strains of interest by electroporation.

All strains were maintained as frozen glycerol stocks, and were sub-cultured into Luria–Bertani medium with appropriate antibiotics (50 μg mL^{-1} kanamycin, 100 μg mL^{-1} ampicillin).

Chemicals and materials for microelectrode preparation

Single layer graphene oxide (SLGO; 0.8 nm thickness, 1–5 μm diameter, 99% purity) was obtained from ACS Materials. Methanol, ascorbic acid and lead acetate were purchased from Fisher Scientific (Atlanta, GA, USA). Cerium (IV) oxide (nanoparticles dispersion, <25 nm particle size, 10 wt% in H_2O) and chloroplatinic acid (8 wt%) were procured from Sigma-Aldrich (St. Louis, MO, USA). Nafion (5% aliphatic alcohol) and *o*-phenylenediamine were acquired from Acros organics (Newark, NJ, USA). Nitric oxide gas (CP grade 99%) was purchased from AirGas (Gainesville, FL, USA).

Nitric oxide donors

MAHMA NONOate and molsidomine were purchased from Sigma-Aldrich (St. Luois, MO, USA). For each compound, 1 mmol L^{-1} stock solutions were prepared in phosphate-buffered saline (PBS), pH 7.3 (PBS, Fisher, Waltham, MA, USA) and aliquots were stored at −80°C. For the assays, serial dilutions were always prepared fresh in ice-cold PBS just before the experiments and used within 5 min of their preparation. The biofilm dispersion potential was tested on polystyrene 96-well plates (Fisher, Waltham, MA, USA).

Biofilm formation and dispersal on plastics

Overnight Luria–Bertani cultures of *Salmonella S.* typhimurium 14028 or *S.* typhimurium 14028 pGFP-ON (with 100 μg mL^{-1} ampicillin) were diluted 1:100 in CFA medium as described previously (Teplitski et al. 2006), and 100 μL of the diluted cultures were aliquoted into wells of 96-well polypropylene and polystyrene plates (Fisher, Waltham, MA, USA). Plates with bacteria were incubated for 24 h or 1 week (well established biofilm) at 37°C inside a Ziploc bag to prevent dehydration.

Measurement of biofilms dispersal

Upon completion of the incubation, the medium with planktonic bacteria was removed and serial dilutions of nitric oxide donors in PBS or CNC (in 200 μL) were added to the wells with biofilms. Dispersal experiments were conducted at 22°C for a time ranging from 1 to 24 h. The dispersion of biofilms treated with the nitric oxide donor dissolved in PBS was measured by staining the remaining biofilms with 1% crystal violet in ethanol, as described previously (O'Toole and Kolter 1998; Merritt et al. 2005).

Biofilms dispersion treated with CNC (or control, CNC + PBS) was measured by directly monitoring the increase of fluorescence of planktonic cells of *S.* typhimurium 14028 pGFP-ON. At the end of the period of exposure, 170 μL of CNC-nitric oxide solution were transferred into a black 96-wells plate and fluorescence was measured by using Victor-2 multimode plate reader with a 485 nm/535 nm excitation/emission filter (Perkin Elmer, Waltham, MA, USA). Increase of fluorescent intensity was used to represent an increase of detached cells.

Luminescence tests

Effects of selected nitric oxide donors on light production by a constitutively luminescent *Salmonella* strain were characterized as indirect assessments of toxicity of the compounds. 1 mL of *S.* typhimurium 14028 pTIM2442 (harboring the *luxCDABE* driven by a strong constitutive phage, Alagely et al. 2011) were washed with BPS and mixed with opportune reagents (BPS, molsidomine, CNC) and grown in black polystyrene plates (Corning, New York, USA). Molsidomine was diluted in PBS (9.89 g L^{-1}) or CNC (Fisher Scientific, Waltham, MA, USA) to final concentrations of 10 μmol L^{-1}. PBS + CNC and PBS alone were used as a control. Luminescence of *S.* typhimurium 14028 pTIM2442 was measured over time using Victor-2 multimode plate reader (Perkin Elmer, Waltham, MA, USA). Each experiment included 12 replicas.

CNC-nitric oxide preparation

1 mmol L^{-1} stock solutions of MAHMA NONOate or molsidomine were prepared in PBS, pH 7.3 (PBS, Fisher, Waltham, MA, USA). Final solutions to 1 μM were prepared by dispersing the nitric oxide stock solution in an opportune volume of CNC. The solution was further vortexed for 30 s and applied of prewashed biofilm formed into polypropylene plates.

Nitric oxide measurement

NO microelectrodes were prepared using the methods in Chaturvedi et al. (2014). Pt/Ir microelectrodes (2 μm tip diameter, 5.1 cm length, 81 μm shaft diameter; MicroProbe, Inc., Gaithersburg, MD, USA) were rinsed in methanol prior to use. Nafion, nanoceria, nanoplatinum, and reduced graphene oxide were deposited based on the methods in Chaturvedi et al. (2014) and Vanegas et al. (2014). Briefly, microelectrodes were platinized via pulsed-sonoelectrodeposition at 10 mV for 60 cycles (each cycle consisted of 1 s electroplating pulse followed by 1 s sonication pulse) in a plating solution of 0.002% lead acetate and 1.44% chloroplatinic acid. Thereafter, electrodes were dip coated for 5 min in a suspension of 10 wt% cerium oxide with 0.8 wt% ascorbic acid and 0.2 wt% SLGO. Electrodes were then dried at 100°C for 20 min. Next, electrodes were dip coated in 5% Nafion and dried at 110°C for 20 min; this step was repeated twice. Finally, an OPD membrane was formed on the tip of the probe by polarizing the electrodes at 900 mV in a solution of 5 mM OPD and 0.1 mM ascorbic acid in PBS (pH 7.4) until a stable current was recorded (less than 2% variability) (Friedemann et al. 1996; Koehler et al. 2008; Porterfield et al. 2001). The working principle for the NO microelectrode is based on oxidation of NO at the surface of the nanocomposite, generating the nitrosonium cation (NO^+) and one free electron based on Eq. 1 below; free electrons are detected as oxidative current versus a reference electrode

$$NO \rightarrow e^- + NO^+ + OH^- \rightarrow HONO \rightarrow NO_2^- + H^+ \quad (1)$$

NO microelectrode calibration and flux measurements

All measurements were performed at a working potential of +500 mV versus a reference electrode composed of a Ag/AgCl wire immersed in 3 M KCl and inserted into a half-cell microelectrode holder (World Precision Instruments, Inc., Sarasota, FL, USA). For electrode calibration, a PBS solution was polarized for 1 h at +900 mV; current output was measured at constant potential while successively injecting NO stock solution prepared based on Porterfield et al. (2001). Calibrated microsensors were used to measure NO flux in the self referencing (SR) modality as described in detail by McLamore and Porterfield (2011). The SR microsensor technique is designed to measure flux within the concentration boundary layer, and involves computer-controlled translation of a microelectrode between two positions separated by a known distance (dX). Differential concentration (dC) is recorded in real time and flux is then calculated using Fick's first law of diffusion ($J = -D \times dC\ dX - 1$). The diffusion coefficient (D) for NO used in all experiments was 2.21×10^{-5} cm^2 s^{-1} based on and Zacharia and Deen (2005).

Statistical analysis

The statistical package JMP (SAS) was used to infer the ANOVA analysis ($p < 0.05$). Tukey means separation analysis was inferred in order to group the means.

Results

Synergistic effect of the association of nitric oxide donors with CNC

The effect of MAHMA NONOate as biofilm dispersal was previously assessed in our recent publication in a range of biofilm-forming microorganisms of industrial and/or clinical significance including *S. enterica* (Marvasi et al. 2014). Our previous efforts were focused on studying young biofilms (maximum 24 h old) and exposure time up to 6 h. In order to focus on more applicative approaches of the nitric oxide donor technology, we tested the nitric oxide donors on well-structured biofilms (1-week old biofilms) and for a shorter contact time (up to 1 h of exposure).

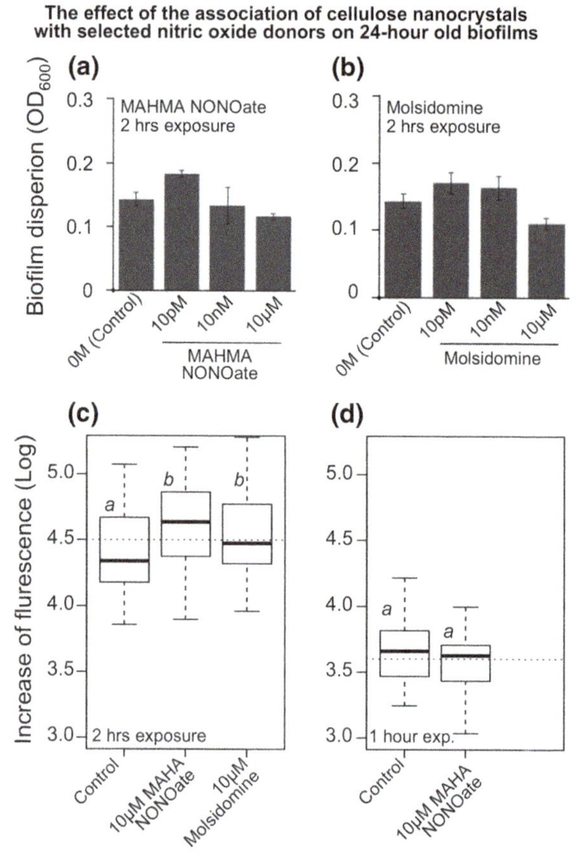

Figure 1 24-h old *Salmonella* biofilm dispersal by MAHMA NONOate dissolved in BPS and CNC. Biofilms were preformed by *S. enterica* typhimurium 14028 for 24 h on polystyrene prior to the treatment with MAHMA NONOate and molsidomine. Contact time at which biofilms were exposed to the nitric oxide donor is listed on each panel. **a**, **b** The decreasing of biofilm formation was measured by using the technique of crystal violet staining. **c**, **d** Fluorescent *Salmonella* 14028 pGFP-ON biofilm was treated with MAHMA NONOate and molsidomine dissolved in CNC. Increase in fluorescence of the planktonic cells was measured at 485 nm/535 nm. The *box-plots* encompass the lower and upper quartiles, *thick lines* within the *box* are the median values, and the *whiskers* indicate the degree of dispersion of the data. Outliers are shown as *dots*. *Dotted line* represents the grand mean. *Letters* above each *box* represent the Tukey means separation: *different letters* are significant different means.

As a first approach, we exposed a 24-h old *Salmonella* biofilm for a minimum of 2 h to different concentrations of MAHMA NONOate dissolved in PBS (10 μM, 10 nM and 10 pM): no significant results were measured (Figure 1a). The experiment was repeated with the nitric oxide donor, molsidomine, observing a similar result (Figure 1b). In further experiments, MAHMA NONOate or molsidomine were dissolved into CNC hydrogel to 10 μM (CNC–NONOate or CNC–molsidomine) to test whether the association of CNC was instrumental in increasing the dispersion potential of the nitric oxide donors within the 2 h. A fluorescent *Salmonella* 14028 pGFP-ON biofilm preformed on polypropylene was treated with 10 μM NONOate–CNC or 10 μM molsidomine–CNC. Fluorescence of the planktonic cells was measured after 2 and 1 h of exposure to the nitric oxide donors. The fraction of cells switching to the planktonic state was significant after 2 h of exposure (Figure 1c), but not significant after 1 h of exposure (Figure 1d).

To determine the viability of *Salmonella* during the exposure to CNC, the luminescence of *Salmonella* 14028 pTIM2442 upon association of CNC and molsidomine was measured. The general metabolic state of the cells was assessed by using the pTIM2442 system (Alagely et al. 2011). No variation in terms of viability was reported during a period of 8 h when CNC exposure was compared with the controls (Additional file 1: Figure S1).

To measure the responsiveness of a well-established biofilm such as a 1-week old biofilm, *Salmonella* 14028 biofilms were exposed to different concentrations of MAHMA NONate dissolved in PBS for 24 and 6 h. 24 h of exposure to MAHMA NONOate reduced the biomass up to 15% (Figure 2a). However, 6 h of exposure did not significantly reduce the biomass (Figure 2b). As before, we tested whether the association of CNC was instrumental in increasing the dispersal potential of the nitric oxide donor. MAHMA NONOate and molsidomine were dissolved into CNC at a concentration of 10 μmol L^{-1}. Interestingly, total fluorescence of the planktonic cells increased significantly in the wells with biofilms treated with both the donors (Figure 2c, d), reflecting an increase in cellular detachment at the tested concentrations. The biofilm exposed for 2 h showed an average of 0.3 log increase in fluorescence when compared with the control (PBS) (Figure 2c). When the same treatment was extended for 6 h, a difference up to 0.6 log was measured (Figure 2d). One-hour exposure was also tested but it was not significant (data not shown).

CNC was also tested on 1-week old biofilms at 4 and 37°C, to determine the effect of temperature on the dispersion potential. No significant differences were determined on 1-week old biofilm when exposed to molsidomine + CNC versus the control CNC + PBS. Biofilms were exposed for 6 h. 22°C seems to be the most appropriate temperature for obtaining a significant dispersal.

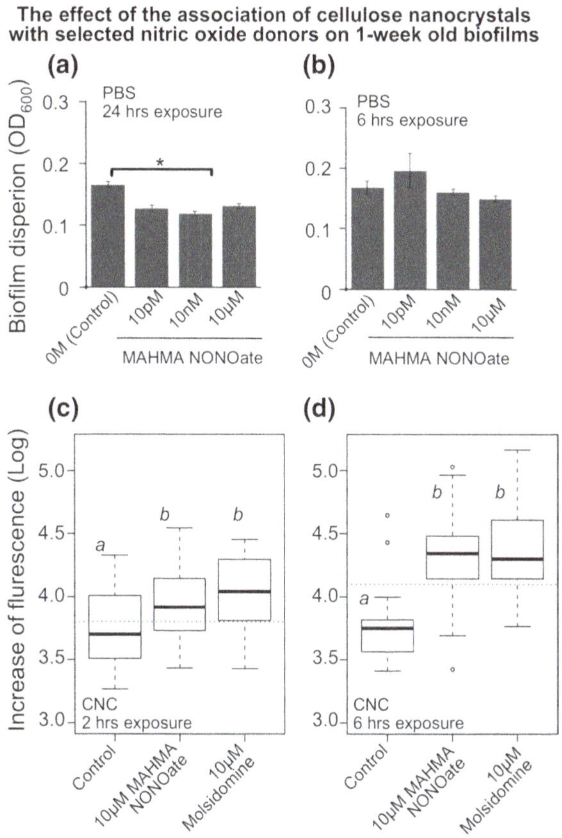

Figure 2 One-week old *Salmonella* biofilm dispersal by MAHMA NONOate dissolved in BPS and CNC. Biofilms were preformed by *S. enterica* typhimurium 14028 for 1-week on polystyrene prior to the treatment with MAHMA NONOate and molsidomine. Contact time at which biofilms were exposed to the nitric oxide donor is listed on each panel. **a**, **b** Decreasing of biofilm formation measured by using the technique of crystal violet staining. **c**, **d** A fluorescent *Salmonella* 14028 pGFP-ON biofilm was treated with MAHMA NONOate or molsidomine dissolved in CNC. Increase in fluorescence of the planktonic cells was measured at 485 nm/535 nm. The *box-plots* encompass the lower and upper quartiles, *thick lines* within the *box* are the median values, and the *whiskers* indicate the degree of dispersion of the data. Outliers are shown as *dots*. *Dotted line* represents the grand mean. *Letters* above each *box* represent the Tukey means separation: *different letters* are significant different means.

Releasing profile of MAHMA NONOate in BPS and CNC

To better understand the synergistic effect of CNC–MAHMA NONOate composites, the diffusion of NO from the hydrogel was studied in the absence of biofilms. Electrodes were first calibrated in buffer solution (Figure 3a, b). The average sensitivity of the NO microelectrodes was 10.5 ± 0.1 pA nM^{-1} within the range of 20 pM–100 nM. The average response time of the electrodes was 0.65 ± 0.24 s.

Microelectrodes were used to directly measure NO flux from CNC hydrogels to better understand the mass transfer under abiotic conditions. A microelectrode was positioned at the surface of a hydrogel immediately after mixing the CNC and MAHMA NONOate while continuously measuring surface concentration and flux. At 25°C, nitric oxide dissociated from the MAHMA NONOate–CNC hydrogel. The surface concentration was 98.8 ± 0.1 μmol L^{-1} and the boundary layer was approximately 400 μm from the surface of the gel (Figure 4). The surface flux of nitric oxide was 0.92 ± 0.01 pmol cm^2 s^{-1}, and the decay profile was modeled using Fick's first law (R = 0.989; Figure 4b) based on the methods in McLamore et al. (2009). As expected, the nitric oxide surface concentration (10.1 ± 0.6 μmol L^{-1}), surface flux (0.36 ± 0.26 pmol cm^2 s^{-1}), and boundary layer thickness (approximately 150 μm) were significantly lower when the MAHMA NONOate was diluted 1:10 in the CNC hydrogel (Figure 4b). For the diluted sample, the

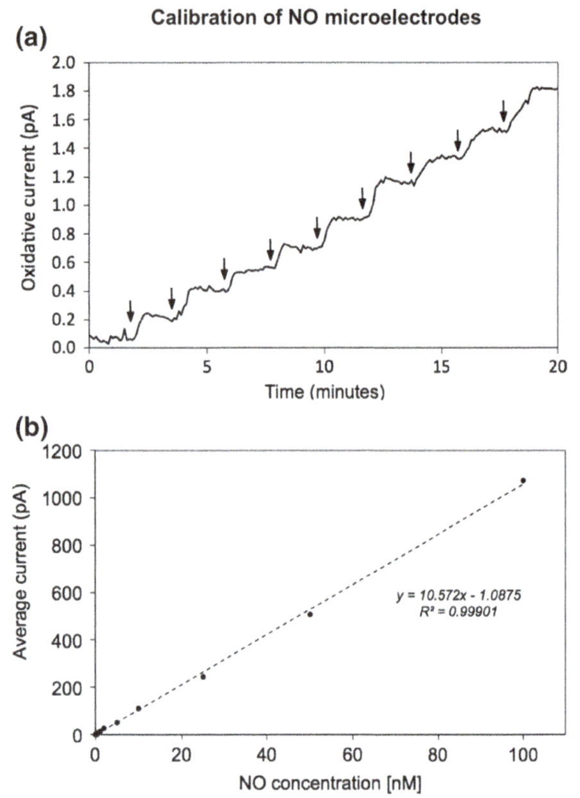

Figure 3 Calibration of NO microelectrodes. **a** Representative real time plot of NO microelectrode. *Vertical arrows* represent injection of NO stock solution. **b** Average output from three replicate microelectrodes (average sensitivity was 10.5 ± 0.1 pA nM^{-1}).

regression coefficient correlating the measured flux profile and the predicted behavior using Fick's first law decreased significantly (R = 0.713; Figure 4b).

Although the exact amount of nitric oxide liberated within biofilms from nitric oxide donors have not yet been established, we measured the release rate of nitric oxide from the CNC hydrogel enriched with MAHMA NONOate. MAHMA NONOate was initially dissolved in PBS to a concentration of 500 mM, and then it was dissolved in CNC at a final concentration of 1 µM. A nitric oxide microelectrode was used to measure the release rate (Figure 5). Our measurements (three replicas) showed that 60% of the nitric oxide is released as gas during the first 23 min (~725 s); the decay constant was 4.1 ± 0.4 h^{-1}. As expected, a 1:10 dilution of MAHMA NONOate in CNC significantly decreased the surface concentration/flux, and the time required for approximately 60% of nitric oxide released as gas (~1125 s) was significantly longer; the decay constant was 2.3 ± 1.1 h^{-1}.

Discussion

Every wet surface is a substrate for microbial biofilms and in food handling facilities biofilms can be particularly problematic. Protected by extracellular polymers, microbes within biofilms are significantly more resistant to chlorine and other disinfectants (Corcoran et al. 2014). As a result, several disinfectants have been used to remove biofilm, however disinfectants fail to completely eradicate biofilms from food contact surface materials

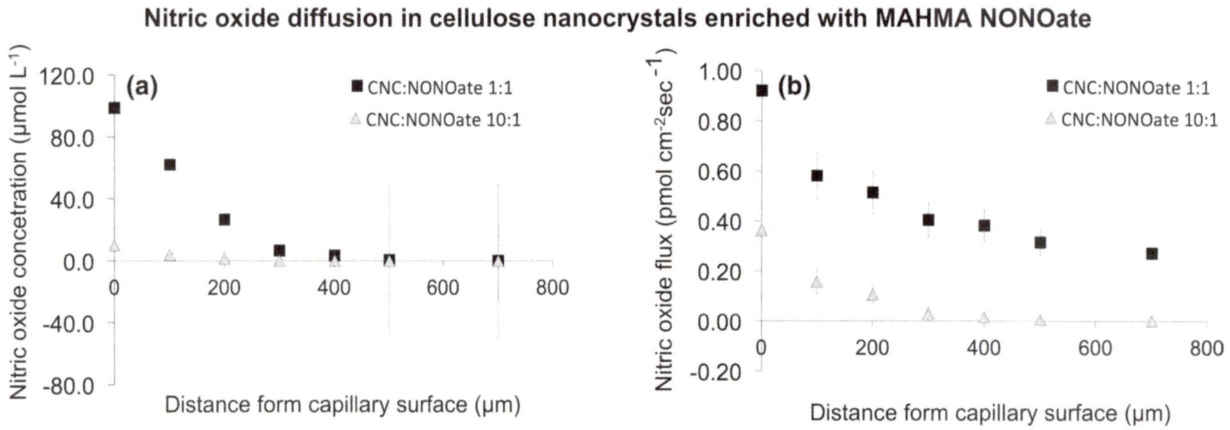

Figure 4 Nitric oxide diffusion in CNC. MAHMA NONOate was mixed with cellulose nanocrystals (*whisker type*) and a hydrogel was formed with a final concentration of 1 mM. A microcapilary was tapered with a tip diameter of ca. 10 µm using a glass puller. Consequently the hydrogel–NONOate was injected into the microcapillary and the microcapillary was placed into a phosphate buffer solution (PBS). A nitric oxide microsensor was immediately used to measure release of NO (both concentration and flux). **a** Nitric oxide concentration profiles from the hydrogel surface. **b** Nitric oxide flux profiles. 10:1 represent a 10 times dilution of the MAHMA NONOate dissolved in CNC. *Error bars* represent the standard error of three replicas.

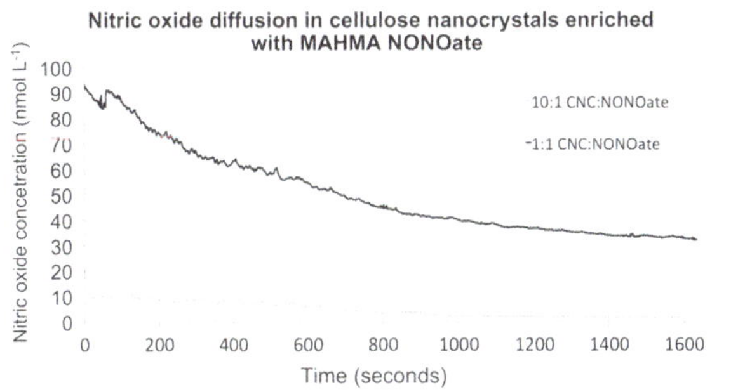

Figure 5 Nitric oxide release profiles of hydrogel-NONOate. Time 0 represents injection of the CNC:NONOate hydrogel into a glass capillary. 10:1 represent a 10 times dilution of the MAHMA NONOate dissolved in CNC. Experiment has been replicated three times. The *graph* represents an average of the three experiments.

(Corcoran et al. 2014). Current research has been instrumental in identify nitric oxide as potential additive to disinfectant (Barraud et al. 2006; Marvasi et al. 2014).

To focus on applicative approaches of the nitric oxide donor technology, we focused on dispersing well-established biofilms in the shortest time possible. To meet this need, a composite of nitric oxide and a hydrogel was used, considering the attractive physicochemical properties of hydrgoels as reviewed by Malmsten (2011). Nitric oxide donors have been combined with a number of hydrogels in preliminary studies (Koehler et al. 2008), but to date no hydrogels have been developed with nitric oxide donors which can potentially be scaled up for commercial applications. In this work, CNC was selected as a candidate for potential large scale application because the material is produced in large batches by a commercial forest based nanomaterial processing facility. In addition, CNC has an ionic charged surface that can facilitate attachment of biofilm EPS (Marvasi et al. 2010). For the first time, we show that association of MAHMA NONOate and molsidomine with CNC improved the biofilm dispersion potential when compared with PBS (Figures 1, 2). CNC–MAHMA NONOate allowed the dispersion of 1-week old biofilm with a contact time of at least 2 h; This can be ascribed to the combination of two different factors: (1) the cells do not have enough time to switch from the biofilm state to the planktonic state in less than 2 h; (2) the nitric oxide diffusion into the biofilm requires a certain amount of time (based on concentration of the nitric oxide donor in the gel). Temperature seems also to play a role on the dispersion potential. In terms of dispersal the exposure to 22°C results to be the most appropriate temperature when compared with 4 and 37°C. We can hypothesize that at 37°C the nitric oxide is completely depleted within a short time. On the opposite at lower temperature the nitric oxide releasing prolife is reduced in CNC. It is well known that properties of CNC change according with the temperature (George et al. 2011).

Due to the porous network, diffusion is the predominant transport process within hydrogels and cell aggregates. We measured the nitric oxide diffusion from CNC, observing that nitric oxide can diffuse up to 500 µm from the capillary opening, which is an important observation since 1-week old biofilms rarely have a thickness greater than 0.5 mm (Paramonova et al. 2007). With reference of the releasing time, our measurements show that 60% of the nitric oxide is released as gas during the first 12 min. In the literature, similar release profiles in PBS were observed with the nitrosothiols *S*-nitroso-*N*-acetylpenicillamine (SNAP), *S*-nitroso-L-glutathione (GSNO) and sodium nitroprusside (Barraud et al. 2009b).

Finally, it is worth mentioning that the improved ability of MAHMA NONOate-CNC association to disperse biofilm may be due to the low-moderate antimicrobial activity of CNC (Azizi et al. 2013), even though in our experimental conditions, CNC did not show any significant antimicrobial activity. In literature other molecules have been associated with CNC to improve its antimicrobial effectiveness: for example, CNC stabilized with ZnO–Ag exhibited greater bactericidal activity against *Salmonella choleraesuis* and *Staphylococcus aureus* compared to cellulose-free ZnO–Ag heterostructure nanoparticles of the same particle size (Azizi et al. 2013). Association of porphyrin (Feese et al. 2011) and silver nanoparticles (Drogat et al. 2011).

Further studies should also be addressed in associating nitric oxide donors with nanoparticle composites with effective antibacterial and biofilm-disrupting properties. To that end several polymers can be used, for example silk fibroin–silver nanoparticle composite showed both an effective antibacterial activity against the methicillin-resistant *S. aureus* and as inhibitor of biofilm formation (Fei et al. 2013).

The mechanical and physical properties of the biocomposites can also be interesting to study in association with nitric oxide donors. It is well known that bacterial adhesion was sensitive to surface roughness and enhanced as the roughness of composite in catheters surfaces (Cheng et al. 2013). For example *Staphylococcus epidermidis* adhesion and growth were markedly higher on rough titanium surfaces than on smooth surfaces (Cheng et al. 2013). Nitric oxide may help in fostering a better dispersion over these surfaces.

To our knowledge, this is the first study that shows the association of nitric oxide donors with CNC as a biofilm dispersant agent. Further studies can determine the association of other additives and further applications in foaming solution, in addition to testing different hydrogels. This, in turn, can implement new sustainable cleaning strategies by expanding the tool-kit of pro-active practices for "Good Agricultural Practices (GAPs), "Hazard Analysis and Critical Control Point" (HACCP) and Cleaning-in-place protocols (CIP).

Additional files

Additional file 1: Figure S1. Luminescence of *Salmonella* 14028 pTIM2442 upon exposure to CNC and molsidomine. General metabolic state of the cells was assessed using the redox-coupled FMNH2/Luciferase produced by a *S.* Typhimurium ATCC14028 strain harboring high copy number plasmid in which the *luxCDABE* operon is under the phage λ promoter (pTIM2442). Concentrations of molsidomine was 10 μM. Combination of molecules to which cultures of *Salmonella* 14028 pTIM2442 were exposed are listed on the figure. Each graph represents the average of 12 replicas.

Authors' contributions

All authors read and approved the final manuscript. MM, ESM conceived experiments. MM wrote the manuscript and ESM helped to draft the manuscript. MM performed the statistical analysis. IAD, DCV, PC carried out the experiments. MM, ESM analyzed data.

Author details

[1] Department of Natural Sciences, School of Science and Technology, Middlesex University, The Burroughs, London NW4 4BT, UK. [2] Soil and Water Science Department, University of Florida, Gainesville, FL, USA. [3] Agricultural and Biological Engineering Department, University of Florida, Gainesville, FL, USA. [4] Food Engineering Department, Universidad del Valle, Cali, Colombia. [5] Department of Mechanical Engineering, University of Colorado, Denver, CO, USA.

Acknowledgements

This research was supported by funding provided by the Florida Tomato Committee, Grant #106486 (MM, University of Florida) and by the UC-Davis Center for Produce Safety (MM, ESM), Grant #2014-308 (University of Florida) and MM start-up Grant from Middlesex University, London.

Compliance with ethical guidelines

Competing interest

The authors declare that they have no competing interests.

References

Ahmed EM (2013) Hydrogel: Preparation, characterization, and applications (2015). J Adv Res. doi:10.1016/j.jare.2013.07.006

Alagely A, Rajamani S, Teplitski M (2011) Luminescent reporters and their applications for the characterization of signals and signal-mimics that alter LasR-mediated quorum sensing. Methods Mol Biol 692:113–130. doi:10.1007/978-1-60761-971-0_9

Azizi S, Ahmad MB, Hussein MZ, Ibrahim NA (2013) Synthesis, antibacterial and thermal studies of cellulose nanocrystal stabilized ZnO–Ag heterostructure nanoparticles. Molecules 18:6269–6280. doi:10.3390/molecules18066269

Barraud N, Hassett DJ, Hwang SH, Rice SA, Kjelleberg S, Webb JS (2006) Involvement of nitric oxide in biofilm dispersal of *Pseudomonas aeruginosa*. J Bacteriol 188:7344–7353. doi:10.1128/JB.00779-06

Barraud N, Schleheck D, Klebensberger J, Webb JS, Hassett DJ, Rice SA et al (2009a) Nitric oxide signaling in *Pseudomonas aeruginosa* biofilms mediates phosphodiesterase activity, decreased cyclic di-GMP levels, and enhanced dispersal. J Bacteriol 191:7333–7342. doi:10.1128/JB.00975-09

Barraud N, Storey MV, Moore ZP, Webb JS, Rice SA, Kjelleberg S (2009b) Nitric oxide-mediated dispersal in single- and multi-species biofilms of clinically and industrially relevant microorganisms. Microb Biotechnol 2:370–378. doi:10.1111/j.1751-7915.2009.00098.x

Barraud N, Kardak BG, Yepuri NR, Howlin RP, Webb JS, Faust SN et al (2012) Cephalosporin-3′-diazeniumdiolates: targeted NO-donor prodrugs for dispersing bacterial biofilms. Angew Chem Int Ed Engl 51:9057–9060. doi:10.1002/anie.201202414

Charville GW, Hetrick EM, Geer CB, Schoenfisch MH (2008) Reduced bacterial adhesion to fibrinogen-coated substrates via nitric oxide release. Biomaterials 29:4039–4044. doi:10.1016/j.biomaterials.2008.07.005

Chaturvedi P, Vanegas DC, Taguchi M, Burrs SL, Sharma P, McLamore ES (2014) A nanoceria–platinum–graphene nanocomposite for electrochemical biosensing. Biosens Bioelectron 58:179–185. doi:10.1016/j.bios.2014.02.021

Cheng M, Yu Q, Sun H (2013) Review: Novel strategies for the prevention and treatment of biofilm related infections. Int J Mol Sci 14:18488–18501. doi:10.3390/ijms140918488

Corcoran M, Morris D, De Lappe N, O'Connor J, Lalor P, Dockery P et al (2014) Commonly used disinfectants fail to eradicate *Salmonella enterica* biofilms from food contact surface materials. Appl Environ Microbiol 80:1507–1514. doi:10.1128/AEM.03109-13

Drogat N, Granet R, Sol V, Memmi A, Saad N, Klein Koerkamp C et al (2011) Antimicrobial silver nanoparticles generated on cellulose nanocrystals. J Nanoparticle Res 13:1557–1562. doi:10.1007/s11051-010-9995-1

Feese E, Sadeghifar H, Gracz HS, Argyropoulos DS, Ghiladi RA (2011) Photobactericidal porphyrin-cellulose nanocrystals: synthesis, characterization, and antimicrobial properties. Biomacromolecules 12:3528–3539. doi:10.1021/bm200718s

Fei X, Jia M, Du X, Yang Y, Zhang R, Shao Z et al. (2013) Green synthesis of silk fibroin-silver nanoparticle composites with effective antibacterial and biofilm-disrupting properties. Biomacromolecules 9;14(12):4483–4488. doi:10.1021/bm4014149

Firoved AM, Wood SR, Ornatowski W, Deretic V, Timmins GS (2004) Microarray analysis and functional characterization of the nitrosative stress response in nonmucoid and mucoid *Pseudomonas aeruginosa*. J Bacteriol 186:4046–4050. doi:10.1128/JB.186.12.4046-4050.2004

Friedemann MN, Robinson SW, Gerhardt GA (1996) *o*-Phenylenediamine-modified carbon fiber electrodes for the detection of nitric oxide. Anal Chem 68:2621–2628. doi:10.1021/ac960093w

George J, Ramana KV, Bawa AS, Siddaramaiah (2011) Bacterial cellulose nanocrystals exhibiting high thermal stability and their polymer nanocomposites. Int J Biol Macromol 48:50–57. doi:10.1016/j.ijbiomac.2010.09.013

Hetrick EM, Shin JH, Paul HS, Schoenfisch MH (2009) Anti-biofilm efficacy of nitric oxide-releasing silica nanoparticles. Biomaterials 30:2782–2789. doi:10.1016/j.biomaterials.2009.01.052

Hrabie JA, Klose JR, Wink DA, Keefer LK (1993) New nitric oxide-releasing zwitterions derived from polyamines. J Org Chem 58:1472–1476. doi:10.1021/jo00058a030

Keefer LK, Nims RW, Davies KM, Wink DA (1996) "NONOates" (1-substituted diazen-1-ium-1,2-diolates) as nitric oxide donors: convenient nitric oxide dosage forms. Methods Enzymol 268:281–293. doi:10.1016/S0076-6879(96)68030-6

Klemm D, Kramer F, Moritz S, Lindstrom T, Ankerfors M, Gray D et al (2011) Nanocelluloses: a new family of nature-based materials. Angew Chem Int Ed Engl 50:5438–5466. doi:10.1002/anie.201001273

Koehler JJ, Zhao J, Jedlicka SS, Porterfield DM, Rickus JL (2008) Compartmentalized nanocomposite for dynamic nitric oxide release. J Phys Chem B 112:15086–15093. doi:10.1021/jp803276u

Malmsten M (2011) Antimicrobial and antiviral hydrogels. Soft Matter 7:8725–8736. doi:10.1039/C1SM05809F

Marvasi M, Visscher PT, Casillas Martinez L (2010) Exopolymeric substances (EPS) from *Bacillus subtilis*: polymers and genes encoding their synthesis. FEMS Microbiol Lett 313:1–9. doi:10.1111/j.1574-6968.2010.02085.x

Marvasi M, Chen C, Carrazana M, Durie I, Teplitski M (2014) Dispersal of biofilms formed by *Salmonella enterica* and *Escherichia coli* O157:H7 by nitric oxide donors. Appl Microbiol Biotech Express 4:42. doi:10.1186/s13568-014-0042-y

McLamore ES, Porterfield DM (2011) Non-invasive tools for measuring metabolism and biophysical analyte transport: self-referencing physiological sensing. Chem Soc Rev 40:5308–5320. doi:10.1039/c0cs00173b

McLamore ES, Porterfield DM, Banks MK (2009) Non-invasive self-referencing electrochemical sensors for quantifying real time biofilm analyte flux. Biotechnol Bioeng 102:791–799. doi:10.1002/bit.22128

Merritt JH, Kadouri DE, O'Toole GA (2005) Growing and analyzing static biofilms. Curr Protoc Microbiol Chapter 1: Unit 1B 1. doi: 10.1002/9780471729259.mc01b01s00

Noel JT, Arrach N, Alagely A, McClelland M, Teplitski M (2010) Specific responses of *Salmonella enterica* to tomato varieties and fruit ripeness identified by in vivo expression technology. PLoS One 5:e12406. doi:10.1371/journal.pone.0012406

O'Toole GA, Kolter R (1998) Initiation of biofilm formation in *Pseudomonas fluorescens* WCS365 proceeds via multiple, convergent signalling pathways: a genetic analysis. Mol Microbiol 28:449–461. doi:10.1046/j.1365-2958.1998.00797.x

Paramonova E, de Jong ED, Krom BP, Can DM, Busscher HJ, Sharma PK (2007) Low-load compression testing: a novel way of measuring biofilm thickness. Appl Environ Microbiol 73:7023–7028. doi:10.1128/AEM.00935-07

Porterfield DM, Laskin JD, Jung SK, Malchow RP, Billack B, Smith PJ et al (2001) Proteins and lipids define the diffusional field of nitric oxide. Am J Physiol Lung Cell Mol Physiol 281:L904–L912

Regev-Shoshani G, Crowe A, Miller CC (2013) A nitric oxide-releasing solution as a potential treatment for fungi associated with tinea pedis. J Appl Microbiol 114:536–544. doi:10.1111/jam.12047

Siro I, Plackett D (2011) Microfibrillated cellulose and new nanocomposite materials: a review. Cellulose 17:459–494. doi:10.1007/s10570-010-9405-y

Srey S, Jahid IK, Ha S (2013) Biofilm formation in food industries: a food safety concern. Food Control 31:572–585. doi:10.1016/j.foodcont.2012.12.001

Teplitski M, Al-Agely A, Ahmer BMM (2006) Contribution of the SirA regulon to biofilm formation in *Salmonella enterica* serovar Typhimurium. Microbiol Sgm 152:3411–3423. doi:10.1099/mic.0.29118-0

Thakur VK, Thakur MK (2014) Recent advances in green hydrogels from lignin: a review. Int J Biol Macromol 72:834–847. doi:10.1016/j.ijbiomac.2014.09.044

Vanegas DC, Taguchi M, Chaturvedi P, Burrs S, Tan M, Yamaguchi H et al (2014) A comparative study of carbon–platinum hybrid nanostructure architecture for amperometric biosensing. Analyst 139:660–667. doi:10.1039/c3an01718d

Wang PG, Cai TB, Taniguchi N (2005) Nitric oxide donors. Wiley, Weinheim

Xiong Y, Liu Y (2010) Biological control of microbial attachment: a promising alternative for mitigating membrane biofouling. Appl Microbiol Biotechnol 86:825–837. doi:10.1007/s00253-010-2463-0

Zacharia IG, Deen WM (2005) Diffusivity and solubility of nitric oxide in water and saline. Ann Biomed Eng 33:214–222. doi:10.1007/s10439-005-8980-9

Zhang L, Mah TF (2008) Involvement of a novel efflux system in biofilm-specific resistance to antibiotics. J Bacteriol 190:4447–4452. doi:10.1128/JB.01655-07

22

Functional screening system for yeast-secreted peptides acting on G-protein coupled receptors

Tomohiro Shigemori, Kouichi Kuroda and Mitsuyoshi Ueda*

Abstract

We established a novel functional screening system for peptides acting on G-protein coupled receptors (GPCRs). Peptides are a promising drug scaffold because of their intermediate molecular size between that of therapeutic small molecules and antibodies. They also offer potential advantages of targeting not only membrane proteins but also intracellular protein–protein interactions. Phage display technology has been used for exploring novel peptides acting on GPCRs, but it is unclear whether the identified peptides functionally modulate targets because the technology selects peptides based on binding ability but not functional activity to targets. In a novel screening system that we established, yeast cells were utilized as a peptide producer while mammalian cells stably producing the receptor for glucagon-like peptide 1 (GLP1R) were used as a biosensor for receptor activation. Three kinds of GLP1R agonists secreted by yeasts were successfully detected for their functional activities without any purification and condensation of those peptides. By applying the functional screening system, we were able to identify GLP1R agonist-secreting yeasts based on GLP1R activation from the cell mixture containing a number of background yeasts that produced non-active control peptides. Further applications of this system would include not only activity evaluation of bioactive peptides without chemical synthesis but also discovery of novel peptides activating druggable GPCRs.

Keywords: Functional screening system, G-protein coupled receptors, Peptides, Yeast, Glucagon-like peptide-1

Introduction

G-protein coupled receptors (GPCRs) are one of critical eukaryotic signal transduction gatekeepers and represent the largest protein family in the human proteome with more than 800 members. They share a common architecture of seven transmembrane helices, and can be classified into five major classes of sequence similarities (Jacoby et al. 2006): rhodopsin receptor family (class A), secretin-like receptor family (class B), glutamate receptor family (class C), frizzled/taste 2 receptor family, and adhesion receptor family. GPCRs recognize a variety of extracellular stimuli, including photons, ions, small molecules, peptides, and proteins; they transmit the resulting extracellular signals across the membrane to elicit intracellular responses. Consequently, GPCRs are involved in physiological and pathophysiological changes in blood pressure, blood sugar, pain, allergies, and so on. Therefore, pharmacological activation or suppression of GPCRs has been effective means to treat various diseases related to GPCR dysregulation.

Peptides are involved in a variety of physiological and pathological processes, and play very important roles in modulating various cell functions such as the absorption of blood glucose into the body through the promotion of insulin secretion in pancreatic β-cells by glucagon-like peptide-1 (GLP1), a peptide hormone that are post-prandially secreted in intestinal L-cells and activates one of class B GPCRs, GLP1 receptor (GLP1R) (Baggio and Drucker 2007). Because of their intermediate molecular size ranging from 0.5 to 5 kDa which is between that of small-molecule drugs and therapeutic monoclonal antibodies, peptides potentially have advantages of easy drug design, high safety, accessibility for protein–protein

*Correspondence: miueda@kais.kyoto-u.ac.jp
Division of Applied Life Sciences, Graduate School of Agriculture, Kyoto University, Sakyo-ku, Kyoto 606-8502, Japan

interactions, targeting of intracellular molecules, and low production cost (Craik et al. 2013).

Phage display is the first innovative technology established by Smith (1985) that allows researchers to prepare and screen a large polypeptide library. However, this methodology is not effective at discovering functional peptides such as activators for GPCRs. This is due to the nature of phage display, whose peptides are screened based only on binding ability to targets, resulting frequently in simple binding peptides without any bioactivity (Chen et al. 2007). Furthermore, the identified peptides on the phage are never a functional, soluble form, potentially leading to dissociation in activity between peptides tied up on the phage as a fusion protein and the secretion form of peptides.

The yeast *Saccharomyces cerevisiae* is suitable as host for production of peptide library; this is due to their abundance in gene manipulation tools, fast growth in a low-cost medium, and protein folding and secretory machinery homologous to that of mammalian cells (Idiris et al. 2010). In this study, we report a novel functional screening method for bioactive peptides acting on GPCRs, which integrated a yeast secretion system and a functional detection system using GPCR-producing mammalian cells. GLP1R produced on mammalian cells was successfully activated by various GLP1R agonistic peptides that were secreted from yeast. We were also able to identify GLP1R agonist-secreting yeasts based on GLP1R activation from the cell mixture containing a number of background yeasts, which produced non-active control peptides, suggesting the effectiveness of our functional screening system to discover novel peptide-based drugs acting on GPCRs.

Materials and methods

Strains and media

Esherichia coli DH5α [F^-, *ΔlacU169* (*φ80lacZΔM15*), *hsdR17* (r_K^-, m_K^+), *recA1, endA1, deoR, thi-1, supE44, gyrA96, relA1*, λ^-] (Toyobo, Osaka, Japan) was used as a host for DNA manipulation. *E. coli* transformants were grown in Luria–Bertani (LB) medium [1% (*w*/*v*) tryptone, 0.5% (*w*/*v*) yeast extract, 1% (*w*/*v*) sodium chloride, and 2% (*w*/*v*) agarose] containing 100 μg/mL ampicillin or kanamycin depending on the plasmids introduced.

Saccharomyces cerevisiae BY4742 (*MATα, his3Δ1, leu2Δ0, lys2Δ0, ura3Δ0*; EUROSCARF, Frankfurt, Germany) was used to construct the yeasts secreting GLP1R agonists, including GLP1, S^2-GLP1 substituted with serine at the position 2 of GLP1, and exendin-4 (Ex4), a naturally occurring peptide found in the saliva of the Gila monster (Furman 2012). Yeast transformants were selected on synthetic dextrose (SDC) solid medium [0.67% (*w*/*v*) yeast nitrogen base without amino acids, 2% (*w*/*v*) glucose, 1% (*w*/*v*) casamino acids, 0.002% (*w*/*v*) adenine, 0.002% (*w*/*v*) L-tryptophan, and 2% (*w*/*v*) agar], and then, the resultant colonies were cultivated in 6-well plate (353046; Thermo Fisher Scientific, Waltham, MA, USA) or 96-well plate (353072; Thermo Fisher Scientific) containing a liquid SDC medium or Dulbecco's modified Eagle Medium (DMEM) (Nacalai Tesque, Kyoto, Japan) at 30°C.

Chinese hamster ovary (CHO) cells (85050302; European Collection of Cell Cultures, Salisbury, UK) were used as a host cell stably producing human GLP1R and cultivated in Ham-F12 (Sigma-Aldrich, St. Louis, MO, USA) containing 10% FBS (Thermo Fisher Scientific) and 400 μg/mL G418 (Nacalai Tesque).

Construction of peptide-secreting yeast

All the primers used in plasmid construction are listed in Table 1. For peptide secretion in yeast, pULS harboring the engineered secretion signal of yeast α-factor, appS4 (Rakestraw et al. 2009), in the downstream of GAPDH promoter, was constructed as follows. The DNA fragment encoding the appS4 (FASMAC, Kanagawa, Japan) in pUC19 was amplified using primers 1 and 2, and was inserted into pULI1 (Miura et al. 2012), which was digested with *EcoR*I and *Xba*I by using In-Fusion (Clontech Laboratories, Inc., Mountain View, CA, USA) to obtain pULS. The DNA fragments encoding GLP1, S^2-GLP1 and Ex4, with or without a FLAG-encoding sequence at the 3′ terminus, were double-stranded from oligonucleotides with the mutual complementary region using DNA polymerase KOD-FX-Neo (Toyobo). The double-stranded DNAs encoding those GLP1 analogues were introduced into the multiple cloning sites of pULS by In-Fusion (Clontech) and named pULS-GLP1, -GLP1FLAG, -S^2-GLP1, -S^2-GLP1FLAG, -Ex4 and -Ex4FLAG, respectively. Yeasts were transformed with those plasmids using Frozen-EZ Yeast Transformation-II kit (Zymo Research, Orange, CA, USA), resulting in GLP1-yeast, GLP1F-yeast, S^2-GLP1-yeast, S^2-GLP1F-yeast, Ex4-yeast, and Ex4F-yeast,

Table 1 Primers used in this study

	Sequence
Primer 1	5′-AAACACACATAAACACCCGGGATG-3′
Primer 2	5′-CAGTCTAGAGGATCCGAATTCTCTTTTATCCAAAGATACC CCTTCTTC-3′
Primer 3	5′-CGAGCTCGGATCGATCGCCACCATGGCCGGCGCCC-3′
Primer 4	5′-TATCTATGCGGCCGCTCAGCTGCAGGAGGCCTG-3′
Primer 5	5′-GAAGAAGGGGTATCTTTGGATAAAAG-3′
Primer 6	5′-CTTGTCATCGTCATCCTTGTAATC-3′

Underlines indicate homologous region to the corresponding plasmids.

respectively. Yeast transformed with pULS (Ctrl-yeast) was used as control.

Construction of human GLP1R-producing CHO

Human GLP1R gene was PCR-amplified from the human brain cDNA library (BioChain, Newark, CA, USA) using DNA polymerase KOD-Plus-Neo (Toyobo) with primers 3 and 4 (Table 1). The DNA fragment coding human GLP1R was inserted into pIRES (Clontech) digested with *Eco*RV and *Bam*HI by using In-Fusion (Clontech), resulting in pIRES-hGLP1R. CHO cells were transfected with pIRES-hGLP1R using Xfect (Clontech), and then, selected with G418 for about 2 weeks to construct a stable cell line producing hGLP1R. Single cell cloning of the resistant cells was conducted by limiting dilution, resulting in GLP1R-CHO.

GLP1R activation assay using GLP1R-CHO

GLP1R-CHO was seeded onto a 96-well plate at 5×10^4 cells and cultured at 37°C for 24 h. After the cells were washed with HANKS buffer (Thermo Fisher Scientific), synthetic GLP1R agonists (GLP1 and Ex4; Peptide Institute, Osaka, Japan. S^2-GLP1; Bachem, Bubendorf, Switzerland) or culture supernatants of GLP1R agonists-secreting yeast were added and incubated at 37°C for 45 min. Then, the cells were lysed with Assay/Lysis buffer (Thermo Fisher Scientific) and the level of cyclic AMP in the cell lysate was determined by using the cAMP-screen® assay (Thermo Fisher Scientific) according to the manufacturer's instructions.

Model screening of Ex4-secreting yeast

Ten yeast cells comprised of Ctrl-yeast and Ex4-yeast in the theoretical ratio of 9:1 were seeded into 16 wells in 96-well plate containing 300 μL of SDC medium and grown for 48 h. Then, the medium was exchanged into 250 μL of DMEM and yeasts were additionally cultivated for 12 h at 30°C. After that, the supernatant was subjected to the GLP1R activation assay mentioned above. Yeasts included in three wells showing or not showing activity were seeded on SDC solid medium to form single colonies. Then, 48 colonies were subjected to colony-direct PCR with primers 5 and 6 (Table 1), and the resultant PCR products were analyzed by agarose gel electrophoresis to identify yeasts with the Ex4 gene.

Results

Construction of stable CHO cells producing human GLP1R

CHO cells producing human GLP1R (GLP1R-CHO) were constructed to detect functional activity of synthetic- or yeast-secreted GLP1R agonists. CHO cells were transfected with pIRES-hGLP1R by lipofection and selected using the neomycin resistance gene on the pIRES vector. The established cells showed cAMP production upon GLP1R activation by three kinds of synthetic GLP1R agonists, GLP1, S^2-GLP1, and Ex4 in a dose-dependent manner, with EC_{50} value of 5.8, 21.5 and 1.4 nM, respectively (Figure 1; Table 2). Therefore, functional human GLP1R was successfully produced in CHO cells.

Medium optimization for GLP1R activation by yeast-secreted peptides and establishment of assay system

We first tested which medium was suitable for yeast growth and evaluation of GLP1R activation in GLP1R-CHO in 6-well plates. Ctrl-yeast and Ex4-yeast were inoculated in a 6-well plate containing SDC medium or DMEM at the initial optical density (OD) which was 600 nm of 0.1 and incubated for 40 h. Then, the growth rates of yeasts and the GLP1R activation in GLP1R-CHO by culture supernatant containing yeast-secreted peptides were evaluated (Figure 2). As a result, SDC gave a higher growth reaching stationary phase at 20 h, showing about 90-fold expansion. On the other hand, yeasts cultivated

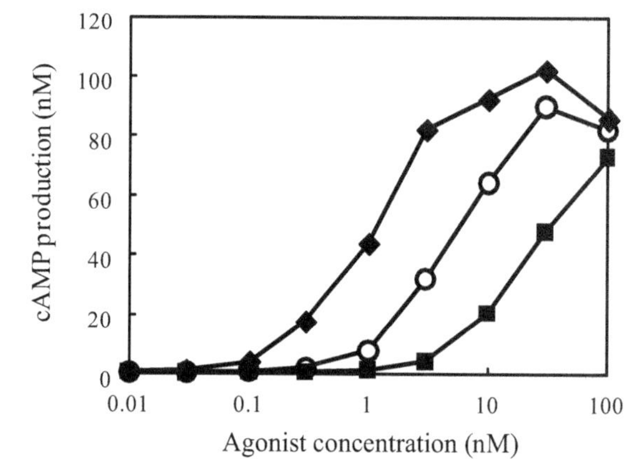

Figure 1 Dose–response curve of GLP1R activation in GLP1R-CHO upon treatment of synthetic GLP1R agonists. GLP1R-CHO cells were exposed to three kinds of GLP1R agonists, GLP1 (*white circle*), S^2-GLP1 (*black square*) and Ex4 (*black diamond*), at increasing concentrations. After 45 min incubation, cells were lysed and cAMP production was determined by ELISA as indicated in "Materials and methods".

Table 2 Amino acid sequences of GLP1R agonists used in this study

Agonists	Amino acid sequence	EC_{50} (nM)
GLP1	HAEGTFTSDVSSYLEGQAAKEFIAWLVKGR	5.8
S^2-GLP1	HSEGTFTSDVSSYLEGQAAKEFIAWLVKGR	22
Exendin4	HGEGTFTSDLSKQMEEEAVRLFIEWLKNGGPS SGAPPPS	1.4

Underlines indicate amino acids different from GLP1.

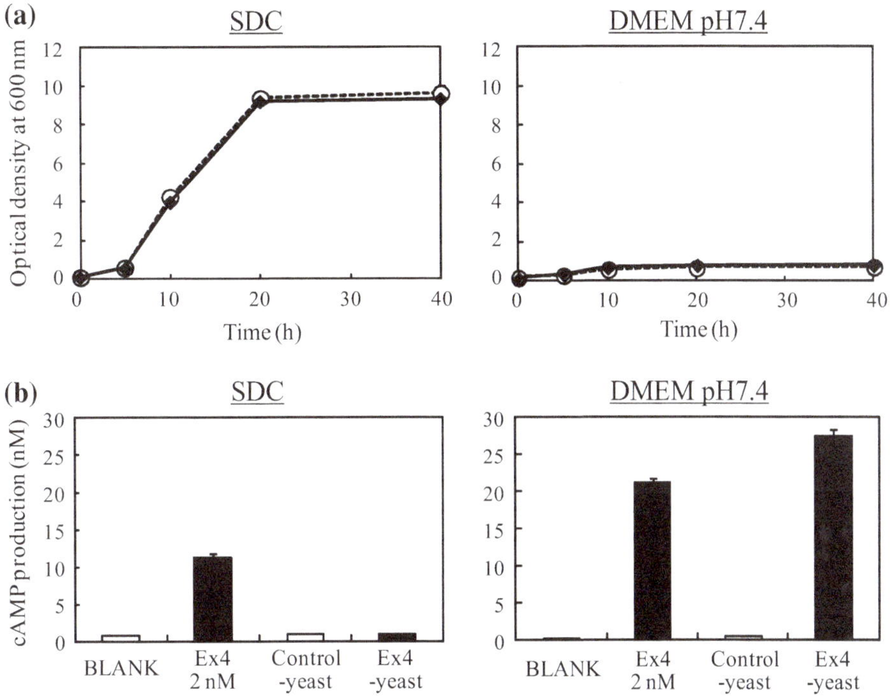

Figure 2 Medium optimization for GLP1R activation assay of yeast-secreted peptides. Yeasts were grown in SDC and DMEM on 6-well plate for 40 h, and then, growth rate (**a**) and GLP1R activation potential of their culture supernatant (**b**) were evaluated by measurement of absorbance at 600 nm and ELISA for cAMP, respectively. For growth curve, *white circle* indicates Ctrl-yeast and *black diamond* shows Ex4-yeast. The data represent the average ± SEM of three independent experiments.

in DMEM showed much lower growth, reaching only the maximal OD_{600} of 0.75 at 20 h (Figure 2a). For GLP1R activation potency after the 40-h cultivation, the culture supernatant of Ex4-yeast grown in SDC showed no activity, although exogenously added Ex4 activated GLP1R by 16-fold, compared with the control. In contrast, Ex4-yeast grown in DMEM strongly induced GLP1R activation by 50-fold compared to the Ctrl-yeast (Figure 2b). These results indicate that SDC medium is suitable for yeast proliferation, while DMEM is excellent for GLP1R activation.

Next, we tried to establish a GLP1R activation assay system combined with peptide-secreting yeast in a 96-well plate format. We first determined the number of Ex4-yeasts required for larger dynamic range in the GLP1R activation. Yeast cells suspended in DMEM were prepared at 4×10^4 to 5×10^6 cells in 96-well plates, and after a 12 h-cultivation at 30°C, a GLP1R activation assay using the culture supernatants was carried out (Figure 3a). The results showed that 4×10^4 yeast cells were enough to detect GLP1R activation, showing 73-fold activation, compared with the control, and 1×10^6 yeast cells provided the largest dynamic range with 224-fold activation of GLP1R. Further, we investigated the cultivation time required for the yeast cell number of 1×10^6 (Figure 3b). Yeast preparations with the initial cell number of 3 were cultivated in SDC medium that showed the highest growth as in Figure 1, and then, yeast cell number was counted at 24, 48, and 72 h cultivation. As results, the cell number per initial cell number exceeded 1×10^6 cells after 48 h cultivation. These results suggested that a single yeast cell can reach 1×10^6 cells, showing the highest dynamic range in GLP1R activation assay when cultivated in SDC for 48 h.

Direct functional assay of various yeast-secreted GLP1R peptide agonists

Based on these optimizations, we constructed a series of workflow for direct functional assay of yeast-secreted peptides on GLP1R, composed of three steps (Figure 4a): cultivation for yeast growth, peptide secretion in yeasts, and GLP1R activation assay using GLP1R-CHO. To demonstrate the effectiveness of the direct functional assay system, we attempted to detect the GLP1R activation by native GLP1, its analogue S^2-GLP1, and Ex4 secreted by yeast. Single colonies of the yeasts secreting GLP1, S^2-GLP1, and Ex4 were inoculated and incubated in a 96-well plate containing SDC for 48 h, and then, the medium was exchanged into DMEM with additional incubation of 12 h, followed by the GLP1R activation assay (Figure 4b). While yeast-secreted Ex4

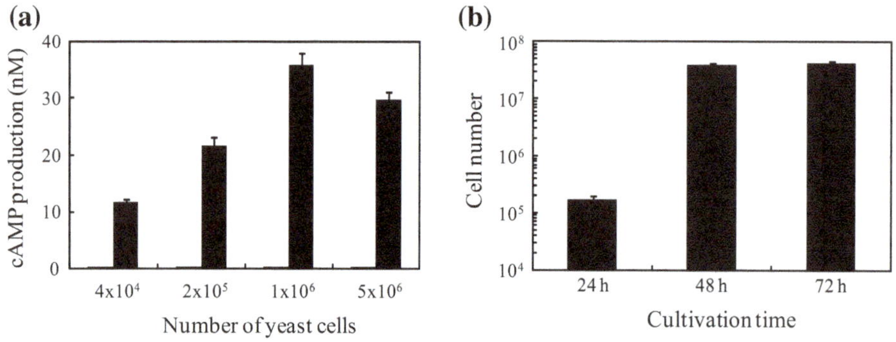

Figure 3 Scale-down assay for GLP1R activation by yeast-secreted peptides. **a** Indicated number of yeast cells, Ctrl-yeast (*left bar*) or Ex4-yeast (*right bar*), were seeded in a 96-well plate containing DMEM and incubated for 12 h. Then, GLP1R activation assay was conducted to find the yeast cell number providing wider dynamic range. The data represent the average ± SEM of three independent experiments. **b** Yeast cultivation time to achieve the target yeast cell number, 1 × 10^6, was evaluated. Yeasts were seeded in a 96-well plate containing SDC at the initial number of 3. After 24-, 48-, and 72-h cultivations, the number of yeasts was determined using hematometer. The data represent the average ± SEM of three independent experiments.

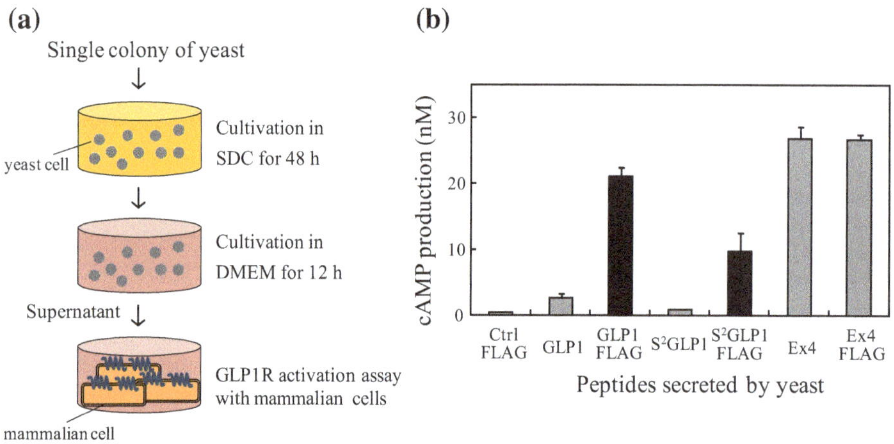

Figure 4 Direct activity detection of various yeast-secreted GLP1R agonists. **a** Schematic illustration for direct activity detection assay was depicted. Single colony of yeasts was inoculated in 96-well plate containing SDC and cultivated for 48 h. After removal of SDC, DMEM was added and incubated for additional 12 h. Then, the culture was subjected to GLP1R activation assay using GLP1R-CHO. **b** Seven kinds of yeasts were tested for GLP1R activation potency according to the assay flow mentioned above. The data represent the average ± SEM of three independent experiments.

showed the highest activation by 59-fold, GLP1 and S^2-GLP1 secreted by yeasts provided only 5.8 and 1.7-fold activation, respectively. Therefore, GLP1 and S^2-GLP1 were considered insufficient given those EC_{50} values are a quarter and one sixteenth of the Ex4 values, respectively. Whereas the N-terminal two residues (His-Ala) in GLP1 are reported to be critical for its biological activity, the C-terminal part is tolerable to various modifications such as fatty acid conjugation or albumin fusion (Baggio et al. 2004; Madsen et al. 2007). Thus, we fused FLAG tag to the C-terminus of GLP1, S^2-GLP1, and Ex4 initially for affinity purification and Western blotting, if needed. Remarkably, when we evaluated the GLP1R activation potency for such yeast-secreted GLP1R agonists with the FLAG tag, GLP1 and S^2-GLP1 activated GLP1R about 10-times higher than those without the FLAG tag, respectively. These results demonstrated that yeast-secreted GLP1 and S^2-GLP1 with the addition of C-terminal FLAG tag were successfully and directly detected without any purification and condensation by using the novel functional detection system established here.

Examination of model screening using our established functional assay system

Finally, we performed a model screening of yeast by applying the direct functional assay system, for example, a screening of Ex4-yeast in the co-presence of excess Ctrl-yeast. A yeast cell mixture including Ctrl-yeast and

Ex4-yeast at a theoretical ratio of 9:1 was prepared, and a GLP1R activation assay was conducted according to the determined workflow (Figure 5a). As a result, we identified three positive wells showing GLP1R activation (Figure 5b). We next investigated the existence of Ex4-yeast in the positive and negative wells by colony-direct PCR after the colony formation of yeasts from both wells. The results indicated the existence of 24% of Ex4-yeast in the positive wells, whereas Ex4-yeast was not identified at all in the negative wells (Figure 5c). These results clearly suggest that our direct functional assay system established was effective in the model screening.

Discussion

In this study, we established the novel functional screening system of yeast-secreted peptides acting on GLP1R. The system directly detected the functional activity of yeast secreted-GLP1R agonists based on activation of GLP1R produced on mammalian cells, without any purification and condensation of yeast-produced peptides. In addition, application of the system enabled identification of agonist-secreting yeast in the model screening.

Binding-based peptide screening strategy as represented by phage display has been the most common method used to discover novel peptide-ligands for certain drug targets, including GPCRs (Molek et al. 2011; Vyroubalova et al. 2006). However, with peptides discovered through this methodology, it is uncertain whether they are biologically active or not. In addition, phage-displayed peptides are distinct from functional peptides in their soluble form. These conventional challenges result in additional chemical synthesis of the identified peptides in soluble form to evaluate their true biological activity, which are time-consuming and quite expensive.

In the first optimization step using Ex4-yeasts as a model for our novel functional assay system, we remarkably found that DMEM buffered at neutral pH designed for mammalian cell culture was very suitable for yeast peptide secretion and GLP1R activation in mammalian cells, even though yeast cells could not grow in the medium. Another remarkable point in this culture system is a protection of the target peptides from degradation by yeast-derived proteases. Because heterologous proteins produced in yeasts could be degraded during several steps, including the intracellular secretory pathways and the post-secreted extracellular environment (Idiris et al. 2010), yeast-derived proteases such as Yps1p and Kex2p, which are most active at a mild acidic condition around pH 5.0, are considered accessible to heterologous proteins, especially when yeasts are incubated in

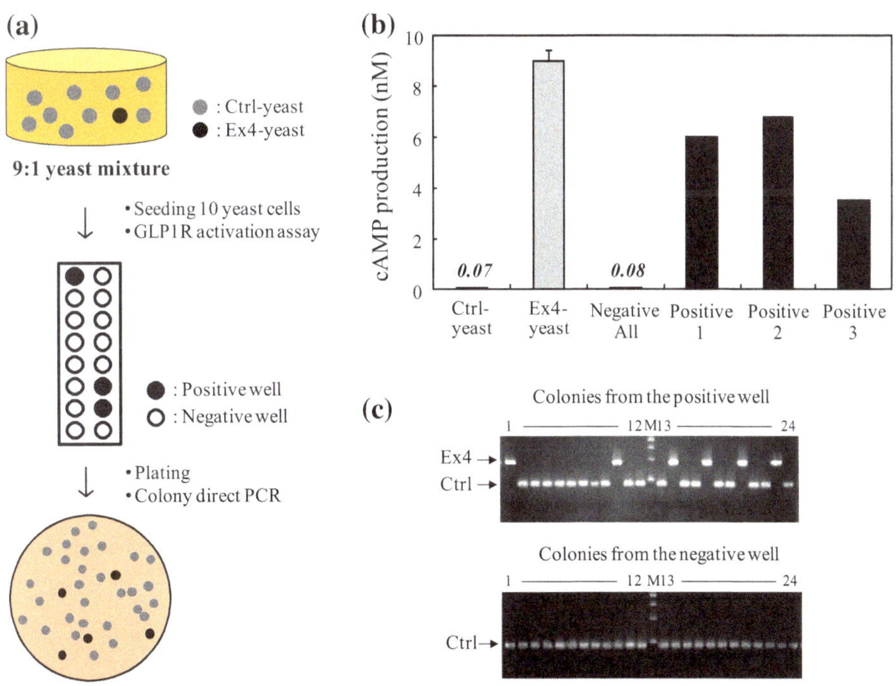

Figure 5 Model screening of Ex4-secreting yeast using the direct activity detection system. **a** Schematic illustration for model screening of Ex4-yeast was indicated. **b** We prepared 16 wells including 9:1 yeast mixture (10 cells in total per well) in the ratio of Ctrl-yeast to Ex4-yeast and conducted GLP1R activation assay according to the workflow indicated above. Only Ctrl-yeast or Ex4-yeast was tested as a control. **c** The existence of Ex4-yeast in the positive and negative wells was investigated by colony-direct PCR after the colony formation of yeasts from both wells (numbering from 1 to 24 means the number of colony picked up). The representative electrophoresis data are shown for colonies from a positive and a negative well, respectively. *M* molecular size marker (New England Biolabs, Ipswich, MA, USA).

SDC medium, which generally has pH 4.5–5.5 (Mizuno et al. 1989; Olsen et al. 1999). In addition, yeasts were a good tool for producing peptides because yeasts swiftly grew in SDC medium; they even in a static 96-well plate setting which would be compatible with high-throughput screening and could easily reach the targeted yeast cell number providing high dynamic range in GLP1R activation assay (Figure 2).

These findings prompted us to conduct the direct functional detection of other GLP1R agonists with weaker activity than Ex4, GLP1, and S^2-GLP1 to be secreted by yeasts. It was difficult to detect their GLP1R activation potency at the high level as expected from the difference of the EC_{50} value of Ex4. Surprisingly, the C-terminus fusion of the FLAG tag with DYKDDDDK in GLP1 and S^2-GLP1 much increased the GLP1R activation potency by about tenfold compared to those without the FLAG tag (Figure 4b). In spite of the fact that the precise mechanisms of the FLAG are still uncertain, the beneficial effect of the FLAG fusion was also observed in somatostatin that is endogenous circular peptide agonist acting on class A GPCR SST receptor (data not shown). Accordingly, the FLAG fusion to isolated peptides would be a promising way to increase their activity.

As shown in Figure 5, we successfully identified the wells including Ex4-yeast by evaluating the GLP1R activation potency, and importantly, the wells that did not show activity did not possess any Ex4-yeast, as confirmed by colony-direct PCR detecting Ex4 gene. This successful model screening of yeasts secreting agonists acting on GLP1R encourages us to carry out an actual screening for novel bioactive peptides through our direct functional detection system.

In conclusion, we successfully established a novel system for direct functional assay for yeast-secreted peptides on GLP1R. This system will be applied not only for biological activity assay of sequenced peptides instead of their chemical synthesis but also discovery of novel bioactive peptides.

Abbreviations

CHO: Chinese hamster ovary; DMEM: Dulbecco's modified Eagle medium; Ex4: exendin-4; GLP-1: glucagon-like peptide-1; GPCR: G-protein coupled receptor.

Authors' contributions

TS designed and carried out the experiments, and drafted the manuscript. KK participated in the experimental design and revised the manuscript. MU contributed to the experimental design and revised the manuscript critically. All authors read and approved the final manuscript.

Acknowledgements

This work was supported by Core Research for Evolutional Science and Technology (CREST) of Japan Science and Technology Corporation (JST).

Compliance with ethical guidelines

Competing interests

The authors declare that they have no competing interests.

References

Baggio LL, Drucker DJ (2007) Biology of incretins: GLP-1 and GIP. Gastroenterology 132:2131–2157

Baggio LL, Huang Q, Brown TJ, Drucker DJ (2004) A recombinant human glucagon-like peptide (GLP)-1-albumin protein (albugon) mimics peptidergic activation of GLP-1 receptor-dependent pathways coupled with satiety, gastrointestinal motility, and glucose homeostasis. Diabetes 53:2492–2500

Chen J, Bai G, Yang Y, Geng P, Cao Y, Zhu Y (2007) Identifying glucagon-like peptide-1 mimetics using a novel functional reporter gene high-throughput screening assay. Peptides 28:928–934

Craik DJ, Fairlie DP, Liras S, Price D (2013) The future of peptide-based drugs. Chem Biol Drug Des 81:136–147

Furman BL (2012) The development of Byetta (exenatide) from the venom of the Gila monster as an anti-diabetic agent. Toxicon 59:464–471

Idiris A, Tohda H, Kumagai H, Takegawa K (2010) Engineering of protein secretion in yeast: strategies and impact on protein production. Appl Microbiol Biotechnol 86:403–417

Jacoby E, Bouhelal R, Gerspacher M, Seuwen K (2006) The 7 TM G-protein-coupled receptor target family. ChemMedChem 1:761–782

Madsen K, Knudsen LB, Agersoe H, Nielsen PF, Thogersen H, Wilken M et al (2007) Structure-activity and protraction relationship of long-acting glucagon-like peptide-1 derivatives: importance of fatty acid length, polarity, and bulkiness. J Med Chem 50:6126–6132

Miura N, Kirino A, Endo S, Morisaka H, Kuroda K, Takagi M, Ueda M et al (2012) Tracing putative trafficking of the glycolytic enzyme enolase via SNARE-driven unconventional secretion. Eukaryot Cell 11:1075–1082

Mizuno K, Nakamura T, Ohshima T, Tanaka S, Matsuo H (1989) Characterization of *KEX2*-encoded endopeptidase from yeast *Saccharomyces cerevisiae*. Biochem Biophys Res Commun 159:305–311

Molek P, Strukelj B, Bratkovic T (2011) Peptide phage display as a tool for drug discovery: targeting membrane receptors. Molecules 16:857–887

Olsen V, Cawley NX, Brandt J, Egel-Mitani M, Loh YP (1999) Identification and characterization of *Saccharomyces cerevisiae* yapsin 3, a new member of the yapsin family of aspartic proteases encoded by the *YPS3* gene. Biochem J 339(Pt 2):407–411

Rakestraw JA, Sazinsky SL, Piatesi A, Antipov E, Wittrup KD (2009) Directed evolution of a secretory leader for the improved expression of heterologous proteins and full-length antibodies in *Saccharomyces cerevisiae*. Biotechnol Bioeng 103:1192–1201

Smith GP (1985) Filamentous fusion phage: novel expression vectors that display cloned antigens on the virion surface. Science 228:1315–1317

Vyroubalova EC, Hartley O, Mermod N, Fisch I (2006) Identification of peptide ligands to the chemokine receptor CCR5 and their maturation by gene shuffling. Mol Immunol 43:1573–1578

Overexpression of *PAD1* and *FDC1* results in significant cinnamic acid decarboxylase activity in *Saccharomyces cerevisiae*

Peter Richard*, Kaarina Viljanen and Merja Penttilä

Abstract

The *S. cerevisiae PAD1* gene had been suggested to code for a cinnamic acid decarboxylase, converting *trans*-cinnamic acid to styrene. This was suggested for the reason that the over-expression of *PAD1* resulted in increased tolerance toward cinnamic acid, up to 0.6 mM. We show that by over-expression of the *PAD1* together with the *FDC1* the cinnamic acid decarboxylase activity can be increased significantly. The strain over-expressing *PAD1* and *FDC1* tolerated cinnamic acid concentrations up to 10 mM. The cooperation of Pad1p and Fdc1p is surprising since the *PAD1* has a mitochondrial targeting sequence and the *FDC1* codes for a cytosolic protein. The cinnamic acid decarboxylase activity was also seen in the cell free extract. The activity was 0.019 µmol per minute and mg of extracted protein. The overexpression of *PAD1* and *FDC1* resulted also in increased activity with the hydroxycinnamic acids ferulic acid, p-coumaric acid and caffeinic acid. This activity was not seen when *FDC1* was overexpressed alone.

An efficient cinnamic acid decarboxylase is valuable for the genetic engineering of yeast strains producing styrene. Styrene can be produced from endogenously produced L-phenylalanine which is converted by a phenylalanine ammonia lyase to cinnamic acid and then by a decarboxylase to styrene.

Keywords: Cinnamic acid; Styrene; Decarboxylase; *S. cerevisiae*; *PAD1*; *FDC1*

Introduction

Enzymes for the decarboxylation of hydroxycinnamic acids such as p-coumaric acid, caffeic acid or ferulic acid have been described. A ferulic acid decarboxylase and the corresponding gene, *fdc*, was identified from *Bacillus pumilus* (Zago et al. 1995). A ferulic acid decarboxylase was also identified from an *Enterobacter* species (Gu et al. 2011a) and a crystal structure obtained (Gu et al. 2011b). A p-coumaric acid decarboxylase was purified from *Lactobacillus plantarum*. The enzyme was inducible and the purified enzyme had a K_M of 1.4 mM and a K_{cat} of 10^3 s^{-1}(Cavin et al. 1997b). The corresponding gene, *pdc*, was cloned and overexpressed in *E. coli* (Cavin et al. 1997a). This protein was crystallised and a structure obtained (Rodríguez et al. 2010). Based on the homology to the *fdc* and *pdc* a hydroxycinnamic acid (phenolic acid) decarboxylase, *pad*, was identified in *Bacillus subtilis* (Cavin et al. 1998). These hydroxycinnamic acid decarboxylases were shown to be active *in vivo* and *in vitro* and the purified enzymes did not require cofactors and were not part of multi subunit enzyme complexes. All these enzymes have in common that they are of bacterial origin and not active with cinnamic acid.

The yeast *Saccharomyces cerevisiae* shows some resistance toward cinnamic acid and the gene that confers this resistance was identified as the *PAD1* (phenylacrylic acid decarboxylase) gene (Clausen et al. 1994). The yeast can convert cinnamic acid to styrene (Figure 1) but also sorbic acid to 1,3-pentadiene however with a lower rate. The deletion of *PAD1* resulted in the inability to convert cinnamic or sorbic acid suggesting that the Pad1p is active with both substrates (Stratford et al. 2007). Overexpression of the *PAD1* gene resulted in strains with increased resistance to cinnamic and ferulic acid (Larsson et al. 2001). A *PAD1* homologue is also present in the mold *Aspergillus niger* where it confers resistance to cinnamic acid and sorbic acid (Plumridge et al. 2008). Recently it was shown in *S. cerevisiae* that not only the *PAD1* but the *PAD1* and the *FDC1* are essential for cinnamic acid,

* Correspondence: Peter.Richard@vtt.fi
VTT Technical Research Centre of Finland Ltd, Tietotie 2, P.O. box 1000, 02044 VTT Espoo, Finland

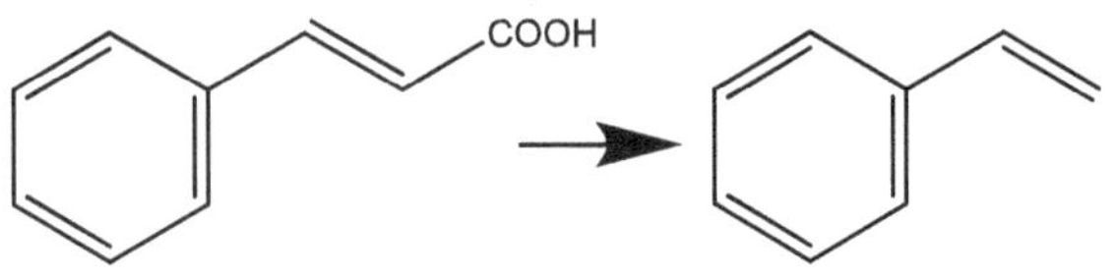

Figure 1 ***Trans*-cinnamic acid is converted in a decarboxylation reaction to styrene.**

ferulic acid or coumaric acid resistance. Deletions in either of these two genes resulted in drastic reduced ability to decarboxylate these acids (Mukai et al. 2010). A possible interpretation of this observation was that the Pad1p and the Fdc1p are two essential subunits of a protein complex. The Pad1p and the Fdc1p are however believed to be in different cellular compartments. The Pad1p has a mitochondrial targeting sequence and the protein is located in the mitochondria (Huh et al. 2003) whereas the Fdc1p is a cytosolic protein.

The *PAD1* and the *FDC1* are clustered, i.e. they are located next to each other on the chromosomes. Such clusters with *PAD1* and *FDC1* homologues are widespread in yeast and filamentous fungi. In filamentous fungi the cluster contains also a transcription factor (Plumridge et al. 2010).

Cinnamic acid decarboxylase is required for the genetic engineering of styrene producing strains. Styrene can be produced from endogenous L-phenylalanine in two steps, by a phenylalanine ammonia lyase to produce cinnamic acid and a cinnamic acid decarboxylase producing styrene (McKenna and Nielsen 2011, McKenna et al. 2014). An efficient cinnamic acid decarboxylase would be desirable for the construction of strains for effective styrene production from biomass.

Materials and methods

Strains and plasmids

Strains: The yeast strain CEN.PK2-1C was obtained from Prof. Friedrich K. Zimmermann (Frankfurt) and used if not otherwise specified.

Plasmids: *PAD1* and *FDC1* were amplified from *S. cerevisiae* by PCR. The *PAD1* was amplified from genomic DNA of the yeast strain S288C (obtained from Dr. G. Shirleen Roeder, Yale), *FDC1* from the CEN.PK2-1C. Through PCR *Eco*RI and *Nhe*I sites were introduced and then ligated to the corresponding sites of p2159 and p2158. The p2159 was derived from the pYX212, a multicopy yeast expression vector with *URA3* marker and *TPI1* promoter, by changing the multiple cloning site as described before (Kuorelahti et al. 2005). The p2158 is derived in the same way from the pYX242, a multicopy yeast expression vector with *LEU2* marker and *TPI1* promoter. The *PAD1* was also amplified by PCR introducing *Xho*I and *Eco*RI sites and then ligated to the pFL60 (Minet et al. 1992), a multicopy yeast expression vector with *URA3* marker and *PGK1* promoter. The *PAD1* was also amplified by PCR without mitochondrial targeting signal. The mitochondrial targeting sequence was identified using the MitoProt software (http://ihg.gsf.de/ihg/mitoprot.html). In one case the first amino acid after the cleavage site was replaced by the start codon so that the amino acid sequence was MVAITG.... In another case the start codon was before the first amino acid resulting in the sequence MVVAITG.... An *FDC1* with mitochondrial targeting sequence was generated by adding the first 60 amino acids of the *PAD1* to the N-terminus of the *FDC1* open reading frame. The plasmids are listed in the Table 1.

Enzyme activity measurement

To prepare a cell free yeast extract yeast cells were collected by centrifugation, washed with water and 1 ml of fresh cell cake was suspended in 1 ml 20 mM sodium phosphate buffer pH 7.0. 1 ml of glass beads (0.4 mm diameter, Sigma) and protease inhibitor (Complete, Roche) were added and the cells were disrupted in two 40s sessions in the Fast Prep (Bio101). The cell extract was centrifuged and the supernatant analysed for cinnamic acid decarboxylase activity. The protein concentration was measured using the Biorad Protein Assay and BSA as a standard. The extract was added to a solution of cinnamic acid in 20 mM sodium phosphate, pH 7.0, and incubated at room temperature. The reaction was quenched by heating the sample to 96°C for 10 minutes.

Cinnamic acid and hydroxy-cinnamic acids were determined by using an analytical UPLC method. The separation of analytes was carried out on an Acquity UPLC BEH C18 1.7 μm column (2.1 × 100 mm) with a Waters Acquity UPLC system including sample manager-FTN, Quaternary solvent manager and PDA eλ detector. Detector was operated on 210-400 nm. The solvents used in gradient elution 0.43 mL/min were A) 5% formic acid and B) acetonitrile. Gradient system was as followed: 0 min 95% A and 5% B; 1.13 min 90% A and 10% B; 5.67 min 60% A and 40% B; 9.00 min 10% A and 90% B; and 10.00 min 90% A and 10% B. On the basis of

Table 1 Table of plasmids used

Plasmid name	Gene expressed	Parent plasmid
p4237	FDC1	p2159 (URA3)
p4517	FDC1	p2158 (LEU2)
p4621	PAD1	pFL60 (URA3)
p4596	PAD1 without targeting (MVAITG…)	pFL60 (URA3)
p4597	PAD1 without targeting (MVVAITG…)	pFL60 (URA3)
p4787	FDC1 with mitochondrial targeting	p2158 (LEU2)

corresponding standards (0-1000 μM) compounds were quantified.

Analysis of styrene by SPME-GC/MS

Samples (300 μl) were analyzed by using SPME (solid phase micro extraction)-GC/MS. Extraction of styrene was done at 80°C for 30 min with preconditioned (300°C, 60 min) 75 μm Carboxen-PDMS fibre (Sulpelco, USA). After extraction the analytes were desorbed during 5 min at 250°C in the splitless injector (flow 14.9 ml/min) of the gas chromatography (Agilent 6890 Series; Palo Alto, CA, USA) combined with a MS detector (Agilent 5973Network MSD, USA) and SPME autosampler (Combipal, Varian Inc., USA). Analytes were separated on BPX5 capillary column of 60 m × 0.25 mm with a phase thickness 1.0 μm (SGE Analytical Science Pty Ltd, Australia). α-Pinene was used as internal standard. The temperature programme started at 60°C with 1 min holding time, then increased 7°C/min up to 100°C, followed by 10°C/min increase up to final temperature 200°C, where the temperature was kept for 4 min. MSD was operated in electron-impact mode at 70 eV, in the full scan m/z 40–550. The ion source temperature was 230°C and the interface was 280°C. Styrene was identified with corresponding standard and by comparing the mass spectra on Palisade Complete 600 K Mass Spectral Library (Palisade Mass Spectrometry, USA).

Results

In vitro activity

It was previously shown that the *PAD1* is responsible for cinnamic acid dehydratase activity and that the overexpression of *PAD1* results in increased cinnamic acid dehydratase activity (Clausen et al. 1994, Larsson et al. 2001) using intact cells of *S. cerevisiae*. In order to test if this activity could be also seen in cell free extracts we analysed the cell free extracts of the CEN.PK strains which were not modified or expressed the *PAD1* gene. The *PAD1* gene was amplified from the *S. cerevisiae* strain S288C since the CEN.PK has a mutation in the *PAD1* gene resulting in a stop codon instead of the tyrosine in position 98. The cell extracts were incubated with 500 μM cinnamic acid at neutral pH and the disappearance of cinnamic acid was followed by UPLC, however no decarboxylase activity was detected. We followed the decrease in cinnamic acid over a period of 5 hours at a protein concentration of 2 g/l but could not detect a decrease. The detection limit for the decrease in concentration was estimated 15 μM. We also tested if the overexpression of *FDC1* would result cinnamic acid decarboxylase activity in the cell free extract however no activity was detected. However when the two cell-free extracts with *PAD1* and the extracts with *FDC1* were mixed, cinnamic acid decarboxylase activity was detected. After 5 hours the cinnamic acid concentration was decreased by 250 μM. This suggested that both enzymes, Pad1p and Fdc1p, are required. We then expressed *PAD1* and *FDC1* in the same yeast strain using two multicopy plasmids. This extract had a much higher cinnamic acid decarboxylase activity. This is summarized in the Figure 2. The initial rate was about 0,019 μmol per minute and mg of extracted protein. We also confirmed that the reaction product was styrene.

In vivo activity

The *PAD1* gene has a mitochondrial targeting sequence of 57 amino acids and the Pad1p is believed to be a mitochondrial protein. The *FDC1* gene however has no targeting sequence and the Fdc1p is believed to be a cytosolic protein. To test if Pad1p and Fdc1p are also active *in vivo* we tested the sensitivity toward cinnamic acid. For that purpose we plated the yeast strain overexpressing the *PAD1* and *FDC1* on selective plates with increasing amounts of cinnamic acid. At concentrations of up to 10 mM cinnamic acid growth was observed. In the control concentrations above 100 μM were toxic (Table 2). The pKa for cinnamic acid is 4.44. Since the medium was not buffered, the cinnamic acid is in its salt form and it is actually cinnamate resistance that was observed.

Mitochondrial targeting

The combination of *PAD1* and *FDC1* seemed to result in cinnamic acid decarboxylase activity although the proteins are targeted to different cellular compartments. We tested if the activity could be increased by targeting the proteins to the same cellular compartment. In one case we targeted both proteins to the mitochondria. For that purpose we added the mitochondrial targeting

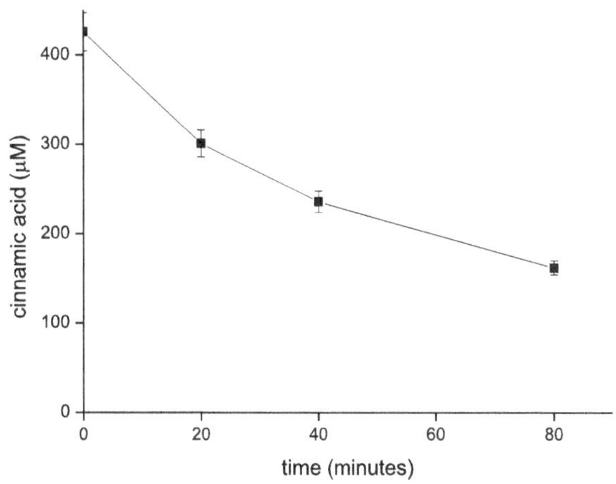

Figure 2 Cinnamic acid decarboxylation with a cell free extract of a yeast strain over-expressing PAD1 and FDC. The protein concentration is 0.32g/l.

Table 2 Growth on increasing cinnamic acid concentrations

Cinnamic acid concentration	H4086	Control
30 mM	-	-
20 mM	-	-
10 mM	+	-
5 mM	+	-
3 mM	+	-
1 mM	+	-
500 μM	+	-
100 μM	+	+

In the strain H4086 PAD1 and FDC1 are expressed from two multicopy plasmids. The control strain harbours the plasmids without open reading frames.

sequence of the *PAD1* to the *FDC1*. In the other case we targeted both to the cytosol. For that we removed the targeting signal from the *PAD1*. In both cases the cell free extract of the yeast strain over-expressing both genes, *PAD1* and *FDC1* with added targeting sequence or *FDC1* and *PAD1* with removed targeting sequence, showed no cinnamic acid decarboxylase activity. Removing or adding mitochondrial targeting signals might affect the protein folding, the lack of activity is therefore not necessarily related to the targeting.

Activity with hydroxy-cinnamic acids

We tested the activity with the following hydroxyl-cinnamic acids in cell free extracts: p-coumaric acid, ferulic acid and caffeic acid. When the *PAD1* or *FDC1* were over-expressed alone we could not detect decarboxylase activity. However, when overexpressing *PAD1* and *FDC1* together we observed activity. This is summarized in the Figure 3. The highest activity was observed with ferulic acid, but this activity was still lower that with cinnamic acid.

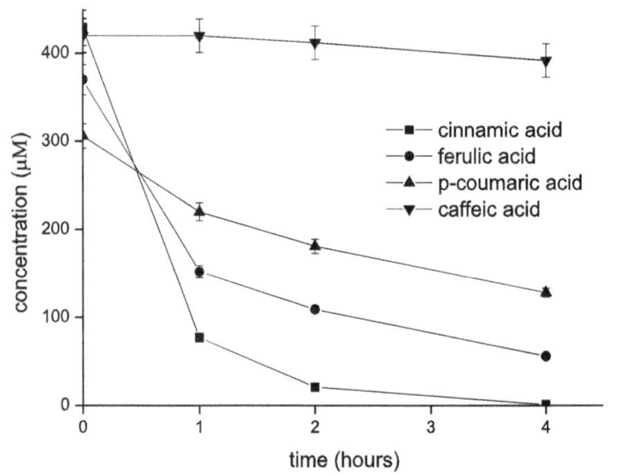

Figure 3 Decarboxylation of the hydroxy cinnamic acids: ferulic acid, p-coumaric acid and caffeic acid. The protein concentration is here 1.6 g/l.

Discussion

Although it is known that yeast and other microorganisms have the ability to convert cinnamic acid to styrene, the enzyme to catalyse this reaction has not been identified. This is mainly because most of the studies were carried in vivo and only a few used a cell free environment (Mukai et al. 2010, Stratford et al. 2012).

In *S. cerevisiae* the *PAD1* was suggested to code for the cinnamic acid decarboxylase. A *PAD1* mutant showed increased sensitivity toward cinnamic acid and expression of *PAD1* repaired the phenotype (Clausen et al. 1994). In the same study permeabilised cells were used to measure cinnamic acid decarboxylase activity. The parent strain showed an activity of 0–3.9 nM/mg dry mass per h and after expression of PAD1 29.4 nM/mg dry mass per h (Clausen et al. 1994). Assuming that 40% of the dry mass is extractable protein this would correspond to about 75 nM per mg extracted protein and hour. When we overexpressed *FDC1* and *PAD1* and analysed the cell extract we estimated the decarboxylase activity to be about 1 mM per mg extracted protein and hour. This corresponds to an increase of activity of about 4 orders of magnitude.

The Padlp and the Fdclp are located in different compartments in *S. cerevisiae.* The Padlp is in mitochondria and the Fdclp in the cytosol. The fact that the overexpression of both proteins results in an increased resistance toward cinnamic acid can be interpreted in different ways. One interpretation is that cinnamic acid is converted to styrene in two steps with an intermediate that is passing through the mitochondrial membrane. Another interpretation is that the proteins are not confined to their compartment. However for the Padlp it was shown that after tagging with the green fluorescence protein the enzyme was exclusively located in the mitochondria (Huh et al. 2003).

No information is available about a possible protein complex including Padlp and Fdclp. In an genome wide survey to identify protein complexes by systematic tagging of open reading frames to identify protein complexes, no protein complexes with Padlp or Fdclp were identified (Gavin et al. 2006).

The physiological role of cinnamic acid decarboxylation is detoxification. Cinnamic acid or cinnamate is more toxic than the reaction product styrene. Cinnamic acid decarboxylation is also of potential biotechnological relevance. L-Phenylalanine can be efficiently converted to cinnamic acid by the action of an L-phenylalanine ammonia lyase (PAL). PAL and cinnamic acid decarboxylase constitute a pathway for styrene production from L-phenylalanine. This had been practiced by McKenna et al. in *E. coli* to produce styrene from D-glucose (McKenna and Nielsen, 2011). In this work the *PAD1* and the *FDC1* of *S. cerevisiae* were expressed in *E. coli* but it was concluded that functional cinnamic acid decarboxylase activity in *E. coli* depends solely upon *FDC1* over-expression and is not dependent on co-expression of *PAD1*. The expression of the *S.cerevisiae PAD1* in *E. coli* did not result in cinnamic acid decarboxylation activity; however the combined expression of *PAD1* and *FDC1* resulted in such activity. The expression of FDC1 alone also resulted in activity suggesting that the *PAD1* is not required for activity; however it cannot be excluded that an *E. coli* protein homologous to *PAD1* would be functional (McKenna and Nielsen, 2011). There are indeed proteins in *E. coli* with homologies to the *PAD1*. One is the *UbiX* that has a role in coenzyme Q biosynthesis (Gulmezian et al. 2007). Another one is a close homologue, *Pad1*, that has been crystallized and a structure determined (Rangarajan et al. 2004), however no function was assigned.

Competing interests

The authors declare that they have no competing interests.

Acknowledgement

This work was supported by the Academy of Finland through the Sustainable Energy (SusEn) programme (Grant 271025).

References

Cavin JF, Barthelmebs L, Diviès C (1997a) Molecular characterization of an inducible p-coumaric acid decarboxylase from *Lactobacillus plantarum*: gene cloning, transcriptional analysis, overexpression in *Escherichia coli*, purification, and characterization. Appl Environ Microbiol 63:1939–1944

Cavin JF, Barthelmebs L, Guzzo J, Van Beeumen J, Samyn B, Travers J-F, Diviès C (1997b) Purification and characterization of an inducible *p*-coumaric acid decarboxylase from *Lactobacillus plantarum*. FEMS Microbiol Lett 147:291–295

Cavin JF, Dartois V, Diviès C (1998) Gene cloning, transcriptional analysis, purification, and characterization of phenolic acid decarboxylase from *Bacillus subtilis*. Appl Environ Microbiol 64:1466–1471

Clausen M, Lamb CJ, Megnet R, Doerner PW (1994) *PAD1* encodes phenylacryil acid decarboxylase which confers resistance to cinnamic acid in *Saccharomyces cerevisiae*. Gene 142:107–112

Gavin AC, Aloy P, Grandi P, Krause R, Boesche M, Marzioch M, Rau C, Jensen LJ, Bastuck S, Dümpelfeld B, Edelmann A, Heurtier MA, Hoffman V, Hoefert C, Klein K, Hudak M, Michon AM, Schelder M, Schirle M, Remor M, Rudi T, Hooper S, Bauer A, Bouwmeester T, Casari G, Drewes G, Neubauer G, Rick JM, Kuster B, Bork P, Russell RB, Superti-Furga G (2006) Proteome survey reveals modularity of the yeast cell machinery. Nat Chem Biol 440:631–636

Gu W, Li X, Huang J, Duan Y, Meng Z, Zhang KQ, Yang J (2011a) Cloning, sequencing, and overexpression in *Escherichia coli* of the *Enterobacter* sp. Px6-4 gene for ferulic acid decarboxylase. Appl Microbiol Biotechnol 89:1797–1805

Gu W, Yang J, Lou Z, Liang L, Sun Y, Huang J, Li X, Cao Y, Meng Z, Zhang KQ (2011b) Structural basis of enzymatic activity for the ferulic acid decarboxylase (FADase) from *Enterobacter* sp. Px6-4. PLoS One 6:e16262

Gulmezian M, Hyman KR, Marbois BN, Clarke CF, Javor GT (2007) The role of UbiX in *Escherichia coli* coenzyme Q biosynthesis. Arch Biochem Biophys 467:144–153

Huh W-K, Falvo JV, Gerke LC, Carroll AS, Howson RW, Weissman JS, O'Shea EK (2003) Global analysis of protein localization in budding yeast. Nature 425:686–691

Kuorelahti S, Kalkkinen N, Penttilä M, Londesborough J, Richard P (2005) Identification in the mold *Hypocrea jecorina* of the first fungal D-galacturonic acid reductase. Biochemistry 44:11234–11240

Larsson S, Nilvebrant N-O, Jönsson LJ (2001) Effect of overexpression of *Saccharomyces cerevisiae* Pad1p on the resistance to phenylacrylic acids and lignocellulose hydrolysates under aerobic and oxygen-limited conditions. Appl Microbiol Biotechnol 57:167–174

McKenna R, Nielsen DR (2011) Styrene biosynthesis from glucose by engineered *E. coli*. Metab Eng 13:544–554

McKenna R, Thompson B, Pugh S, Nielsen DR (2014) Rational and combinatorial approaches to engineering styrene production by *Saccharomyces cerevisiae*. Microb Cell Fact 13:123

Minet M, Dufour ME, Lacroute F (1992) Complementation of *Saccharomyces cerevisiae* auxotrophic mutants by *Arabidopsis thaliana* cDNAs. Plant J 2:417–422

Mukai N, Masaki K, Fujii T, Kawamukai M, Iefuji H (2010) *PAD1* and *FDC1* are essential for the decarboxilation of phenylacrylic acid in *Saccharomyces cerevisiae*. J Biosci Bioeng 109:564–569

Plumridge A, Stratford M, Lowe KC, Archer DB (2008) The weak-acid preservative sorbic acid is decarboxylated and detoxified by a phenylacrylic acid decarboxylase, PadA1, in the spoilage mold *Aspergillus niger*. Appl Environ Microbiol 74:550–552

Plumridge A, Melin P, Stratford M, Novodvorska M, Shunburne L, Dyer PS, Roubos JA, Menke H, Stark J, Stam H, Archer DB (2010) The decarboxylation of the weak-acid preservative, sorbic acid, is encoded by linked genes in *Aspergillus spp*. Fungal Genet Biol 47:683–692

Rangarajan ES, Li Y, Iannuzzi P, Cygler M, Matte A (2004) Crystal structure of *Escherichia coli* crotonobetainyl-CoA: carnitine CoA-transferase (CaiB) and its complexes with CoA and carnitinyl-CoA. Biochemistry 44:5728–5738

Rodríguez H, Angulo I, DelasRivas B, Campillo N, Páez JA, Muñoz R, Mancheño JM (2010) *p*-Coumaric acid decarboxylase from *Lactobacillus plantarum*: Structural insights into the active site and decarboxylation catalytic mechanism. Proteins 78:1662–1676

Stratford M, Plumridge A, Archer DB (2007) Decarboxylation of sorbic acid by spoilage yeasts is associated with the *PAD1* gene. Appl Environ Microbiol 73:6534–6542

Stratford M, Plumridge A, Pleasants MW, Novodvorska M, Baker-Glenn CAG, Pattenden G, Archer DB (2012) Mapping the structural requirements of inducers and substrates for decarboxylation of weak acid preservatives by the food spoilage mould *Aspergillus niger*. Int J Food Microbiol 157:375–383

Zago A, Degrassi G, Bruschi CV (1995) Cloning, sequencing, and expression in *Escherichia coli* of the *Bacillus pumilus* gene for ferulic acid decarboxylase. Appl Environ Microbiol 61:4484–4486

24

Engineering of *Saccharomyces cerevisiae* for the production of poly-3-D-hydroxybutyrate from xylose

Anders G Sandström, Alejandro Muñoz de las Heras, Diogo Portugal-Nunes and Marie F Gorwa-Grauslund*

Abstract

Poly-3-D-hydroxybutyrate (PHB) is a promising biopolymer naturally produced by several bacterial species. In the present study, the robust baker's yeast *Saccharomyces cerevisiae* was engineered to produce PHB from xylose, the main pentose found in lignocellulosic biomass. The PHB pathway genes from the well-characterized PHB producer *Cupriavidus necator* were introduced in recombinant *S. cerevisiae* strains already capable of pentose utilization by introduction of the fungal genes for xylose utilization from the yeast *Scheffersomyces stipitis*. PHB production from xylose was successfully demonstrated in shake-flasks experiments, with PHB yield of 1.17 ± 0.18 mg PHB g^{-1} xylose. Under well-controlled fully aerobic conditions, a titer of 101.7 mg PHB L^{-1} was reached within 48 hours, with a PHB yield of 1.99 ± 0.15 mg PHB g^{-1} xylose, thereby demonstrating the potential of this host for PHB production from lignocellulose.

Keywords: Polyhydroxyalkanoate; PHB; Xylose; *Saccharomyces cerevisiae*; Metabolic engineering; Biopolymer

Introduction

Petroleum-based plastics are essential elements of our contemporary society, and will remain so for a foreseeable future. Still, the world is standing at the threshold of an era marked by scarce fossil-based resources (Lambert et al. 2014; Foley et al. 2011), which called for a transition towards the production of bio-based bulk chemicals from renewable feedstock (European Parliament CotEU 2009). Among candidate feedstock for bio-based polymer production, lignocellulosic biomass such as corn stover, wheat straw or sugar cane bagasse, are amongst the most promising ones (Gírio et al. 2010). Another readily available lignocellulosic feedstock is the spent sulfite liquor (SSL), one of the main by-products from the pulp and paper industry (Novy et al. 2013).

However, up to now, the bio-based polymers that are entering the market as replacement for their petroleum-based counterparts are produced from simple sugars, such as glucose, or food crop residuals such as glycerol. Polyhydroxyalkanoates (PHAs) are some of the most promising biodegradable biopolymers for a sustainable future (Philip et al. 2007). Poly-(*R*)-3-hydroxybutyrate (PHB), that is naturally produced by several bacterial species as an energy storage depot, typically under conditions of nitrogen starvation (Suriyamongkol et al. 2007; Madison and Huisman 1999), is one of the most studied PHAs. Numerous hosts have been suggested as PHA producers, such as natural PHB-producing bacterial species or engineered *Escherichia coli.* Nevertheless *E. coli* suffers from regular phage infections in industrial production settings (Lian et al. 2014), while the natural PHB-producing bacterial species are not well adapted to harsh environments, such as growth in SSL. Instead baker's yeast *Saccharomyces cerevisiae,* that is known for its tolerance to acidic and inhibitor-rich conditions, is considered one of the favored species for converting the sugars found in SSL into bulk chemicals of interest – e.g. PHB (Demeke et al. 2013). SSLs are commonly rich in xylose, a pentose sugar that is not naturally consumed by *S. cerevisiae*. Therefore, efficient xylose-metabolizing *S. cerevisiae* strains have been developed (Hahn-Hägerdal et al. 2007; Kim et al. 2013).

The potential of engineered *S. cerevisiae* as a producer of PHAs from pure hexose sugars has previously been

* Correspondence: Marie-Francoise.Gorwa@tmb.lth.se
Division of Applied Microbiology, Department of Chemistry, Lund University, PO Box 124, Lund SE-221 00, Sweden

explored by several groups. First, only the polyhydroxyalkanoate synthase gene from the PHB producer *Cupriavidus necator* (formerly known as *Ralstonia eutropha* (Vandamme and Coenye 2004)) was overexpressed (Leaf et al. 1996) and PHB was produced by the available cytosolic 3-hydroxybutyryl-CoA, derived from the fatty acid β–oxidation pathway. Several strategies aiming at increasing the available pool of precursors have been tested in *S. cerevisiae* in order to increase the yield of PHB, for example addition of the *C. necator* β-ketothiolase and acetoacetyl-CoA reductase genes (Carlson and Srienc 2006; Breuer et al. 2002), enzymes for boosting ethanol assimilation and acetyl-CoA synthesis (Kocharin et al. 2012) and addition of the phosphoketolase pathway and non-phosphorylating $NADP^+$-dependent glyceraldehyde-3-phosphate dehydrogenase for increasing the NADPH pool (Kocharin et al. 2013). Here, we demonstrate that *S. cerevisiae* can also be engineered to produce PHB from D-xylose.

Materials and methods

Strains, media and culture conditions

Plasmids and yeast strains used in this study are listed in Table 1. *E. coli* strain NEB5-α (New England Biolabs) was used for sub-cloning of plasmid DNA and further propagation. Lysogeny broth (LB) (5 g L^{-1} yeast extract, 10 g L^{-1} peptone, 5 g L^{-1} NaCl, pH 7.0) medium was used for routine culturing of *E. coli* and 50 mg L^{-1} ampicillin was added to LB when needed. Bacterial transformants were selected on solid LB plates (15 g L^{-1} agar), supplemented with ampicillin (50 mg L^{-1}), for 16 h at 37°C. Cultures of vector-carrying *E. coli* were recovered from 25% glycerol stocks stored at −80°C and grown in liquid LB medium, supplemented with ampicillin, for 14–16 h at 37°C and 180 rpm in an orbital shaker.

Yeast strains were recovered from 20% glycerol stocks stored at −80°C, by streaking on solid YPD plates (10 g L^{-1} yeast extract, 20 g L^{-1} peptone, 20 g L^{-1} glucose, 15 g L^{-1} agar) for two days at 30°C. Yeast cultures were grown in liquid YPD when preparing the strain for transformation for 14–16 h, or less when required, at 30°C and 180 rpm in an orbital shaker.

Engineered yeast strains were selected on solid YNB medium (6.7 g L^{-1} Yeast Nitrogen Base without amino acids (Becton, Dickinson and Company, USA) supplemented with 20 g L^{-1} glucose or xylose and 15 g L^{-1} agar). Leucine was added for complementation at a concentration of 30 mg L^{-1} when required.

All chemicals were purchased from Sigma-Aldrich if not otherwise stated.

PHB pathway expression vector construction

The PHB pathway genes *PhaA*, *PhaB1*, and *PhaC1* from *C. necator* (Poehlein et al. 2011) were codon optimized towards *S. cerevisiae* and synthesized (Eurofins Genomics, Germany). The codon-optimized genes have been deposited with following accession numbers; *PhaA* [GenBank: KP681582], *PhaB1* [GenBank:KP681583], *PhaC1* [GenBank: KP681584]. The three coding regions were inserted in-between the strong constitutive promoter-terminator pairs of the genes *TEF1*, *TPI1*, *GPM1* (Sun et al. 2012), respectively. The promoters and terminators were amplified directly from the genome of *S. cerevisiae* CEN.PK113-7D (Nijkamp et al. 2012), and the PHB pathway genes were amplified by PCR from respective vector. The fragments were amplified with Phusion Hotstart II polymerase (Thermo Scientific, USA) using the following PCR program: initial denaturation at 98°C for 30 s, 30 cycles of denaturation at 98°C for 10 s, annealing at 65°C (−0.5°C/cycle) for 30 s and elongation at 72°C for 15 s or 1 min, and a final elongation step for 10 min at 72°C. The three purified promoter, coding sequence and terminator fragments were assembled by overlap extension PCR (OE-PCR). Equimolar fragment amount were mixed, and the following overlap-extension program was used; 98°C for 30 s, then 16 cycles at 98°C for 10 s, 68°C (−0.5°C/cycle) for 30 s, 72°C for 1 min and then hold at 12°C for the addition of amplification primers (50 μM each), whereafter the program continued at 98°C for 30 s, then 20 cycles at 98°C

Table 1 Plasmids and strains used in this study

Plasmids	Relevant genotype	Reference
YIplac128	*LEU2*	(Gietz and Akio 1988)
YIpOB8	*pTDH3-XYL1-tADH1; pPGK1-XYL2-tPGK1; URA3*	(Runquist et al. 2010)
YIpAGS2	YIplac128; *pTEF1-PhaA-tTEF1; pGPM1-PhaB1-pGPM1; pTPI1-PhaC1-tTPI1; LEU2*	This study
Yeast strains	**Relevant genotype**	**Reference**
TMB 3043	CEN.PK2-1C; *gre3Δ; his3::pPGK1-XKS1-tPGK1, HIS3; tal1::pPGK1-TAL1-tPGK1; tkl1::pPGK1-TKL1-tPGK1; rki1::pPGK1-RKI1-tPGK1; rpe1::pPGK1-RPE1-tPGK1; ura3, leu2*	(Karhumaa et al. 2005)
TMB 4440	TMB 3043; *ura3*::YIpOB8; *leu2*	This study
TMB 4443	TMB 4440; *leu2*::YIpAGS2	This study
TMB 4444	TMB 4440; *leu2*::YIplac128	This study

for 30 s, 68°C (−0.5°C/cycle) for 30 s, 72°C for 1 min, and then a final elongation at 72°C for 10 min. The primers were designed such that the flanking ends of the generated fragments overlapped with the neighboring regions, enabling the use of In-Fusion cloning (Clontech, Takara-Bio, Japan). Linear vector fragments of YIplac128 (Gietz and Akio 1988) were generated by double-digestion with Fast Digest *Cfr*9I (*Xma*I) and *Ssp*DI (*Kas*I) (Thermo Scientific, USA). The digested vector, and the three promoter-gene-teminator fragments were mixed in equimolar amounts and were assembled by In-Fusion cloning, according to the supplier's instruction. High efficiency competent NEB5-α cells were transformed according to the supplier's instructions. Plasmids were propagated, a diagnostic restriction digestion was performed and the cloned region was fully sequenced. The oligonucleotides used for the PCRs are listed in Table 2.

Yeast strain engineering

S. cerevisiae TMB 3043 (Karhumaa et al. 2005) was transformed with the *Eco*RV-linearized vector YIpOB8 (Runquist et al. 2010), generating the strain TMB 4440. The yeast cells were transformed using the high-efficiency Li-Ac method (Gietz and Schiestl 2007), and plated on YNB-xylose (20 g L^{-1})-plates, supplemented with leucine (30 mg L^{-1}). The strain was subsequently transformed with *Eco*91-linearized YIpAGS2 or YIp128 – to form the PHB-producing strain TMB 4443 or the corresponding control strain TMB 4444 (Table 1). Transformed strains were validated by amplifying stretches of the integrated cassette by PCR, using extracted genomic DNA as template and primers found in Table 2.

Aerobic cultivations

For the shake flasks experiments, well-aerated baffled shake flasks were used. Pre-cultures were started by transferring an isolated colony into 5.0 mL of defined buffered YNB-xylose (13.4 g L^{-1} Yeast nitrogen base, 50 g L^{-1} xylose, 50 mM potassium hydrogen phthalate, pH 5.5) in 50 mL conical centrifuge tube, shaking at 30°C at 180 rpm overnight. 50 mL buffered YNB-xylose was inoculated with the pre-culture to a starting OD_{620} of 0.05, in a 250 mL baffled shake flask. The cultures were grown at 30°C at 180 rpm for at least 82 hours. Optical density during growth was monitored at 620 nm using an Ultrospec 2100 Pro spectrophotometer (Amersham Biosciences Corp., USA). Samples were extracted for PHB quantification, OD_{620}, cell dry weight measurements, and metabolite analysis, and were done in technical duplicates. Maximum specific growth rate was calculated by the analysis of the exponential growth phase, using a linear regression of the natural logarithm derived from the OD_{620}.

For the experiments in bioreactors, unbuffered YNB-xylose (13.4 g L^{-1} Yeast nitrogen base, 50 g L^{-1} xylose) was used as media. Well-controlled Multifors 1.2 L vessels (Infors, Switzerland) were sparged with 1.0 vvm air at 30°C and 300 rpm. A culture volume of 300 mL was inoculated with YNB-xylose freshly grown cells to a

Table 2 Primers used in the present study

Name	Amplification target	Sequence (5′ to 3′)
TEF1p_f	*S. cerevisiae TEF1* promoter	CTAGAGGATCCCCGGGTATAGCTTCAAAATGTTTCTAC
TEF1p_r		CACGTCAGTCATTTTGTAATTAAAACTTAGATTAG
PhaA_f	Codon optimized *PhaA* ORF from *C. necator*	GTTTTAATTACAAAATGACTGACGTGGTGATTG
PhaA_r		CACGTCAGTCATTTTGTAATTAAAACTTAGATTAGATTG
TEF1t_f	*S. cerevisiae TEF1* terminator	GTTGAAAGAAAGTAAGGAGATTGATAAGACTTTTC
TEF1t_r		GATATAGAGCTCAGATAGCGCCGATCAAAG
TPI1p_f	*S. cerevisiae TPI1* promoter	TCTGAGCTCTATATCTAGGAACCCATCAG
TPI1p_r		CAATCCTTTGCGTCATTTTTAGTTTATGTATGTGTTTTTG
PhaB1_f	Codon optimized *PhaB1* ORF from *C. necator*	CATACATAAACTAAAAATGACGCAAAGGATTGCGTATG
PhaB1_r		GATAATATTTTTATATAATTATATTATTAACCCATGTGTAAGCCAC
TPI1t_f	*S. cerevisiae TPI1* terminator	CACATGGGTTAATAATATAATTATATAAAAATATTATCTTCTTTTC
TPI1t_r		CTAACTAGTTATATAACAGTTGAAATTTGGATAAG
GPM1p_f	*S. cerevisiae GPM1* promoter	TATATAACTAGTTAGTCGTGCAATGTATGAC
GPM1p_r		CTTTACCAGTGGCCATTATTGTAATATGTGTGTTTGTTTG
PhaC1_f	Codon optimized *PhaC1* ORF from *C. necator*	CATATTACAATAATGGCCACTGGTAAAGG
PhaC1_r		CATTCATTCTTCAGACTTATGCCTTTGCTTTCACATAC
GPM1t_f	*S. cerevisiae GPM1* terminator	GAAAGCAAAGGCATAAGTCTGAAGAATGAATGATTTG
GPM1t_r		TACCGCATCAGGCGCCATATTCGAACTGCCCATTCAG

starting OD_{620} of 0.05. Silicon antifoam RD emulsion (Dow Corning, USA), 3 M KOH and 3 M H_2SO_4 were added as needed in order to avoid the excess of foaming and to keep the pH at 5.5, respectively. Two independent biological replicates were performed for each condition.

Analysis of biomass and metabolites

Cell dry weight was measured in triplicate by filtering a known volume (3.0 to 5.0 mL) of the culture through a pre-weighed nitrocellulose filter with 0.45 μm pore size (PALL, USA). The filters were washed, dried and weighed after equilibrating to room temperature in a desiccator. Biomass was correlated to OD_{620} by a single point calibration, based on the final measuring time point. Concentrations of glucose, xylose, xylitol, glycerol, acetate and ethanol were analyzed by HPLC (Waters, USA). Metabolites were separated using an Aminex HPX-87H ion exchange column (Bio-Rad, USA) at 45°C with 5 mM H_2SO_4 as mobile phase, at a flow rate of 0.6 mL min^{-1}. All compounds were detected with a RID-10A refractive index detector (Shimadzu, Japan). Concentrations of metabolites were calculated from an external seven-point standard calibration curve. Samples were analyzed in technical duplicates.

PHB quantification

The method used relied on the quantitative conversion of PHB to crotonic acid catalyzed by hot concentrated sulfuric acid (Law and Slepecky 1961). A culture volume of 0.5 or 1.0 mL was harvested and transferred to a 1.5 mL microcentrifuge tube. The harvested cells were centrifuged for 5 min at 6000 × *g* and the supernatant was discarded by careful aspiration (FTA-1, Biosan, Latvia). The pellet was washed with 1.0 mL ultrapure H_2O and resuspended by vortexing, centrifuged for 5 min at 6000 × *g* and the supernatant was discarded. This wash step was repeated two times. H_2SO_4 (0.5 mL, 96% v/v) was added to the pellet. The pellet was resuspended by careful pipetting, and the open-lid tubes were transferred to a heat block (Grant QBD1, Grant Instruments, UK), and incubated at 95°C for 1 h. The resulting solution was diluted 20 times by a serial dilution and further analyzed by HPLC, using the same conditions as for the analysis of metabolites, and quantified by comparing to an external calibration curve of crotonic acid. Commercially available PHB (#363502, Sigma-Aldrich) was used as method control and was processed similarly as pelletized cells. Samples were analyzed in technical duplicates.

Results

Construction of the expression vector carrying the PHB pathway genes

The *C. necator* PHB pathway, consisting of a β-ketothiolase (acetyl-CoA acetyltransferase) (encoded by *PhaA*), an acetoacetyl-CoA reductase (encoded by *PhaB1*), and a PHB synthase (encoded by *PhaC1*) (Figure 1A), was chosen for PHB production in *S. cerevisiae* as it has been successfully used in the past (see e.g. (Carlson and Srienc 2006;

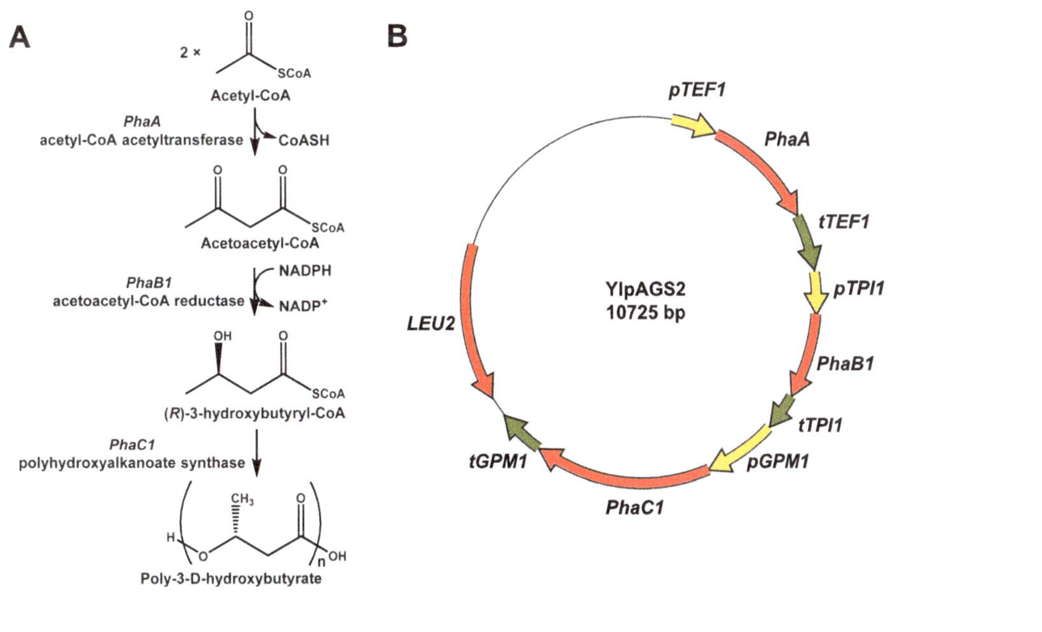

Figure 1 The PHB metabolic pathway, consisting of A) metabolite structures, involved enzymes, cofactors and B) the genes and regulatory sequences in the integrative vector YIpAGS2 in a schematic representation.

Kocharin et al. 2012; Breuer et al. 2002)). The genes, ordered from MWG Eurofins, were codon optimized for *S. cerevisiae*, which notably led to lower GC-content (47–52% GC) than the native bacterial genes (63–68% GC). All three genes were cloned behind a strong constitutive promoter (*pTEF1*, *pTPI1* and *pGPM1* for *PhaA*, *PhaB1* and *PhaC1*, respectively) into a single large vector, named YIpAGS2 (Figure 1B), based on YIplac128 which has a *LEU2* gene for complementation of auxotrophic strains. The entire pathway was sequenced and validated.

Strain engineering

S. cerevisiae strain TMB 3043 (Karhumaa et al. 2005), that has been engineered for increased flux from xylulose to glycolysis, through the overexpression of xylulokinase and the non-oxidative pentose phosphate pathway genes, was transformed with YIpOB8 carrying the *Scheffersomyces stipitis* xylose genes encoding xylose reductase (XR) and xylitol dehydrogenase (XDH) (Bengtsson et al. 2009), generating strain TMB 4440.

TMB 4440 was further transformed with the constructed linearized YIpAGS2 that carries *C. necator* PHB pathway gene or with a linearized integrative control vector (YIplac128), generating strains TMB 4443 (PHB+) and TMB 4444 (PHB-), respectively.

Evaluation of PHB production from xylose

The generated strains were compared aerobically in well-stirred baffled flasks to evaluate their PHB production capacity. The strains were grown with D-xylose as sole carbon source, until the cultures reached late stationary phase. Metabolites were analyzed and cell dry weight calculated. Representative growth plots for the control strain TMB 4444 and the TMB 4443 strain expressing the PHB pathway genes are displayed in Figure 2A and 2B, respectively.

As expected, both strains were able to grow on xylose. Xylitol, acetate and ethanol were the main by-products (Table 3). However strain TMB 4443 (PHB+) generated

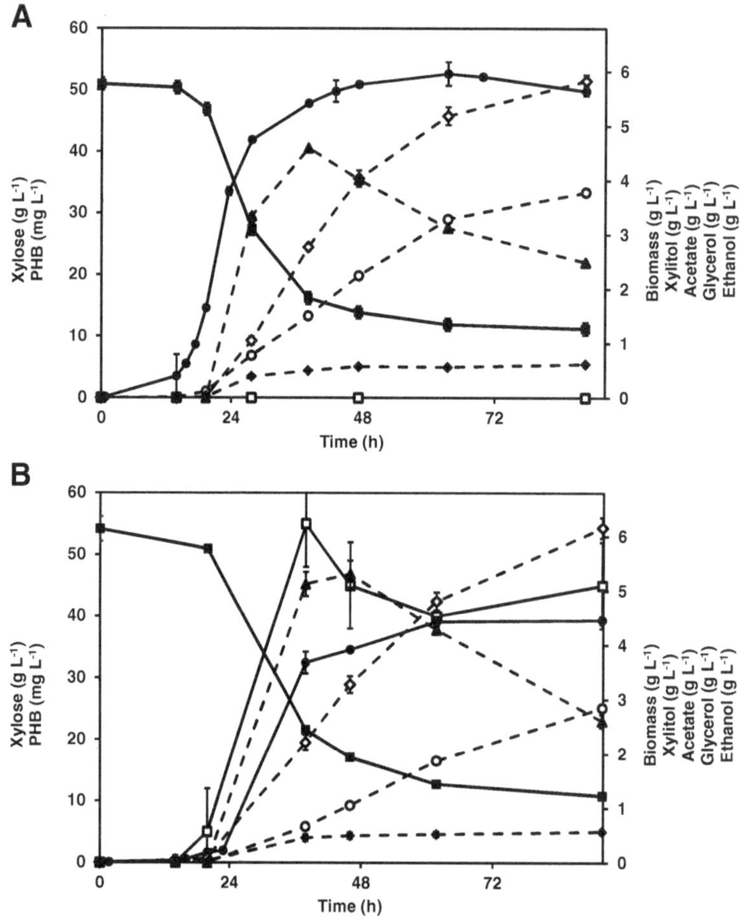

Figure 2 Representative aerobic growth and metabolite profiles from recombinant *S. cerevisiae* strains A) TMB 4444 (Control strain, PHB-) and B) TMB 4443 (PHB+), with xylose defined medium in buffered shake flask cultures. *Legend*: xylose(■); biomass(●); glycerol(♦); ethanol(▲); PHB(□); xylitol(○); acetate(◊).

Table 3 Physiological results of the strains grown aerobically on xylose on shake flasks or bioreactors

Strain	Growth condition	Growth rate μ_{max} (h^{-1})	Yields						PHB titer (mg/L)	Final PHB content (% CDW)	Xylose/sugars consumed (%)
			Y_{sx} (g/g Xyl)	Y_{sEtOH} (g/g Xyl)	Y_{sAc} (g/g Xyl)	$Y_{sXylitol}$ (g/g Xyl)	Y_{sGly} (g/g Xyl)	Y_{sPHB} (mg/g Xyl)			
TMB 4443	Shake flask	0.231 ± 0.01	0.11 ± 0.02	0.10 ± 0.05	0.14 ± 0.0	0.05 ± 0.03	0.01 ± 0.0	1.17 ± 0.18	45.0 ± 3.54	1.04 ± 0.03	76.8 ± 4.6
TMB 4444	Shake flask	0.249 ± 0.01	0.14 ± 0.0	0.06 ± 0.0	0.14 ± 0.0	0.10 ± 0.0	0.02 ± 0.0	0.0 ± 0.0	0.0 ± 0.0	0.0 ± 0.0	80.1 ± 3.0
TMB 4443	Bioreactor	0.184 ± 0.01	0.41 ± 0.02	0.004 ± 0.0	0.0 ± 0.0	0.028 ± 0.0	0.0 ± 0.0	1.99 ± 0.15	101.7 ± 7.1	0.49 ± 0.06	100 ± 0

The values reported are calculated for aerobic cultivations on defined media with xylose (50 g L^{-1}) as carbon source. The specific growth rate is calculated for the exponential phase. Yields and titers are calculated from a single time point in the early stationary phase. Reported values represent the mean ± SD of at least two independent cultivations performed. Y_{sx}: biomass yield on xylose, Y_{sEtOH}: yield of ethanol on xylose, Y_{sAc}: yield of acetate on xylose, $Y_{sXylitol}$: yield of xylitol on xylose, Y_{sGly}: yield of glycerol on xylose, Y_{sPHB}: yield of PHB on xylose. PHB titer: The volumetric PHB titer. PHB/CDW: The PHB component as percentage of total cell dry weight.

less biomass and xylitol and more ethanol than the control strain TMB 4444. This was accompanied by PHB production that appeared to be concomitant with biomass formation. In the early exponential phase, PHB content rose more rapidly than the biomass titers under these shake flask conditions (Figure 2B). PHB yield peaked at 1.17 ± 0.33 mg PHB g^{-1} xylose consumed (Table 3) at the end of the exponential phase and further ethanol consumption did not result in additional PHB accumulation (Figure 2B).

As acetate levels were high and xylose utilization stopped before depletion, PHB production was also evaluated in another set-up using well-aerated and pH-controlled bioreactor for strain TMB 4443 (PHB+). Under these conditions, full xylose consumption was observed within 48 hours (Figure 3) and biomass accumulation was very high in comparison to the shake flask-experiments. Also, acetate, ethanol and xylitol production was significantly reduced. At the end of the exponential phase (48 hours), maximum biomass accumulation was reached and cellular PHB reached 0.49 ± 0.06% PHB/CDW, corresponding to a PHB yield of 1.99 ± 0.15 mg PHB g^{-1} xylose (Figure 3, Table 3).

Discussion

In the present study, PHB production from xylose was demonstrated using recombinant *Saccharomyces cerevisiae.* PHB production occurred under exponential growth phase, concomitant with the accumulation of ethanol and other by-products. The maximal PHB yield from xylose (1.99 ± 0.15 mg PHB g^{-1} xylose, or 2.79 Cmmol PHB $Cmol^{-1}$ xylose), that was achieved in well-aerated bioreactors, was comparable to reported non-xylose engineered strains harboring equivalent *C. necator* PHB pathways growing on glucose (Table 4), thereby demonstrating the potential of using a pentose carbon source, and more generally C5-rich lignocellulosic substrates for PHB production. One of the possible reasons might be

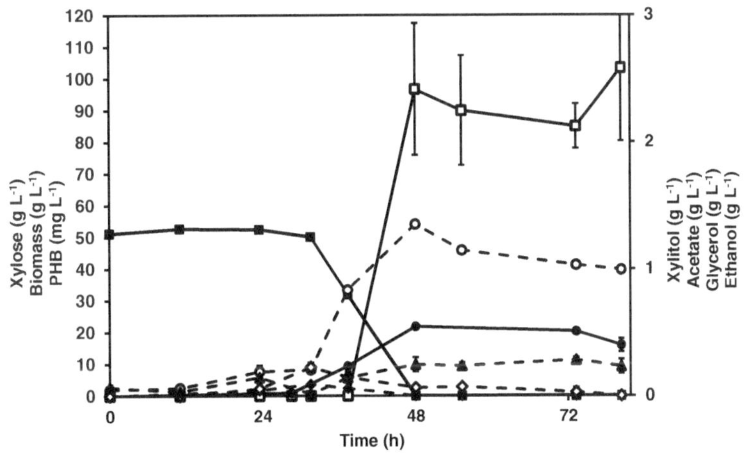

Figure 3 Representative fully aerobic batch growth and metabolite profile from recombinant *S. cerevisiae* strain TMB 4443 (PHB+) in xylose defined medium at pH 5.5 using a well-controlled bioreactor. *Legend*: xylose(■); biomass(●); glycerol(♦); ethanol(▲); PHB(□); xylitol(○); acetate(◊).

Table 4 Comparison of selected PHB producing *S. cerevisiae* strains harbouring the *C. necator* PHB pathway, consuming different carbon sources

Strain	Genotype	Carbon source	Growth conditions	PHB yield/carbon source		PHB content (mg/g CDW)	Reference
				(mg/g)	(Cmmol/Cmol)		
SCKK005	PHB pathway	Glucose	Shake flask	~1.5[a]	2.1[a]	4.02	(Kocharin et al. 2013)
SCK006	PHB pathway, *ADH2*, *ALD6*, *acs*^{L641P}, *ERG10*	Glucose	Shake flask	~5[a]	7[a]	15.89	(Kocharin et al. 2013)
SCK032	PHB pathway, *xpkA*, *ack*, *acs*^{L641P}	Glucose	Shake flask	~9[a]	12.5[a]	27.86	(Kocharin et al. 2013)
SCK034	PHB pathway, *gapN*, *ADH2*, *ALD6*, *acs*^{L641P}, *ERG10*	Glucose	Shake flask	~7[a]	9.8[a]	27.52	(Kocharin et al. 2013)
SCKK006	PHB pathway, *ADH2*, *ALD6*, *acs*^{L641P}, *ERG10*	Glucose	Aerobic, chemostat bioreactor, dilution rate 0.10 h^{-1}	2.79	3.67	5.59	(Kocharin and Nielsen 2013)
Strain RKS	GAL1-10-induced PHB pathway	Galactose/Glucose	Aerobic, batch bioreactor	13.5[b]	18.8[b]	75	(Carlson and Srienc 2006)
TMB4443	PHB Pathway, *XYL1*, *XYL2*, *XKS*, PPP↑, *gre3Δ*	Xylose	Shake flask	1.17	1.63	10.4	This study
TMB4443	PHB Pathway, *XYL1*, *XYL2*, *XKS*, PPP↑, *gre3Δ*	Xylose	Aerobic, batch bioreactor	1.99	2.79	4.9	This study

[a]Recalculated from graph presented in the respective references, based on total consumption of carbon sources.
[b]Calculated based on total consumption of carbon sources.

that xylose conversion to pyruvate does not involve the oxidative pentose phosphate pathway that is the main source of NADPH inside the cell (Bruinenberg et al. 1983). As a consequence more carbon may be diverted towards the $NADP^+$-catalyzed formation of acetate that is a direct precursor for PHB synthesis.

The final titer reached a total of 45.0 ± 3.5 mg L^{-1} of PHB for TMB 4443 (PHB+) in shake flask conditions, and increased to 101.7 ± 7.1 mg L^{-1} of PHB in well-aerated bioreactor. By-product accumulation was much lower in the fully aerobic and pH-controlled bioreactor set-up than in the shake-flask experiment,

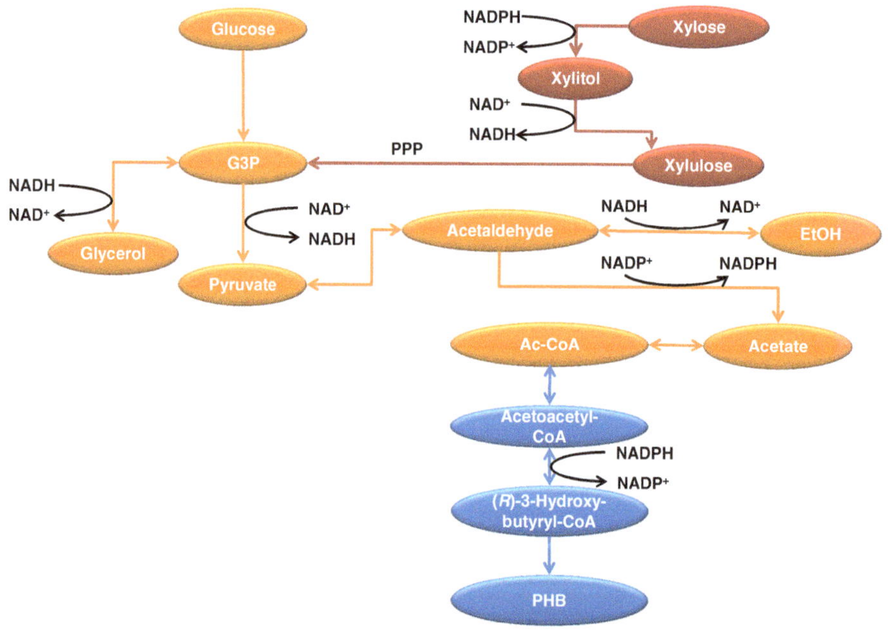

Figure 4 A condensed view of the glycolytic pathway steps, as well as the XR and PHB pathway steps that are relevant for the PHB accumulation in the constructed strain. Abbreviations: EtOH, ethanol; Ac-CoA, acetyl coenzyme A; G3P, glyceraldehyde 3-phosphate; PPP, pentose phosphate pathway The parts marked in red are the introduced xylose assimilation pathway and the overexpressed PPP. The parts marked in blue are the components of the PHB metabolic pathway.

indicating increased respiratory metabolism that directly translated into higher biomass. This may explain the positive impact on PHB production as more cells were available to synthesize PHB while less carbon was dissipated into reduced byproducts. This observation was made despite the fact that low levels of acetate, that is required for PHB production (Figure 4), were generated. In contrast, unusually high $NADP^+$-catalyzed acetate levels were observed in the shake flask experiments. In that case, the limited respiration may trigger higher availability of acetaldehyde precursor; this, combined with the need for additional routes for NADPH generation for biomass synthesis and for xylose utilization via the NAD(P)H-dependent xylose reductase, could explain the dramatic increase in acetate production. It may also explain why xylose consumption stopped before completion as pH decreased over time (data not shown). The accumulation of acetate also indicates that, under these conditions, PHB generation may rather be controlled by reactions downstream of acetate (Figure 4). This includes the acetyl-CoA synthesis that has been shown to be limited by the availability of ATP and by the efficiency of the acetyl-CoA synthetase (Shiba et al. 2007), or the acetoacetyl-CoA reductase-catalyzed reduction that could be limited by NADPH shortage. Overall, the results indicate that a threshold in oxygenation may be necessary to provide sufficient resources for biomass synthesis while enabling sufficient formation and further conversion of the acetate precursor.

In this report PHB production from xylose was demonstrated in *S. cerevisiae.* Our results also point towards the importance of aeration on the synthesis of the PHB pathway precursor and cofactors. To further enhance the PHB accumulation process aspects such as redox balance and the acetyl-CoA node have to be considered.

Abbreviations

Acetyl-CoA: Acetyl coenzyme A; NAD^+/H: Nicotinamide adenine dinucleotide; $NADP^+$/H: Nicotinamide adenine dinucleotide phosphate; PCR: Polymerase chain reaction; OE-PCR: Overlap extension PCR; XR: Xylose reductase.

Competing interests

The authors declare that they have no competing interests.

Authors' contributions

AGS participated in the design of the study and its coordination, carried out molecular genetic engineering, strain constructions and drafted the manuscript. AMH carried out molecular genetic engineering and strain constructions. DPN carried out all the growth assays, bioreactor experiments and HPLC method optimization and metabolite analysis. MGG conceived the study, participated in the design of the study and its coordination and helped to draft the manuscript. All authors read and approved the final manuscript.

Acknowledgements

We would like to thank Nathasha Rizza, Henrik Almqvist, Gunnar Lidén, Basti Bergdahl, Ed van Niel, Biotrend S.A. and SilicoLife Lda for valuable input. The EU Commission (FP7 project BRIGIT, Contract nr 311935) is acknowledged for their financial support.

References

Bengtsson O, Hahn-Hägerdal B, Gorwa-Grauslund MF (2009) Xylose reductase from *Pichia stipitis* with altered coenzyme preference improves ethanolic xylose fermentation by recombinant *Saccharomyces cerevisiae*. Biotechnol Biofuels 2. doi:910.1186/1754-6834-2-9

Breuer U, Terentiev Y, Kunze G, Babel W (2002) Yeasts as producers of polyhydroxyalkanoates: genetic engineering of *Saccharomyces cerevisiae*. Macromol Biosci 2(8):380–386. doi:10.1002/1616-5195(200211)2:8<380::aid-mabi380>3.0.co;2-x

Bruinenberg PM, Vandijken JP, Scheffers WA (1983) A theoretical analysis of NADPH production and consumption in yeasts. J Gen Microbiol 129 (APR):953–964

Carlson R, Srienc F (2006) Effects of recombinant precursor pathway variations on poly [(R)-3-hydroxybutyrate] synthesis in *Saccharomyces cerevisiae*. J Biotechnol 124(3):561–573. doi:10.1016/j.jbiotec.2006.01.035

Demeke MM, Dietz H, Li Y, Foulquie-Moreno MR, Mutturi S, Deprez S, Den Abt T, Bonini BM, Liden G, Dumortier F, Verplaetse A, Boles E, Thevelein JM (2013) Development of a D-xylose fermenting and inhibitor tolerant industrial *Saccharomyces cerevisiae* strain with high performance in lignocellulose hydrolysates using metabolic and evolutionary engineering. Biotechnol Biofuels 6(1):89, 1754-6834-6-89 [pii] 10.1186/1754-6834-6-89

European Parliament CotEU (2009) Promotion of the use of energy from renewable sources. http://eur-lex.europa.eu/legal-content/EN/ALL/?uri=CELEX:32009L0028

Foley JA, Ramankutty N, Brauman KA, Cassidy ES, Gerber JS, Johnston M, Mueller ND, O'Connell C, Ray DK, West PC, Balzer C, Bennett EM, Carpenter SR, Hill J, Monfreda C, Polasky S, Rockström J, Sheehan J, Siebert S, Tilman D, Zaks DP (2011) Solutions for a cultivated planet. Nature 478(7369):337–342. doi:10.1038/nature10452

Gietz RD, Akio S (1988) New yeast-*Escherichia coli* shuttle vectors constructed with in vitro mutagenized yeast genes lacking six-base pair restriction sites. Gene 74(20):527–534

Gietz R, Schiestl R (2007) Quick and easy yeast transformation using the LiAc/SS carrier DNA/PEG method. Nat Protoc 2(1):35–37. doi:10.1038/nprot.2007.14

Gírio F, Fonseca C, Carvalheiro F, Duarte L, Marques S, Bogel-Lukasik R (2010) Hemicelluloses for fuel ethanol: a review. Bioresour Technol 101(13):4775–4800. doi:10.1016/j.biortech.2010.01.088

Hahn-Hägerdal B, Karhumaa K, Fonseca C, Spencer-Martins I, Gorwa-Grauslund MF (2007) Towards industrial pentose-fermenting yeast strains. Appl Microbiol Biotechnol 74(5):937–953. doi:10.1007/s00253-006-0827-2

Karhumaa K, Hahn-Hägerdal B, Gorwa-Grauslund MF (2005) Investigation of limiting metabolic steps in the utilization of xylose by recombinant *Saccharomyces cerevisiae* using metabolic engineering. Yeast 22(5):359–368. doi:10.1002/yea.1216

Kim S, Park Y-C, Jin Y-S, Seo J-H (2013) Strain engineering of *Saccharomyces cerevisiae* for enhanced xylose metabolism. Biotechnol Adv. doi:10.1016/j.biotechadv.2013.03.004

Kocharin K, Nielsen J (2013) Specific growth rate and substrate dependent polyhydroxybutyrate production in *Saccharomyces cerevisiae*. AMB Express 3(1):18. doi:10.1186/2191-0855-3-18

Kocharin K, Chen Y, Siewers V, Nielsen J (2012) Engineering of acetyl-CoA metabolism for the improved production of polyhydroxybutyrate in *Saccharomyces cerevisiae*. AMB Express 2(1):52. doi:10.1186/2191-0855-2-52

Kocharin K, Siewers V, Nielsen J (2013) Improved polyhydroxybutyrate production by *Saccharomyces cerevisiae* through the use of the phosphoketolase pathway. Biotechnol Bioeng. doi:10.1002/bit.24888

Lambert JG, Hall CAS, Balogh S, Gupta A, Arnold M (2014) Energy, EROI and quality of life. Energy Policy 64:153–167. doi:10.1016/j.enpol.2013.07.001

Law J, Slepecky R (1961) Assay of poly-beta-hydroxybutyric acid. J Bacteriol 82:33–36

Leaf T, Peterson M, Stoup S, Somers D, Srienc F (1996) *Saccharomyces cerevisiae* expressing bacterial polyhydroxybutyrate synthase produces poly-3-hydroxybutyrate. Microbiology 142(Pt 5):1169–1180. doi:10.1099/13500872-142-5-1169

Lian J, Si T, Nair NU, Zhao H (2014) Design and construction of acetyl-CoA overproducing *Saccharomyces cerevisiae* strains. Metab Eng 24:139–149. doi:10.1016/j.ymben.2014.05.010

Madison LL, Huisman GC (1999) Metabolic engineering of poly(3-hydroxyalkanoates): from DNA to plastic. Microbiol Mol Biol Rev: MMBR 63(1):21–53

Nijkamp JF, van den Broek M, Datema E, de Kok S, Bosman L, Luttik MA, Daran-Lapujade P, Vongsangnak W, Nielsen J, Heijne WH, Klaassen P, Paddon CJ, Platt D, Kötter P, van Ham RC, Reinders MJ, Pronk JT, de Ridder D, Daran J-MMC (2012) De novo sequencing, assembly and analysis of the genome of the laboratory strain *Saccharomyces cerevisiae* CEN.PK113-7D, a model for modern industrial biotechnology. Microb Cell Fact 11:36. doi:10.1186/1475-2859-11-36

Novy V, Krahulec S, Longus K, Klimacek M, Nidetzky B (2013) Co-fermentation of hexose and pentose sugars in a spent sulfite liquor matrix with genetically modified *Saccharomyces cerevisiae*. Bioresour Technol 130:439–448

Philip S, Keshavarz T, Roy I (2007) Polyhydroxyalkanoates: biodegradable polymers with a range of applications. J Chem Technol Biotechnol 82. doi:10.1002/jctb.1667

Poehlein A, Kusian B, Friedrich B, Daniel R, Bowien BC (2011) Complete genome sequence of the type strain *Cupriavidus necator* N-1. J Bacteriol 193(18):5017. doi:10.1128/jb.05660-11

Runquist D, Hahn-Hägerdal B, Bettiga M (2010) Increased ethanol productivity in xylose-utilizing *Saccharomyces cerevisiae* via a randomly mutagenized xylose reductase. Appl Environ Microbiol 76(23):7796–7802. doi:10.1128/aem. 01505-10

Shiba Y, Paradise EM, Kirby J, Ro DK, Keasing JD (2007) Engineering of the pyruvate dehydrogenase bypass in Saccharomyces cerevisiae for high-level production of isoprenoids. Metab Eng 9(2):160–168. doi:10.1016/j.ymben.2006.10.005

Sun J, Shao Z, Zhao H, Nair N, Wen F, Xu J-H, Zhao H (2012) Cloning and characterization of a panel of constitutive promoters for applications in pathway engineering in *Saccharomyces cerevisiae*. Biotechnol Bioeng 109(8):2082–2092. doi:10.1002/bit.24481

Suriyamongkol P, Weselake R, Narine S, Moloney M, Shah S (2007) Biotechnological approaches for the production of polyhydroxyalkanoates in microorganisms and plants - a review. Biotechnol Adv 25(2):148–175. doi:10.1016/j.biotechadv.2006.11.007

Vandamme P, Coenye T (2004) Taxonomy of the genus *Cupriavidus*: a tale of lost and found. Int J Syst Evol Microbiol 54(Pt 6):2285–2289. doi:10.1099/ijs. 0.63247-0

Permissions

All chapters in this book were first published in AMBJ, by Springer; hereby published with permission under the Creative Commons Attribution License or equivalent. Every chapter published in this book has been scrutinized by our experts. Their significance has been extensively debated. The topics covered herein carry significant findings which will fuel the growth of the discipline. They may even be implemented as practical applications or may be referred to as a beginning point for another development.

The contributors of this book come from diverse backgrounds, making this book a truly international effort. This book will bring forth new frontiers with its revolutionizing research information and detailed analysis of the nascent developments around the world.

We would like to thank all the contributing authors for lending their expertise to make the book truly unique. They have played a crucial role in the development of this book. Without their invaluable contributions this book wouldn't have been possible. They have made vital efforts to compile up to date information on the varied aspects of this subject to make this book a valuable addition to the collection of many professionals and students.

This book was conceptualized with the vision of imparting up-to-date information and advanced data in this field. To ensure the same, a matchless editorial board was set up. Every individual on the board went through rigorous rounds of assessment to prove their worth. After which they invested a large part of their time researching and compiling the most relevant data for our readers.

The editorial board has been involved in producing this book since its inception. They have spent rigorous hours researching and exploring the diverse topics which have resulted in the successful publishing of this book. They have passed on their knowledge of decades through this book. To expedite this challenging task, the publisher supported the team at every step. A small team of assistant editors was also appointed to further simplify the editing procedure and attain best results for the readers.

Apart from the editorial board, the designing team has also invested a significant amount of their time in understanding the subject and creating the most relevant covers. They scrutinized every image to scout for the most suitable representation of the subject and create an appropriate cover for the book.

The publishing team has been an ardent support to the editorial, designing and production team. Their endless efforts to recruit the best for this project, has resulted in the accomplishment of this book. They are a veteran in the field of academics and their pool of knowledge is as vast as their experience in printing. Their expertise and guidance has proved useful at every step. Their uncompromising quality standards have made this book an exceptional effort. Their encouragement from time to time has been an inspiration for everyone.

The publisher and the editorial board hope that this book will prove to be a valuable piece of knowledge for researchers, students, practitioners and scholars across the globe.

List of Contributors

Andrea Brenna
Pasteur Cenci Bolognetti Foundation, c/o Department of Biology and Biotechnology "Charles Darwin", Sapienza University, Piazzale A. Moro 5, 00185 Rome, Italy

Barbara Montanini
Department of Life Sciences, Biochemistry and Molecular Biology Unit, Laboratory of Functional Genomics and Protein Engineering, University of Parma, 43124 Parma, Italy

Eleonora Muggiano
Pasteur Cenci Bolognetti Foundation, c/o Department of Biology and Biotechnology "Charles Darwin", Sapienza University, Piazzale A. Moro 5, 00185 Rome, Italy

Marco Proietto
Pasteur Cenci Bolognetti Foundation, c/o Department of Biology and Biotechnology "Charles Darwin", Sapienza University, Piazzale A. Moro 5, 00185 Rome, Italy

Patrizia Filetici
Institute of Biology and Molecular Pathology, CNR, 00185 Rome, Italy

Simone Ottonello
Department of Life Sciences, Biochemistry and Molecular Biology Unit, Laboratory of Functional Genomics and Protein Engineering, University of Parma, 43124 Parma, Italy

Paola Ballario
Pasteur Cenci Bolognetti Foundation, c/o Department of Biology and Biotechnology "Charles Darwin", Sapienza University, Piazzale A. Moro 5, 00185 Rome, Italy
Institute of Biology and Molecular Pathology, CNR, 00185 Rome, Italy

Masayoshi Sakaguchi1
Department of Applied Chemistry, Kogakuin University, 2,665-1 Nakano-cho, Hachioji, Tokyo 192-0015, Japan

Kanae Osaku
Department of Applied Chemistry, Kogakuin University, 2,665-1 Nakano-cho, Hachioji, Tokyo 192-0015, Japan

Susumu Maejima
Department of Applied Chemistry, Kogakuin University, 2,665-1 Nakano-cho, Hachioji, Tokyo 192-0015, Japan

Nao Ohno
Department of Applied Chemistry, Kogakuin University, 2,665-1 Nakano-cho, Hachioji, Tokyo 192-0015, Japan

Yasusato Sugahara
Department of Applied Chemistry, Kogakuin University, 2,665-1 Nakano-cho, Hachioji, Tokyo 192-0015, Japan

Fumitaka Oyama
Department of Applied Chemistry, Kogakuin University, 2,665-1 Nakano-cho, Hachioji, Tokyo 192-0015, Japan

Masao Kawakita
Department of Applied Chemistry, Kogakuin University, 2,665-1 Nakano-cho, Hachioji, Tokyo 192-0015, Japan
Center for Medical Research Cooperation, Tokyo Metropolitan Institute of Medical Science, 2-1-6 Kamikitazawa Setagaya-ku, Tokyo 156-8506, Japan

Jasper Sloothaak
Microbial Systems & Synthetic Biology, Laboratory of Systems and Synthetic Biology, Wageningen University, Dreijenplein 10, NL-6703 HB, Wageningen, Netherlands

Mike Schilders
Microbial Systems & Synthetic Biology, Laboratory of Systems and Synthetic Biology, Wageningen University, Dreijenplein 10, NL-6703 HB, Wageningen, Netherlands

Peter J Schaap
Microbial Systems & Synthetic Biology, Laboratory of Systems and Synthetic Biology, Wageningen University, Dreijenplein 10, NL-6703 HB, Wageningen, Netherlands

Leo H de Graaff
Microbial Systems & Synthetic Biology, Laboratory of Systems and Synthetic Biology, Wageningen University, Dreijenplein 10, NL-6703 HB, Wageningen, Netherlands

Alma Lorena Almaraz-Delgado
Unidad Profesional Interdisciplinaria de Biotecnología, Instituto Politécnico Nacional, Av. Acueducto SN Col. Barrio la Laguna Ticoman, 07340 Mexico City, México

José Flores-Uribe
Unidad Profesional Interdisciplinaria de Biotecnología, Instituto Politécnico Nacional, Av. Acueducto SN Col. Barrio la Laguna Ticoman, 07340 Mexico City, México

Víctor Hugo Pérez-España
Unidad Profesional Interdisciplinaria de Biotecnología, Instituto Politécnico Nacional, Av. Acueducto SN Col. Barrio la Laguna Ticoman, 07340 Mexico City, México

Edgar Salgado-Manjarrez
Unidad Profesional Interdisciplinaria de Biotecnología, Instituto Politécnico Nacional, Av. Acueducto SN Col. Barrio la Laguna Ticoman, 07340 Mexico City, México

Jesús Agustín Badillo-Corona
Unidad Profesional Interdisciplinaria de Biotecnología, Instituto Politécnico Nacional, Av. Acueducto SN Col. Barrio la Laguna Ticoman, 07340 Mexico City, México

Ryota Moriuchi
Department of Environmental Life Sciences, Graduate School of Life Sciences, Tohoku University, Sendai 980-8577, Japan
The United Graduate School of Agricultural Science, Gifu University 1-1 Yanagido, Gifu 501-1193, Japan

Hiroki Tanaka
Department of Environmental Life Sciences, Graduate School of Life Sciences, Tohoku University, Sendai 980-8577, Japan

Yuki Nikawadori
Department of Environmental Life Sciences, Graduate School of Life Sciences, Tohoku University, Sendai 980-8577, Japan

Mayuko Ishitsuka
Department of Environmental Life Sciences, Graduate School of Life Sciences, Tohoku University, Sendai 980-8577, Japan

Michihiro Ito
Department of Environmental Life Sciences, Graduate School of Life Sciences, Tohoku University, Sendai 980-8577, Japan
Consolidated Research Institute for Advanced Science and Medical Care, Waseda University, 2-2 Wakamatsu-cho, Shinjuku, Tokyo 162-8480, Japan

Yoshiyuki Ohtsubo
Department of Environmental Life Sciences, Graduate School of Life Sciences, Tohoku University, Sendai 980-8577, Japan

Masataka Tsuda
Department of Environmental Life Sciences, Graduate School of Life Sciences, Tohoku University, Sendai 980-8577, Japan

Jiri Damborsky
Loschmidt Laboratories, Department of Experimental Biology and Research Centre for Toxic Compounds in the Environment RECETOX, Faculty of Science, Masaryk University, Kamenice 5/A13, 625 00 Brno, Czech Republic

Zbynek Prokop
Loschmidt Laboratories, Department of Experimental Biology and Research Centre for Toxic Compounds in the Environment RECETOX, Faculty of Science, Masaryk University, Kamenice 5/A13, 625 00 Brno, Czech Republic

Yuji Nagata
1Department of Environmental Life Sciences, Graduate School of Life Sciences, Tohoku University, Sendai 980-8577, Japan

Elena Volodina
Institut für Molekulare Mikrobiologie und Biotechnologie, Westfälische Wilhelms-Universität Münster, Corrensstraße 3, D-48149 Münster, Germany

Alexander Steinbüchel
Institut für Molekulare Mikrobiologie und Biotechnologie, Westfälische Wilhelms-Universität Münster, Corrensstraße 3, D-48149 Münster, Germany
King Abdulaziz University, Jeddah, Saudi Arabia

Narin Kirikyali
Division of Food Sciences, School of Biosciences, University of Nottingham, Sutton Bonington Campus, Loughborough LE12 5RD, UK

Jonathan Wood
Division of Food Sciences, School of Biosciences, University of Nottingham, Sutton Bonington Campus, Loughborough LE12 5RD, UK

Ian F Connerton
Division of Food Sciences, School of Biosciences, University of Nottingham, Sutton Bonington Campus, Loughborough LE12 5RD, UK

Shunsuke Aburaya
Division of Applied Life Sciences, Graduate School of Agriculture, Kyoto University, Sakyo-ku, Kyoto, Japan

Kohei Esaka
Division of Applied Life Sciences, Graduate School of Agriculture, Kyoto University, Sakyo-ku, Kyoto, Japan

Hironobu Morisaka
Division of Applied Life Sciences, Graduate School of Agriculture, Kyoto University, Sakyo-ku, Kyoto, Japan
Kyoto Integrated Science and Technology Bio-Analysis Center, Shimogyo-ku, Kyoto, Japan

Kouichi Kuroda
Division of Applied Life Sciences, Graduate School of Agriculture, Kyoto University, Sakyo-ku, Kyoto, Japan

Mitsuyoshi Ueda
Division of Applied Life Sciences, Graduate School of Agriculture, Kyoto University, Sakyo-ku, Kyoto, Japan
Kyoto Integrated Science and Technology Bio-Analysis Center, Shimogyo-ku, Kyoto, Japan

Jessica Zampolli
Department of Biotechnology and Biosciences, University of Milano-Bicocca, Piazza della Scienza 2, 20126 Milano, Italy
Department of Earth and Environmental Sciences, University of Milano-Bicocca, Milano, Italy

Elena Collina
Department of Earth and Environmental Sciences, University of Milano-Bicocca, Milano, Italy

Marina Lasagni
Department of Earth and Environmental Sciences, University of Milano-Bicocca, Milano, Italy

Patrizia Di Gennaro
Department of Biotechnology and Biosciences, University of Milano-Bicocca, Piazza della Scienza 2, 20126 Milano, Italy

Yuya Mitsudome
Department of Biochemistry and Applied Biosciences, University of Miyazaki, 1-1 Gakuen, Kibanadai-Nishi, Miyazaki 889-2192, Japan

Mamiko Takahama
Department of Biochemistry and Applied Biosciences, University of Miyazaki, 1-1 Gakuen, Kibanadai-Nishi, Miyazaki 889-2192, Japan

Jun Hirose
Department of Applied Chemistry, University of Miyazaki, 1-1 Gakuen, Kibanadai-Nishi, Miyazaki 889-2192, Japan

Naoto Yoshida
Department of Biochemistry and Applied Biosciences, University of Miyazaki, 1-1 Gakuen, Kibanadai-Nishi, Miyazaki 889-2192, Japan

Mine Gungormusler-Yilmaz
Department of Biosystems Engineering, Faculty of Engineering, University of Manitoba, Manitoba, Winnipeg, MB R3T 5 V6, Canada.
Department of Bioengineering, Ege University, Izmir, Turkey

Dmitry Shamshurin
Manitoba Centre for Proteomics and Systems Biology, University of Manitoba, Winnipeg, Manitoba, Canada

Marine Grigoryan
Manitoba Centre for Proteomics and Systems Biology, University of Manitoba, Winnipeg, Manitoba, Canada

Marcel Taillefer
Department of Microbiology, University of Manitoba, Winnipeg, Manitoba, Canada

Victor Spicer
Department of Physics and Astronomy, University of Manitoba, Winnipeg, Manitoba, Canada

Oleg V Krokhin
Department of Physics and Astronomy, University of Manitoba, Winnipeg, Manitoba, Canada

Richard Sparling
Department of Microbiology, University of Manitoba, Winnipeg, Manitoba, Canada

David B Levin
Department of Biosystems Engineering, Faculty of Engineering, University of Manitoba, Manitoba, Winnipeg, MB R3T 5 V6, Canada

Gaurav Singh Kaira
Biotechnological Applications, G. B. Pant Institute of Himalayan Environment and Development, Kosi-Katarmal, Almora 263 643, Uttarakhand, India

Kusum Dhakar
Biotechnological Applications, G. B. Pant Institute of Himalayan Environment and Development, Kosi-Katarmal, Almora 263 643, Uttarakhand, India

Anita Pandey
Biotechnological Applications, G. B. Pant Institute of Himalayan Environment and Development, Kosi-Katarmal, Almora 263 643, Uttarakhand, India

Jan Dines Knudsen
Division of Applied Microbiology, Department of Chemistry, Lund University, PO Box 124, SE 221 00 Lund, Sweden

Magnus Carlquist
Division of Applied Microbiology, Department of Chemistry, Lund University, PO Box 124, SE 221 00 Lund, Sweden

Marie Gorwa-Grauslund
Division of Applied Microbiology, Department of Chemistry, Lund University, PO Box 124, SE 221 00 Lund, Sweden

Peter R Chenaux
Department of Biology, Queen's University, Kingston, ON K7L 3 N6, Canada

Narisa Lalji
Department of Biology, Queen's University, Kingston, ON K7L 3 N6, Canada

Daniel D Lefebvre
Department of Biology, Queen's University, Kingston, ON K7L 3 N6, Canada

Dorit Eberhardt
Genetics of Prokaryotes, Faculty of Biology & Center for Biotechnology, Bielefeld University, Universitätsstraße 25, 33615 Bielefeld, Germany

Jaide V K Jensen
Genetics of Prokaryotes, Faculty of Biology & Center for Biotechnology, Bielefeld University, Universitätsstraße 25, 33615 Bielefeld, Germany

Volker F Wendisch
Genetics of Prokaryotes, Faculty of Biology & Center for Biotechnology, Bielefeld University, Universitätsstraße 25, 33615 Bielefeld, Germany

Songsak Wattanachaisaereekul
Pilot Plant Development and Training Institute, King Mongkut's University of Technology Thonburi, 49 Soi Thianthale 25, Bangkhunthian-Chaithale Rd., Thakham, Bangkhunthian, Bangkok 10150, Thailand

Anuwat Tachaleat
Pilot Plant Development and Training Institute, King Mongkut's University of Technology Thonburi, 49 Soi Thianthale 25, Bangkhunthian-Chaithale Rd., Thakham, Bangkhunthian, Bangkok 10150, Thailand

Juntira Punya
Bioresource Research Unit, National Center for Genetic Engineering and Biotechnology, 113 Thailand Science Park, Phahonyothin Rd., Khlong Nueng, Klong Luang, Pathum Thani 12120, Thailand

Rachada Haritakun
Bioresource Research Unit, National Center for Genetic Engineering and Biotechnology, 113 Thailand Science Park, Phahonyothin Rd., Khlong Nueng, Klong Luang, Pathum Thani 12120, Thailand

Chollaratt Boonlarppradab
Bioresource Research Unit, National Center for Genetic Engineering and Biotechnology, 113 Thailand Science Park, Phahonyothin Rd., Khlong Nueng, Klong Luang, Pathum Thani 12120, Thailand

Supapon Cheevadhanarak
Pilot Plant Development and Training Institute, King Mongkut's University of Technology Thonburi, 49 Soi Thianthale 25, Bangkhunthian-Chaithale Rd., Thakham, Bangkhunthian, Bangkok 10150, Thailand
School of Bioresources and Technology, King Mongkut's University of Technology Thonburi, 49 Soi Thianthale 25, Bangkhunthian-Chaithale Rd., Thakham, Bangkhunthian, Bangkok 10150,Thailand

Ayrat M Ziganshin
Department of Microbiology, Kazan (Volga Region) Federal University, ul. Kremlyovskaya 18, Kazan 420008, The Republic of Tatarstan, Russia

Elvira E Ziganshina
Department of Microbiology, Kazan (Volga Region) Federal University, ul. Kremlyovskaya 18, Kazan 420008, The Republic of Tatarstan, Russia

James Byrne
Geomicrobiology, Center for Applied Geosciences, University of Tübingen, Tübingen D-72076, Germany

Robin Gerlach
Department of Chemical and Biological Engineering, Center for Biofilm Engineering, Montana State University, Bozeman, Montana 59717, USA

Ellen Struve
Geomicrobiology, Center for Applied Geosciences, University of Tübingen, Tübingen D-72076, Germany

Timur Biktagirov
Institute of Physics, Kazan (Volga Region) Federal University, Kazan 420008, The Republic of Tatarstan, Russia

Alexander Rodionov
Institute of Physics, Kazan (Volga Region) Federal University, Kazan 420008, The Republic of Tatarstan, Russia

Andreas Kappler
Geomicrobiology, Center for Applied Geosciences, University of Tübingen, Tübingen D-72076, Germany

Seitaro Goto
Product Development Laboratory, NH Foods Ltd., Chikusei, Ibaraki 308-0042, Japan

Jun Kawamoto
Institute for Chemical Research, Kyoto University, Uji, Kyoto 611-0011, Japan

Satoshi B Sato
Research Center for Low Temperature and Materials Sciences, Kyoto University, Kyoto 606-8501, Japan

Takashi Iki
Product Development Laboratory, NH Foods Ltd., Chikusei, Ibaraki 308-0042, Japan

Itaru Watanabe
Product Development Laboratory, NH Foods Ltd., Chikusei, Ibaraki 308-0042, Japan

Kazuyuki Kudo
Product Development Laboratory, NH Foods Ltd., Chikusei, Ibaraki 308-0042, Japan

Nobuyoshi Esaki
Institute for Chemical Research, Kyoto University, Uji, Kyoto 611-0011, Japan

Tatsuo Kurihara
Institute for Chemical Research, Kyoto University, Uji, Kyoto 611-0011, Japan

Angelina M M Basso
Department of Molecular Pathology, University of Brasilia, Brasilia, DF, Brazil
Laboratory of Plant-Pest Interaction, Embrapa - Genetic Resources and Biotechnology, Brasília, DF, Brazil

Patrícia B Pelegrini
Laboratory of Plant-Pest Interaction, Embrapa - Genetic Resources and Biotechnology, Brasília, DF, Brazil

Fernanda Mulinari
Laboratory of Plant-Pest Interaction, Embrapa - Genetic Resources and Biotechnology, Brasília, DF, Brazil
Pioneer Union for Social Insertion - UPIS, Planaltina, DF, Brazil

Michelle C Costa
Laboratory of Plant-Pest Interaction, Embrapa - Genetic Resources and Biotechnology, Brasília, DF, Brazil

Antonio B Viana
Laboratory of Plant-Pest Interaction, Embrapa - Genetic Resources and Biotechnology, Brasília, DF, Brazil
Catholic University of Brasilia, Brasilia, DF, Brazil

Luciano P Silva
Laboratory of Mass Spectrometry, Embrapa - Genetic Resources and Biotechnology, Brasília, DF, Brazil

Maria Fatima Grossi-de-Sa
Laboratory of Plant-Pest Interaction, Embrapa - Genetic Resources and Biotechnology, Brasília, DF, Brazil
Catholic University of Brasilia, Brasilia, DF, Brazil

Yoko Asahi
Department of Restorative Dentistry and Endodontology, Osaka University Graduate School of Dentistry, 1-8 Yamadaoka, Suita, Osaka 565-0871, Japan.

Jiro Miura
Division for Interdisciplinary Dentistry, Osaka University Graduate School of Dentistry, 1-8 Yamadaoka, Suita, Osaka 565-0871, Japan

Tetsuya Tsuda
Department of Applied Chemistry, Graduate School of Engineering, Osaka University, Suita, Osaka 565-0871, Japan

Susumu Kuwabata
Department of Applied Chemistry, Graduate School of Engineering, Osaka University, Suita, Osaka 565-0871, Japan

Katsuhiko Tsunashima
Department of Material Science, Wakayama National College of Technology, 77 Noshima, Nada-cho, Gobo, Wakayama 644-0023, Japan

Yuichiro Noiri
Department of Restorative Dentistry and Endodontology, Osaka University Graduate School of Dentistry, 1-8 Yamadaoka, Suita, Osaka 565-0871, Japan

Takao Sakata
Osaka University Research Center for Ultra-high-voltage Electron Microscopy, Suita, Osaka565-0871, Japan

Shigeyuki Ebisu
Department of Restorative Dentistry and Endodontology, Osaka University Graduate School of Dentistry, 1-8 Yamadaoka, Suita, Osaka 565-0871, Japan

Mikako Hayashi
Department of Restorative Dentistry and Endodontology, Osaka University Graduate School of Dentistry, 1-8 Yamadaoka, Suita, Osaka 565-0871, Japan

Massimiliano Marvasi
Department of Natural Sciences, School of Science and Technology, Middlesex University, The Burroughs, London NW4 4BT, UK
Soil and Water Science Department, University of Florida, Gainesville, FL, USA

Ian A Durie
Soil and Water Science Department, University of Florida, Gainesville, FL, USA

Eric S McLamore
Agricultural and Biological Engineering Department, University of Florida, Gainesville, FL, USA

Diana C Vanegas
Agricultural and Biological Engineering Department, University of Florida, Gainesville, FL, USA
Food Engineering Department, Universidad del Valle, Cali, Colombia

Prachee Chaturvedi
Department of Mechanical Engineering, University of Colorado, Denver, CO, USA

Tomohiro Shigemori
Division of Applied Life Sciences, Graduate School of Agriculture, Kyoto University, Sakyo-ku, Kyoto 606-8502, Japan

Kouichi Kuroda
Division of Applied Life Sciences, Graduate School of Agriculture, Kyoto University, Sakyo-ku, Kyoto 606-8502, Japan

Mitsuyoshi Ueda
Division of Applied Life Sciences, Graduate School of Agriculture, Kyoto University, Sakyo-ku, Kyoto 606-8502, Japan

Peter Richard
VTT Technical Research Centre of Finland Ltd, Tietotie 2, P.O. box 1000, 02044 VTT Espoo, Finland

Kaarina Viljanen
VTT Technical Research Centre of Finland Ltd, Tietotie 2, P.O. box 1000, 02044 VTT Espoo, Finland

Merja Penttilä
VTT Technical Research Centre of Finland Ltd, Tietotie 2, P.O. box 1000, 02044 VTT Espoo, Finland

Anders G Sandström
Division of Applied Microbiology, Department of Chemistry, Lund University, PO Box 124, Lund SE-221 00, Sweden

Alejandro Muñoz de las Heras
Division of Applied Microbiology, Department of Chemistry, Lund University, PO Box 124, Lund SE-221 00, Sweden

Diogo Portugal-Nunes
Division of Applied Microbiology, Department of Chemistry, Lund University, PO Box 124, Lund SE-221 00, Sweden

Marie F Gorwa-Grauslund
Division of Applied Microbiology, Department of Chemistry, Lund University, PO Box 124, Lund SE-221 00, Sweden